MTP International Review of Science

Magnetic Resonance

MTP International Review of Science

Publisher's Note

The MTP International Review of Science is an important new venture in scientific publishing, which we present in association with MTP Medical and Technical Publishing Co. Ltd. and University Park Press, Baltimore. The basic concept of the Review is to provide regular authoritative reviews of entire disciplines. We are starting with chemistry because the problems of literature survey are probably more acute in this subject than in any other. As a matter of policy, the authorship of the MTP Review of Chemistry is international and distinguished; the subject coverage is extensive, systematic and critical; and most important of all, new issues of the Review will be published every two years.

In the MTP Review of Chemistry (Series One), Inorganic, Physical and Organic Chemistry are comprehensively reviewed in 33 text volumes and 3 index volumes, details of which are shown opposite. In general, the reviews cover the period 1967 to 1971. In 1974, it is planned to issue the MTP Review of Chemistry (Series Two), consisting of a similar set of volumes covering the period 1971 to 1973. Series Three is planned for 1976, and so on.

The MTP Review of Chemistry has been conceived within a carefully organised editorial framework. The over-all plan was drawn up, and the volume editors were appointed, by three consultant editors. In turn, each volume editor planned the coverage of his field and appointed authors to write on subjects which were within the area of their own research experience. No geographical restriction was imposed. Hence, the 300 or so contributions to the MTP Review of Chemistry come from many countries of the world and provide an authoritative account of progress in chemistry.

To facilitate rapid production, individual volumes do not have an index. Instead, each chapter has been prefaced with a detailed list of contents, and an index to the 13 volumes of the MTP Review of Physical Chemistry (Series One) will appear, as a separate volume, after publication of the final volume. Similar arrangements will apply to the MTP Review of Organic Chemistry (Series One) and to subsequent series.

Butterworth & Co. (Publishers) Ltd.

Physical Chemistry Series One

Consultant Editor
A. D. Buckingham
Department of Chemistry
University of Cambridge

Volume titles and Editors

1 THEORETICAL CHEMISTRY
Professor W. Byers Brown, *University of Manchester*

2 MOLECULAR STRUCTURE AND PROPERTIES
Professor G. Allen, *University of Manchester*

3 SPECTROSCOPY
Dr. D. A. Ramsay, F.R.S.C., *National Research Council of Canada*

4 MAGNETIC RESONANCE
Professor C. A. McDowell, F.R.S.C., *University of British Columbia*

5 MASS SPECTROMETRY
Professor A. Maccoll, *University College, University of London*

6 ELECTROCHEMISTRY
Professor J. O'M Bockris, *University of Pennsylvania*

7 SURFACE CHEMISTRY AND COLLOIDS
Professor M. Kerker, *Clarkson College of Technology, New York*

8 MACROMOLECULAR SCIENCE
Professor C. E. H. Bawn, F.R.S., *University of Liverpool*

9 CHEMICAL KINETICS
Professor J. C. Polanyi, F.R.S., *University of Toronto*

10 THERMOCHEMISTRY AND THERMODYNAMICS
Dr. H. A. Skinner, *University of Manchester*

11 CHEMICAL CRYSTALLOGRAPHY
Professor J. Monteath Robertson, F.R.S., *University of Glasgow*

12 ANALYTICAL CHEMISTRY — PART 1
Professor T. S. West, *Imperial College, University of London*

13 ANALYTICAL CHEMISTRY — PART 2
Professor T. S. West, *Imperial College, University of London*

INDEX VOLUME

Physical Chemistry
Series One

Consultant Editor
A. D. Buckingham

MTP International Review of Science

Volume 4

Magnetic Resonance

Edited by **C. A. McDowell, F.R.S.C.**
University of British Columbia

Butterworths · London
University Park Press · Baltimore

THE BUTTERWORTH GROUP

ENGLAND
Butterworth & Co (Publishers) Ltd
London: 88 Kingsway, WC2B 6AB

AUSTRALIA
Butterworth Pty Ltd
Sydney: 586 Pacific Highway 2067
Melbourne: 343 Little Collins Street, 3000
Brisbane: 240 Queen Street, 4000

NEW ZEALAND
Butterworth of New Zealand Ltd
Wellington: 26–28 Waring Taylor Street, 1

SOUTH AFRICA
Butterworth & Co (South Africa) (Pty) Ltd
Durban: 152–154 Gale Street

ISBN 0 408 70265 6

UNIVERSITY PARK PRESS

U.S.A. and CANADA
University Park Press Inc
Chamber of Commerce Building
Baltimore, Maryland, 21202

Library of Congress Cataloging in Publication Data

McDowell, Charles A.
  Magnetic resonance

  (Physical chemistry, series one, v. 4) (MTP international review of science)

  1. Magnetic resonance–Addresses, essays, lectures.
QD453.2.P58 Vol. 4 [QC762] 541′.3′08s [538′.3]
ISBN 0–8391–1018–9   72–4332

First Published 1972 and © 1972
MTP MEDICAL AND TECHNICAL PUBLISHING CO. LTD.
Seacourt Tower
West Way
Oxford, OX2 OJW
and
BUTTERWORTH & CO. (PUBLISHERS) LTD.

Filmset by Photoprint Plates Ltd., Rayleigh, Essex
Printed in England by Redwood Press Ltd., Trowbridge, Wilts
and bound by R. J. Acford Ltd., Chichester, Sussex

Consultant Editor's Note

The MTP International Review of Science is designed to provide a comprehensive, critical and continuing survey of progress in research. The difficult problem of keeping up with advances on a reasonably broad front makes the idea of the Review especially appealing, and I was grateful to be given the opportunity of helping to plan it.

This particular 13-volume section is concerned with Physical Chemistry, Chemical Crystallography and Analytical Chemistry. The subdivision of Physical Chemistry adopted is not completely conventional, but it has been designed to reflect current research trends and it is hoped that it will appeal to the reader. Each volume has been edited by a distinguished chemist and has been written by a team of authoritative scientists. Each author has assessed and interpreted research progress in a specialised topic in terms of his own experience. I believe that their efforts have produced very useful and timely accounts of progress in these branches of chemistry, and that the volumes will make a valuable contribution towards the solution of our problem of keeping abreast of progress in research.

It is my pleasure to thank all those who have collaborated in making this venture possible – the volume editors, the chapter authors and the publishers.

Cambridge   A. D. Buckingham

Preface

This volume was initially planned to include work proceeding in e.s.r., n.m.r., radiofrequency and Mössbauer spectroscopy. As all of these are very active fields in which a large number of publications appear each year, it is obvious that some selectivity had to be exercised in deciding what material should be included. For the present volume it was decided that an attempt should be made to cover developments in those areas in which, during the past five years, there had been advances of considerable significance. Limitations of space caused some modifications to the preliminary list of topics. It should also be mentioned that because the volume forms part of the MTP International Review of Science, some effort was made to give the work a truly international flavour. This latter factor admittedly did to some extent influence the decisions as to what to include and who to ask to write about a particular topic, but in general it is felt that this has been beneficial to the volume as a whole.

It is, of course, the natural hope of all editors that readers will find the choice of the topics, the authors selected, and the material covered, to be interesting and worthwhile. Later issues of the volumes on Magnetic Resonance will include material such as multiple resonance techniques; high resolution n.m.r. in solids; Fourier transform techniques; ENDOR studies on free radicals in solution; and many other interesting topics which regrettably have had to be omitted this time.

Vancouver   C. A. McDowell

Contents

1
Nuclear Spin Relaxation in Gases

M. BLOOM
University of British Columbia, Vancouver *

*Address during 1971–72 academic year: Laboratoire de Physique des Solides, Université Paris-Sud, 91-Orsay, France.

1.1 INTRODUCTION

1.1.1 Relationship between molecular spectroscopy and nuclear spin relaxation in gases

1.1.1.1 Review of electric dipole spectroscopy and correlation functions

When a system of molecules is subjected to an oscillating electric field $E_x \cos \omega t$, the coupling between the electric dipole moment $\sum_{i=1}^{N} \mu_i$ of the system of N molecules and the electric field causes transitions between states $|a>$ and $|b>$ of the system at a rate proportional to $|<a| \sum_{i=1}^{N} \mu_{ix} |b>|^2$ if the angular frequency ω of the oscillating field is tuned to the frequency $(E_a - E_b)/\hbar$ corresponding to the energy difference between the states. For small values of E_x, the induced polarisation P_x of the molecular system is linear in E_x, i.e.

$$P_x = \chi'(\omega)E_x \cos \omega t + \chi''(\omega)E_x \sin \omega t \qquad (1.1)$$

The power absorbed from the oscillating electric field is proportional to the out-of-phase component $\chi''(\omega)$ of the electric susceptibility. At low densities, when the frequency of the molecular collisions becomes small, the frequencies of most of the allowed transitions are very close to the single-molecule transition frequencies. Under these conditions, $\chi''(\omega)$ is closely related to the spectral density $J(\omega)$ of the correlation function $G(t)$ of the single-molecule electric dipole moment operator μ_x [1]

$$\chi''(\omega) \propto J(\omega) = \int_{-\infty}^{\infty} e^{-i\omega t} \; G(t)dt \qquad (1.2)$$

where

$$G(t) = \text{Real} <\mu_x(t)\mu_x> \; = \; \text{Real} <e^{iHt/\hbar}\mu_x e^{-iHt/\hbar}\mu_x> \qquad (1.3)$$

where the brackets $< \; >$ denote an average over an equilibrium ensemble of molecular systems and equation (1.3) defines the time-dependent operator $\mu_x(t)$ in terms of the Hamiltonian H of the N-molecule system. In the absence of collisions (*the free-molecule approximation*), the correlation function of $\mu_x(t)$ can be expressed in terms of the eigen-values ε_n and the eigen-functions $|n>$ of the free molecule as

$$[G(t)]_{\text{free}} = \text{Re} \sum_{m,n} P_m <m|\mu_x(t)|n><n|\mu_x|m> \qquad (1.4)$$

$$= \sum_{m,n} P_m |<m|\mu_x|n>|^2 \cos \omega_{mn}t = \sum_{m,n} G_{mn} \cos \omega_{mn}t$$

In this case, the spectral density of $G(t)$ consists of a superposition of δ-functions at the frequencies $\omega_{mn} = (\varepsilon_m - \varepsilon_n)/\hbar$ of intensity $G_{mn} = P_m |<m| \mu_x|n>|^2$ where P_m is the Boltzmann factor given by

$$P_m = \frac{e^{-\varepsilon_m/kT}}{\sum_m e^{-\varepsilon_m/kT}} \qquad (1.5)$$

The effect of collisions can be described in terms of the 'reduced correlation function' $g(t)$ defined by

$$G(t) = [G(t)]_{\text{free}}g(t) \tag{1.6}$$

so that

$$J(\omega) = \tfrac{1}{2}\sum_{m,n} G_{mn}[j(\omega - \omega_{mn}) + j(\omega + \omega_{mn})] \tag{1.7}$$

where

$$j(\omega - \omega_{mn}) = \int_{-\infty}^{\infty} e^{-i(\omega - \omega_{mn})t}g(t)\mathrm{d}t \tag{1.8}$$

Clearly, $g(0) = 1$. The effect of collisions is to limit the lifetime of the molecules in any state $|m>$ and to cause $g(t)$ to approach zero at long times. At very low densities, the effect of the collisions is often well approximated by an exponential function of time characterised by the correlation time τ_c,

$$g(t) = \exp(-|t|/\tau_c) \tag{1.9}$$

Under these conditions, the molecular absorption lines at the frequencies ω_{mn} have a Lorentzian shape and are of width τ_c^{-1}, i.e.

$$j(\omega - \omega_{mn}) = \frac{2\tau_c}{1 + (\omega - \omega_{mn})^2\tau_c^2} \tag{1.10}$$

An illustration of a pure rotational spectrum for linear molecules such as

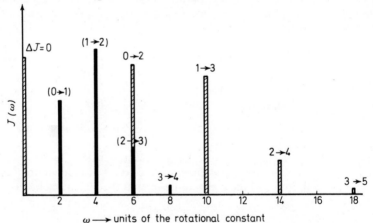

Figure 1.1 Schematic illustration of the positive frequency part of the pure rotational spectral densities of a dilute gas of diatomic molecules. The electric dipole spectrum ($\Delta J = 1$) and the 'nuclear spin relaxation spectrum' associated with the dipolar coupling between the nuclear magnetic moments $\Delta J = 0,2$ are superposed as solid and hatched lines respectively. The relative intensities are appropriate to HD gas at 300 K. For spin–rotation interactions, only contributions for $\Delta J = 0$ are obtained

HCl which have permanent electric dipole moments is shown in Figure 1.1 as solid lines. In this case, the states $|n>$ and $|m>$ connected by the electric dipole moment operator are characterised by the rotational angular momentum quantum numbers $m,n = J,J'$, and they satisfy the selection rule $J - J' = \pm 1$.

1.1.1.2 Nuclear spin relaxation spectroscopy—a thought experiment

In 'electric dipole spectroscopy', a uniform oscillating electric field is coupled to the molecules by means of the molecular electric dipole moment. The resonant transitions in the dilute gas are defined by the spectral density of the electric dipole moment operator as discussed in the previous section. In principle, it would be possible to couple to the molecular rotation by means of the nuclear spins on the molecule since the nuclear spins interact with the molecular rotation. For example, in a molecule having two nuclei of gyro-magnetic ratios γ_1 and γ_2 and spins I_1 and I_2, respectively, two interactions by which the spins are coupled to the molecule are the spin–rotation and dipolar interactions[2, 3]. *The spin–rotation interaction* Hamiltonian is given by

$$H_{sr} = 2\pi\hbar(C_1 I_1 + C_2 I_2) \cdot K \tag{1.11}$$

where C_1 and C_2 are spin–rotation coupling constants for nuclei 1 and 2, respectively and the rotational operator K reduces to the angular momentum operator for diatomic molecules.

The dipolar interaction Hamiltonian is given by

$$H_{dip} = \frac{\gamma_1 \gamma_2 \hbar}{r^3}\left[I_1 \cdot I_2 - \frac{3(I_1 \cdot r)(I_2 \cdot r)}{r^2}\right] \tag{1.12}$$

$$= A_d \hbar \sum_{\mu = -2}^{+2} (-1)^\mu T_2(I_1, I_2) Y_{2, -\mu}(\theta, \phi)$$

where $r = (r, \theta, \phi)$ is the vector joining the two spins in spherical polar coordinates, $A_d = (24\pi/5)^{1/2} \gamma_1 \gamma_2 \hbar / r^3$ and the interaction has been expressed as the scalar product of the *spherical harmonics* of rank two $Y_{2, -\mu}$ and the second-rank tensor formed from the nuclear spin operators I_1 and I_2 in the usual way[4].

We imagine the following nuclear magnetic resonance (n.m.r.) experiment; firstly, suppose that a static magnetic field $H_0 \hat{z}$ is applied along the z-axis where $\hat{z}$ is a unit vector along the z-axis. Then, the nuclear spins 1 and 2 precess about the z-axis at their Larmor frequencies $\omega_0^{(1)} = -\gamma_1 H_0$ and $\omega_0^{(2)} = -\gamma_2 H_0$ respectively. Now, imagine that a rotating r.f. field of magnitude H_1 is applied at right angles to $H_0 \hat{z}$ at the frequency $\omega_0^{(1)}$ or $\omega_0^{(2)}$ for a time t_1 or t_2 such that $\gamma_1 H_1 t_1 = \pi/2$ or $\gamma_2 H_1 t_2 = \pi/2$. The effect of this '$\pi/2$ pulse' is to rotate the nuclear spins through an angle of 90 degrees so that, if the nuclear spins undergoing n.m.r. are initially orientated along the z-axis, they are orientated in the x–y plane after the pulse and are 90 degrees out of phase with the rotating field. Finally, the phase of the r.f. field is shifted by 90 degrees to make the orientation of the r.f. field co-linear with the nuclear spins. The process we have just described is called 'spin-locking' and is commonly performed using n.m.r.[5].

The net effect of the above operations is to maintain I_1 or I_2 in a state of precession about the z-axis at the frequency $\omega_0^{(1)}$ or $\omega_0^{(2)}$, respectively. The coupling to the molecule by the spin–rotation or dipolar interaction, respectively, can now produce transitions between the molecular rotational states in complete analogy with 'electric-dipole spectroscopy'. For spin–rotation interactions, the spectrum is given by the spectral density $J_1(\omega)$ of the correlation functions of $\mathbf{J}(t)$, while for dipolar interactions the spectrum is $J_2(\omega)$, the spectral density of $Y_{2,-\mu}(\theta)(t),\phi(t))$. The selection rules for those two spectra are $J'-J=0$ and $J'-J=0,\pm 2$, respectively, *for a diatomic molecule* as illustrated in Figure 1.1 as hatched lines.

A simpler type of n.m.r. experiment than the spin-locking experiment described above is the measurement of the approach to equilibrium of the nuclear spin system. For a large class of nuclear spin systems, the approach of the z-component of the nuclear magnetisation to its equilibrium value M_0, is well approximated by an exponential function of time

$$M_z(t) = M_0 - (M_0 - M_z(0))\exp(-t/T_1) \qquad (1.13)$$

The relaxation time T_1 is called the 'longitudinal' or 'spin–lattice' relaxation time. The relaxation rate T_1^{-1} is a measure of the rate at which energy is exchanged between the nuclear spin system and the lattice, i.e. the rate at which transitions between nuclear magnetic (Zeeman) energy levels are induced by the spin–lattice interactions. Since the spin–lattice interactions for a dilute molecular gas are usually *intra*-molecular spin-dependent interactions such as the *spin–rotation* and *dipolar* interactions, T_1^{-1} is also governed by spectral densities such as $J_1(\omega_0)$ and $J_2(\omega_0)$. The precise relationship between T_1^{-1} and the spectral densities has been developed quite rigorously in the general theory of magnetic resonance[2]. Therefore, measurements of T_1^{-1} as a function of the nuclear Larmor frequency ω_0 in a dilute gas should give a maximum whenever ω_0 is tuned to a rotational frequency associated with a pair of molecular levels between which the spin–lattice interactions have non-zero matrix elements. No direct measurements of 'nuclear spin relaxation molecular spectroscopy' have ever been performed – and for good reason: typical rotational frequencies fall in the infrared. The largest Larmor frequencies possible in the highest conceivable magnetic fields correspond to hundreds of MHz. In practice, the information obtained from measurements of T_1 in dilute molecular gases is closely related to that obtained using the molecular spectroscopy technique known as 'non-resonant absorption'.

1.1.2   Relationship between T_1 and non-resonant absorption in molecular gases

1.1.2.1   The T_1 minimum

Since the rotational frequencies are usually much greater than ω_0, the spin–lattice relaxation in dilute gases is determined primarily by that part of the spin–lattice interaction which couples free molecular states having the same energy. The effect of collisions is to distribute the spectral density of the *intra*-molecular spin-dependent interactions over a range of frequencies of order τ_c^{-1} about the pure rotational frequencies where τ_c is the correlation time for collisions which interrupt the molecular rotation or cause molecular reorientation. Since T_1^{-1} is governed by the intensity of $J_1(\omega)$ and/or $J_2(\omega)$ at the nuclear transition frequencies, a plot of T_1 v. density ρ gives a characteristic minimum in the vicinity of the density at which $\tau_0^{-1} \simeq \omega_0$. This is illustrated in Figure 1.2. In the dilute gas $\tau_c^{-1} \propto \rho$. At very low densities

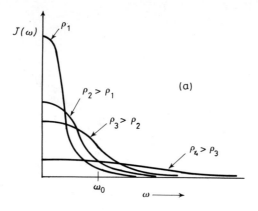

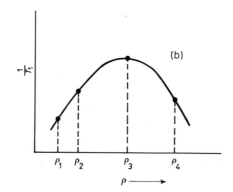

Figure 1.2 (a) Illustration of the $\Delta J = 0$ term of the spectral density $J_1(\omega)$ or $J_2(\omega)$ at several different densities. (b) Schematic plot of the relaxation rate of the nuclear spins in a dilute molecular gas as a function of density. The relative relaxation rates associated with the spectral densities in (a) are shown

$\omega_0 \tau_c \gg 1$ and $J(\omega_0)$ increases with increasing density thereby increasing the relaxation rate T_1^{-1}. At high densities $\omega_0 \tau_c \ll 1$ and $J(\omega_0)$ decreases with increasing density.

The first observation of the T_1 minimum in a dilute molecular gas was performed by Hardy[6] on H_2, and his measurements at 78 K are shown in Figure 1.3. The special significance of Hardy's work has been reviewed in detail elsewhere[7]. In the case of H_2, the spin–rotation and dipolar coupling constants are known accurately from molecular beam experiments. As may be seen from Figure 1.3, the minimum value of T_1 is sensitive to these coupling constants, and the agreement between $(T_1)_{min}$ and theory is excellent. For other molecules in which the measurement of spin–rotation coupling constants by molecular beam techniques is more difficult, they may be deter-

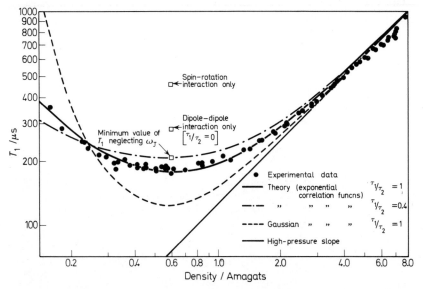

Figure 1.3 Plot of T_1 as a function of density in the H_2 gas at 77 K as measured by Hardy
(From Hardy, W. N.[6], by courtesy of the National Research Council, Ottawa)

mined from the measurement of the T_1 minimum as will be discussed later. Hardy's measurements also demonstrated that the spectral densities $J_1(\omega)$ and $J_2(\omega)$ were Lorentzian, i.e. that they had the form of equation (1.10) and the ratio of the correlation times τ_1 for $J(t)$ and τ_2 for $Y_{2\mu}(\theta(t),\phi(t))$ was consistent with the value of 0.6 predicted for the $J = 1$ state of H_2.

1.1.2.2   *Non-resonant absorption*

The phenomenon of non-resonant absorption[8, 9] essentially corresponds to 'electric-dipole molecular spectroscopy' associated with collisional broadening of the $\omega_{mn} = 0$ terms in equation (1.10). Most absorption experiments have been carried out at microwave frequencies[8] though it is not difficult to measure the density dependence of the dispersion term $\chi'(\omega)$ at frequencies[10] in the 1–100 MHz range commonly used in T_1 measurements. Non-resonant absorption does not give interesting information in the case of diatomic molecules since the only non-zero matrix elements of the molecular electric dipole moment operator correspond to $\Delta J = \pm 1$. For symmetric-top molecules, however, the matrix elements $<JKM\,|\,\boldsymbol{\mu}\,|\,J'K'M'>$ are non-zero for $J'-J = 0,\pm 1$ and $M'-M = 0,\pm 1$. The dependence of $\chi''(\omega)$ on density is very similar to that of T_1^{-1} and gives information on the strength of the molecular electric dipole moment and on the correlation time for molecular re-orientation as determined by the correlation function of $Y_{1\mu}(\theta(t),\phi(t))$, where θ,ϕ denotes the orientation of the molecular symmetry axis.

1.2　EXPERIMENTAL METHODS

1.2.1　The spin-echo method

The most common method used to study nuclear spin relaxation in gases has been the 'spin-echo' technique[2, 11]. In addition to measuring T_1, the spin-echo technique is the best available method for measuring the 'transverse' or 'spin–spin' relaxation time T_2 and the spin diffusion constant D. The spin-

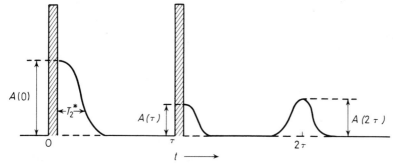

Figure 1.4　Schematic illustration of an oscilloscope display of a spin-echo experiment involving a two-pulse sequence. The two pulses are applied at $t = 0$ and τ, respectively. The maximum amplitudes of the free induction decay following each pulse is indicated by $A(0)$ and $A(\tau)$, respectively, and the echo amplitude by $A(2\tau)$

echo technique is illustrated schematically in Figure 1.4. In this illustration, two r.f. pulses of length t_w such that $\gamma H_1 t_w = \pi/2$ are applied at times $t = 0$ and τ, respectively. The pulses each rotate the nuclear magnetisation by 90 degrees. Free induction signals of maximum amplitude $A(0)$ and $A(\tau)$ are observed after the first and second pulses, respectively. In addition, a characteristic echo of maximum amplitude $A(2\tau)$ appears at $t = 2\tau$. If the z-component of the nuclear magnetisation relaxes towards its equilibrium value exponentially as given by equation (1.13), then it is easily shown that

$$A(0) - A(\tau) = A(0)\exp(-\tau/T_1) \tag{1.14}$$

Furthermore, if the spatial motion of the molecules in the inhomogeneous magnetic field is adequately described by the solution to the diffusion equation, the echo amplitude is given by[2, 11–14]

$$A(2\tau) \propto \exp\left[-\frac{2\tau}{T_2} - \frac{2}{3}\gamma^2 G^2 D\tau^3 \right] \tag{1.15}$$

where it is assumed that the time-independent magnetic field has a gradient G which is constant over the sample volume. The decay time T_2^* of the free induction signals (see Figure 1.4) is inversely proportional to the gradient and provides a method of measuring G.

Since the spin-echo technique was first introduced by Hahn[11], the method has been refined in many ways, e.g. by Carr and Purcell[12], Clark[15] and others. We do not attempt to review the various pulse trains and detection methods in current use — a measure of the extent to which the technique has become standardised is that several commercial spin-echo spectrometers of sufficiently high quality to perform most feasible measurements of interest are now available. We confine ourselves here to a brief discussion of some of the experimental factors associated with the applicability of n.m.r. to molecular gases. A more detailed discussion has been given by Hardy[6, 16] and Clark[15].

1.2.2   Signal-to-noise considerations

The free precession of the nuclear spins gives rise to an induced voltage in a coil whose axis is perpendicular to the field $H_0 = H_0 \hat{z}$. The maximum voltage $A(0)$ is proportional to the nuclear Larmor frequency and the equilibrium magnetisation, which is given by Curie's law[2]

$$A(0) \propto \omega_0 M_0 = (\gamma H_0) \frac{N \gamma^2 \hbar^2 <I(I+1)> H_0}{3 kT} \tag{1.16}$$

where N is the number of molecules per cm^3 and $<I(I+1)>$ is the mean squared value of the nuclear spin per molecule. It is important to produce a rotation of the magnetisation by the desired angle, usually 90 or 180 degrees, in a time $t_w \ll T_2^*$. This requirement can be the factor which limits the size of the sample used, since T_2^* decreases with increasing sample size, while the power required to achieve a given value of H_1 increases. Typical values of t_w attainable using standard methods with samples having volume of about $10 \; cm^3$ range down to about $1 \; \mu s$.

Often the same coil used to transmit the r.f. power is used to pick up the free precession signal, though some workers prefer to use a crossed-coil system to minimise the direct r.f. signal picked up by the amplifier from the transmitter. In any case, it is necessary to match the pick-up coil to the amplifier for optimum signal-to-noise. When this is properly done, the noise at the input to the r.f. amplifier is determined by the thermal noise in the sample coil. In general, it is possible to satisfy this condition at room temperature using conventional techniques. The r.f. amplifier and detector system of free induction signals of the type indicated schematically in Figure 1.4 is necessarily broad-band. In order to improve the signal-to-noise ratio, the pulse sequence is repeated many times, the signal sampled each time by means of a suitable gated detector and the information stored in digital form or by charging a condenser in a circuit having a long time constant. The time between successive sampling sequences must be much greater than T_1 while the sampling time must be $\approx T_2^*$. Therefore, the signal-to-noise ratio increases as T_2^*/T_1 increases. In practice, the signal-to-noise can be made very high after a few minutes of sampling for typical proton and fluorine resonances in gases at densities $\rho \gtrsim 1$ Amagat, which is the number density at standard temperature and pressure (1 Amagat $= 2.69 \times 10^{19}$ molecules cm^{-3} for an ideal gas). At low densities, and especially below the T_1 minimum, the signal-to-noise decreases very rapidly with decreasing ρ. Not only is $N \propto \rho$, but T_1 increases and T_2 decreases as ρ decreases[6]. Since T_2 generally becomes less than the value of T_2^* determined by the homogeneity of the magnet at low densities, the use of a very homogeneous magnet does not generally give a very significant improvement in signal-to-noise at very low densities. So far, the lowest density at which T_1 measurements have been reported is 0.006 Amagats of CH_4 gas at room temperature[56]. This corresponds to a density of about 2×10^{17} molecules cm^{-3} or 8×10^{17} spins cm^{-3} and may be taken as a rough indication of the lowest density at which practical measurements are feasible. Each measurement at 0.006 Amagats took about 1 hour to carry out.

At the present time, most of the new developments in the improvement of signal-to-noise involve the use of computers for data processing and control of the experiments. It seems to the author that the 'state of the art' in this regard has not been adequately described in the literature at the time of writing. Therefore, in order to help anyone intending to learn the experimental techniques now being used, some of the active laboratories in the study of nuclear spin relaxation in gases are listed below.

Table 1.1   Some of the active laboratories in the study of nuclear spin relaxation in gases

Name	Department	University
R. L. Armstrong	Physics	Toronto
M. Bloom	Physics	British Columbia
H. Y. Carr	Physics	Rutgers
P. S. Hubbard	Physics	North Carolina
M. Lipsicas	Belfer Graduate School of Science	Yeshiva
J. W. Riehl	Chemistry	Pennsylvania State
N. J. Trappeniers K. O. Prins	van der Waals Laboratorium	Amsterdam
J. S. Waugh	Chemistry	M.I.T.

1.2.3 Impurity effects

The effect of small amounts of molecular impurities on nuclear spin relaxation in molecular gases of low density is usually negligible. The reason for this is that because of the large separation of the molecules at low densities, the mean squared value of the intermolecular spin-dependent interactions in dilute gases is always much smaller than that of the intra-molecular interactions, even for paramagnetic impurities. For monatomic gases, however, even small amounts of paramagnetic impurities can play an important role in spin–lattice relaxation as will be discussed in Section 1.7.

In many molecular gases, T_1 can be as long as the time taken for a molecule to reach the walls of the container. Since the time taken for a molecule to reach the walls is proportional to the density, relaxation due to collisions with the walls can be difficult to distinguish from relaxation in the pure gas above the T_1 minimum in which case T_1 is also proportional to the density. As will be discussed in Section 1.6.2, there is now some experimental evidence that wall relaxation may be playing a role in proton spin relaxation in the CHF_3 and CH_3F gases.

1.3 REVIEW OF THE THEORY OF NUCLEAR SPIN RELAXATION IN DILUTE MOLECULAR GASES NEGLECTING PERMUTATION SYMMETRY OF THE NUCLEI

1.3.1 General remarks

It is often convenient to express the relaxation rate for nuclear spin systems in molecular gases, liquids and solids as the sum of three contributions

$$1/T_1 = R_A + R_B + R_C \tag{1.17}$$

where R_A is the relaxation rate due to *intra*-molecular dipolar and quadrupolar interactions, i.e. spin–lattice interactions of the type given by equation (1.12) in which the spin operators transform as irreducible tensors of the second rank and the lattice operators are also second rank tensors involving the rotational coordinates of the molecule, R_B is the relaxation rate due to *inter*-molecular dipolar interactions and R_C is the relaxation rate due to spin–rotation interactions. For diatomic molecules, the spin–rotation interaction for a given nucleus is characterised by a single coupling constant as given by equation (1.11) but more parameters are required for polyatomic molecules as will be discussed later.

The effective strengths of the *intra*-molecular interactions, which are simply the hyperfine interactions of the individual molecules, are independent of density. By contrast, the effective strength of the *inter*-molecular dipolar interactions is proportional to the density. Thus, except for such systems as monatomic gases in which R_A and R_C are zero, one can always make the approximation $R_B \ll R_A + R_C$ at low densities. In this section, we shall summarise the theoretical expressions which have been developed for R_A and R_C for dilute molecular gases. Before doing so, it should be pointed out that, though equations (1.17) and (1.13) are sufficient to interpret almost all of the experimental data available at the present time, a more general theory may be required for future experiments. *Intra*-molecular interactions other than those associated with R_A and R_C may manifest themselves in molecular gases[2], interference effects must sometimes be taken into account[17] and the approach to equilibrium can be non-exponential[18].

In the general theory of nuclear spin relaxation[2], R_A, R_B and R_C are expressed in terms of the spectral density of the correlation functions of the lattice operators of the spin–lattice interaction; the spectral density is evaluated at frequencies corresponding to the energy differences between nuclear spin states coupled by the spin–lattice interaction. The basic formulae obtained for a system of molecules each containing S equivalent nuclei of spin I are as follows[2,3]

$$(R_A)_{\text{dipolar}} = \frac{S-1}{6} A_d^2 I(I+1)[J_2(\omega_0) + 4J_2(2\omega_0)] \tag{1.18}$$

with a similar expression for quadrupolar interactions[3] and

$$R_C = 4\pi^2 C^2 J_1(\omega_0) \qquad (1.19)$$

where A_d is the dipolar coupling constant for a pair of spins separated by $r = (r,\theta,\phi)$ as defined in equation (1.12), $J_2(\omega)$ and $J_1(\omega)$ are spectral densities of the correlation functions of $Y_{20}(\theta,\phi)$ and of K_z respectively. In equations (1.18) and (1.19), ω_0 stands for the *difference* between the nuclear and molecular Larmor frequencies, the difference between these frequencies appearing because the molecular hyperfine interactions couple states satisfying $\Delta M_I + \Delta M_J = 0$. Though the molecular Larmor frequency is usually much less than the nuclear Larmor frequency, its influence on T_1 has been observed in the gas[6] (see Figure 1.13). In the derivation of equations (1.18) and (1.19), effects associated with the Bose–Einstein or Fermi–Dirac statistics of the nuclear spins have not been included explicitly. The effect of the symmetry or anti-symmetry of the molecular wave functions with respect to permutation of a pair of nuclear spins is to introduce correlations between the nuclear spins and also between the nuclear spin and rotational states. Such correlations lead to departures from the predictions of equations (1.18) and (1.19), which we shall discuss later in connection with the interpretation of the experimental results.

1.3.2   Low-density approximations

We distinguish between two types of simplifying approximations which can be made at low densities. The first of these, which has been adequately discussed elsewhere[7, 19, 88, 89] is that at sufficiently low densities the influence of three-body and high-order collisions can be neglected. In most cases of interest, this results in the widths of the rotational levels being proportional to density at constant temperature. A *second* type of low density approximation can be made when the collision frequency is much less than any of the rotational frequencies. Since the rotational frequencies are usually much greater than ω_0, the contribution to T_1^{-1} of the matrix elements of the spin–lattice interaction between rotational states of different energies is quenched. To illustrate this effect, the correlation functions of the lattice operators have been approximated as follows[3, 20]

$$G_i(t) = [G_i(t)]_{\text{free}} g_i(t) \qquad (1.20)$$

where $i = 1$ and 2 refer to the correlation functions of K_z and Y_{20}, respectively. The free molecule correlation function $[G_i(t)]_{\text{free}}$ can be easily expressed in terms of the 'intensities' $G_{i,mn}$ associated with the molecular states $|m\rangle$ and $|n\rangle$ in analogy with equation (1.4). Denoting the sum of all the zero-frequency terms ($\omega_{mn} = 0$) by $G_{i,0}$, we write

$$J_i(\omega_0) = \int_{-\infty}^{\infty} e^{-i\omega_0 t} G_i(t) dt = \sum_{m,n} \int_{-\infty}^{\infty} e^{-i\omega_0 t} G_{i,mn} \cos(\omega_{mn}t) g_i(t) dt$$

$$= \tfrac{1}{2} \sum_{m,n} G_{i,mn}[j_i(\omega_0 - \omega_{mn}) + j_i(\omega_0 + \omega_{mn})]$$

$$\simeq G_{i,0} j_i(\omega_0) \text{ at low densities} \qquad (1.21)$$

where $j_i(\omega)$ is related to $g_i(t)$ by equation (1.8) and the approximation made in the last line of equation (1.21) is only valid at low densities (where $\omega_{mn}\tau_i \gg 1$ for the case of exponential correlation functions as in equations (1.9) and (1.10)). The effective strength of the ith interaction is thus quenched by a factor

$$q_i = G_{i,0}/G_i(0) \tag{1.22}$$

For diatomic molecules at high temperatures, it is easily shown (see Section 1.3.3) that $q_1 = 1$ and $q_2 = 0.25$. In general, the lattice operators associated with R_A tend to be quenched much more in the gas than those associated with R_C, while the opposite is true in dense fluids and solids[21]. The reason for this is that J is a good quantum number for free molecules while in a dense fluid or solid, crystalline electric fields of low symmetry remove the rotational degeneracy thus quenching J in a manner similar to the quenching of the orbital angular momentum of electrons in solids[22].

Explicit expressions have been derived for $[G_1(t)]_{\text{free}}$ and $[G_2(t)]_{\text{free}}$ for diatomic, spherical-top and symmetric-top molecules. These expressions are required to interpret the experimental data and are summarised in the next section.

1.3.3 Explicit expressions for correlation functions of the lattice operators at low densities

1.3.3.1 Diatomic molecules

For diatomic molecules, $\mathcal{H}_{\text{sr}}$ is given by (1.11) with $K = J$ so that

$$[G_1(t)]_{\text{free}} = G_1(0) = \;<J_z^2> \;= \tfrac{1}{3}<J(J+1)> \;= \tfrac{1}{3}\sum_J P_J J(J+1) \tag{1.23}$$

where

$$P_J = \frac{(2J+1)e^{-J(J+1)\alpha}}{\sum_J (2J+1)e^{-J(J+1)\alpha}}, \quad \alpha = \frac{\hbar B}{kT} = \frac{T_0}{T} \tag{1.24}$$

where B and T_0 are the molecular rotational constant in units of $\hbar$ and k, respectively.

At high temperatures, the sum in equation (1.23) may be well approximated by an integral[23] and evaluated to give $<J(J+1)> \simeq T/T_0$; hence

$$[G_1(t)]_{\text{free}} \simeq \frac{1}{3}\frac{T}{T_0} \tag{1.25}$$

At low temperatures, $P_J \ll 1$ for $J \neq 0$, so that

$$[G_1(t)]_{\text{free}} \simeq 2e^{-2\alpha} \tag{1.26}$$

The expression for $[G_2(t)]_{\text{free}}$ for diatomic molecules has been shown to be[24]

$$[G_2(t)]_{\text{free}} = G_2(0)\sum_J P_J \sum_{q=\pm 2, 0} [C(J2,J+q;00)]^2 \cos[(2J+1+q)qBt] \tag{1.27}$$

where $G_2(0) = 1/4\pi$ and $C(\quad)$ is a Clebsch–Gordan coefficient. The sum over J has been evaluated in the high-temperature limit in closed form by O'Reilly[24], but it is clear from equation (1.26) that since $[C(J2J;00)]^2$ approaches $\frac{1}{4}$ as $J \to \infty$, $q_2 \simeq \frac{1}{4}$ at high temperatures. Also *at low temperatures*

$$q_2 \simeq [C(121;00)]^2 P_1 = \frac{6}{5}e^{-2T_0/T} \tag{1.28}$$

1.3.3.2 Spherical-top molecules

The spin–rotation interaction for each nucleus in a spherical-top molecule is characterised by two coupling constants $C_{\parallel}$ and $C_{\perp}$ corresponding to the two principal values of the spin–rotation tensor. For example, in tetrahedral molecules such as CH_4, $C_{\parallel}$ and $C_{\perp}$ for a given proton characterise the spin–rotation energy when J is parallel and perpendicular to the C—H bond, respectively. Thus, when equation (1.11) is used for a spherical-top molecule, CK for a given nucleus is given in spherical tensor notation by[3, 25, 26]

$$CK_m = (-1)^m C_a J_m - (\tfrac{2}{3})^{1/2} C_d \sum_{\mu} (-1)^{\mu} C(112;m\mu) D^2_{0,-(m+\mu)} J_{\mu} \tag{1.29}$$

where $D^2_{0,-(m+\mu)}$ is a rotation matrix[3, 26] and

$$C_a = \tfrac{1}{3}(C_{\parallel} + 2C_{\perp}), \quad C_d = C_{\perp} - C_{\parallel} \tag{1.30}$$

Hubbard[25] and Blicharski[27] first arrived at expressions for T_1 due to spin–rotation interactions in *liquids* composed of spherical-top molecules. They found that the effective value of the spin–rotation coupling constant was $(C_a^2 + \tfrac{2}{9}C_d^2)^{\frac{1}{2}}$. Though their theories involved some simplifying assumptions, Dong[28] has recently shown by a direct calculation that *at high temperatures*

$$G_1(0) = \tfrac{1}{3}(C_a^2 + \tfrac{2}{9}C_d^2)\langle J(J+1)\rangle \tag{1.31}$$

where

$$\langle J(J+1)\rangle = 3/2\alpha \tag{1.32}$$

for spherical-top molecules. Bloom et al.[3] extended the theory to *dilute gases* composed of spherical and symmetric-top molecules. Their result for spherical-top molecules *at high temperatures* may be summarised in terms of the value of the quenching factor (see equation (1.21))

$$q_1 = \frac{C_a^2 + \tfrac{4}{45}C_d^2}{C_a^2 + \tfrac{2}{9}C_d^2}, \quad G_{i,0} = \frac{1}{2\alpha}(C_a^2 + \tfrac{4}{45}C_d^2) \tag{1.33}$$

Similarly, *at high temperatures*, $q_2 = 0.2$ for spherical-top molecules. We do not give any results for the low temperature limit because it seems impossible to study gases composed of spherical-top or symmetric-top molecules at low enough temperatures to satisfy the low temperature limit. The reason for this is that at low temperatures the vapour pressure of all gases composed of spherical-top molecules is too small to give an observable n.m.r. signal.

1.3.3.3 Symmetric-top molecules

For symmetric-top molecules of the type CX_3Y, the spin–rotation interaction for the Y nucleus, which is located on the molecular symmetry axis, is again characterised by two coupling constants and it is a straightforward matter to calculate the effective value of the spin–rotation coupling constant. This has been done[3, 20, 29] but since the result is quite complicated and will not be used here explicitly, we do not reproduce it. It should be noted, however, that the last terms in equations (9) and (11) of Reference 20 should be multiplied by a factor of $\frac{1}{9}$, though equations (18) and (19) of that paper which are based on equation (11), are correct.

The quenching factor q_2 for symmetric-top molecules is quite interesting and gives rise to easily observable effects. Suppose that the vector joining two nuclei which interact via the dipolar interaction, *or* the vector representing the axis of symmetry of the electric field gradient for a nucleus experiencing a quadrupolar interaction, makes an angle θ_0 with the axis of symmetry of the molecule. Then the quenching factor is given by[3]

$$q_2 = \left(\frac{3\cos^2\theta_0 - 1}{2}\right)^2 < C(J2J;KO)^2 > + \tfrac{3}{4}\sin^4\theta_0 < C(J2J;1,-2)^2 >_{K=1}$$

(1.34)

where contributions to the second term are only obtained for $K = 1$, K being the quantum number giving the projection of J on the molecular symmetry axis. The physical interpretation of the first term in equation (1.34) is simple. In the dilute gas, the symmetry axis of the molecule precesses many times about J between collisions thus resulting in an averaging of the spin–lattice interaction for a tensor of rank 2 by an amount $P_2(\cos\theta_0)$. The second term in equation (1.34) arises from the degeneracy of the $K = \pm 1$ states which are coupled by the tensor of rank 2 interaction.

1.4   EXPERIMENTAL RESULTS FOR LOW-DENSITY GASES OF DIATOMIC MOLECULES

1.4.1   The effects of permutation symmetry—'nuclear spin symmetry species'

All studies of n.m.r. in gases thus far have involved only the ground electronic states of molecules. For a homonuclear diatomic molecule in the ground electronic state, it can be shown[30] that the operation of permutating the identical nuclei is formally equivalent to the parity operation. Since the molecular hyperfine interactions are all invariant with respect to the parity operation, they do not mix nuclear spin states which are even or odd with respect to permutation. The product of the spin and the vibration–rotation wave functions are, of course, even or odd depending on whether the nuclei obey Bose–Einstein or Fermi–Dirac statistics, respectively. The states of homonuclear diatomic molecules are thus rigorously separated into two 'nuclear spin symmetry species' called *ortho* or *para* corresponding to the species of greater or lesser nuclear spin statistical weight, respectively. It is well known that the rate of conversion between *ortho*- and *para*-H_2 and D_2 due to collisions in the pure gas is extremely slow so that it is possible to do experiments on *ortho*- and *para*-hydrogen mixtures of any composition. Presumably this could also be true of other molecules such as N_2 or Cl_2.

However, no one has ever produced gaseous samples of any molecule other than H_2 or D_2 in which the ratio of the concentrations of the different nuclear spin symmetry species is different from the high-temperature equilibrium value.

One anticipates that the *ortho-* and *para-* species of a gas of homonuclear diatomic molecules will relax at different rates thus giving a relaxation curve which involves the sum of two exponentials. However, the basic theoretical approach described in Section 1.3 is still valid providing one calculates T_1 for each spin symmetry species separately. Clearly, $[\mathscr{H}_{sr}, I] = 0$ for diatomic molecules in which $\gamma_1 = \gamma_2$ (see equation (1.11)) so that the expression for R_C is the same for the *ortho-* and *para-*species. The corresponding result for R_A is not so obvious. For H_2, only *ortho-*H_2 contributes to the nuclear magnetisa-tion, since $I = 1$ for *ortho-*H_2 and $I = 0$ for *para-*H_2. For molecules such as D_2 and N_2 composed of nuclei of spin 1, however, the *ortho-*species has six nuclear spin states arising from $I = 0$ and 2, while the *para-*species has 3 states from $I = 1$. For nuclei with $I > \frac{1}{2}$, the inequality $(R_A)_{quad} \gg R_C$ is usually satisfied and this is true of D_2 and N_2, in particular. It may be shown by calculating the transition rates among the states $| I,M_I >$ and identifying the nuclear magnetisation of N molecules of a given species with $N \sum\limits_{I,M_I}$ $M_I P_{I,M_I}$, where P_{I,M_I} is the probability that a molecule of a given species is in the state $| I,M_I >$, that the magnetisation of the *ortho-* and *para-*species of molecules such as N_2 and D_2 each relax exponentially with a relaxation rate

$$(R_A)_{quad} = \left(\frac{3\pi}{20}\right)\left(\frac{eqQ}{\hbar}\right)^2 \left[J_2(\omega_0) + 4 J_2(2\,\omega_0)\right] \qquad (1.35)$$

where $J_2(\omega)$ for the *ortho-* and *para-*species are evaluated using an ensemble of molecules restricted to even and odd values of J, respectively. At low temperatures, the *ortho-*species should relax much more slowly than the *para-*species as has been observed for D_2 by Hardy[16], while at high temper-atures the relaxation rates should be almost identical as has been verified by Speight and Armstrong[31] for N_2. In establishing equation (1.35), it was found that the transitions between the $I = 0$ and 2 states of the *ortho-*species do play an essential role. It would be interesting to investigate whether a result similar to equation (1.35), which is identical to that obtained by Abragam for $I_1 = I_2 = I$ (see equation (138) of Reference 2), holds for $(R_A)_{quad}$ in general. When spin symmetry is neglected, Abragam (Reference 2, p.134) has shown that the relaxation should be non-exponential in the long correlation time limit, i.e. at densities below the T_1 minimum (see also Ref-erence 18).

1.4.2   The use of nuclear spin relaxation in H_2 gas to study intermolecular forces

One of the most important applications of the study of nuclear spin relaxation in gases has been the study of intermolecular interactions, especially the anisotropic intermolecular potential which gives rise to transitions between the molecular states $| JM_J >$. The H_2 molecule has been especially useful as a probe for the anisotropic potential because a very large fraction of the *ortho*-H_2 molecules are in the $J = 1$ state at temperatures below about 250 K. Under these conditions, the reduced correlation functions $g_i(t)$ are each characterised by a single time constant τ_i and the experimental result can be used in a simple way to evaluate the velocity averaged cross-section σ_i defined in terms of the velocity dependent cross-section $\sigma_i(v)$ by

$$\frac{1}{\tau_i} = \rho\overline{\sigma_i(v)v} = \rho\sigma_i\bar{v} \qquad (1.36)$$

where v is the relative speed of two molecules and the bar represents an average over a Maxwell–Boltzmann velocity distribution. It would be useful at this time to have a critical review of the potentialities and limitations of the n.m.r. method of obtaining intermolecular forces from measurements of σ_i, especially as compared with other techniques, but we shall not attempt to supplement the already existing discussions on this topic[1, 7, 32, 33]. Instead the history of the development of the theory on this subject is briefly surveyed below and the most important recent experimental work in intermolecular forces using T_1 measurements in dilute H_2 gas is briefly outlined in Section 1.4.3.

The original theory of nuclear spin relaxation in H_2 gas was due to Schwinger (Reference 2, pages 316–321), who restricted himself to low temperatures where all of the *ortho*-H_2 molecules are in the ground state. Schwinger's theory was extended to include higher rotational states by Needler and Opechowski[34] and their model was generalised by Freed[35] using the Redfield method. Freed calculated, in detail, the consequences of the 'strong collision limit'. In order to understand the interesting (and at that time puzzling) dependence of T_1 on *ortho*-H_2 and temperature[7, 36, 37], the intermolecular potential had to be taken into account in an explicit and well-defined way. This was achieved first using the methods of statistical mechanics and the 'constant acceleration approximation'[7, 19, 32] and later by Gordon[38] using a classical kinetic theory approach. Gordon has also discussed the close relationship between nuclear spin relaxation and other types of measurements[1].

During the last few years two important theoretical developments have taken place. One is the development of a quantum-mechanical Boltzmann equation appropriate for the treatment of nuclear spin relaxation in gases by Chen and Snider[39]. The second is the carrying out of a detailed scattering calculation for realistic forms of both the isotropic and anisotropic parts of the H_2–He potential using the 'distorted wave' and 'close-coupling' approximations[40].

1.4.3   Hydrogen—recent experimental work

1.4.3.1 H_2

All T_1 studies thus far performed with a view to obtaining information on intermolecular potentials have been done at densities well above the T_1 minimum, in which case $T_1 \propto \rho$. Though the expression for $g_i(t)$ becomes fairly complicated when higher rotational states of *ortho*-H_2 are populated[7], the weakness of the anisotropic interactions in the case of H_2–H_2 and H_2–He (and other inert gas atoms) collisions enables a precise determination of the relationship between T_1/ρ and a small number of parameters associated with cross-sections for transitions of *ortho*-H_2 between different rotational states[7]. A major theoretical weak point at the present time is in the treatment of *inelastic* collisions. This point has been emphasised recently[41] in the interpretation of measurements of T_1/ρ in normal-H_2 (75% *ortho*-H_2) up to 738 K. The conclusion was reached that although most inelastic collision cross-sections are too small to be important in H_2–H_2 collisions, certain inelastic collisions between *ortho*-H_2 and *para*-H_2 molecules do play an important role; e.g. those in which the change in the rotational quantum number $(J,J') \Leftrightarrow (J'',J''')$ gives a partial compensation of the change of rotational energy as in the cases $(1,2) \Leftrightarrow (3,0)$ or $(1,4) \Leftrightarrow (3,2)$. The questions raised

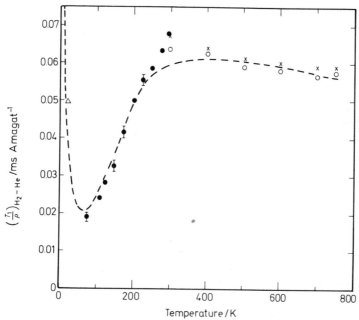

Figure 1.5 Plot of T_1/ρ as a function of temperature for H_2 infinitely diluted in helium gas. The experimental data are derived from extrapolations to 100% helium concentration of measurements of T_1/ρ as a function of helium concentration: $\triangle$, Reference 44; $\bullet$, Reference 43; $\circ$, Reference 42 (quadratic extrapolation); x Reference 42 (linear extrapolation). The significance of the theoretical curve is discussed in the text

in this paper could be resolved by studying ultrasonic relaxation in *ortho–para*-H_2 mixtures or by systematically studying the T_1 minimum in H_2 gas as a function of temperature.

In the case of T_1 studies of H_2 diluted in helium gas, the inelastic collisions should not play an important role and an explicit relationship can be derived[7, 42] between T_1/ρ and the cross-section for transitions between *any* degenerate magnetic substates M_J of H_2. Results have been obtained for T_1/ρ as a function of temperature between liquid N_2 and room temperature by Riehl et al.[43], from room temperature to 730 K by Lalita[42] and, much earlier, at 20 K by Bloom[44]. They are plotted in Figure 1.5. Remarkable agreement is obtained with the results of the calculations of Kinsey et al.[40] using an a

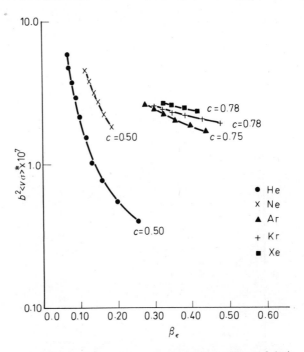

Figure 1.6 Plots of the temperature dependence of $\langle \sigma v \rangle$ for collisions between H_2 and different rare gas atoms derived from measurements of T_1/ρ. The significance of the theoretical fits has been discussed by Foster[45]. The parameters b and c are measures of the strength of short range and long range terms in the anisotropic potential. $\beta = 1/kT$ and ε is the Lennard–Jones parameter for the interacting pair (From Foster, K. R.[45], by courtesy of the author)

priori H_2–He potential. However, J. W. Riehl (private communication) has found on repeating the early measurement of Bloom at 20 K that he obtains a value of T_1/ρ which is much smaller than that shown in Figure 1.5. Further work is needed to clarify this point (see also Ref. 116).

A very complete study of T_1/ρ in H_2–Ne, H_2–Ar, H_2–Kr and H_2–Xe mixtures as a function of temperature between 165 K and 265 K has just been completed by Foster[45]. In this thesis, Dr Foster makes an exhaustive and critical comparison between the different theoretical approaches to the interpretation of T_1 data[7, 40, 42, 43], and discusses the relationship between the observed attractive part of anisotropic H_2–rare gas interaction and that predicted by molecular polarisability measurements. A summary of Foster's results analysed to give $\langle \sigma v \rangle$ as a function of temperature for all the systems studied by him is given in Figure 1.6.

1.4.3.2 **HD**

T_1/ρ for the proton and deuteron in HD gas was measured by Hardy[16] between 30 and 313 K. These results are shown in Figure 1.7 together with a value previously obtained[44] at 20.5 K. These results illustrate the predictions of equations (1.18)–(1.28) beautifully. At low temperatures, most of the temperature dependence of T_1 for both isotopes is associated with the activation energy for the $J = 1$ state. Since the relaxation rate for the proton $T_1^{-1}(P)$ is associated with R_C and that of the deuteron $T_1^{-1}(D)$ with R_A, and since the spin–rotation and dipolar plus quadrupolar coupling constants

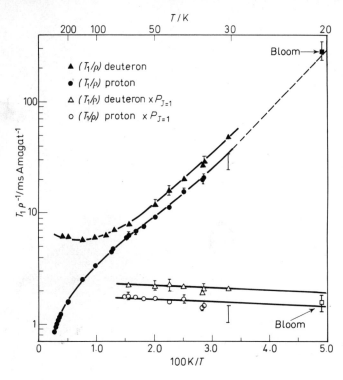

Figure 1.7 Plots of T_1/ρ v. $100/T$ for protons and deuterons in HD gas. $P_{J=1}$ is the probability that an HD molecule is in the $J = 1$ state (From Hardy, W. N.[16], by courtesy of the author)

are known for HD[46], the observed ratio of $T_1(P)/\rho/T_1(D)/\rho \simeq 1.33$ *at low temperatures* can be interpreted to give τ_2/τ_1 if one assumes that $g_i(t) = \exp(-t/T_i)$. The result obtained is $0.9 < \tau_2/\tau_1 < 1.07$ which is consistent with the value of unity to be expected if the dominant process for molecular reorientation involves a transition from the $J = 1$ to the $J = 0$ state. Finally, the fact that *at high temperatures* $T_1(P)/\rho$ decreases while $T_1(D)/\rho$ does not, is a consequence of the fact that higher rotational states are populated at high temperatures and that $G_{1,0} \propto \langle J(J+1) \rangle$ while $G_{2,0} \propto \langle J(J+1)/(2J-1)(2J+3) \rangle$ according to equations (1.23) and (1.27), respectively.

1.4.3.3 D_2

Hardy has also measured the approach to equilibrium of the deuteron spin system in D_2 [16]. As expected, two relaxation times were observed in D_2 gas, one associated with *ortho*-D_2 molecules ($I = 0,2; J = 0,2,4,...$) and the other with *para*-D_2 ($I = 1; J = 1,3,5...$). The ratio of the *amplitudes* of the two components was found to be 4.9 in agreement with the predicted value of 5 for normal-D_2. A plot of T_1/ρ for *ortho*- and *para*-D_2 as a function of temperature is shown in Figure 1.8.

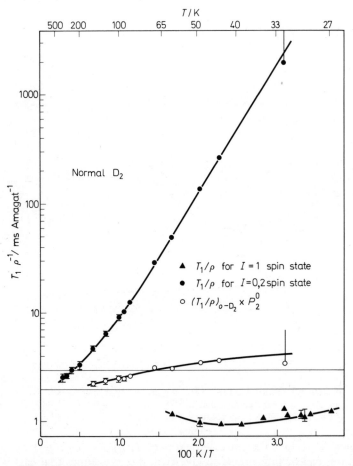

Figure 1.8   Plots of T_1/ρ v. $100/T$ for the $I = 1$ spin states of para-D_2 and the $I = 0,2$ spin states of *ortho*-D_2 as measured in normal-D_2. P_2^0 is the probability that an *ortho*-D_2 molecule is in the state $J = 2$
(From Hardy, W. N.[16], by courtesy of the author)

As may be seen from this plot, $(T_1/\rho)_{ortho} > (T_1/\rho)_{para}$ over the entire range. At low temperatures, most of the variation with temperature of $(T_1/\rho)_{ortho}$ is associated with the activation energy for the population of the $J = 2$ state as expected theoretically. Furthermore, the difference between $(T_1/\rho)_{ortho}$ and $(T_1/\rho)_{para}$ becomes small at high temperatures as predicted by the theory. The results given in Figure 1.8 have not yet been analysed systematically in order to obtain insight on intermolecular potentials. From that point of view, it would seem fruitful to extend these measurements to study the dependence of $(T_1/\rho)_{ortho}$ and $(T_1/\rho)_{para}$ on the ratio of *ortho-* to *para-*D_2. There is an advantage to studying D_2 in that the rate of molecular reorientation can be obtained from the even as well as the odd rotational states.

1.4.4 F_2

T_1/ρ has been measured for F_2 gas at temperatures of 222 K, 290 K and 333 K by Courtney and Armstrong[47] using data over the density range from 4 to 15 Amagats. The spin–rotation coupling constant of $C = 157$ kHz is so large for F_2 that $R_C \gg R_A$. A common signature of $T_1^{-1} = R_C$ is that $T_1/\rho \propto T^{-3/2}$. For $g_1(t) = e^{-t/\tau_1}$, equations (1.19) to (1.25) give

$$\frac{1}{T_1} = R_C = \frac{8\pi^2 C^2 T \tau_1}{3 T_0} \tag{1.37}$$

It may also be shown that for weak collisions in which the isotropic potential may be adequately approximated by a hard sphere potential, $\sigma(v) \propto v^{-2}$ so that equation (1.36) gives $\sigma_1 \propto T^{-1}$ or $\tau_1 \propto T^{1/2}/\rho$ thus yielding a $T^{-3/2}$ law for T_1/ρ. This was the experimental result obtained by Courtney and Armstrong, who derived an effective cross-section for molecular reorientation due to F_2–F_2 collisions of $\sigma_1 = 12.4 \, \text{Å}^2$ at 290 K from their data.

1.4.5 N_2

As discussed in Section 1.4.1, Speight and Armstrong[31] showed in a careful experiment on N_2 gas that the approach to equilibrium of the nitrogen nuclear spin system is exponential to a very good approximation, thus demonstrating that the *ortho-*N_2 and *para-*N_2 spin species have almost identical relaxation times as expected from the theory of Section 1.3.3 at high temperatures. The experiment on N_2 gas was carried out between 145 and 380 K and at pressures between 50 and 700 atm. It is of unusual interest to us for another reason; the second type of low density approximation discussed in Section 1.3.2 was observed to be violated in a clear-cut way as the density was increased.

In the case of N_2, $(R_A)_{quad} \gg (R_A)_{dipolar}$, R_C so that $T_1^{-1} \simeq (R_A)_{quad}$ to a very good approximation. Speight and Armstrong used equations (1.18) and (1.20) with $g_2(t) = \exp(-|t|/\tau_2)$, equation (1.21) *without the low density approximation* and equation (1.27) to obtain the following result at high temperatures in the short correlation time limit $\omega_0 \tau_2 \ll 1$

$$\frac{1}{T_1} = (R_A)_{quad} \simeq \frac{3}{32}\left(\frac{eqQ}{\hbar}\right)^2 \tau_2[1 + 3pe^p E_1(p)] \tag{1.38}$$

where

$$E_1(p) = \int_p^\infty \frac{e^{-x}}{x}\, dx$$

and

$$\frac{1}{p} = \frac{16BkT}{\hbar}\tau_2^2 = 16\langle J^2\rangle (B\tau_2)^2 \qquad (1.39)$$

The dimensionless parameter p is unity when the mean squared splitting of rotational states satisfying $\Delta J = \pm 2$ (i.e. states connected by the dipolar interaction) is equal to the width of the rotational states due to collisions.

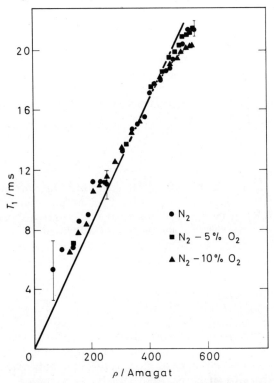

Figure 1.9 Plot of T_1 as a function of density in pure nitrogen and in nitrogen–oxygen mixtures at 293 K (From Speight, P. A. and Armstrong, R. L.[31], by courtesy of the National Research Council, Ottawa)

As $p \to \infty$. $p e^p E_1(p) \to 0$ in which case the quadrupolar interactions are partially quenched, while as $p \to 0$, $pE_1(p) \to 1$. This corresponds to all the rotational lines overlapping with each other in this simple model, in which case the quenching of R_A no longer occurs. It will be recalled from the discussion of equation (1.27) that the quenching factor $q_2 \simeq 1/4$ at high temperatures.

The experimental results for T_1 as a function of density at room temperature are shown in Figure 1.9. Also shown on the same graph are measurements of T_1 v. ρ for samples containing 5% and 10% O_2 so as to demonstrate that the density dependence in the lower part of the density range studied was not due to contributions to the relaxation from paramagnetic impurities. The main point to be made is that the data cannot be fitted by a straight line passing through the origin. The reason for this is that the second term

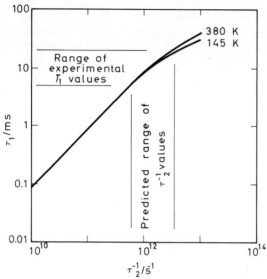

Figure 1.10 Plot of the theoretical dependence of T_1 on τ_2^{-1} for N_2 according to equation (1.38). The range of experimental values obtained in gaseous N_2 and the corresponding values of τ_2^{-1} are indicated
(From Speight, P. A. and Armstrong, R. L.[31], by courtesy of the National Research Council, Ottawa)

in equation (1.38) is not negligible in the region of densities studied as may be seen from a plot of T_1 $v. \tau_2^{-1}$ based on equation (1.38) shown in Figure 1.10. The range of experimental T_1 values falls in range of τ_2^{-1} in which the departure from linearity is appreciable.

1.4.6 HCl, HBr and HI

T_1/ρ has been measured for HCl between 228 K and 338 K, for HBr between 22 K and 360 K and for HI between 293 K and 337 K by Tward and Armstrong[48]. In each case, they found that $T_1/\rho \propto T^{-3/2}$ and that the relaxation is by spin–rotation interactions, as discussed for F_2 earlier. The cross-sections satisfy $\sigma_1 \propto T_0^{-1}$ roughly, and have a value of 98 Å for HCl at 263 K, 103 Å for HBr at 316 K and 129 Å for HI at 314 K, each of these temperatures being roughly in the middle of the temperature range studied. These very large cross-sections are not surprising for molecules having large electric dipole moments. It does raise questions, however, about the criteria for obtaining $R_C \propto T^{+3/2}$ which were mentioned earlier. In each of these cases, the collisions leading to molecular reorientation do not seem to be predominantly weak since σ_1 is greater than the kinetic cross-section. The meaning of the word 'weak' in 'weak collisions' may be ambiguous here. In the case of H_2 which involves small values of J, the statement that the collisions are weak implies that the probability that a molecule in a state $|J,M_J\rangle$ performs a transition to any other state $|J'M'_J\rangle$ in a single collision is small. In the case of HCl, HBr or HI, however, $\langle J(J+1)\rangle \gg 1$ at room temperature which makes it more appropriate to use a semi-classical theory of relaxation. In the semi-classical theory of nuclear spin relaxation in gases as formulated by Gordon[38], the results obtained for R_C are the same as given by equations (1.19), (1.21), (1.23), (1.25), (1.26) and (1.36) with cross-section σ_1 in equation (1.36) expressed in terms of $\langle(\Delta J)^2\rangle$, which is the mean squared change in J per collision for an impact parameter b as follows[38]

$$\sigma_1 = \frac{1}{2\langle J^2\rangle}\int \langle(\Delta J)^2\rangle\, 2\pi b\, db \qquad (1.40)$$

The condition required to obtain a $T^{\frac{3}{2}}$ law is that the collision integral in equation (1.40) be independent of temperature in this formulation[48]. In my opinion, the weak collision approximation which is implicit in this semi-classical treatment is that $\langle(\Delta J)^2\rangle \ll \langle J^2\rangle$ for the range of impact parameters which contribute appreciably to σ_1.

More recently, an independent study of T_1 in gaseous HCl has been carried out[48a] between 304 K and 423 K. In addition, measurements have been made on HCl–argon mixtures and the cross-section σ_1 for HCl–HCl and HCl–argon collisions have been obtained. This represents the first serious study of T_1 in gas mixtures for a polar molecule and the results have already been interpreted[48b] using 'exact collisional calculations'. The main interest in the HCl–argon system is that it is an example of a relatively well-defined system in which perturbation methods cannot be readily applied. Students of the art of measuring and interpreting T_1 in gases will be interested in studying critically the way in which some important departures from linearity in the dependence of T_1 in density were handled when these interesting results were being obtained[48a].

1.5   EXPERIMENTAL RESULTS FOR LOW-DENSITY GASES OF SPHERICAL-TOP MOLECULES

1.5.1   The T_1 minimum in spherical-top molecules containing fluorine nuclei

In general $R_C \gg R_A$ for fluorine nuclei in molecular gases so that we can put $T_1^{-1} \simeq R_C$. In order to apply the theory of Section 1.3 to experiment, some assumption has to be made about the form of the reduced correlation function $g_1(t)$ in equation (1.20), since there exists no theoretical prediction on this point. If one assumes, as is customary that $g_1(t) = \exp(-|t|/\tau_1)$, then (1.19)–(1.21), (1.33) and (1.10) give the following result for spherical-top molecules in dilute gases[3, 20]

$$\frac{1}{T_1} = R_C = \frac{4\pi^2}{\alpha}C_{\text{eff}}^2\frac{\tau_1}{1+\omega_0^2\tau_1^2} \qquad (1.41)$$

where $\alpha = T_0/T$ and the 'effective spin–rotation coupling constant' C_{eff} for spherical-top molecules in dilute gases is given by

$$C^2_{eff} = C^2_a + \tfrac{4}{45}C^2 \qquad (1.42)$$

To test the *ad hoc* assumption about the form of $g_1(t)$, it is necessary to measure T_1 as a function of ρ near the T_1 minimum. This was first achieved by Dong[20] for CH_4, CF_4, CHF_3 and CH_3F. Recently, Courtney and Armstrong[49] have studied CF_4 with improved accuracy and have also obtained

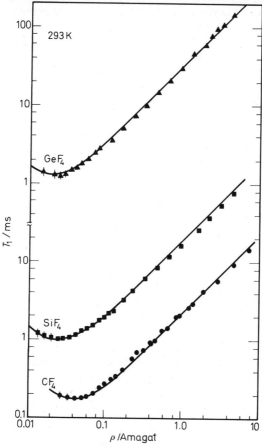

Figure 1.11 Plots of T_1 as a function of density for the fluorine spins in CF_4, SiF_4 and SF_6 showing the T_1 minimum in these gases (Courtney, J. A. and Armstrong, R. L., private communication). The results of Table 1.2 are based on these data
(From Courtney, J. A., Ph.D Thesis, University of Toronto, by courtesy of the author)

results for SiF_4, GeF_4 and SF_6. Their results for the tetrahedral molecules at 28 MHz are shown in Figure 1.11. The theoretical curve is a two-parameter fit using equations (1.36) and (1.41), the two unknown parameters being C_{eff}

and σ_1. As may be seen from Figure 1.11, the fit is good in each case which indicates that the relaxation theory of Section 1.3 in which nuclear spin symmetry is neglected is probably adequate for the interpretation of fluorine spin relaxation. The spin–rotation parameters extracted from the data of Figure 1.11 by Courtney and Armstrong are shown in Table 1.2. The values of C_a and C_d obtained from molecular beam measurements on these systems[50] are shown for comparison. In the molecular beam experiments, the beam

Table 1.2   Spin–rotation constants in spherical-top molecules
(From Courtney, J. A., Ph.D. Thesis, University of Toronto, by courtesy of the author)

Molecule	C_{eff}^2/kHz^2†	C_a^2/kHz^2*	C_d^2/kHz^2‡	C_d^2/kHz^2*
CF_4	47.6 ± 1.1	46.5 ± 1.1	12.5^{+17}_{-12}	<200
SiF_4	6.30 ± 0.22	5.86 ± 0.10	4.95 ± 3.60	<9
GeF_4	4.02 ± 0.10	3.53 ± 0.06	5.50 ± 1.80	<9
SF_6	22.9 ± 1.2	27.1 ± 4.1	(<65)	—

*From molecular beam experiments – Reference 50 and private communications from I. Ozier
†From T_1 minimum
‡Deduced from * and † by use of equation (1.42)

consists of molecules distributed over a very large number of rotational states because of the small values of the rotational parameter for these molecules. Since the nuclear magnetic resonance associated with the different rotational states in the beam are not resolved, the values of C_a and C_d are obtained from an analysis of the molecular beam n.m.r. line shape. As may be seen from the values of C_a and C_d obtained by Ozier et al.[50], this procedure yields quite precise values of C_a, but not of C_d. Therefore, the values of C_{eff} obtained from the measurements of the T_1 minimum can usually be used with the molecular beam measurements of C_a to give a value of C_d better than that obtainable from the molecular beam measurements alone. The derived values of C_d following this procedure are also shown in Table 1.2.

In the case of CH_4, it has proven to be possible to do molecular beam measurements on molecules in well-defined rotational states[51] thus leading to precise values of both C_a and C_d. However, it is necessary to take into account the permutation symmetry of protons in CH_4 in order to interpret the T_1 measurements, as discussed in the next two sections.

1.5.2   Nuclear spin symmetry considerations in CH₄ gas

When the permutation symmetry of the four protons in CH_4 is taken into account the centrifugal distortion energy partially removes the degeneracy of the states characterised by the J quantum number[52]. It may be shown that the non-degenerate states of a J manifold may be classified according to the irreducible representations of the tetrahedral group. In the case of CH_4, there is a one-to-one correspondence between the total nuclear spin per molecule I and the irreducible representations. The *meta* (A), *ortho* (T) and *para* (E) spin symmetry species have $I = 2, 1$ and 0 respectively.

Since the spin–rotation interaction is linear in the nuclear spin operators the only non-zero matrix elements satisfy the selection rules $\Delta I = 0, \pm 1$. Therefore, in addition to a large term at zero frequency in $[G_1(t)]_{\text{free}}$ corresponding to matrix elements of the spin–rotation interaction between states $A \leftrightarrow A$ and $T \leftrightarrow T$, there exist terms at non-zero frequencies corresponding to certain non-zero $A \leftrightarrow T$, $T \leftrightarrow E$ and $T \leftrightarrow T$ matrix elements between states within the same J manifold[53]. Most of the intensity of the centrifugal distortion spectrum of $J_1(\omega)$ in CH_4 at room temperature is in the vicinity of 200 MHz.

From the properties of the spin–rotation interaction[26], we know that the C_a (scalar) term involves only zero frequency matrix elements while the C_d (tensor) term *does* connect states whose energies differ by the centrifugal distortion energy. A simple extrapolation of the theory given in Section 1.3 and equation (1.41) and (1.42) then gives the following expression for R_C which should be valid at high temperatures under conditions that the relaxation is exponential

$$R_C = \frac{4\pi^2}{\alpha}\left[C_a^2 \frac{\tau_1}{1+\omega_0^2\tau_1^2} + \frac{4}{45}C_d^2 \sum_k \frac{F_k\tau_k}{1+(\omega_0-\omega_k)^2\tau_k^2}\right] \qquad (1.43)$$

where the F_k satisfy the normalisation condition $\Sigma_k F_k = 1$ and give a measure of the sum of squares of the matrix elements of the tensor part of the spin–rotation interaction between states having energy difference $\hbar\omega_k$. It is clear that the problem of spin–lattice relaxation in CH_4 gas at low densities is closely related to the rate of conversion between nuclear spin symmetry species. In fact the conversion rate has been calculated for CH_4 gas by Curl et al.[54] in terms of the molecular hyperfine interactions. Ozier and Yi[55], commenting on their paper, pointed out the importance of the centrifugal distortion splittings which had been neglected by Curl et al. in their calculation and computed the conversion rate in the dilute gas taking these splittings into account.

1.5.3 The T_1 minimum in CH_4 gas

It would be very strange if the influence on nuclear spin symmetry and the centrifugal distortion splitting did not manifest themselves in a study of the T_1 minimum of CH_4 since the characteristic spectrum associated with the C_d contributions is predicted to lie in the radio-frequency regime. In the first study of the T_1 minimum of CH_4, no evidence for the centrifugal distortion splitting was detected[20], but a more careful experiment performed recently[56] has revealed the characteristic effects of the distortion splitting on the proton spin relaxation. The experimental plot of T_1^{-1} v. ρ at 30 MHz is shown in Figure 1.12. The two curves show the predicted density dependence using the theory of Section 1.3 with the known value[51] of C_{eff} and the predictions for $C_d = 0$. At the highest densities, the rotational levels are broadened by an amount larger than the average centrifugal distortion frequency so that the theory of Section 1.3 may be used in this regime. At the density of the T_1 minimum (or T_1^{-1} maximum) the broadening of the rotational levels is much less than the average centrifugal distortion and the terms involving C_d are almost completely quenched. The difference between T_1^{-1} and the predicted T_1^{-1} for $C_a = 0$ is directly associated with the centrifugal distortion spectrum. The fact that it peaks at a density about six times that at which the T_1 minimum occurs is consistent with the average centrifugal distortion splitting being in the vicinity of 200 MHz.

In an interesting theoretical paper, McCourt and Hess[57] have shown how to incorporate the theory of nuclear spin relaxation in spherical-top molecules

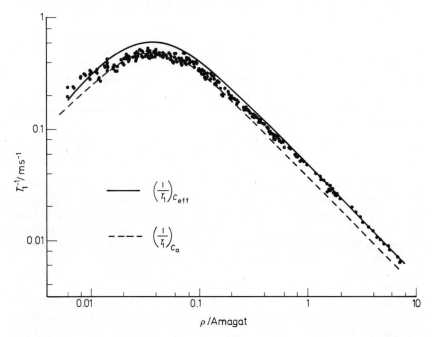

Figure 1.12 Plot of the experimental values of $1/T_1$ v. density in CH_4 gas. The significance of the comparison with the theoretical curves is discussed in the text
(From Beckman, P. A. *et al.*[56], by courtesy of the National Research Council, Ottawa)

such as CH_4 into the Boltzmann equation approach developed by Chen and Snider[39]. Some of their comments on the role of nuclear spin symmetry in CH_4 have been strongly criticised by Ozier[53] so that some of the numerical results are incorrect[58]. These points are not, however, directly related to the main arguments on the kinetic theory of transport and relaxation processes which McCourt and Hess stressed in their paper.

1.5.4   Studies of the temperature dependence of T_1/ρ

The temperature dependence of T_1/ρ in the short correlation time region has been measured over the temperature range of 112–800 K for CH_4 [3, 59], 112–295 K for CD_4 [3], 244–371 K for CF_4 [60] and 231–337 K for SiF_4 [60]. In the cases of CH_4, CF_4 and SiF_4, the relaxation is governed by spin–rotation interactions and in each case it was found that $T_1/\rho \propto T^{-3/2}$ within experimental error. In the case of CD_4 the relaxation is due to quadrupolar inter-actions, and, as may be seen from Figure 1.13, T_1/ρ varied with temperature more slowly than $T^{-\frac{1}{2}}$, which would be the expected power law for the same model which gives $T^{-\frac{3}{2}}$ for spin–rotation interactions. Thus we must conclude that the ratio τ_1/τ_2 is temperature dependent in the case of methane, a result which is not yet explained in detail[3]

An interesting result obtained by Lalita[59] is that T_1/ρ for CH_4 diluted in helium gas varies according to $T^{-0.8}$ instead of the usual $T^{-\frac{3}{2}}$ for spin–rotation interactions. This result could be due to the short range nature of the CH_4–He intermolecular potential. However, in view of the recently discovered influence of the centrifugal distortion splitting on the density dependence of T_1 in CH_4, discussed in the previous section, these results must be re-examined to see whether the temperature dependence of the average centrifugal distortion splitting could account for the unusual temperature dependence of T_1/ρ. Lalita's measurements were carried out between room temperature and 100 K and the lowest density at which measurements could be made decreased with increasing temperature because of signal-to-noise considerations, thus introducing the possibility of a systematic effect on the apparent T_1/ρ due to the centrifugal distortion splitting.

1.6 EXPERIMENTAL RESULTS FOR LOW-DENSITY GASES OF SYMMETRIC-TOP AND ASYMMETRIC-TOP MOLECULES

1.6.1 The deuterated methane system—effects of molecular geometry

The most striking prediction from the simple theory of Section 1.3 concerning nuclear spin relaxation in symmetric-top molecules is that important geometrical effects are predicted for R_A, but not for R_C. The reason for this is that the dominant contribution to R_C is unusually the scalar interaction $C_a I \cdot J$ which is not affected by the precession of the molecular symmetry axis about J between collisions, while the second-rank tensor type of interaction which contributes to R_A is averaged by an amount $P_2(\cos \theta_0)$ where θ_0 is the angle between the molecular symmetry axis and the particular molecule-fixed vector associated with the second-rank tensor interaction. The study of T_1/ρ v. temperature for the proton and deuteron spin systems in CH_4, CH_3D, CH_2D_2, CHD_3 and CD_4 has verified the existence of this geometrical effect[3]. In these molecules, the protons relax primarily due to spin–rotation interactions so that $T_1^{-1} \simeq R_C$ while the deuterons relax due to quadrupolar interactions so that $T_1^{-1} \simeq R_A$. For the protons, it was found that $T_1/\rho \propto T^{-\frac{3}{2}}$ for all the molecules while for the deuterons T_1/ρ varies somewhat more slowly than $T^{-\frac{1}{2}}$ as discussed in Section 1.5.4.

The small differences in T_1/ρ values for the proton spins found experi-mentally[3] for the different molecules $CH_{4-n}D_n$ can be understood in terms of the variation of $\langle J(J+1) \rangle$ with the rotational constants, but the much larger differences in T_1/ρ for the deuteron spins, which are shown in Figure 1.13, cannot be understood without taking into account the molecular geometry as specified by equation (1.34). When this is done, and it is assumed that the electric field gradient tensor in each case is axially symmetric about the C–D axis, which must surely be a good approximation, then the ratio of T_1/ρ for CD_4, CHD_3 and CH_3D are completely explained. Although no attempt has been made to interpret T_1/ρ for the asymmetric-top CH_2D_2 system, the methods described in Section 1.3 should be directly applicable.

At the time that the experiments described above were carried out, the value of the quadrupolar coupling constant was not well known, the only estimate being $eqQ/\hbar \simeq 2\pi \times 100$ MHz from a molecular beam experiment[61]. By assuming that $\tau_2 \lesssim \tau_1 \lesssim 3\tau_2$ on the basis of general theoretical considerations[3] for molecules in states of large J it was found that the following limits on $eqQ/\hbar$ could be placed,

$$2\pi \times 130 \text{ kHz} < eqQ/\hbar < 2\pi \times 200 \text{ kHz}$$

Since then the quadrupolar coupling constant has been measured to be[62]

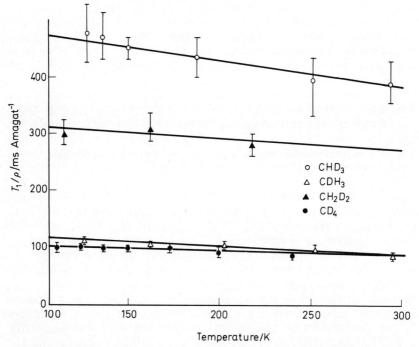

Figure 1.13   Plots of T_1/ρ v. temperature for the deuteron spins in CH_3D, CH_2D_2, CHD_3 and CD_4.
(From Bloom, M. *et al.*[3], by courtesy of the National Research Council, Ottawa)

$eqQ/\hbar = 2\pi \times (191.48 \pm 0.77)$ kHz in agreement with these limits. The fact that the correct value of $eqQ/\hbar$ is near the top of the range allowed by the T_1 measurements indicates that $\tau_1 \approx \tau_2$ for methane gas.

Another interesting result was that the relative values of τ_1 and τ_2 for the different molecules $CH_{4-n}D_n$ were consistent with the intermolecular interactions being independent of n within experimental error. The fact that the centre of mass is not coincident with the geometrical centre for methane molecules other than the spherical-top molecules CH_4 and CD_4 does not seem to produce large changes in the cross-section for molecular reorientation. This is not so for the analogous system of H_2, D_2 and HD, where it has been found[16] that the HD cross-section for reorientation is much greater than H_2 or D_2.

1.6.2 The fluorinated methane system—evidence for 'steps' in the density dependency of T_1/ρ

The T_1 minimum of the fluorine resonance in CHF_3 and the proton resonance in CHF_3 was measured by Dong and Bloom[20]. In addition, the observed density dependence of T_1/ρ for the proton resonance was observed to undergo 'steps'[63], i.e. in the case of CH_3F, T_1/ρ was roughly constant up to a density of about 25 Amagats at room temperature, then decreased over a range of about 10 Amagats by about 30% to a new roughly constant value. The same type of behaviour was observed for the proton resonance in HF_3, while the T_1/ρ for the fluorine resonance in each of these systems was independent of density over the entire density range.

At the time of writing this review, there is strong evidence[64, 65] that some, and possibly all of these effects can be classified as 'impurity effects' of the second type discussed in Section 1.2.3, though the possibility of the existence of 'steps' even in the pure gases is not yet completely excluded either experimentally or theoretically[39, 66]. A discussion on a possible physical origin of the 'steps' will be given in Section 1.8 in which the passage to the dense gas is discussed. Here, we shall briefly summarise the experimental situation at the present time.

Courtney and Armstrong[64] obtain the same value of T_1/ρ as Dong and Bloom[63] at the higher densities for CHF_3 but no evidence of a T_1 minimum at 0.04 Amagats. The study of the T_1 minimum for fluorine resonances, in which the value of T_1 at the minimum is quite short, has some pitfalls. The very short value of T_2, which decreases still more at densities below the T_1 minimum, makes it necessary to exercise unusual care in the processing of the very weak signals to avoid systematic errors. The Toronto group has identified a spurious T_1 minimum in BF_3 in their own earlier work[64, 67].

The proton T_1/ρ values in CH_3F recently obtained at Toronto[64] and M.I.T.[65] are slightly higher than those reported and the decrease in T_1/ρ above 25 Amagats is less abrupt than reported by the U.B.C. group[63]. However the difference between the density dependence of T_1/ρ for protons and fluorine remains. Much larger values of T_1/ρ are now being observed for the protons in CHF_3 at low densities in all these laboratories indicating that wall relaxation effects have been playing an important role. Whether the 'steps' themselves are due to wall relaxation has not yet been established, but work is in progress at U.B.C., Toronto and M.I.T. on this problem. It may be added that in Dong's later unpublished work[68], he found a correlation between the 'steps' in T_1/ρ and 'steps' in the proton–fluorine nuclear Overhauser effect, indicating that the enhanced relaxation was at least due to the 'unquenching' of the dipolar interactions, which make no contribution to the proton spin relaxation at densities below the 'steps'.

1.6.3 PH$_3$ and NH$_3$

Measurements on T_1/ρ for the protons in PH$_3$ at 289 and 330 K indicate that $T_1/\rho \propto T^{-\frac{3}{2}}$ and that the relaxation is governed by spin–rotation interactions[69]. Similar remarks can be made of NH$_3$ [64].

1.6.4 BF$_3$

Hinshaw and Hubbard[70] have measured T_1 as a function of density at 270, 300 and 350 K for ^{11}B and ^{19}F nuclei in BF$_3$ gas. Measurements of the fluorine T_1 had been reported previously for this system[64]. The ^{11}B relaxation is governed by quadrupolar interactions and the fluorine spin relaxation by spin–rotation interactions. Part of the motivation for the more recent study was to look for the type of 'steps' in T_1/ρ described in Section 1.6.2. No such behaviour was found. However, the density dependence of T_1 for the boron spins was different than for the fluorine spins. The bending of the T_1 v. ρ curve exhibited at 300 K as the density was increased from a few Amagats to 80 Amagats is similar to the results obtained by Speight and Armstrong[31] in N$_2$ and discussed in some detail in Section 1.4.5. Over the same density range, the fluorine T_1 was linear in ρ. It is likely that the density dependence of $(R_A)_{\text{quad}}$ in BF$_3$ could be interpreted in a similar way to that described in Section 1.4.5 for N$_2$.

1.7 RELAXATION DUE TO INTERMOLECULAR SPIN-DEPENDENT INTERACTIONS

1.7.1 Monatomic gases

Nuclear spin relaxation in monatomic gases is completely due to intermolecular spin-dependent interactions, i.e. $T_1^{-1} = R_B$ (see equation (1.17)). For nuclei of spin $\frac{1}{2}$ such as ^{3}He and ^{129}Xe, it was first believed that the dominant spin–lattice interaction would always be the dipolar interaction for which the 'constant acceleration approximation' gives[19]

$$\frac{1}{T_1} = (R_B)_{\text{dipolar}} = \frac{2\gamma^4 \hbar^2 I(I+1)}{a^2} \left(\frac{\pi m}{kT} \right)^{\frac{1}{4}} \rho \tag{1.44}$$

where the atoms have mass m and diameter a, and the number density of the gas is ρ. For ^{3}He, the result obtained at NTP is $T_1 \simeq 5 \times 10^6$ s $\simeq 60$ days while the result for ^{129}Xe is almost two orders of magnitude larger. It is not surprising, in view of this large value of T_1, that there exists no published measurement of $(R_B)_{\text{dipolar}}$ in pure monatomic gases at this time.

Measurements of T_1 v. density have been made on ^{3}He gas in the vicinity of 4 K [71] and higher temperatures[72–74]. In each case the spins relax via collisions with the surface of the sample holder or by diffusion into the material comprising the surface. The surface relaxation rate R_S is easily distinguished from R_B through its density dependence. Whereas $R_B \propto \rho$, $R_S \propto \rho^{-1}$ since it is limited by the rate of diffusion of the atoms to the surface. Most of the measurements[72-74] on ^{3}He gas have been made using optical pumping methods in which case it is also possible to observe an interesting type of nuclear spin relaxation due to diffusion in the inhomogeneous magnetic field[75]. This process is important at low pressures and small magnetic field.

Measurements of R_B in pure ^{3}He gas at 4.2 K have recently been obtained for the first time[76]. At these temperatures, it is possible to carry out a precise partial waves scattering analysis of the cross-section for nuclear spin transitions using only a few partial waves. Shizgal has recently out such an analysis[77] using the best available potential for the He–He interaction and it will be interesting to compare the experimental results with this theory.

In the case of ^{129}Xe, a new relaxation mechanism has been found by Carr and his co-workers[78, 79] which is stronger by several orders of magnitude than the dipolar interaction. The interaction responsible for this relaxation has been conclusively shown to be associated with a *transient* spin–rotation coupling between the nuclear spins and the angular momentum of a pair of colliding atoms[80]. The predicted relationship between T_1 and the pressure-dependent chemical shift arising from the same interaction[81] has been verified experimentally[79].

In addition to ^{129}Xe, T_1 and the chemical shift have been measured for ^{131}Xe and ^{83}Kr which relax due to a collision-induced quadrupolar interaction. The experimental results have been reviewed elsewhere[82] and are well understood[83].

1.7.2   Influence of paramagnetic impurities

Relatively small concentrations of paramagnetic molecules can contribute appreciably to $(R_B)_{dipolar}$ because of their relatively large magnetic moments. Just as (1.44) follows from equations (77) and (115) of chapter 8 of Reference 2, a similar formula for dipolar interactions with paramagnetic molecules follows from the corresponding equations (88) and (118), i.e.

$$(R_B)_{dipolar} = -\frac{4\,\gamma^2\gamma_e^2\hbar^2 S(S+1)}{a^2}\left(\frac{2\pi\mu}{kT}\right)^{\frac{1}{2}}c\rho \qquad (1.45)$$

where the magnetic moment μ_e and electronic spin S of the paramagnetic molecule are assumed to be related by $\mu_e = \gamma_e\hbar S$, a is the effective diameter of a solvent plus impurity molecule pair, μ is their reduced mass and c is the molar concentration of paramagnetic molecules.

The first systematic study of nuclear spin relaxation by paramagnetic impurities in molecular gases was carried out by Johnson and Waugh[84] for mixtures of CH_4 and O_2 and it was found that equation (1.45) was approximately valid for this case. Some interesting insight into the limitations of equation (1.45), however, was provided by an investigation of the CH_4–O_2 and CH_4–NO systems by Siegel and Lipsicas[85].

When the structure of the paramagnetic molecule must be taken into account, i.e. coupling between the electronic spin and the molecular rotation, it has been customary to replace $\gamma_e^2\hbar^2 S(S+1)$ in equation (1.45) by an effective magnetic moment $\langle\mu_{eff}^2\rangle$, (see equation (117), chapter 8 of Reference 2) and to associate $\langle \mu_{\mathrm{eff}}^2 \rangle$ with the effective magnetic moment required to properly interpret the static susceptibility of a paramagnetic molecular system[86]. The experimentally measured ratio of $\langle \mu_{\mathrm{eff}}^2 \rangle$ for NO to that of O_2 is found to be smaller according to T_1 measurements[85] than given by measurements of static susceptibility. Lipsicas and Siegel have explained this effect in terms of the relatively large coupling of the electronic spin and molecular rotation in NO as compared with O_2. It is well known that the contribution of the off-diagonal matrix elements of the electronic magnetic moment operator between states differing in energy by ΔE is appreciable if $\Delta E \lesssim kT$. In order to contribute appreciably to R_B however, a more restrictive condition has to be satisfied, namely $\Delta E \lesssim \hbar/\tau_{\mathrm{coll}}$, where $\tau_{\mathrm{coll}} \simeq (\mu a^2/kT)^{\frac{1}{2}}$ is the duration of a single collision. If this condition is violated, as it is in NO but not as much in O_2 in the experiments of Siegel and Lipsicas[85], the effectiveness of the off-diagonal elements of $\boldsymbol{\mu}_e$ is reduced considerably.

The study of T_1 in monatomic gases with small amounts of paramagnetic molecular impurities could be used to check these structural effects on $\langle \mu_{\mathrm{eff}}^2 \rangle$, but the available data[87] are not yet sufficient to make such a check.

1.8 HIGH-DENSITY GASES

1.8.1 Some general comments on the passage from low to high densities

As discussed in Sections 1.3.2 and 1.4.5, two distinct types of effects must be taken into account at high densities. Account must be taken of the effect of three-body and higher-order collisions and, in addition, the effect of overlapping rotational levels must be properly treated.

The effect of higher order collisions has been treated theoretically in two ways. An expansion of T_1 in powers of density, applicable for any form of the intermolecular potential, has been made with the help of the constant acceleration approximation[19]. At the moment, this theory can only be applied to monatomic gases or to systems such as molecular hydrogen in which a small number of rotational states are populated. An alternative approach which uses the Enskog kinetic theory of dense gases has been suggested by Gordon[88] and has been found to fit the available data for H_2 at 195 and 298 K for a hard sphere *inter*-molecular potential[89]. The hard sphere model does not fit the data at 77 K which were taken up to higher densities[89]; nor do the same experimented results fit a theory[91] designed for liquid densities[92]. In general, the experiments which have been carried out thus far at densities at which 3-body collisions are important are not adequate to provide a *critical* test of the available theories, taking into account the effects of a realistic potential.

In the theoretical discussion of Section 1.3, it was implicitly assumed that the effect of collisions is simply to broaden the rotational levels of the molecules. This would correspond, for example, to $g_i(t)$ in equation (1.20) being a monatomically decreasing function of time. In dense liquids or solids, the *form* of the correlation functions is very different from equation (1.20). Often, the molecules can be pictured as being orientated for very long times.

as compared with the rotational periods, within a narrow range of solid angle and undergoing reorientational jumps in a stochastic manner. The correlation function $G_2(t)$ for such behaviour is often well approximated by $G_2(t) = (1/4\pi) \exp(-t/\tau_2)$, i.e. the molecules behave as though their rotational spectrum has collapsed, to speak in the language of Section 1.3.

The collapse of rotational spectra of various sorts – infrared and Raman[88, 93, 94], non-resonant absorption[88, 95], etc. – is already signalled in gases which are dense enough for the molecular collision frequency to be of the order of the rotational frequencies. The physical origin of the collapse lies in the fact that when a molecule undergoes a collision-induced transition from a given rotational state to another, other rotational states undergo *correlated* transitions. Some of these other states may be connected to the given state by the transition operator associated with the type of spectroscopy being carried out. The resultant transfer of the transition moment from one pair of states to another can, and usually, does, give rise to a type of motional narrowing or collapse of the spectrum. It was thought that this mechanism was responsible for the steps in T_1/ρ observed by Dong[63, 68] for CHF_3 and CH_3F. However, as discussed in Section 1.6.2, there is now considerable doubt as to the validity of these results.

1.8.2 Separation of the contributions to the nuclear spin relaxation rate of the inter- and intra-molecular interactions

In high density gases, it is generally not possible to say in advance what is the relative importance of R_A, R_B and R_C in nuclear spin relaxation. Usually, for nuclei of $I > \frac{1}{2}$ in a molecule in which the electric field gradient is not zero at the nuclear site, the *intra*-molecular quadrupolar relaxation is dominant, i.e. $R_A \gg R_B$, R_C. However, for nuclei with spin $\frac{1}{2}$ it is generally necessary to separate the different contributions experimentally. Many papers have been written about this problem in connection with liquids and no attempt will be made to review the entire literature here. The most precise and clear-cut work of this type involving the passage from the low-density gas to the high-density gas and liquid is that on the proton T_1 in methane and its deuterated modifications by Trappeniers and his co-workers[96, 97]. Their study involves measurements of the proton T_1 in CH_4, CH_3D, CH_2D_2, CHD_3 and on mixtures of CH_4–CD_4. In addition, use is made of the relationship between R_B and the diffusion constant D, together with independent measurements[98] of D. One of the main results in these papers is that the molecular reorientation in liquid methane is not well described by rotational diffusion[25], a result which may also hold for other nearly spherical molecular liquids[21]. In fact, the theoretical approach given in Section 1.3 here seems to give a reasonably good description of relaxation in dense methane gas.

1.8.3 Studies near the critical point

The properties of fluids near their critical points have a special interest at this time[99] so that it is not surprising that some attempt has been made to study this region using n.m.r. methods. It has been pointed out[100] that, because of the relatively short-ranged nature of the *inter*-molecular interactions responsible for molecular reorientation, striking effects associated with T_1 are *not* expected to occur near the critical point. This has been borne out by studies in ethane[100], methane[96, 97], ethylene[101] and sulphur hexafluoride[102].

The spin-echo method can also be used to study the diffusion constant D in fluids, as discussed in Section 1.2.1. One report of an 'anomaly' in D near the critical point has been reported in ethane[100, 103]. In similar studies of methane[98], sulphur hexafluoride[102] and xenon[104], no such anomaly has been found and a general attitude of scepticism has been expressed about the validity of the interpretation of the ethane experiments[100, 103]. In view of the tricky nature of these experiments, this seems to be the most sensible attitude to take at this time until the experiment has been repeated. At this time, the only clearly established critical behaviour using n.m.r. methods involves a detailed study of the density dependence of the chemical shift in xenon near the critical point[105].

1.9 MISCELLANEOUS TOPICS

Future interest in nuclear spin relaxation in gases is likely to involve the establishment of connections with other types of measurements in a more explicit manner than has been done thus far and in the development of interesting applications to other systems. A few illustrative examples are included in this section. No attempt has been made in this discussion to be exhaustive.

1.9.1 Dependence of transport properties of gases on magnetic and electric fields—The Senftleben–Beenakker effect

When two molecules collide the (usually small) non-spherically-symmetric part of the intermolecular potential gives rise to a (usually small) correlation between the orientation of each molecule and its final momentum. Thus, under normal conditions, the transport coefficients are determined by collisions between orientated molecules. In the presence of a magnetic field H, however, the molecules precess at an angular frequency $\omega_J = \gamma_J H$, where γ_J is the gyromagnetic ratio of the molecule. As the magnetic field is increased from low values ($\omega_J \tau_c \ll 1$) to high values ($\omega_J \tau_c \gg 1$), the correlation between orientation and momentum is made smaller and smaller in the mean time τ_c between collisions. The resultant change in the transport coefficients, though small, is detectable. It has a characteristic dependence on $\omega_J \tau_c$ or, equivalently, on H/ρ. This dependence of the transport coefficients on H/ρ was first detected for paramagnetic molecules by Senftleben in 1930, and for diamagnetic molecules by Beenakker in 1962. A concise review of the extensive literature on this subject has been given recently by Beenakker and McCourt[106]. It is clear that the information obtained from T_1 measurements in molecular gases is similar to that obtained from a study of the Senftleben–Beenakker effect.

1.9.2 Electron spin relaxation in molecular gases

The e.s.r. spectrum of molecular gases is considerably more complicated than the n.m.r. spectrum, which probably accounts for the comparatively small amount of experimental work which has been done on electron spin *relaxation* in gases. Gordon[107] has discussed some aspects of the theory of electron spin relaxation in light gaseous molecules such as NO, OH, O_2, etc. and references to experimental work may be found in his paper. One basic difference between nuclear and electron spin relaxation discussed by Gordon is that in the latter case the coupling between the electron spin and the molecular angular momentum is sometimes so large that the electron spin adiabatically follows collisional changes in the molecular rotation. Information on electron spin relaxation in a gas composed of heavier molecules — $(CF_3)_2NO$ — has been obtained from a study of the e.s.r. line width as a function of pressure[108].

1.9.3 Bubbles of hydrogen gas in LiH crystals

When LiH solid is irradiated with γ-rays, it has been found that small particles of lithium metal are produced and that these are accompanied by small bubbles of H_2 gas under high pressure which causes the LiH crystals to swell. The proton magnetic resonance signals of the H_2 gas have been studied and an interesting, though not yet completely explained, dependence of T_2 on bubble size and *crystal orientation in the magnetic field* has been observed[109]. In the same experiments, T_1 is almost independent of bubble size and crystal orientation. The behaviour of T_2 is thought to be due to interaction with nuclei in LiH solid near the surface of the H_2 bubbles. Similar effects had been observed previously in a study of CH_4 gas in a container in which the dimensions of the boundary could be varied to quite small values[110]. The effect described here is an unusual type of surface effect. It is clear from Section 1.7.1 and from other studies[114] that measurements of nuclear spin relaxation in other systems could be applied to the study of surface processes.

1.9.4 van der Waals molecules

Some atoms or molecules which do not interact chemically to any appreciable degree can form loosely bound states due to the attractive van der Waals interaction. Recently, some beautiful studies of the bound states of H_2 and D_2 with rare gas atoms have been made by Welsh and his collaborators using spectroscopic techniques[111]. Interpretation of these experiments gives precise information on the anisotropic *inter*-molecular potential which is of interest in the interpretation of nuclear spin relaxation times. A systematic study of the temperature and density dependence of dielectric relaxation and T_1 in these systems should give independent information on the structural parameters associated with the bound states. van der Waals molecules composed of alkali and rare gas atoms have also been studied using the method of optical pumping and a detailed theory of the electron spin relaxation of the alkali atoms due to these molecules has been given[112, 113, 115]. This theory is directly applicable to the evaluation of the correlation functions of rotational angular momentum operators of rotating molecules such as H_2 involved in van der Waals molecules. The theory could be used, therefore, as the basis of a theory of nuclear spin relaxation in systems containing van der Waals molecules.

Acknowledgements

I am grateful to R. L. Armstrong and P. A. Beckmann for some very helpful and critical remarks about points of the original manuscript. I also appreciate the helpful comments and information received from many people including H. J. Bernstein, G. Birnbaum, J. S. Blicharski, J. Courtney, J. Deutch, K. R. Foster, R. G. Gordon, W. Hardy, P. S. Hubbard, E. P. Jones, M. Lipsicas, A. A. Maryott, T. Nakamura, R. E. Norberg, T. Oka, J. G. Powles, K. O. Prins, H. A. Resing, J. Riehl, B. C. Sanctuary, M. Schwab, B. Shizgal, R. F. Snider, J. S. Waugh, C. G. Wade and D. E. Woessner.

References

1. Gordon, R. G. (1968). *Advances in Magnetic Resonance*, Vol. 3, 1, 4–10 (Waugh, J. S. editor). (New York: Academic Press)
2. Abragam, A. (1961). *Nuclear Magnetism*. (London: Oxford University Press)
3. Bloom, M., Bridges, F. B. and Hardy, W. N. (1967). *Can. J. Phys.*, **45**, 3533
4. Tinkham, M. (1964). *Group Theory and Quantum Mechanics*, 129. (New York: McGraw-Hill)
5. Goldman, M. (1970). *Spin Temperature and Nuclear Magnetic Resonance in Solids*, pp. 33–35. (London: Oxford University Press)
6. Hardy, W. N. (1966). *Can. J. Phys.*, **44**, 265
7. Bloom, M. and Oppenheim, I. *Advances in Chemical Physics*, Vol. 12, 549 (Hirschfelder, J. O., editor). (New York: Wiley)
8. Birnbaum, G. (1966). *Phys. Rev.*, **150**, 101
9. Gordon, R. G. (1966). *J. Chem. Phys.*, **45**, 1635
10. Clark, R. B. (1971). *M.Sc. Thesis, University of British Columbia* (unpublished)
11. Hahn, E. L. (1950). *Phys. Rev.*, **80**, 980
12. Carr, H. Y. and Purcell, E. M. (1954). *Phys. Rev.*, **94**, 630
13. Torrey, H. C. (1956). *Phys. Rev.*, **104**, 563
14. Muller, B. H. and Bloom, M. (1960). *Can. J. Phys.*, **38**, 1318
15. Clark, W. G. (1964). *Rev. Sci. Instr.*, **35**, 316
16. Hardy, W. N. (1964). *Ph.D. Thesis, University of British Columbia* (unpublished)
17. Blicharski, J. S. (1969). *Acta Phys. Polon.*, **36**, 211
18. Hubbard, P. S. (1970). *J. Chem. Phys.*, **53**, 987
19. Oppenheim, I. and Bloom, M. (1961). *Can. J. Phys.*, **39**, 845
20. Dong, R. Y. and Bloom, M. (1970). *Can. J. Phys.*, **48**, 793
21. Bloom, M. (1967). *Proc. 14th Colloque Ampere*, p. 65 (Blinc, R., editor). (Amsterdam: North-Holland)
22. Slichter, C. P. (1963). *Principles of Magnetic Resonance*, 181. (New York: Harper and Row)
23. Birnbaum, G. (1957). *J. Chem. Phys.*, **27**, 360
24. O'Reilly, D. E. (1968). *J. Chem. Phys.*, **49**, 5416
25. Hubbard, P. S. (1963). *Phys. Rev.*, **131**, 1155
26. Yi, P., Ozier, I. and Anderson, C. H. (1968). *Phys. Rev.*, **165**, 92
27. Blicharski, J. S. (1963). *Acta Phys. Polon.*, **24**, 817
28. Dong, R. Y., private communication
29. Blicharski, J. S. (1968). *Physica*, **39**, 161
30. Landau, L. D. and Lifshitz, E. M. (1959). *Quantum Mechanics*, 261, 262, 291–294. (London: Pergamon Press)

31. Speight, P. A. and Armstrong, R. L. (1969). *Can. J. Phys.*, **47,** 1475
32. Deutch, J. M. and Oppenheim, I. (1966). *Advances in Magnetic Resonance,* Vol. 2, 225 (Waugh, J. S., editor). (New York: Academic Press)
33. Waugh, J. S. and Johnson, C. S. (1962). *Discuss. Faraday Soc.,* **34,** 191
34. Needler, G. T. and Opechowski, W. (1961). *Can. J. Phys.,* **39,** 870
35. Freed, J. H. (1964). *J. Chem. Phys.,* **41,** 7
36. Lipsicas, M. and Bloom, M. (1961). *Can. J. Phys.,* **39,** 881
37. Lipsicas, M. and Hartland, A. (1963). *Phys. Rev.,* **131,** 1187
38. Gordon, R. G. (1966). *J. Chem. Phys.,* **44,** 228
39. Chen, F. M. and Snider, R. F. (1967). *J. Chem. Phys.,* **46,** 3937; (1968). **48,** 3185; (1969) **50,** 4802
40. Kinsey, J. L., Riehl, J. W. and Waugh, J. S. (1968). *J. Chem. Phys.,* **49,** 5269
41. Lalita, K., Bloom, M. and Noble, J. D. (1969). *Can. J. Phys.,* **47,** 1355
42. Lalita, K. and Bloom, M. (1971). *Can. J. Phys.,* **49,** 1018
43. Riehl, J. W., Kinsey, J. L., Waugh, J. S. and Rugheimer, J. H. (1968). *J. Chem. Phys.,* **49,** 5276
44. Bloom, M. (1957). *Physica,* **23,** 237
45. Foster, K. R. (1971). *Ph.D. Thesis, University of Indiana* (unpublished)
46. Ramsey, N. F.(1956). *Molecular Beams.* (London: Oxford University Press)
47. Courtney, J. A. and Armstrong, R. L. (1970). *J. Chem. Phys.,* **52,** 2158
48. Tward, E. and Armstrong, R. L. (1967). *J. Chem. Phys.,* **47,** 4068
48a. Leonardi-Cattolica, A. M., Prins, K. O. and Waugh, J. S. (1971). *J. Chem. Phys.,* **54,** 769
48b. Gordon, R. G., private communication. Professor Gordon writes that the results on the exact collisional calculations which reproduce the T_1 measurements on the HCl–argon system will be included in Walter Neilson's Harvard Ph.D. Thesis.
49. Courtney, J. A. and Armstrong, R. L., to be published
50. Ozier, I., Crapo, L. M. and Lee, S. S. (1968). *Phys. Rev.,* **172,** 63
51. Yi, P., Ozier, I. and Ramsey, N. F. (1971). *Phys. Rev.,* to be published; Yi, P., Ozier, I., Khosla, A. and Ramsey, N. F. (1967). *Bull. Amer. Phys. Soc.,* **12,** 509
52. Hecht, K. T. (1960). *J. Molec. Spectrosc.,* **5,** 355, 390
53. Ozier, I. (1971). *Z. Naturforsch,* **26a,** 1212
54. Curl, R. F., Jr., Kaspar, J. V. V. and Pitzer, K. S. (1967). *J. Chem. Phys.,* **46,** 3220
55. Ozier, I. and Yi, P. (1967). *J. Chem. Phys.,* **47,** 5458
56. Beckmann, P. A., Bloom, M. and Burnell, E. E. (1971). *Can. J. Phys.,* to be published. Beckmann, P. A. (1971). *M.Sc. Thesis, University of British Columbia* (unpublished)
57. McCourt, F. R. and Hess, S. (1970). *Z. Naturforsch,* **25a,** 1169
58. McCourt, F. R. and Hess, S. (1971). *Z. Naturforsch,* **26a,** 1234
59. Lalita, K. and Bloom, M. (1971). *Chem. Phys. Lett.,* **8,** 285
60. Armstrong, R. L. and Tward, E. (1968). *J. Chem. Phys.,* **48,** 332
61. Anderson, C. H. and Ramsey, N. F. (1966). *Phys. Rev.,* **149,** 14
62. Wofsy, S. C., Muenter, J. S. and Klemperer, W. (1970). *J. Chem. Phys.,* **53,** 4005
63. Dong, R. Y. and Bloom, M. (1968). *Phys. Rev. Lett.,* **20,** 981
64. Courtney, J. A. and Armstrong, R. L., private communication
65. Pausak, S. and Waugh, J. S., private communication
66. Sanctuary, B. C. (1971). *Ph.D. Thesis, University of British Columbia* (unpublished)
67. Armstrong, R. L. and Hanrahan, T. A. J. (1968). *J. Chem. Phys.,* **49,** 4777
68. Dong, R. Y. (1969). *Ph.D. Thesis, University of British Columbia* (unpublished)
69. Armstrong, R. L. and Courtney, J. A. (1969). *J. Chem. Phys.,* **51,** 457
70. Hinshaw, W. S. and Hubbard, P. S. (1971). *J. Chem. Phys.,* **54,** 428
71. Luszczynski, K., Norberg, R. E. and Opfer, J. E. (1962). *Phys. Rev.,* **128,** 186
72. Gamblin, R. L. and Carver, T. R. (1965). *Phys. Rev.,* **138,** A 946
73. Fitzsimmons, W. A., Tankersley, L. L. and Walters, G. K. (1969). *Phys. Rev.,* **179,** 156
74. Timsit, R. S., Daniels, J. M. and May, A. D. (1971). *Can. J. Phys.,* **49,** 560
75. Shearer, L. D. and Walters, G. K. (1965). *Phys. Rev.,* **139,** A 1398
76. Chapman, R. and Richards, M., private communication
77. Shizgal, B., private communication
78. Streever, R. L. and Carr, H. Y. (1961). *Phys. Rev.,* **121,** 20
79. Hunt, E. R. and Carr, H. Y. (1963). *Phys. Rev.,* **130,** 2302
80. Torrey, H. C. (1963). *Phys. Rev.,* **130,** 2306; Bloom, M., Oppenheim, I. and Torrey, H. C., *Can. J. Phys.,* **42,** 70

81. Adrian, F. J. (1964). *Phys. Rev.*, **136**, A 980
82. Brinkman, D. (1968). *Helv. Phys. Acta*, **41**, 367
83. Adrian, F. J. (1965). *Phys. Rev.*, **138**, A 403
84. Johnson, C. S., Jr. and Waugh, J. S. (1961). *J. Chem. Phys.*, **35**, 2020
85. Siegel, M. M. and Lipsicas, M. (1970). *Chem. Phys. Lett.*, **6**, 259. See also the proceedings of the 4th International Magnetic Resonance Conference, Israel, August 1971, to be published in *J. Mag. Res.*
86. Van Vleck, J. H. (1932). *Electric and Magnetic Susceptibilities.* (London: Oxford University Press)
87. Karra, J. S. and Kemmerer, G. E., Jr. (1970). *Phys. Lett.*, **33A**, 105
88. Gordon, R. G. (1966). *J. Chem. Phys.*, **45**, 1649
89. Gordon, R. G., Armstrong, R. L. and Tward, E. (1968). *J. Chem. Phys.*, **48**, 2655
90. Tward, E. and Armstrong, R. L. (1968). *Can. J. Phys.*, **46**, 331
91. Deutch, J. M. and Oppenheim, I. (1966). *J. Chem. Phys.*, **44**, 2843
92. Lipsicas, M. and Hartland, A. (1966). *J. Chem. Phys.*, **44**, 2839; Miller, C. E. and Lipsicas, M. (1968). *Phys. Rev.*, **176**, 273
93. Alekseyev, V., Grasiuk, A., Ragulsky, I., Sobelman, I. and Faizulov, F. (1968). *IEEE J. Quant. El.*, **QE-4**, 654
94. Alekseyev, V. and Sobelman, I. (1968). *Acta Phys. Polon.*, **34**, 579
95. Ben-Reuven, A. (1966). *Phys. Rev.*, **141**, 34
96. Gerritsma, C. J., Oosting, P. H. and Trappeniers, N. J. (1971). *Physica*, **51**, 381
97. Oosting, P. H. and Trappeniers, N. J. (1971). *Physica*, **51**, 395
98. Oosting, P. H. and Trappeniers, N. J. (1971). *Physica*, **51**, 418
99. Kadanoff, L. P., Götze, W., Hamblen, D., Hecht, R., Lewis, E. A. S., Palciauskas, V. V., Rayl, M., Swift, J., Aspnes, D. and Kane, J. (1967). *Rev. Mod. Phys.*, **39**, 395
100. Bloom, M. (1965). *Nat. Bur. Std. U.S. Misc. Publ.*, **273**, 178
101. Trappeniers, N. J. and Prins, K. O. (1967). *Physica*, **33**, 435
102. Tison, J. K. and Hunt, E. R. (1971). *J. Chem. Phys.*, **54**, 1526
103. Noble, J. D. and Bloom, M. (1965). *Phys. Rev. Lett.*, **14**, 250
104. Ehrlich, R. S. and Carr, H. Y. (1970). *Phys. Rev. Lett.*, **25**, 341
105. Stacey, L. M., Pass, B. and Carr, H. Y. (1969). *Phys. Rev. Lett.*, **23**, 1424
106. Beenakker, J. J. M. and McCourt, F. R. (1970). *Ann. Rev. Phys. Chem.*, **21**, 47
107. Gordon, R. G. (1967). *J. Chem. Phys.*, **46**, 448
108. Schaafsma, T. J. and Kivelson, D. (1968). *J. Chem. Phys.*, **49**, 5235
109. Souers, P. C., Imai, T., Blake, T. S., Penpraze, R. M. and Leider, H. R. (1970). *J. Phys. Chem. Solids*, **31**, 1461
110. Wayne, R. C. and Cotts, R. M. (1966). *Phys. Rev.*, **151**, 264
111. McKellar, A. R. W. and Welsh, H. L. (1971). *J. Chem. Phys.*, **55**, 595
112. Bouchiat, C. C., Bouchiat, M. A. and Pottier, L. C. (1969). *Phys. Rev.*, **181**, 144
113. Bouchiat, C. C. and Bouchiat, M. A. (1970). *Phys. Rev.*, **A2**, 1274
114. Resing, H. (1971). *Proc. 8th Colloquium on NMR Spectroscopy,* to be published. (Berlin: Springer-Verlag)
115. Hartmann, F. and Hartmann-Boutron, F. (1970). *Phys. Rev.*, **A2**, 1885
116. Pinto-Vega, S., Foster, K. R. and Rugheimer, J. H. (1972). *J. Chem. Phys.*, **56**, 678

2
N.M.R. Studies of Molecular Motion in Solids

P. S. ALLEN
University of Nottingham

2.1   INTRODUCTION

In this discussion an outline will be given (Sections 2.2–2.4) of the nuclear resonance methods currently employed for investigating molecular motion in solids. In addition, Sections 2.5 and 2.6 contain a number of examples of the use of these methods in a wide variety of molecular-motion problems. The remainder of this introduction is to set the stage by sketching a model for molecular motion in solids and by presenting the nuclear interactions[1, 2] which enable the resonance technique to be employed as a method of observation for such motion.

It will be assumed throughout that the following model is applicable to random molecular reorientations in the solid state. In a solid the inter- and intra-molecular forces constrain the molecules and their internal groups to certain equilibrium orientations, which are related by the symmetry operations for the molecule. In the equilibrium orientations the molecule, or its internal group, may exist in a number of torsional oscillator energy states. At energies near the bottom of the hindering barrier the states will be essentially harmonic oscillator in character, whereas approaching the top of the barrier the energy states will develop more and more free rotor character, until at energies well above the potential barriers the wave functions will be entirely free rotor functions. A typical reorientational jump is then the excitation of the hindered rotor from its ground torsional oscillator state, to a state at the top of the barrier in which it is able to rotate before falling back into a torsional oscillator state, where it remains until further excited.

It is assumed that these energy states are strongly coupled to the lattice modes, so that a Boltzmann distribution of populations, which is characterised by the lattice temperature, exists among them. An alternative view of this Boltzmann distribution, which is useful from a nuclear resonance point of view, is to imagine that it represents the mean residence time spent by an individual hindered rotor in each of its energy states. Therefore the residence time of a molecule in one of its equilibrium orientations is much longer than the time it takes to change orientations.

In n.m.r. however, the important time is the mean time *between* changes of orientation, i.e. between excitations to the top of the barrier, which is usually called τ_c. For a two level system τ_c would equal the residence time in the ground state, but in general τ_c is greater because the ground state residence time is curtailed by transitions to other excited states. It is still to be expected that the temperature dependence of τ_c will reflect the Boltzmann character of the distribution, namely

$$\tau_c = \tau_0 \exp (E_A/kT) \qquad (2.1)$$

where one may identify E_A, the activation energy, with the energy separation of the ground state and states near the top of the barrier, and where τ_0 is determined largely though not solely by the lifetime of the states at the top of the barrier. An alternative discussion of equation (2.1), the Arrhenius relation, may be found elsewhere[3].

The object of molecular motion studies using nuclear resonance is first to ascertain which motional processes occur and secondly to measure the temperature dependence of their rate $1/\tau_c$, so that an estimate of the magnitude of the potential barriers hindering such motion can be made by means of equation (2.1). The success of such an enterprise relies on one being able to interpret the observable nuclear resonance data in terms of the molecular motions. This is feasible since the nuclear interactions in a solid, with their angular and radial dependencies, are quite sensitive to nuclear motion. The interactions which are of importance to the investigation of molecular motion are the internuclear magnetic dipole–dipole interaction, the nuclear electric quadrupole interaction with the surrounding electric field gradient and occasionally, the nuclear spin–rotation interaction with the rotational angular momentum of the molecule. Furthermore, in many cases of interest one or more of these interactions can be neglected and therefore it is often possible to refer to the solids of interest as either dipolar or quadrupolar.

The Hamiltonians for the important interactions in solids may be written as follows for a system of like spins,

Zeeman $\qquad H_{\mathbf{Z}} = -\gamma\hbar \sum_i \mathbf{H}_{\text{app}} . \mathbf{I}_i$ $\qquad\qquad\qquad\qquad\qquad$ (2.2)

Dipolar $\qquad H_{\mathbf{D}} = \gamma^2\hbar^2 \sum_{i<j} \left\{ \dfrac{\mathbf{I}_i . \mathbf{I}_j}{r_{ij}^3} - \dfrac{3(\mathbf{I}_i . \mathbf{r}_{ij})(\mathbf{I}_j . \mathbf{r}_{ij})}{r_{ij}^5} \right\}$

$\qquad\qquad\qquad = \gamma^2\hbar^2 \sum_{i<j} \dfrac{1}{r_{ij}^3} \cdot \{A_{ij} + B_{ij} + C_{ij} + D_{ij} + E_{ij} + F_{ij}\}$ $\qquad$ (2.3)

where

$$A_{ij} = \{1 - 3\cos^2\theta_{ij}\} I_{iz} I_{jz}$$

$$B_{ij} = -\tfrac{1}{4}\{1 - 3\cos^2\theta_{ij}\} \cdot \{I_{i+} I_{j-} + I_{i-} I_{j+}\}$$

$$C_{ij} = -\tfrac{3}{2}\{\sin\theta_{ij}\cos\theta_{ij}\exp(-i\phi_{ij})\}\{I_{iz} I_{j+} + I_{i+} I_{jz}\}$$

$$D_{ij} = C_{ij}^*$$

$$E_{ij} = -\tfrac{3}{4}\{\sin^2\theta_{ij}\exp(-2i\phi_{ij})\} I_{i+} I_{j+}$$

$$F_{ij} = E_{ij}^*$$

r_{ij}, θ_{ij} and ϕ_{ij} denote the coordinates of the vector between spins i and j and the asterisk denotes the Hermitian conjugate. Terms A and B commute with H_z, whereas terms C, D, E and F do not. For a nucleus of quadrupole moment, Q, in an axially symmetric electric field of gradient $\dfrac{\partial^2 V}{\partial \zeta^2}$ we have,

Quadrupolar $\qquad H_Q = \dfrac{eQ}{4I(2I-1)} \cdot \dfrac{\partial^2 V}{\partial \zeta^2} \cdot \{3I_\zeta^2 - I^2\}$ $\qquad\qquad$ (2.4)

Finally the interaction with the molecular angular momentum J may be written,

Spin–rotation $\qquad\qquad\qquad H_{SR} = -I.C.J$ $\qquad\qquad\qquad\qquad$ (2.5)

where C is the spin–rotation interaction tensor.

Random molecular motion imparts a time dependence to the coordinates in equation (2.3) and (2.4) and to the angular momentum in equation (2.5). Therefore the total Hamiltonian can be written as the sum of the two parts

$$H = H_0 + H'(t)$$

where H_0 is time independent and includes for example H_z and the time independent parts of H_D and H_Q. $H'(t)$ on the other hand contains those parts of the Hamiltonian which are functions of time. The shape and structure of the nuclear resonance line is fixed by H_0 and the relaxation behaviour depends on $H'(t)$. In many experiments $H_0 \gg H'(t)$. H_0 then determines the static energy states and the fluctuations included in $H'(t)$ are able to promote transitions between them. Such a situation may be classified as a *weak collision* case. An entirely different situation arises if $H'(t) > H_0$. Now the total Hamiltonian is predominantly time dependent, and the term *strong collision* is used as a description. The relaxation behaviour of nuclei is markedly different in the two situations and as a consequence each situation is accorded a separate section in the following discussion.

2.2   INVESTIGATIONS USING THE RESONANCE SPECTRUM

2.2.1   Second moments in dipolar solids

2.2.1.1   Introduction

In a rigid 'dipolar' lattice an exact solution of the nuclear resonance line shape problem is not available[4, 5]. Instead, in order to obtain an exactly calculable observable, one turns to the method of moments[4, 6]. This is a technique which enables one to compare exactly certain properties of the experimental and theoretical resonance lines, without solving explicity for the energy eigenstates of the spin system. The property of the resonance line which has most relevance in dipolar solids which are rigid, or which exhibit molecular motion, is the second moment, M_2. This may be defined in terms of the normalised resonance line shape function $f(\omega)$ as follows:

$$M_2 = \int_{-\infty}^{\infty} (\omega - \omega_0)^2 \, f(\omega) \, d\omega \qquad (2.6)$$

where $f(\omega)$ is centred on ω_0.

In a rigid lattice, where the relative positions of the nuclear spins do not change in time, the second moment has a value governed by terms A and B in H_D. Abragam[4], by relating the line shape function to the behaviour of the x magnetisation, shows that M_2 may be written in terms of the dipolar Hamiltonian according to

$$M_2 = \frac{-\mathrm{Tr}\{[H_{D_{sec}}, I_x]^2\}}{\mathrm{Tr}\{I_x^2\}} \qquad (2.7)$$

where

$$H_{D_{sec}} = \sum_{i<j} \frac{\gamma^2 \hbar^2}{r_{ij}^3} \cdot \{A_{ij} + B_{ij}\}$$

Terms C to F in H_D are omitted because their net effect is to cause unobservable contributions to the second moment from satellite lines at frequencies $\omega = 0$ and $2\omega_0$. Evaluation of the traces in equation (2.7) leads to the familiar Van Vleck expression[4,7] for the second moment in terms of the microscopic coordinates, i.e.

$$M_2 = \tfrac{3}{4} \gamma^4 \hbar^2 I(I+1) \sum_j \frac{(1 - 3\cos^2 \theta_{ij})^2}{r_{ij}^6} \qquad (2.8)$$

Thus M_2 is a measure of the strength and distribution of the local dipolar magnetic fields.

If the spins are in relative motion, the local field experienced by one nucleus due to its neighbours will fluctuate in time, since r_{ij} and θ_{ij} will depend on time. Only the time-average local field, taken over a time which may or may not be long compared with the duration of the fluctuation, will be observed in a second-moment or line-shape measurement. This average local field which reflects the time average of terms A and B in H_D may then, if the averaging time is much longer than the fluctuation time, be smaller than the static value. Hence the distribution of local fields may be much smaller when the spins are in motion and this results in a smaller line width and second moment of the resonance line.

The question which immediately arises here is, what is the criterion which determines whether or not the fluctuation time is much smaller than the averaging time? The order of magnitude of the averaging time can be seen if one considers the system in a frame of reference rotating at the Larmor frequency. In this frame[8], the only 'static' field seen by a nucleus is the local dipolar field, if the radiofrequency field H_1 is small. Consequently a nuclear magnetic moment will precess around this field with a frequency which is the order of the local field expressed in frequency units. If the fluctuation time is short compared with this precessional period, the nucleus will only experience the time averaged local field and will modify its local field precession accordingly.

Remembering that the time-average local field $\sim$ line width $\sim (M_2)^{\frac{1}{2}}$, and assuming that one can characterise the motional fluctuations by a correlation time τ_c, one may write the criterion for the line having narrowed as

$$(M_2)^{\frac{1}{2}} . \tau_c \ll 1 \tag{2.9}$$

where M_2 is in frequency units.

It should be emphasised here that it is only the 'observable' second moment of the resonance line in the region of the Larmor frequency ω_0, which is reduced by molecular motion[9]. Strictly speaking the second moment as computed from equation (2.7) is independent of molecular motion. However the effect of such motion is to transfer some contributions of the resonance line out into the wings, so as to make them unobservable, even though in

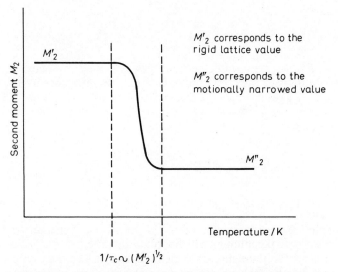

Figure 2.1 The temperature dependence of the second moment, M_2, in some imaginary dipolar solid which exhibits a single motional process.

principle they still contribute to the second moment. As a consequence one has to be careful to use the time average $H_{D,\,sec}$ to compute the second moment of the observable resonance line, so that one may purposely omit all the unobservable contributions in the wings.

The temperature dependence of the observable second moment therefore acts as an indicator of the motion exceeding a threshold frequency region of $(M_2)^{\frac{1}{2}}$. This is illustrated in Figure 2.1. If, for example, the rotational jumps are characterised by a frequency $(1/\tau_c) \ll (M_2)^{\frac{1}{2}}$, then the local field appears to be the full rigid lattice value and the second moment is that corresponding to the rigid lattice (M'_2). As the rate of reorientation increases (corresponding to an increase in temperature) the system changes to a situation where $(1/\tau_c) \gg (M_2)^{\frac{1}{2}}$, the local field develops a much smaller time-average value and the observable second moment reduces to M''_2, corresponding to the smaller local field.

2.2.1.2   Identification of the type of motion

The detailed calculation of second moments makes use of equation (2.8) together with the molecular and crystal structures. In all but the simplest of molecular solids such a calculation can be quite tedious and, in addition, may be hindered by a lack of knowledge of the crystal structure. In this latter

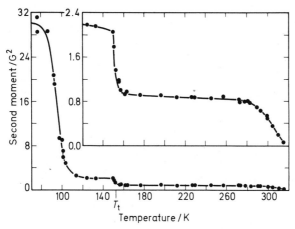

Figure 2.2   The temperature dependence of the proton second moment in hexamethylethane. The inset is an exploded view of the data from the transition temperature upwards.
(From Smith, G. W.[15] by courtesy of the American Institute of Physics)

case one has to make intelligent estimates of the intermolecular contribution to the total second moment. In order to facilitate a second-moment calculation, M_2 is usually divided into its main contributions e.g. inter- and intramolecular, and inter- and intra-group, and these contributions are determined in turn. Such a calculation is quite sensitive to the structure because of its dependence on $1/r^6$. Consequently the rigid plateau value M'_2 gives a good test of these structures in the rigid lattice. To explain M''_2 one needs to postulate the existence of certain types of motion and to calculate the observable second moments for molecules undergoing such motions. Agreement between experiment and a calculated second moment is good evidence for that postulated motion. *Thus the second moment can provide a test of which molecular motions are taking place at a frequency above the line-width threshold frequency.*

The modification to equation (2.8), necessary for each of the contributions to M_2 when a molecular solid exhibits one or more types of motion, are well illustrated in the papers of Smith[10–15]. He makes use of the ideas suggested by Gutowsky and Pake[16] and by Andrew and Eades[17].

For example in hexamethylethane[15] (see Figure 2.2) Smith was able to establish that (i) below 80 K the lattice is essentially rigid and no molecular motions occur at a rate sufficiently high to affect the nuclear resonance line shape; (ii) above 90 K the C_3 reorientation of the t-butyl groups proceeds at a rate which soon exceeds 10^4 Hz; (iii) at the phase transition at 152.5 K there is an abrupt increase in the rate of general molecular reorientation to a value in excess of 10^4 Hz; (iv) and finally beyond 270 K the rate self-diffusional jumps becomes large enough to cause appreciable further narrowing of the resonance line. These data on hexamethylethane have been confirmed independently[18].

Further work making use of this technique will be discussed in Section 2.6.

2.2.1.3   *Extraction of an activation energy for the motion*

A further use to which second-moment or line-width data is often put, is to try to obtain τ_c as a function of temperature from the transition region, and so extract an activation energy by means of equation (2.1). Such an exercise must be undertaken with caution[177].

The expression which is used is

$$\frac{1}{\tau_c} = \alpha . \delta H / \tan\left[\frac{\pi}{2} . (\delta H^2 - B^2)/(C^2 - B^2)\right] \qquad (2.10)$$

where α is constant, δH is the line-width, B is the fully narrowed δH and C is the unnarrowed δH. This expression was developed from the original treatment of Bloembergen, Purcell and Pound[19] and some of the limitations of its derivation are discussed by Abragam[9] and by Kubo and Tomita[20]. The line-width is used here rather than the second moment, because the determination of the latter in the transition region becomes very uncertain, if not meaningless. The reason for this is that as contributions to the resonance line are transferred into the wings with increasing temperature, the value of the measured second moment will depend on how far into the wings one can follow the resonance line. It depends in fact on the available signal-to-noise ratio. Hence the line-width is a less ambiguous parameter to measure in the transition region. However, equation (2.10) still assumes that there is no change in line shape accompanying the observed change in line-width and this is often not true. In addition the accuracy of this method deteriorates as δH approaches B, since the uncertainty introduced by the term $(\delta H^2 - B^2)$ increases. Finally the range of correlation times which are obtained is fairly small, two orders of magnitude or less. Consequently, when one remembers the uncertainties in τ_c, one sees that there could be considerable ambiguity in the slope of an Arhennius diagram.

This method of determining E_A does not compare too favourably with its determination from the spin–lattice relaxation time, as outlined in Section 2.3.1.

2.2.1.4   The experimental problem of saturation

Although measurements of the second moment can[21–23, 176] and have[24–26] been made using pulsed n.m.r. techniques without any fear of saturation, the vast majority of second-moment measurements are still being made using continuous wave (c.w.) techniques. C.W. measurements are subject to the possible distortion of the resonance line due to saturation, and it is not always apparent from the literature that adequate steps have been taken to ensure its absence.

Goldman[27] has recently highlighted the problem of saturation for experimenters using continuous wave spectrometers with lock-in detection. From his analysis he is able to show how, if non-negligible saturation exists, the output from a lock-in detector differs from the pure derivative of either the absorption or the dispersion signal. Furthermore, and this is possibly of greater relevance to the previously published second-moment work, he points out that a linear dependence of the amplitude v_1 of the lock-in absorption signal on the r.f. level H_1 is not necessarily a guarantee of the absence of saturation, even at H_1 levels well below those corresponding to maximum observable v_1. In other words even when the system is well saturated, the lock-in absorption signal may be proportional to H_1, but in this case it yields a bell-shaped curve which is different from the true line shape. As a consequence a common method of c.w. experimenters to establish if no saturation exists, i.e. the proportionality of v_1 to H_1 is shown to be invalid. One must therefore adopt other methods to monitor the threshold of saturation. For example the amplitude of the lock-in dispersion signal at the exact centre of the resonance line does saturate normally and may therefore be used to determine the maximum safe H_1 level for absorption line measurements.

2.2.2   Resonance spectrum in quadrupolar solids

2.2.2.1   Introduction

The preceding section dealt with the effects of nuclear motion on the resonance spectrum from a solid in which the dipole–dipole interaction was the dominant interaction between the spins and their surroundings. Molecular motion may also be detected from the spectra of quadrupolar solids, due to its effect on the nuclear quadrupole splittings.

Experiments using this technique are most profitably done on single crystals and so in the subsequent discussion mono-crystalline samples will be assumed unless it is stated otherwise. Furthermore, such work is almost exclusively confined to deuteron magnetic resonance (d.m.r.), because of the convenient ratio of dipolar to quadrupolar interaction energies for the deuteron. Consequently the illustrative examples chosen below are in line with this trend.

The Hamiltonian for a high-field nuclear resonance experiment in a quadrupolar solid is, from equations (2.2) and (2.4),

$$H = H_z + H_Q$$
$$= -\gamma \hbar H . I + \frac{eQ}{4I(2I-1)} . \frac{\partial^2 V}{\partial \zeta^2} . (3I_\zeta^2 - I^2) \qquad (2.11)$$

The ζ direction specified in the quadrupolar Hamiltonian is that corresponding to the direction of the principal component of the electric field gradient tensor. Thus for high magnetic field experiments it is convenient to transform[28] to a new set of axes (x,y,z) defined with the direction of the applied magnetic field along the z axis. The transformed Hamiltonian then leads, in a first order approximation, to the following nuclear energy levels

$$E_m = -\gamma \hbar . H . m + \frac{e^2 qQ}{4I(2I-1)} . \tfrac{3}{2} (3\cos^2\theta - 1)(m^2 - I(I+1)/3) \qquad (2.12)$$

where θ is the angle between the ζ and z axes. Hence in a static monocrystalline quadrupolar lattice, the quadrupole interaction of a nucleus of spin $I > \tfrac{1}{2}$ splits the nuclear Zeeman resonance line into $2I$ components for each non-equivalent nuclear site. The widths of these components are, to a first order, determined by the weaker dipolar interaction and the inhomogeneity in the quadrupole interaction. Their separation is a function of the quadrupole coupling constant, and the angle θ, and may be written for the satellite lines $(m-1) \leftrightarrow m$ and $-m \leftrightarrow -(m-1)$

$$\Delta\omega = \tfrac{3}{2} . \frac{e^2 qQ}{I(2I-1)\hbar} . (3\cos^2\theta - 1)(m - \tfrac{1}{2}) \qquad (2.13)$$

Thus, for example, in a deuterated monohydrate containing static equivalent D_2O molecules one usually observes two pairs of lines, since the spin quantum number of the deuteron is 1 and since eq and θ will differ from one deuteron in the water molecule to the other. With deuteron magnetic resonance one is therefore able to investigate more complicated hydrates than with proton magnetic resonance. The reason for this is that the quadrupole splittings, and the differences in quadrupole splittings, are large compared with the line-width and so one may observe lines from each type of deuteron site. With proton magnetic resonance on the other hand all the protons contribute to the same line and one is only able to extract averages over all the non-equivalent sites.

With the onset of molecular motion a quadrupole nucleus may sample, as a function of time, sites of different eq and θ. When the jump rate, $1/\tau_c$, between such non-equivalent sites (a and b) is much less than the difference in their corresponding quadrupole splittings $(\Delta\omega_a - \Delta\omega_b)$, then the observed spectrum will contain the individual spectra of sites a and b. However, if $1/\tau_c \gg (\Delta\omega_a - \Delta\omega_b)$, the spectrum will be that corresponding to a single time-averaged quadrupole splitting. For the previous example of a monohydrate, the onset of rapid 180 degree flips of the D_2O molecule about its DOD bisector would cause the two pairs of lines to coalesce into one pair, since both deuterons would now experience the same time-averaged environment.

2.2.2.2 Identification of the type of motion

The identification of a particular type of motion from the quadrupole spectrum involves the observation of the variation with temperature of the number of pairs of lines, their relative intensities, their relative splittings and their rotation patterns. The following examples illustrate the use of this technique.

The linear jumping motion of a deuteron back and forth along a hydrogen bond, so important in the study of hydrogen-bonded ferroelectrics, may be detected using deuteron magnetic resonance. Bjorkstam[29] has observed that in the ferroelectric phase of KD_2PO_4, the deuteron spectrum is composed of two pairs of lines corresponding to the non-equivalent deuteron sites in a hydrogen bond. At the critical temperature these two pairs collapse into one with an intermediate quadrupole coupling constant and which is consistent with the onset of rapid jumping of the deuteron back and forth along the bond in the paraelectric phase. The rotation patterns of the spectra in the two phases confirm an averaging of the directions of the principal axes in the paraelectric phase.

The rotational motion of the ND_4^+ ion has been observed by Genin and O'Reilly[30] in deuterated leconite ($NaND_4SO_4 \cdot 2D_2O$). At room temperature they observe a single central line due to the ND_4^+ group. The single line is a result of the rapid isotropic tumbling of the ND_4 tetrahedron causing a zero time average for eq. As the sample is cooled to 77 K the central line disappears and two pairs of satellites emerge with the intensity ratio 3:1. The weak pair is assigned to the axial deuteron and the intense pair to the three equatorial deuterons of an ND_4 group undergoing rapid C_3 reorientations. These assignments were confirmed by a calculation of the relative quadrupolar splittings to be expected for these two types of deuterons.

The transition from isotropic reorientation to C_3 reorientation to no effective motions has been observed by O'Reilly et al.[31] in hydronium perchlorate, but these experiments were made on a polycrystalline sample and the results are not quite so 'clean' as with single crystals.

2.2.2.3 Extraction of an activation energy for the motion

When a quadrupole nucleus exhibits jumping between sites of different quadrupolar splitting, the observed spectral pattern may change with increasing temperature due to the averaging of the quadrupolar splitting over the non-equivalent sites. Examples of this have been given in the preceding sub-section. The temperature dependence of the change in the spectrum, which usually consists of the formation of fewer pairs of lines, provides a method of estimating the mean jump frequency between the non-equivalent sites and hence E_A by use of equation (2.1).

Considering the deuterated monohydrate example used before, the low-temperature spectrum consists of two pairs of lines. As the 180 degree flip frequency of the D_2O increases, the pairs of lines broaden, coalesce and reappear as one sharp pair consistent with the time-averaged quadrupolar splitting in the rapid-motion limit. The general theory of the destruction of line structure due to the motion of the spins was developed by Anderson[32] and has also been described by Kubo[33] and by Abragam[34]. This theory may be used to correlate the width of these lines with the jump frequency. For the simple case considered here, one has in the low jump-frequency region $(\Delta\omega_a - \Delta\omega_b) \gg 1/\tau_c$, the width δ of the component lines given by

$$\delta = \delta_D + 1/\tau_c$$

where δ_D is the residual dipolar width. In the high-temperature region of rapid jumping $(\Delta\omega_a - \Delta\omega_b) \ll 1/\tau_c$,

$$\delta = \delta_D + \frac{1}{2}\left\{\frac{\Delta\omega_a - \Delta\omega_b}{2}\right\}^2 \cdot \frac{1}{\tau_c}$$

Figure 2.3 illustrates the spectral pattern in such an example, where we have assumed for simplicity that the time-averaged quadrupole splitting is the exact mean of the individual splittings.

Soda and Chiba[35] have used this technique successfully to extract the

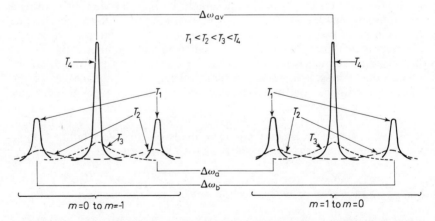

Figure 2.3 An illustration of the partial destruction of the quadrupolar structure of a deuteron magnetic resonance line from the heavy water molecules in some imaginary mono-hydrate. The change in spectrum is due to the temperature dependence of the rate of 180 degree flips of the D_2O molecules. T_1 to T_4 denotes an increase in temperature

separate activation energies of the five non-equivalent water molecules in deuterated cupric sulphate pentahydrate. The theory mentioned above may be extended[34] to encompass systems with more than two quadrupolar splittings and unequal jump frequencies.

2.3 INVESTIGATIONS USING RELAXATION: THE WEAK COLLISION CASE

A study of the static resonance spectra, although useful in ascertaining the nature of any molecular motion, does not provide a quantitative measure of the frequency of the motional jumps $(1/\tau_c)$ over a wide range of values. Such information however can often be obtained from the temperature dependence of the nuclear spin–lattice relaxation.

2.3.1 Spin–lattice relaxation in high fields in dipolar solids

2.3.1.1 Introduction

A theoretical description of nuclear spin–lattice relaxation occurring in high magnetic fields and due to a time dependent dipole–dipole interaction can be found in many places and at a number of different levels. The original treatment of Bloembergen, Purcell and Pound[19] is outlined in the book by Andrew[36], while an introductory description of the derivation using density matrix formalism is given by Slichter[37]. Abragam[38] is a standard work of reference and other discussions have been given by Hubbard[39], Kubo and Tomita[20], Argyres and Kelley[40], Redfield[41] and by Noack[42]. The following paragraph outlines the physical picture underlying such a theoretical derivation.

Thermally activated reorientations of molecular groups will produce a time dependence of the coordinates in the dipolar Hamiltonian and correspondingly cause fluctuations in the local magnetic field at any nuclear site. Such a fluctuating field may be Fourier analysed into a series of components at different frequencies. The component at the resonance frequency will be able to stimulate transitions between the nuclear Zeeman states. These transitions come about through the action of terms C to F in the dipolar Hamiltonian. Because the thermal bath which is activating the reorientations is characterised by an infinite heat capacity and a Boltzmann distribution among its states, it will preferentially absorb energy from an excited nuclear spin system and will therefore result in an equalisation of the temperatures of the bath and the spins, i.e. in nuclear spin–lattice relaxation.

Restricting this introduction to a simple two spin relaxation case which results in an exponential approach to equilibrium with the lattice, one may define a single characteristic relaxation time T_1 for the system. If a spin temperature is maintained within the system, T_1 is related[43] to the probability per unit time, W_{mn}, of a transition between spin states $|m>$ and $|n>$ by

$$\frac{1}{T_1} = \frac{1}{2} \sum_{m,n} W_{mn} (E_m - E_n)^2 \Big/ \sum_n E_n^2 \qquad (2.14)$$

where E_m and E_n are the energies of states $|m>$ and $|n>$. The transition probabilities[44] occurring in the summation will obviously depend upon the magnitude of the spin part of the dipolar matrix elements $<m|H_D^{spin}|n>$, but in addition they will also depend on the spectral representation of the fluctuations in the spatial part of the dipolar Hamiltonian. This is because it is only those fluctuations at certain specific frequencies which are effective in causing relaxation. Therefore it is necessary to have a measure of the intensity of the fluctuations at those frequencies. The assumption that the fluctuations may be represented by a stationary random process results[44] in the following expression for W_{mn}

$$W_{mn} = |<m|H_D^{spin}|n>|^2 J(\omega_{mn}) \tag{2.15}$$

where $J(\omega_{mn})$ is the spectral density of the dipolar fluctuations at ω_{mn}. For this two spin dipolar relaxation there are two contributions, one at frequency ω_0 from fluctuations in the terms C and D of H_D and another at frequency $2\omega_0$ from fluctuations in the terms E and F of H_D. Thus equation (2.14) becomes on evaluation of the matrix elements in equation (2.15)

$$\frac{1}{T_1} = \frac{3}{2} \frac{\gamma^4 \hbar^2}{N} \frac{I(I+1)}{N} \sum_{i \neq j} \{J_{ij}^{(1)}(\omega_0) + J_{ij}^{(2)}(2\omega_0)\} \tag{2.16}$$

where we have generalised to a system in which each spin may interact with several identical spins and where N is the number of spins.

The calculation of the spectral densities is probably the most difficult part of the problem and for this reason it is here that the most crude assumptions are usually made. The precise form of the spectral densities will depend on the nature of the molecular motion and on the molecular and crystal structures. In many cases much of this information is not available, and only occasionally[45-47, 107] are attempts made to calculate the spectral densities

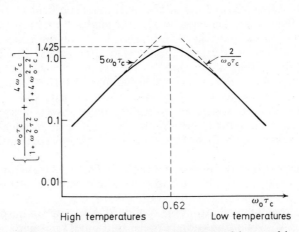

Figure 2.4 An illustration of the dependence of the sum of the spectra density functions at ω_0 and $2\omega_0$ on the correlation time τ_c.

explicitly. In the absence of an explicit calculation, it is usually assumed that the fluctuations can be described by an exponential correlation function, with a characteristic correlation time τ_c (the correlation function is the Fourier transform of the spectral density). Such an assumption is not unreasonable, in view of the fact that several realistic models lead to a correlation function of this form, e.g. any two valued randomly jumping function[48]. More details and examples are given in Reference 41, Section Vc and in Reference 42, Table 3.

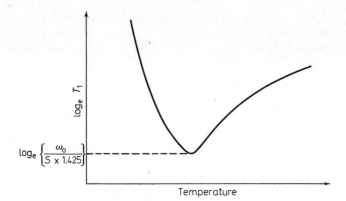

Figure 2.5 A schematic diagram of the temperature dependence of the spin–lattice relaxation time T_1, from a single-phase dipolar solid which exhibits a single thermally activated motional process.

Once the assumption of an exponential correlation function has been made, the functional dependence of the spectral densities on τ_c is fixed, leaving a coefficient which may be lumped together with the nuclear constants in equation (2.16) to form a constant S, the strength of the relaxation interaction, i.e.

$$\frac{1}{T_1} = \frac{S}{\omega_0}\left[\frac{\omega_0\tau_c}{(1+\omega_0^2\tau_c^2)} + \frac{4\omega_0\tau_c}{(1+4\omega_0^2\tau_c^2)}\right] \qquad (2.17)$$

where the factor in brackets is the correlation time dependence of the spectral densities. The overall term involving the correlation time has the form illustrated in Figure 2.4. If the temperature dependence of τ_c follows a relation of the form $\tau_c = \tau_0 \exp(E_A/RT)$, then the temperature dependence of T_1 will contain a single minimum as shown in Figure 2.5.

Since it is not always easy to calculate S explicitly because of the inter- and intra-molecular interactions, recourse is sometimes made to using S as an adjustable parameter when fitting experiment and theory. This allows the fitting to be done at the minimum in the temperature dependence of T_1, and one is therefore using the experiment to determine the strength of the relaxation interaction. Such a method of analysing the T_1 data is frequently useful but at the same time it must be borne in mind that it may obscure some interesting physics. The correct evaluation of the spectral densities will almost certainly alter the depth of the minimum from that predicted by the simple assumptions outlined above, and in addition it may change the position of the minimum slightly from $\omega_0\tau_c = 0.62$.

In order to study nuclear spin–lattice relaxation experimentally one usually turns the nuclear magnetisation from its equilibrium value of M_0 in the applied field direction through some angle θ, and then observes its return to equilibrium. Such a process may be described by the equation

$$M_z(t) - M_0 = (\cos \theta - 1)M_0 \exp \{-t/T_1\} \qquad (2.18)$$

where $M_z(t)$ is the magnitude of the nuclear magnetisation along the applied field direction at a time t after the θ pulse.

Thus if $\theta = 180$ degrees, the recovery R of the magnetisation is given by

$$\ln\{R\} = \ln\left\{\frac{M_0 - M_z(t)}{2M_0}\right\} = -\frac{t}{T_1} \qquad (2.19)$$

and the straight line graph of $\ln\{R\}$ against t will enable one to extract T_1. If T_1 is obtained over a wide range of temperature, which includes the minimum, S may be fitted to the minimum and $\omega_0\tau_c$ corresponding to each T_1 obtained from equation (2.17). It is then a simple matter to plot $\ln(\omega_0\tau_c)$ versus reciprocal temperature according to equation (2.1) in order to extract both E_A and τ_0. Consequently these constants for the motional process may be determined from the temperature dependence of τ_c over many orders of magnitude.

The description given above refers to the situation where one has random uncorrelated motion of the spins, which is describable by an exponential correlation function with a single correlation time, and which results in a return of the nuclear magnetisation to thermal equilibrium which can be described by a single exponential. Such a situation does not always exist in practice. Differences may arise, for example, because the return to thermal equilibrium is not exponential or because the motion may not be characterised by a single correlation time. These two cases will be treated in a little more detail in the following sub-sections.

2.3.1.2 Non-exponential relaxation

Deviations from simple exponential relaxation may be due to a number of possible causes. For example, they may arise[49] because of the correlated motions of more than two nuclei in a reorienting molecular group (e.g. a methyl group or an ammonium ion). Alternatively their source may be the superposition of a number of exponential decay processes due to the existence of a multiphase system. In addition non-exponential relaxation results[50, 51] from nuclei of $I > 1$ in a non-cubic environment because of the unequal spacing of energy levels. However, within a discussion of molecular motion in solids the most pertinent of these examples is probably the first and therefore a little space will be devoted to its description.

In solids, where a significant contribution to the relaxation may come from the modulation of the intragroup dipolar interactions within a reorienting internal group, the relaxation may be the sum of a number of exponentials, none of which dominates the others. The reason for this departure from exponential behaviour has been adequately explained by Hilt and Hubbard[49]. The calculation of the nuclear magnetic relaxation involves certain correlation functions of each dipole–dipole interaction with itself (auto-correlations) and with other dipole–dipole interactions (cross-correlations). If the cross-correlation terms are omitted, the calculated relaxation of the longitudinal component of the nuclear magnetisation may be a simple exponential decay, namely equation (2.18). If the cross-correlation terms are included in the calculation, the longitudinal relaxation is found to be the sum of more than one decaying exponential. For the case of the intragroup relaxation of a

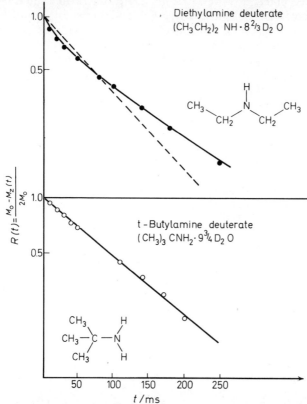

Figure 2.6 A reproduction of the recovery of the longitudinal proton magnetisation following a 180 degree r.f. pulse in the heavy ice clathrates of diethylamine and t-butylamine. The t axis refers to both parts of the Figure.

rigid three-spin triangle undergoing hindered rotations, the equation equivalent to equation (2.18) is

$$M_z(t) - M_0 = (\cos\theta - 1)M_0 \sum_{j=1}^{4} C_j \exp(-tq_j/T') \qquad (2.20)$$

where C_j and q_j are both functions of $\omega_0\tau_c$ and of β, the angle between the axis of rotation and the magnetic field, and T' is a measure of the strength of the relaxation interaction.

$$\frac{1}{T'} = \left\{\frac{\gamma^2\hbar}{r^3}\right\}^2 \cdot \frac{1}{\omega_0} \qquad (2.21)$$

where r is the internuclear vector. Hilt and Hubbard[49] have tabulated C_j and q_j in equation (2.20) for various values of $\omega_0\tau_c$ and β to allow analysis of the experimental data in an analogous[52] way to that outlined in Section 2.3.1.1. The net effect of the cross-correlations is to shift slightly the position of most efficient relaxation from $\omega_0\tau_c = 0.62$ to $\omega_0\tau_c = 0.68$, and to modify the maximum efficiency.

It should be noted, however, that for intergroup or intermolecular interactions, the cross-correlations are negligible and produce no non-exponential character in the relaxation. Thus if the overall relaxation is dominated by intercontributions, then even for hindered triangular rotors one may not observe any non-exponential character. Intra- and inter-group effects are illustrated in Figure 2.6, where the relaxation curves are from data[53] on diethylamine and t-butylamine molecules encaged in heavy ice clathrates, so as to remove intermolecular effects. In diethylamine the methyl groups are well separated, the relaxation is predominantly intragroup and largely non-exponential. In t-butylamine on the other hand, intergroup methyl interactions dominate the relaxation and result in an essentially exponential decay. Unfortunately, with non-exponential relaxation from reorientating molecular groups, most practical cases fall somewhere between the extreme of a relaxation totally determined by intragroup relaxation and the extreme of a relaxation unaffected by cross-correlations. Consequently in these cases the theory of Hilt and Hubbard[49] is not totally applicable.

2.3.1.3 The absence of a unique correlation time

The presence of more than one correlation time may arise in at least two different ways.

First, one may have, for example, two independent types of motion pro-

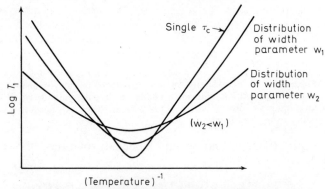

Figure 2.7 A diagramatic representation of the effect of a distribution of correlation times on the temperature dependence of the spine–lattice relaxation time T_1

ceeding at the same temperature but at different rates and each contributing towards the efficiency of the spin–lattice relaxation. One example[14, 15, 18] of this type is the t-butyl group, which may exhibit both methyl group rotation and t-butyl rotation. In these cases the shape of the temperature dependence of T_1 and also M_2 will depend on the difference in barrier heights for the two competing motions, and since this varies from sample to sample each case has to be treated individually.

Secondly, and this case is one which has been considered in some detail in the literature[42, 54, 55], there may exist a distribution of correlation times, all due to the same type of motion but resulting from a distribution of environments within the sample. Such a distribution may come about because of static crystal inhomogeneities or from thermal fluctuations. The observable symptom of such a distribution is that the shape of the minimum in T_1 is broader and flatter bottomed than the shape predicted by equation (2.17). Figure 2.7 illustrates the effect of a symmetric distribution of correlation times on the temperature dependence of T_1.

In order to account for the broad minimum quantitatively, the procedure is to generalise equation (2.17) as follows:

$$\frac{1}{T_1} = S \left\{ \int_0^\infty \frac{\tau G(\tau) d\tau}{(1 + \omega_0^2 \tau^2)} + 4 \int_0^\infty \frac{\tau G(\tau) d\tau}{(1 + 4\omega_0^2 \tau^2)} \right\} \qquad (2.22)$$

where $G(\tau)$ is a density function describing the distribution of correlation times

$$\int_0^\infty G(\tau) d\tau = 1$$

It is then necessary to postulate the density function $G(\tau)$ and obtain the form of $(1/T_1)$ from equation (2.22). The density functions usually chosen are those which have been found useful in dielectric studies of molecular motion. In general it is possible to conclude from the references quoted that the limiting slopes of a $\ln T_1$ v. 1/temperature diagram do not always provide an unambiguous determination of E_A. For example, for the symmetric Fuoss–Kirkwood and Cole–Cole distributions, the limiting slopes are (wE_A/k) where w is a parameter determined by the width of the distribution ($w \leqslant 1$ and $w = 1$ corresponds to a δ-function distribution). Hence one may underestimate E_A by neglecting the width of the distribution. In the case of an asymmetric Cole–Davidson distribution, $\ln T_1$ is asymmetric with respect to the reciprocal of temperature. The high-temperature limiting slope gives E_A unambiguously, whereas the low-temperature slope is again modified by the width of the distribution. A detailed fit of the experimental data to a postulated distribution function is probably the best way to obtain E_A, but it is very tedious.

2.3.2   Spin–lattice relaxation by spin–rotation interactions

The spin–rotation interaction although commonplace in gases and liquids[56–58] has been postulated in very few solids[59–65]. The reason for postulating this interaction in the solid has been the observation of a spin–lattice relaxation time with a temperature dependence and a frequency independence characteristic of a fluid whose dominant relaxation mechanism is spin–rotation.

In simple terms the interaction comes about because a rotating molecule may be thought of as a rotating charge distribution which is able to produce a magnetic field at a nuclear site. Since the rotations of the molecule will no doubt be interrupted, the magnetic field so produced will be a random function of time and the fluctuation of its interaction with the nucleus will form a possible relaxation mechanism. It should be noted that for the spin–rotation interaction it is the fluctuation in the angular velocity which is important, in contrast to the dipolar interaction where the fluctuations in coordinates are instrumental in relaxation.

The Hamiltonian for the interaction may be written as equation (2.5)

$$H_{SR} = -I.C.J.$$

where I is the nuclear spin, J is the molecular angular momentum operator, and C is the spin–rotation interaction tensor. The spin–rotation contribution to T_1 may be estimated using the perturbation theory approach outlined in Section 2.3.1.1. The magnitude of this contribution hinges on the calculation of the correlation functions for the rotational angular momentum, which in turn depends upon the model chosen for molecular rotation. Since many of the results derived for liquids can be taken over into mobile solids, the rotational diffusion model of rotation is sometimes used to determine the correlation functions and this leads to the expression,

$$\frac{1}{T_{1,SR}} = \frac{2}{3} \cdot \frac{I_0 kT}{\hbar^2} \cdot (2C_\perp^2 + C_{||}^2) \cdot \tau_j \qquad (2.23)$$

where I_0 is the moment of inertia of the molecule, $C_\perp$ and $C_{||}$ are the diagonal components of the spin–rotation interaction tensor and τ_j may be thought of as the mean time between changes in angular velocity. For spherical-top molecules which satisfy the rotation diffusion equation τ_j is related to τ_c, the correlation time for the angular orientation of the molecule, by

$$\tau_c . \tau_j = I_0/6kT \qquad (2.24)$$

and since τ_c is known to be thermally activated the temperature dependence of $T_{1,SR}$ may be obtained.

However, for solids one does not really expect the rotational diffusion model to be perfectly valid. Rigny and Virlet[60], in a search for a more realistic model of a solid, based their analysis on the transient rotation model of Gutowsky et al.[56,57]. They proposed that both the libration and the interrupted rotation of a molecule could lead to an observable spin–rotation interaction. They proceeded to calculate a correlation function for the angular momentum with components from librations and interrupted rotations, from which it followed that in plastic phases where interrupted rotations dominate

$$\frac{1}{T_{1,SR}} = \frac{2}{3} \cdot \frac{I_0 . kT}{\hbar^2} \cdot (2C_\perp^2 + C_{||}^2) \cdot \frac{\tau_2^2}{(\tau_1 + \tau_2)} \qquad (2.25)$$

τ_2 is the mean duration of a rotational jump and τ_1 is the mean interval between such jumps. Unfortunately this analysis is not followed up with a discussion of the temperature dependence of the characteristic times occurring in equation (2.25) or of the absolute magnitude of the spin–rotation relaxation.

The current state of affairs seems to be therefore that experimental relaxation data have been obtained for the metal hexafluorides[59, 60], the ammonium ion[61], P_4 [62], PH_3 [63], PF_3 [64] and HBr [65] which are similar to those obtained in liquids which exhibit the spin–rotation interaction. However, our quantitative understanding of these results is somewhat uncertain[66], because of reservations about the acceptability of the rotational models.

2.3.3 Spin–lattice relaxation in the rotating frame in dipolar solids

As the temperature decreases and $\omega_0 \tau_c$ becomes larger than unity for a particular type of motion, T_1 will increase and may in fact become difficult to measure experimentally because it is so long. In such circumstances it is often more profitable to measure $T_{1\rho}$, the characteristic decay time of the nuclear magnetisation along the resonant radiofrequency field in the rotating frame[67]. Such a relaxation mechanism samples the spectral density of the local field fluctuations at low frequencies ($\sim \gamma H_1$) in addition to those in the region of the Larmor frequency ($\sim \gamma H_{app}$) which are important in the decay of the longitudinal magnetisation in the laboratory frame. Consequently rotating-frame relaxation is more efficient for infrequent reorientations whose power spectrum is concentrated at low frequencies.

In such an experiment one needs to satisfy the condition that $\gamma H_{app} \gg \gamma H_1 \gg \gamma H_L$, ($H_L$ is the local dipolar field) so that at resonance the effective field in the rotating frame is H_1, which is greater than H_L. Under these circumstances it is still possible to treat the problem using a perturbation theory approach; the time-dependent dipolar Hamiltonian is the perturbation. This has been done by Lowe[46, 47, 68] and by Jones[69]. It turns out that there are now finite matrix elements of part B of the perturbing Hamiltonian H_D, which pick out the spectral density of the fluctuations at $2\omega_1 = 2\gamma H_1$, in addition to the finite matrix elements of C, D, and E, F which depend on the spectral densities at ω_0 and $2\omega_0$ respectively. In a similar fashion to equation (2.16) therefore, one obtains,

$$\frac{1}{T_{1\rho}} = \frac{3}{2} \cdot \frac{\gamma^4 \hbar^2 I(I+1)}{N} \sum_{i \neq j} \left\{ \frac{5}{2} J_{ij}^{(1)}(\omega_0) + \frac{1}{4} J_{ij}^{(2)}(2\omega_0) + \frac{1}{4} J_{ij}^{(0)}(2\omega_1) \right\} \quad (2.26)$$

For the simple case of a two-spin interaction, modulated by a random motion of the spins which is itself describable by an exponential correlation function, Jones[69] obtains a relation for $T_{1\rho}$ which is analogous to equation (2.17) for T_1, i.e.

$$\frac{1}{T_{1\rho}} = S \left\{ \frac{5}{2} \frac{\tau_c}{(1 + \omega_0^2 \tau_c^2)} + \frac{\tau_c}{(1 + 4\omega_0^2 \tau_c^2)} + \frac{3}{2} \frac{\tau_c}{(1 + 4\omega_1^2 \tau_c^2)} \right\} \quad (2.27)$$

According to equation (2.27) the temperature dependence of $T_{1\rho}$ should have a similar form to that for T_1, but with its minimum at a lower temperature than the T_1 minimum and coalescing with the T_1 curve in the high temperature ($\omega_0 \tau_c \ll 1$) limit. Consequently equation (2.27) can be used in the same way as equation (2.17) to obtain E_A and τ_0 for a reorientation process.

Look and Lowe[46] have treated the specific example of the water molecules of crystallisation in gypsum and have calculated the correct correlation functions for such a system, and as a consequence obtain a slightly different form of equation (2.27) from that given above. However, the general form of the temperature dependence is not changed and Figure 2.8 illustrates a comparison of $T_{1\rho}$ at two different H_1 values and also of T_1.

It is to be anticipated that each of the problems which arise in the interpretation of T_1 will affect the rotating-frame relaxation, e.g. non-exponential

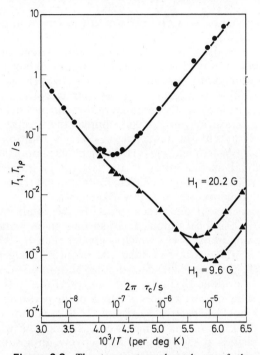

Figure 2.8 The temperature dependence of the spin–lattice relaxation time in both the laboratory frame (T_1, ●) and the rotating frame ($T_{1\rho}$, ▲) for gypsum. The rotating frame relaxation is illustrated at two different values of the r.f. field H_1
(From Look, D. C. and Lowe, I. J.[46], by courtesy of the American Institute of Physics)

relaxation or correlation-time distributions. However there is as yet no published work relating them to $T_{1\rho}$.

2.4   INVESTIGATIONS USING RELAXATION: THE STRONG COLLISION CASE

When the jump rate of a motional process becomes very low, the techniques outlined in Section 2.3 are no longer able to provide a convenient measure of that jump rate. This may be illustrated by considering infrequent re-orientational jumps in a dipolar solid. If the infrequent jumps only affect the system Hamiltonian to a small extent, e.g. dipolar changes in high-field n.m.r., then, as one has seen above, perturbation theory can be used and the results one obtains for T_1 etc. are equivalent to the infrequent rotation limit of equation (2.17), i.e.

$$\frac{1}{T_1} \sim \left\{\frac{\omega_D}{\omega_0}\right\}^2 \cdot \frac{1}{\tau_c} \tag{2.28}$$

where $\omega_D = \gamma^2 \hbar / r^3$ is a measure of the dipolar interaction strength, $\omega_0 = \gamma H_{app}$, a measure of the Zeeman interaction and $(\omega_D/\omega_0) \ll 1$. Consequently the infrequent rotational jumps constitute an inefficient longitudinal relaxation mechanism.

On the other hand, one can observe the effects of this infrequent motion by devising an experiment in which the rotational jumps have an appreciable effect on the system Hamiltonian. However it will not be possible to analyse such an experiment using perturbation theory and a different approach must be adopted, the so-called 'strong collision' approach; the following subsections contain examples of its use. Both the topics discussed in this section have been dealt with in the articles on spin thermodynamics by Jeener[70] and by Goldman[71].

2.4.1 Dipolar relaxation by infrequent motional jumps

Slichter and Ailion[72] have shown that very infrequent motional jumps may be detected in dipolar solids if one observes the rate of change of order due to the infrequent jumps within the dipolar system itself. For there to be a significant amount of order in the dipolar system, the local fields must form a large part of the total field and hence one must conduct an essentially low-field experiment. In such a low-field experiment, a single rotational jump will produce a large change in the dipolar Hamiltonian and as a result a large change in the total Hamiltonian. Consequently the problem of explaining the destruction of the dipolar order must be treated from the 'strong collision' point of view.

Slichter and Ailion base their theory on two assumptions. The first is that since in low fields the Zeeman and dipolar systems couple strongly, one may suppose that they establish a common spin temperature between jumps. This limits the range of validity of their theory to $\tau > T_2$. Secondly, the sudden approximation is assumed, which supposes that all spins have the same orientation just after the jump as they did just before. On this basis Slichter and Ailion are able to show that the characteristic decay time for the dipole order T_{1D} is given by

$$\frac{1}{T_{1D}} = \frac{2(1-p)}{\tau} \tag{2.29}$$

where τ is the mean time between motional jumps and p is a factor introduced to account for any correlation between the local field before and after the jump. The factor p will thus depend on the details of the motion and it has been worked out for various types of reorientation[72] and numerous types of diffusion[73, 74]. Equation (2.29) illustrates how efficiently the motional jumps destroy the dipolar order, i.e. $T_{1D} \simeq \tau$.

One now has the problem of observing the decay of dipolar order. Ailion and Slichter[73] adopted the technique of converting Zeeman order to dipolar order by an adiabatic demagnetisation in the rotating frame (ADRF) to a field $H_1 \simeq H_L$. If molecular motion provides the only spin–lattice relaxation mechanism, the decay of the magnetisation in the resonant rotating frame will then be due to the destruction of order in the dipolar system and the relaxation time T_1 of the magnetisation in that frame is given by[72]

$$\frac{1}{T_{1\rho}} = \frac{1}{T_{1D}} \cdot \left\{ \frac{H_L^2}{H_1^2 + H_L^2} \right\} \tag{2.30}$$

Hence one can obtain the temperature dependence of τ from the temperature dependence of T_1. Ailion and Slichter[73] verified their theory by observing the very infrequent diffusional jumps of Li atoms in lithium metal. Resing[3] has used the same technique to perform a comparative analysis of slow molecular motions in solid adamantane and solid hexamethylenetetramine. His results will be discussed in Section 2.6.1.

Jeener and Broekaert[75] have suggested an alternative method of measuring T_{1D}, using a three pulse (θ, ϕ) sequence, where θ is the pulse length and ϕ is its r.f. phase. The sequence is $(\pi/2, 0) - t_1 - (\pi/4, \pi/2) - t_2 - (\pi/4, \pi/2)$ where $t_1 \lesssim T_2$ and $T_2 \ll t_2 \lesssim T_{1D}$. The first two pulses prepare the system in a state of partial dipolar order and the third pulse inspects the dipolar order remaining after a time t_2. The dipolar signal is detected coherently with phase $\phi = 0$. Such a method suffers from the disadvantage that the degree of dipolar order which it can produce is only about half of that produced in an ideal ADRF. However, this is offset by the advantage that for long T_{1D} the sample is not subject to the heating of a long r.f. pulse. The three pulse method has been used successfully by Virlet and Rigny[76] to measure the ultra slow rotational jumps in MoF_6.

A more recent proposal is that of Haeberlen et al.[77] who present another sequence of r.f. pulses which transfers nuclear spin energy from the laboratory-frame Zeeman reservoir to the rotating-frame Zeeman and dipolar reservoirs. Their sequence comprises an initial $\pi/2$ pulse, which is followed immediately with a spin-locking pulse whose amplitude is reduced adiabatically from the H_1 of the $\pi/2$ pulse to the low value of H_1 in which the relaxation is to be observed. At a suitable interval of time after the spin-locking pulse a $\pi/4$ inspection pulse is applied in phase with it. The Zeeman signal appears after the spin-locking pulse and the dipolar signal after the inspection pulse. Examples of the use of this technique to investigate large aromatic molecules are given in reference 78.

2.4.2   Pure quadrupole relaxation by infrequent motional jumps

Alexander and Tzalmona[79] have drawn attention to another case in which the strong collision approach must be adopted. In pure nuclear quadrupole resonance in molecular crystals the energy states are determined by the quadrupole Hamiltonian (equation (2.4)). The direction and/or the magnitude of the principal components of the electric field gradient tensor occurring in this Hamiltonian may change markedly from one nuclear site in the molecule to another. If the nuclei exchange positions due to infrequent molecular reorientations then a large sudden change will occur in the total Hamiltonian. As a consequence the problem of spin–lattice relaxation due to the infrequent reorientations must be treated by a strong collision method. The specific example considered by Alexander and Tzalmona is the ^{14}N spin–lattice relaxation by molecular reorientations in hexamethylenetetramine. Each molecular reorientation causes a corresponding reorientation of the gradients of the electric field, and leads[79] to the following relation for the spin–lattice relaxation time, T_1,

$$\frac{1}{T_1} \simeq \frac{1}{\tau_c} \qquad (2.31)$$

where τ_c is of the order of the time required for the number of reorientations to equal the number of molecules. Equation (2.31) is very similar to equation (2.29) for the decay of the dipolar order due to strong collisions, i.e. $T_1 \simeq \tau_c$. It must be emphasised that the mechanism described by this strong collision model is not a resonant mechanism, whereby a Fourier component of an oscillating field induces transitions between the stationary energy states produced by a large static Hamiltonian. The 'static' Hamiltonian itself is changing in this case. This non-resonant character is reflected in the absence of a minimum in the correlation time dependence of T_1. Alexander and Tzalmona were able to extract from their T_1 data the temperature dependence of τ_c over the range $5 \times 10^{-3}\,\text{s} < \tau_c < 2\,\text{s}$.

Tzalmona[80] has extended the analysis to anisotropic molecular reorientations in piperazine in which one possible molecular reorientation is that around the molecular twofold axis. It is highly hindered and therefore well suited to investigation by a strong collision experiment. Such a reorientation will contribute unequally to the spin–lattice relaxation of each of the ^{14}N pure quadrupole resonance transitions. Tzalmona was able to measure T_1 for each of these transitions, establish that the molecular motion is indeed twofold reorientation and estimate its activation energy at 7 kcal mol^{-1}.

2.5   TUNNELLING EFFECTS AT LOW TEMPERATURES

The semi-classical theory of nuclear resonance which works well at intermediate and high temperatures, may begin to lose its credibility at very low temperatures for certain simple molecular rotors such as methyl groups, ammonium ions, methane and ammonia molecules, etc. A fundamental tenet of the above treatment of nuclear resonance is that the changes in line width and relaxation times are determined by the variation in the rate of random molecular reorientation. As a consequence there should be consistency between various nuclear resonance observables with regard to their predicted random molecular jump rate. Such a consistency which is present at higher temperatures can break down at low temperatures. There are many low temperature observations[25, 26, 81–83] of methyl groups for which the second moment is much less than its rigid-lattice value and consistent with random reorientations of the methyl group at a rate in excess of 50 kHz. In addition to this however there exist T_1 data[25, 26, 81-83] on the same compounds which at the same temperatures indicate that $1/\tau_c \ll 50$ kHz and predict that the resonance line should have broadened. It turns out that this inconsistency arises because the semi-classical method of interpreting narrowed lines is no longer valid at low temperatures. To resolve the dichotomy one needs to take account of the quantisation of the hindered molecular rotor.

2.5.1   The effect on the line shape

In a crystalline hindering potential field a symmetric molecular group has a number of torsional oscillator energy states. Furthermore[84] these energy states are split into a number of sub-states. The nature of the splittings is determined by the symmetry of both the reorienting group and the crystalline field. The magnitude of the splittings is a function of the magnitude of this hindering potential.

The torsional oscillator wave function for each of the substates has the property that it transforms according to an irreducible representation of

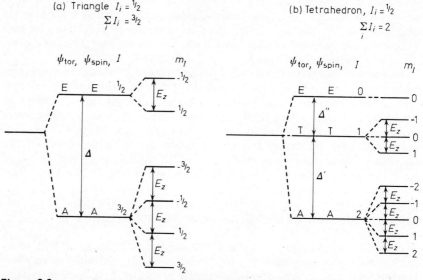

Figure 2.9   A diagram representing the tunnelling and Zeeman energy splittings of (a) a triangle of spins ($I_i = \frac{1}{2}$) hindered by a threefold potential barrier and of (b) a tetrahedron of spins ($I_i = \frac{1}{2}$) in a tetrahedral crystal field. The symmetry species of ψ_{torsion} and ψ_{spin} are tabulated together with the spin number for each state

the symmetry group to which the hindered rotor system belongs, i.e. there exists a substate for each irreducible representation. An important point[85] now is that the assignment of a nuclear spin state to each of the torsional substates is not arbitrary, but is determined by the Pauli principle and the symmetry of the system. In order not to violate the Pauli principle the product wavefunctions $\psi_{torsion}\psi_{spin}$ must transform according to the A representation and the multiplication table for the symmetry group determines which products satisfy this criterion. Figure 2.9 illustrates schematically the torsional ground state energy configuration for (i) a triangle of spins[86] $(I_i = \frac{1}{2})$ hindered

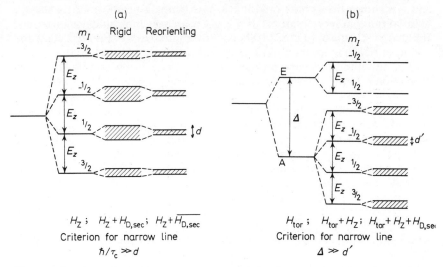

Figure 2.10 An illustration of the energy states of a triangle of spins according to (a) equation (2.33) and to (b) equation (2.32). The Hamiltonians used are shown below their corresponding energy level structure. The spread in energy of the Zeeman levels represents the dipolar broadening in a polycrystalline sample

by a threefold potential barrier and (ii) a tetrahedron[87] of spins $(I_i = \frac{1}{2})$ in a tetrahedral field.

The torsional energy splittings of the ground state can vary as a function of a barrier height from a few hundred millidegrees to negligibly small for hindered triangles and from the order of ten Kelvin degrees to negligibly small for tetrahedra. Thus at very low temperatures when only the torsional ground state is appreciably populated, the nuclear resonance line shape will be determined by an energy level scheme calculated from matrix elements such as

$$< \psi_{torsion}\, \psi_{spin} \,|\, H_{tor} + H_Z + H_{D,\,sec} |\, \psi_{torsion}\, \psi_{spin} > \qquad (2.32)$$

rather than

$$< \phi_{spin} \,|\, H_Z + \overline{H_{D,sec}} \,|\, \phi_{spin} > \qquad (2.33)$$

as was assumed in the semi-classical theory. The bar in equation (2.33) denotes the time average. These two approaches to the calculation are illustrated for a triangular group in Figure 2.10. On the basis of equation (2.32), it may be shown[88] that if $\Delta > d'$ (barriers less than about 2.5 kcal mol^{-1}), the second moment for a methyl group is one quarter of its rigid-lattice value, independent of any random reorientations. A criterion different from equation (2.9) is therefore obtained for a narrowed line, i.e.

$$\Delta > d'$$ (2.34)

and it is shown that a narrowed line at low temperatures is not a guarantee of rapid random reorientations. If $\Delta < d'$, i.e. for very high hindering barriers, the energy scheme of Figure 2.10(b) goes over into that[89] of Figure 2.10(a) and a rigid-lattice line occurs if random reorientations are absent.

In addition to the second moment, the line shape has been calculated[90] as a function of Δ for a triangular group. Such line shapes have been tested experimentally[91] on 1,1,1-trichloroethane and methyl trichlorosilane molecules embedded in perchloropropene glass. For the low barrier methyl trichlorsilane, line shape and relaxation data are consistent in their prediction of a hindering barrier. However the interpretation of the 1,1,1-trichloroethane results is complicated somewhat by intermolecular barrier effects.

The nuclear resonance line shape of a tetrahedron, characteristic of its ground-state energy splittings, has also been calculated[92]. Although line structure has been observed[93-95] for the NH_4^+ ion in a number of ammonium salts, the spectra are not usually in direct correspondence to the simple energy diagram of Figure 2.9. The distortion of the crystal field from pure tetrahedral, with its subsequent splitting of the T state, may possibly explain these departures.

The distortion of the crystal field from pure tetrahedral has also been invoked[96] to account for the measurements of nuclear susceptibility in solid methane. The static nuclear susceptibility which is proportional to $<I(I+1)>$, will obviously be sensitive to which of the torsional substates are populated and in what proportions, i.e. on the Δ' and Δ'' of Figure 2.9 relative to kT. It was found that at 4.2 K the nuclear susceptibility of CH_4 could best be explained by inclusion of a trigonal distortion of the crystal field, which lifts the degeneracy of the T state and shifts a third of the T state contribution to an energy much higher than 4.2 K. A search[97] for similar effects in silane produced no departure from a Curie law down to liquid helium temperatures. This result indicates that the separation of the substates of different spin symmetry species is much less for SiH_4 than it is for CH_4, which is consistent with a larger crystal field for SiH_4.

2.5.2 The effect on spin–lattice relaxation

In addition to its effect on the magnetic resonance line shape, the splitting and symmetrisation of the torsional energy states also causes marked differences in the temperature dependence of T_1 from that predicted by a semi-classical theory. Much work to date has centred on the triangular methyl group[26, 83, 98, 99]. For example, relaxation data on methyl reorientation in the methylbenzenes established the existence of multiple T_1 minima, where only one was expected. Initial attempts to explain the multiple minima in terms of non-equivalent methyl sites within the molecules were soon discounted by deuteriation studies and by crystal structure effects[100]. It was then pointed out[101] that a modulation of the tunnel splitting of the torsional ground state could produce a further maximum in the temperature dependence of the nuclear spin–lattice transition probability. However conceptual difficulties arose concerning the proposed mechanics of such a modulation.

To overcome such difficulties Clough[102, 103] has proposed the introduction of a third parameter into the symmetry condition. He suggests that the product $\psi_v \psi_{torsion} \psi_{spin}$ should transform according to the A representation, rather than just $\psi_{torsion} \psi_{spin}$, where ψ_v 'involves space coordinates associated with the hindering barrier'. Such a suggestion leads to the existence of two new sets of energy states, corresponding to the E_a and E_b symmetry species of ψ_v. As a result many new transitions are possible, which opens the way for a modulation of the ground-state splitting.

An alternative method of overcoming the conceptual difficulty mentioned above is to note that in the first torsional excited state of a methyl group, the splitting is larger than for the ground state and also the A and E states are reversed. If one then considers only symmetry-conversion transitions caused by the dipolar interaction, the thermal excitations to the first excited torsional state produce the required modulation of the energy splitting. This in turn leads to the possibility of a second maximum in the nuclear transition probability when a time average of the tunnel splitting is of the order of the Larmor frequency. The second maximum will only be resolved over a limited range of barrier heights. Outside this range, the effect of transitions to the first excited torsional state is to lead to a change in the slope of the temperature dependence of T_1. In either case the temperature dependence of T_1 at low temperatures is governed by thermal excitations to the first excited torsional state and not by excitations to the top of the barrier.

Wallach[104, 105] has independently noted the shortcomings of the first proposed model and has approached the problem from a more formal position by attempting to formulate the probability of transitions involving nuclear spin states, the torsional states and the phonon states. He obtains the result that for symmetry conversion transitions due to the dipolar interaction the probability of nuclear relaxational transitions is proportional to the rate of transitions amongst the torsional states of the same symmetry. Hence at low temperatures where only the two lowest torsional states need to be considered, the nuclear transition probability is proportional to the transition rate between the ground state and first excited torsional state, a similar result to that suggested in the previous paragraph. The formal approach of Wallach has also been adopted by Haupt[106] who arrives at essentially the same conclusion with regard to the temperature dependence of T_1. Namely, that it is determined by thermal excitations to the first excited torsional state.

The area in which these three calculations are likely to differ, and therefore the area to which experimental effort should be devoted, is in their predicted frequency dependence for T_1. In addition, the relative efficiencies of relaxation displayed in the classical reorientation limit on the one hand and in the excited torsional state limit on the other are also likely to vary between treatments.

The semi-classical theory has also been proved inadequate to explain the numerical values of T_1 in solid methane[107] at low temperatures. In the words of the authors, before their data can be properly interpreted 'more quantitative calculations of energy levels of methane in a crystal field are required and the role of lattice vibrations in causing molecular reorientations must be elucidated'.

2.6 A SURVEY OF RECENTLY PUBLISHED WORK

The nuclear magnetic resonance methods outlined in Sections 2.2 and 2.4 continue to be exploited in a wide variety of molecular motion studies. The following survey of some of the work published in the last few years is not intended to be exhaustive, but it is hoped that the survey contains a number of the more important papers. References to earlier work can be found in the reviews by Powles[108, 110], by Andrew[109, 111, 112] and by Slichter[113].

2.6.1 Nearly spherical molecules

A characteristic of a number of solids composed of molecules which are nearly spherical in shape is that they exhibit a *plastic* phase before melting. Their bulk plastic properties are closely related to the freedom of molecular movement on a microscopic scale and this molecular freedom has been investigated in many cases using n.m.r. spectroscopy.

The globular molecules pentaerythritol[13], 2-methyl-2-propanethiol[14] and hexamethylethane[15, 16] have recently been the subject of nuclear resonance investigations. Each of these solids displays general molecular reorientation and diffusion in its plastic phase. In addition the last two exhibit internal group reorientation which tends to dominate the nuclear resonance parameters in lower-temperature solid phases. Smith[13] has compared the properties of pentaerythritol with those of other globular molecules such as adamantane, which in its turn has received detailed attention from others. Resing[3] for example has studied the differences in molecular freedom between the plastic crystal adamantane and the non-plastic hexamethylenetetramine (HMT). He was able to show that in the plastic phase of adamantane the diffusion constant is at least 10^4 times greater than in HMT. In addition the freedom of general reorientation is greater in both phases of adamantane than it is in HMT. A point which should not be glossed over here is the fact that the three[3, 79, 114] independent investigations of molecular reorientation in HMT are not really in agreement over the magnitude of the activation energy. Adamantane has also been used[115] as a comparison for the work on carboranes. The *ortho* isomer of carborane exhibits general reorientation and diffusion in the solid state which is characteristic of plastic crystals. In a similar investigation[116] it was found that 2-chloro-2-methylpropane and 2,2-dimethylpropanol display general molecular reorientation and diffusion before melting whereas trimethylchlorosilane and 3,3-dimethyl-2-butane and 2,2-dimethylpropane nitrile do not. The previous nuclear resonance work on the symmetric molecules cyclohexane, perfluorocyclohexane and neopentane has been extended[117] into the rotating frame. Such an extension confirms the existence of general reorientation and diffusion in the high-temperature phase of each substance.

The metal hexafluorides also form spherically shaped molecules which are expected to be distinctly mobile in their cubic phase which exists just below the melting point. Virlet and Rigny have shown this to be the case for the hexafluorides of Mo and W [60] and of S, Se and Te [118]. They have also measured the infrequent reorientation rate of MoF_6 in its orthorhombic phase at low temperatures.

White phosphorus (P_4) exists in a plastic cubic (α) phase between its melting point and 196.3 K, where it is transformed into a crystalline (β) phase with a symmetry lower than tetragonal. In the plastic phase the P_4 molecule is highly mobile, performing both general reorientation and self diffusion. T_1 in this phase is dominated by the modulation of the anisotropic chemical shift. However Boden and Folland [62] were able to make a separation of the dipolar, spin–rotation and anisotropic chemical shift contributions to the relaxation rate. From an interpretation of the spin–rotation contribution in terms of the transient rotation model [57], they conclude that the molecule is executing 120 degree jumps about its C_{3v} axes. General molecular reorientation still persists in the β-phase where it promotes spin–lattice relaxation by modulating the anisotropic chemical shift and the dipolar interaction.

2.6.2   Small molecules

The small molecule which has received most attention is undoubtedly the water molecule, primarily because the hydrates provide an extensive supply of interesting materials. Excluding the ferroelectric hydrates, the following materials have been studied by means of deuterium magnetic resonance (d.m.r.), cupric sulphate pentahydrate [35], calcium bromate monohydrate [119] and potassium oxalate monohydrate [120]. In each of these the nuclear resonance behaviour is dominated by the 180 degree flip motion about the DOD bisector. In the copper salt, Soda and Chiba were successful in observing the spectra of each individual water molecule of crystallisation. This enabled them to measure a separate activation energy for the 180 degree flips of each water molecule. The variation was from 7.5 to 12.5 kcal mol^{-1}. The work of Soda and Chiba is a detailed application of the d.m.r. method to deuterated hydrates and it contains much structural information in addition to the data on water motion.

Solid water in its own right has been the subject of a definitive investigation by Barnaal and Lowe [24, 121] who maintain that high frequency proton tunnelling can be excluded on the basis of their second moment results and obtain from their relaxation data an activation energy of (14.1 ± 0.1) kcal mol^{-1} for proton jumping, without actually being able to identify explicitly the detailed nature of the jumping process.

A similar molecule to water in its molecular structure is H_2S [167, 178]. Hydrogen sulphide exists in three solid phases and in its low-temperature phase (III) it is found to execute 90 degree flips around its dipole axis. At higher temperatures (phase (II) and (I)) the rate of 90 degree flips is much greater, direction reversing reorientations are observed for the molecular dipole and molecular diffusion is also detected.

The hydrogen halides [122], HCl [123, 124], HBr [65, 125] and HI form an interesting group of solids because they too exist in a number of crystalline phases. In their high-temperature phase (solid (I)) general molecular reorientation and diffusion were detected in each halide. Indeed molecular reorientation is so free as to produce a detectable spin–rotation interaction in the bromide. The intermediate temperature phases of the bromide and iodide display both two-plane and three-plane reorientation, while in their low temperature phase (solid (III)) only single plane reorientation is observed. Evidence was also found for single plane reorientation of HCl in its ferroelectric phase (solid (II)).

Molecular solids such as hydrogen, nitrogen and fluorine are easily investigated by nuclear resonance. Solid hydrogen, however, because of the fundamental effect on its nuclear resonance behaviour of the quantisation of its rotational states, requires a discussion beyond the scope of this brief survey. An analysis of the present state of knowledge of the properties of solid hydrogen can be found in the articles by Brooks-Harris[126, 127]. The nuclear resonance investigation of solid nitrogen on the other hand[128, 179] may be considered from a semi-classical point of view. In the h.c.p. β-phase the nitrogen molecules exhibit a hindered reorientation about the c-axis at an angle very close to $54° 44'$. This has the effect of reducing the average quadrupole coupling constant by three orders of magnitude from that present in the f.c.c. α-phase, where the only motion to affect the nuclear resonance observables is molecular libration. The molecules of solid fluorine, in contrast, behave in a much more cavalier fashion initially. They are reported[129] to display rapid translational diffusion and rapid anisotropic reorientation in their plastic β-phase. However, on cooling through the transition at 45.5 K a very dense low-symmetry α-phase is produced in which, to account for the $T_{1\rho}$ data, a 15 degree 'tilt' motion is invoked relative to the (001) direction.

Two small molecules with C_{3v} symmetry are ammonia[130, 131] and phosphine[63]. Neither of these has been thoroughly investigated at very low temperatures to look for quantum effects in the spin–lattice relaxation. However, the second moment of ammonia has been observed[131] at liquid helium temperatures and is much less than the rigid-lattice value as would be expected from the reasoning given in Section 2.5.1. Throughout the solid range studied C_3 reorientation of the ammonia molecule dominates the proton spin–lattice relaxation time. Phosphine on the other hand seems to be much more mobile. It is suggested[63] that a plastic phase exists below the melting point, in which both rapid general molecular reorientation and a slow diffusion are manifest. Such a conclusion is based on the observation of ^1H, ^2H and ^{31}P relaxation in the laboratory frame and ^1H relaxation in the rotating frame.

2.6.3 Reorientation of internal groups

2.6.3.1 Three-spin triangle

The reorientating methyl group occurs in many molecular solids and has been the subject of much investigation. The second moment and spin–lattice relaxation time of some of the globular molecules mentioned in Section 2.6.1, e.g. 2-methyl-2-propanothiol[12] and hexamethylethane[15, 18] are dominated by methyl reorientation at temperatures lower than their plastic crystal phase. The mobility of methyl groups has also been documented for certain alkanes[132], ketones[133], odd- and even-numbered sodium and lithium soaps[134], methylamines[135] and methylamine complexes[136, 137], in addition to the work on methylbenzenes which was noted in Section 2.5. Solid boron trifluoride–amine complexes have revealed[138] the reorientation of their triangular groups and the dynamics of the NH_3 reorientation has been observed in glycine[139] and in certain amino acids[140].

In the solid state a number of acid hydrates exist as oxonium salts and one may observe the motion of the H_3O^+ oxonium ion. Such ions have been shown to reorientate around their threefold axes in perchloric acid[141] and its deuteriated analogue[31] and in hydrated gallium sulphate[142].

2.6.3.2   Four-spin tetrahedron

The ammonium ion is a very mobile group which is amenable to nuclear resonance study. Since a number of ammonium salts are ferroelectric they will be referred to in Section 2.6.4 rather than here. Of the remainder, the ease with which the group reorientates has been demonstrated in the ammonium halides NH_4Cl[143], NH_4Br[144, 145] and NH_4I[61], and in $(NH_4)_2SF_6$[146]. Ammonium bromide exemplifies the halides in that it displays an order–disorder transition (tetragonal to cubic) at 234.5 K, below which it is suggested that there exist domains defined by the orientation of tetragonal axes. From the nuclear relaxation data, it has been established that the activation energy for NH_4^+ ion reorientation changes from $3.64 \, \text{kcal mol}^{-1}$ to $4.95 \, \text{kcal mol}^{-1}$ on cooling through the transition. A lower-temperature order–order transition has also been studied[147] in NH_4Br.

The characteristics of the phosphonium halides are somewhat different from those of the ammonium halides. For example, whereas the activation energy for NH_4^+ decreases significantly in going from chloride to bromide to iodide, there is no such observed[148] marked trend within the phosphonium halides. The activation energies[148, 149] for the PH_4^+ ion reorientation in the phosphonium halides are all larger than for the corresponding ammonium halide. Such differences are attributed to the existence of non-electrostatic repulsive forces in the phosphonium salts.

Isoelectronic with the ammonium halides are the alkali metal borohydrides. Since the BH_4^- ion has a tetrahedral geometry, these borohydrides provide a complimentary extension of the work on ammonium halides. As was to be expected BH_4^- ion reorientations are active in the solid state of KBH_4, $NaBH_4$ and $LiBH_4$, where their activation energies have been measured[150, 151] in the cubic phases of all three compounds and in the tetragonal phase of $NaBH_4$.

Ammonium tetrafluoroborate[152] provides a solid in which there are two four-spin tetrahedra per molecule. Of the two, the NH_4^+ ion provides a motional narrowing of the nuclear resonance line at the lowest temperature and is therefore presumably the more mobile. Nevertheless isotropic reorientation of the BF_4^- ions is rapid enough to dominate the ^{19}F nuclear resonance at 170 K. Furthermore it has been demonstrated[153] in a series of measurements with the following cations Na, NO, K, Rb, ND_4, NH_4, Cs and NO_2, that the activation energy for BF_4^- tumbling decreases smoothly with increasing cation radius.

2.6.3.3   Six-spin octahedron

In order to explain the nuclear resonance results in both $(NH_4)_2SiF_6$[146] and certain iron-group fluosilicate hexahydrates[154], the hindered reorientation of the SiF_6 octahedron has been invoked. At temperatures above room temperature the reorientation rate of the ammonium tetrahedra in $(NH_4)_2SiF_6$ becomes too great to produce effective nuclear spin–lattice relaxation. The reorientating SiF_6 group takes over at these higher temperatures, where it is now suggested that the protons are relaxed by inter-ionic dipolar interactions and the fluorine nuclei by intra-ionic interactions.

The reorientations of the BiF_6 octahedron in $KBiF_6$ have proved particularly useful[155], in that they lead to a sufficient narrowing of the ^{19}F resonance line to allow its indirect spin–spin structure to be observed. From the static resonance spectrum a $^{209}Bi-^{19}F$ coupling of 2.7 kHz has been measured in solid $KBiF_6$.

2.6.3.4   Eight- and nine-spin groups

The hindered reorientation of UF_8 groups in Na_3UF_8 is clearly evident[156] from the temperature dependence of the ^{19}F resonance line profile. On cooling from above room temperature the line broadens from a symmetric narrow line to become distinctly asymmetric due to the anisotropic ^{19}F chemical shift. By extracting the field independent dipolar contribution to M_2, it was possible to observe the temperature dependence of that dipolar part exhibit a normal M_2 transition as the hindered reorientation rate decreased. About 9 kcal mol^{-1} was given as an estimated activation energy.

The ReH_9^- ions in K_2ReH_9 occupy two non-equivalent crystallographic sites, with two markedly different hindering barriers. The second-moment data[157] established that both types of ions undergo the same motional process, while the double minimum in T_1 yielded the activation energies of 2.4 kcal mol^{-1} and 6.0 kcal mol^{-1} for the two non-equivalent sites.

2.6.4   Ferroelectrics

The subject of hydrogen-bonded ferroelectrics was thoroughly reviewed by Blinc[158] in 1968. Since then a continuation of the work has proceeded along similar lines. In ferroelectric materials there are often a number of molecular motions which occur simultaneously and not all of which are critically involved in the ferroelectric transition. Nuclear resonance has been used as a technique to identify which motions exist and to observe the temperature dependence of their rate, in an attempt to discover what motional processes are affected by the onset of ferroelectricity. For example, in the paraelectric phase of KH_2PO_4-type hydrogen-bonded ferroelectrics, two types of molecular motion occur at widely different rates. A fast motion, which displays a discontinuous decrease in rate as the sample is cooled through the critical point, has been identified as the jumping of a proton between two off-centre positions with the O—H····O bonds. The freezing in of such a motion has been observed[159] from the d.m.r. spectra in the deuteriated forms of both phosphates and arsenates. A very infrequent motion on the other hand, which appears to play no part in the onset of ferroelectricity, corresponds to an exchange of hydrogen atoms between hydrogen bonds. This motion has been identified[160] as a reorientation of the H_2PO_4 groups from the temperature dependence of a rotating-frame relaxation time. The activation energy for this reorientation increases with increasing cation size. Similar results occur for the $NaH_3(SeO_3)_2$ type of crystals. The deuteron dynamics in the antiferroelectric materials $NH_4D_2PO_4$, (ADP)[161], resemble strongly those in KDP with respect to the deuteron jumping along the hydrogen bond. In addition, however, ADP exhibits reorientation of its NH_4^+ ions, the activation energy of which changes at T_c.

Ferroelectric ammonium salts often display an active reorientation of their NH_4^+ ions both above and below T_c; however below T_c the ions are often distorted from perfect tetrahedra and locked as some angle to a crystal plane, even though they are still reorientating. NH_4SO_4 and $(ND_4)2BeF_4$ are examples of this sort of behaviour[162].

Leconite[30], $NaNH_4SO_4 \cdot 2H_2O$, contains both NH_4^+ ions and water molecules. Because of the invariance in the orientation of their interproton vectors on transforming to the ferroelectric phase, it is suggested that the water molecules are not involved in the ferroelectric transition. Furthermore, since the NH_4^+ ions appear, from the d.m.r. results, to be undistorted in the ferroelectric phase and since a change in activation energy (which heralds a change in lattice parameters) is observed at T_c, it is proposed that the ferroelectric transition in this material is of the displacive type.

The proton T_1 in both diglycine nitrate and tris-sarcosine calcium chloride is found[163] to be dominated by the reorientation of threefold groups which is unaffected by the ferroelectric transition. It is from the temperature dependence of the second moment that evidence is obtained to the effect that the ferroelectric transition is accompanied by the freezing of the flipping motion of molecular units much larger than the threefold groups. Thiourea[164] is a material in which the whole molecule executes 180 degree flips about its C—S axis in the paraelectric phase. $LiNbO_3$ on the other hand[165] allows only the motion of its lithium atom. Below 550 K this is confined to a to and fro motion between normal and interstitial sites through a loose triangle of oxygen atoms, whereas above 550 K it appears to blossom into full-scale translational diffusion.

2.6.5 Miscellaneous

Of the molecular motion studies which do not fit neatly into the above sections, it should be mentioned that hindered reorientations about an axis perpendicular to their 'plane' have been detected and studied in both cyclohexene[166] and piperazine[80].

The succinonitrile molecule $(CH_2CN)_2$ is far from spherical yet it shows a marked degree of mobility in its high-temperature phase (I). The high-field spin–lattice relaxation is dominated[167] by molecular rotation, while self-diffusion may be detected from the rotating-frame relaxation[168]. The interpretation of both sets of data is complicated by what is thought to be the existence of both *trans* and *gauche* forms of the molecule in phase (I). Indeed because of the proposed molecular flexibility, it is suggested that the self-

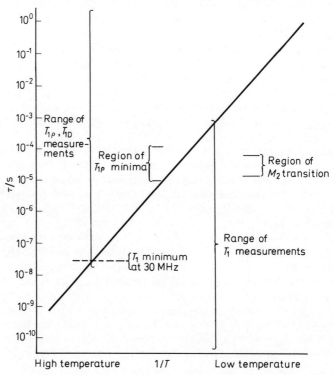

Figure 2.11 A schematic diagram illustrating the useful range of the nuclear resonance observables in a typical dipolar solid containing protons.

diffusion process may not be characterised by a single τ_c. In the low-temperature phase (II) 'no large-scale molecular motion' was established[167].

The T_1 data[169] from anhydrous sodium and lithium soaps above room temperature are made difficult to interpret by the distributions of correlation times for the methylene group oscillations. However it is possible to conclude from the data that the even-numbered soaps pack closer than do the odd-numbered ones.

An area which continues to expand is the investigation of the mobility of guest molecules in clathrate cages. Such investigations give information about the cage structure on the one hand and about the effects on molecular mobility of changing the crystal-field environment on the other. A system which provides excellent isolation of the guest molecules, from a proton resonance point of view, is the heavy ice clathrate[170, 171]. A further advantage of this system is that it is able to act as host to a wide variety of guest molecules[172-175].

2.7 SUMMARY

The object of Section 2.2 was to outline how the temperature dependence of the steady-state nuclear resonance spectrum enables one to identify which molecular motions are active in the region of tens of kHz. Such information is complementary to the more extensive quantitative data which may be obtained from relaxation time measurements, as indicated in Sections 2.3 and 2.4. By means of relaxation-time measurements reliable estimates of the frequency of the motional jumps may be made, typically over ten orders of magnitude, i.e. 10^9–10^{-1} Hz. Figure 2.11 is a schematic illustration of the useful range of each n.m.r. observable for a single reorientational mechanism in some imaginary dipolar solid containing protons.

The fact that such diverse motional rates can be measured enables one to estimate the hindering barriers for a wide variety of motions, from the most cavalier translational diffusion to the more restrained overall rotations of the larger molecules. Because the relaxation times are so sensitive to τ_c, and furthermore since τ_c is a rapid function of both barrier height and temperature, the temperature dependence of relaxation times is an excellent detector of any changes in barrier due to phase changes, etc. In addition spatial distributions of barrier heights can, as a result of the distribution of correlation times which they produce, be detected from the temperature dependence of the relaxation time. The survey presented in Section 2.6 is intended to direct attention to some of the examples of molecular motion studied by nuclear resonance over the past few years.

Acknowledgements

The author would like to thank his colleagues Professor E. R. Andrew, Dr. S. Clough, Dr. W. Derbyshire and Dr. P. Mansfield for many stimulating and helpful discussions.

References

1. Abragam, A. (1961). *Principles of Nuclear Magnetism* (Oxford: Clarendon Press)
2. Slichter, C. P. (1964). *Principles of Magnetic Resonance* (New York: Harper and Row)
3. Resing, H. A. (1969). *Mol. Cryst. Liquid Cryst.*, **9**, 101
4. Reference 1, Chapter 4
5. Mansfield, P. (1971). *Progress in Nuclear Magnetic Resonance Spectroscopy*, Vol. 8, 41, (J. W. Emsley, J. Feeney and L. H. Sutcliffe, editors). (Oxford: Pergamon Press)
6. Reference 2, Chapter 3
7. Van Vleck, J. H. (1948). *Phys. Rev.*, **74**, 1168
8. Reference 2, Chapter 2
9. Reference 1, Chapter 10
10. Smith, G. W. (1962). *J. Chem. Phys.*, **36**, 3081

11. Smith, G. W. (1965). *J. Chem. Phys.*, **42**, 4229
12. Smith, G. W. (1965). *J. Chem. Phys.*, **43**, 4325
13. Smith, G. W. (1969). *J. Chem. Phys.*, **50**, 3595
14. Smith, G. W. (1969). *J. Chem. Phys.*, **51**, 3569
15. Smith, G. W. (1971). *J. Chem. Phys.*, **54**, 174
16. Gutowsky, H. S. and Pake, G. E. (1950). *J. Chem. Phys.*, **18**, 162
17. Andrew, E. R. and Eades, R. G. (1953). *Proc. Roy. Soc.*, **A218**, 537
18. Chezeau, J. M., Dufourcq, J. and Strange, J. H. (1971). *Mol. Phys.*, **20**, 305
19. Bloembergen, N., Purcell, E. M. and Pound, R. V. (1948). *Phys. Rev.*, **73**, 679
20. Kubo, R. and Tomita, K. (1954). *Proc. Phys. Soc. Japan*, **9**, 888
21. Reference 1, Chapter 4
22. Powles, J. G. and Strange, J. H. (1963). *Proc. Phys. Soc. (London)*, **82**, 6
23. Mansfield, P. (1965). *Phys. Rev.*, **137**, A961
24. Barnaal, D. E. and Lowe, I. J. (1967). *J. Chem. Phys.*, **46**, 4800
25. Allen, P. S. and Cowking, A. (1967). *J. Chem. Phys.*, **47**, 4286
26. Allen, P. S. and Cowking, A. (1968). *J. Chem. Phys.*, **49**, 789
27. Goldman, M. (1970). *Spin Temperature and Nuclear Magnetic Resonance in Solids*, Chapter 5. (Oxford: Clarendon Press)
28. Reference 1, Chapter 7
29. Bjorkstam, J. L. (1967). *Phys. Rev.*, **153**, 599
30. Genin, D. J. and O'Reilly, D. E. (1969). *J. Chem. Phys.*, **50**, 2842
31. O'Reilly, D. E., Peterson, E. M. and Williams, J. M. (1971). *J. Chem. Phys.*, **54**, 96
32. Anderson, P. W. (1954). *J. Phys. Soc. Japan*, **9**, 316
33. Kubo, R. (1962). *Fluctuations, Relaxation and Resonance in Magnetic Systems*, 23. (D. Ter Haar, editor). (Edinburgh: Oliver and Boyd)
34. Reference 1, Chapter 10
35. Soda, G. and Chiba, T. (1969). *J. Chem. Phys.*, **50**, 439
36. Andrew, E. R. (1954). *Nuclear Magnetic Resonance* (Cambridge: Cambridge Univ. Press
37. Reference 2, Chapter 5
38. Reference 1, Chapter 8
39. Hubbard, P. S. (1961). *Rev. Mod. Phys.*, **33**, 249
40. Argyres, P. N. and Kelley, P. L. (1964). *Phys. Rev.*, **134**, A98
41. Redfield, A. G. (1965). *Advances in Magnetic Resonance*, Vol. 1, 1. (J. S. Waugh, editor). (New York: Academic Press)
42. Noack, F. (1971). *N. M. R. Basic Principles and Progress*, Vol. 3, 83. (P. Diehl, E. Fluck and R. Kosfeld, editors). (Berlin: Springer-Verlag)
43. Reference 2, Chapter 5
44. Reference 1, Chapter 5D
45. Eisenstadt, M. and Redfield, A. G. (1963). *Phys. Rev.*, **132**, 635
46. Look, D. C. and Lowe, I. J. (1966). *J. Chem. Phys.*, **44**, 2995
47. Look, D. C. and Lowe, I. J. (1966). *J. Chem. Phys.*, **44**, 3437
48. Reference 2, Appendix C
49. Hilt, R. L. and Hubbard, P. S. (1964). *Phys. Rev.*, **134**, A392
50. Andrew, E. R. and Tunstall, D. P. (1961). *Proc. Phys. Soc. (London)*, **78**, 1
51. Hubbard, P. S. (1970). *J. Chem. Phys.*, **53**, 985
52. Baud, M. F. and Hubbard, P. S. (1968), *Phys. Rev.*, **170**, 384
53. Allen, P. S., Khanzada, A. W. and McDowell, C. A., to be published
54. Connor, T. M. (1964). *Trans. Faraday. Soc.*, **60**, 1574
55. Noack, F. and Preissing, G. (1969). *Z. Naturforsch.*, **24a**, 143
56. Gutowsky, H. S., Lawrenson, I. J. and Shimonura, K. (1961). *Phys. Rev. Lett.*, **6**, 349
57. Brown, R. J. C., Gutowsky, H. S. and Shimonura, K. (1963). *J. Chem. Phys.*, **38**, 76
58. Hubbard, P. S. (1963). *Phys. Rev.*, **131**, 1155
59. Blinc, R. and Lahajnar, G. (1967). *Phys. Rev. Lett.*, **19**, 685
60. Rigny, P. and Virlet, J. (1969). *J. Chem. Phys.*, **51**, 3807
61. Sharp, A. R. and Pintar, M. M. (1970). *J. Chem. Phys.*, **53**, 2428
62. Boden, N. and Folland, R. (1971). *Mol. Phys.*, **21**, 1123
63. Sawyer, D. W. and Powles, J. G. (1971). *Mol. Phys.*, **21**, 83
64. Allen, P. S. and Dunell, B. A. To be published
65. Norris, M. O., Strange, J. H., Powles, J. G., Rhodes, M., Marsden, K. and Krynicki, K. (1968). *J. Phys. C.: Solid State Phys.*, **1**, 422

66. Bloom, M. (1967). *Proceedings of XIV Colloque Ampere,* 65. (R. Blinc, editor). (Amsterdam: North Holland)
67. Redfield, A. G. (1955). *Phys. Rev.,* **98,** 1787
68. Look, D. C., Lowe, I. J. and Northby, J. A. (1966). *J. Chem. Phys.,* **44,** 3441
69. Jones, G. P. (1966). *Phys. Rev.,* **148,** 332
70. Jeener, J. (1965). *Advances in Magnetic Resonance,* Vol. 1, 205. (J. S. Waugh, editor). (New York: Academic Press)
71. Reference 27, Chapter 3D
72. Slichter, C. P. and Ailion, D. C. (1964). *Phys. Rev.,* **135,** A1099
73. Ailion, D. C. and Slichter, C. P. (1965). *Phys. Rev.,* **137,** A235
74. Ailion, D. C. and Ho, P. P. (1968). *Phys. Rev.,* **168,** 662
75. Jeener, J. and Broekaert, P. (1967). *Phys. Rev.,* **157,** 232
76. Virlet, J. and Rigny, P. (1970). *Chem. Phys. Lett.,* **4,** 501
77. Haeberlen, U., Stehlik, D. and Lauer, O. (1971). *Proceedings of XVI Colloque Ampere,* 664. (I. Ursu, editor). (Bucharest: Academy of the Socialist Republic of Rumania)
78. Hausser, K. H., Lauer, O., Schuch, H. and Stehlik, D. (1971). *Proceedings of XVI Colloque Ampere,* 83. (I. Ursu, editor). (Bucharest: Academy of the Socialist Republic of Rumania)
79. Alexander, S. and Tzalmona, A. (1965). *Phys. Rev.,* **138,** A845
80. Tzalmona, A. (1968). *J. Chem. Phys.,* **50,** 366
81. Eades, R. G., Jones, G. P., Llewellyn, J. P. and Terry, K. W. (1967). *Proc. Phys. Soc. (London),* **91,** 124
82. Eades, R. G., Jones, T. A. and Llewellyn, J. P. (1967). *Proc. Phys. Soc. (London),* **91** 632
83. Jones, G. P., Eades, R. G., Terry, K. W. and Llewellyn, J. P. (1968). *J. Phys. C: Solid State Phys.,* **1,** 415
84. King, H. F. and Hornig, D. F. (1966). *J. Chem. Phys.,* **44,** 4520
85. Wilson, E. B. (1935). *J. Chem. Phys.,* **3,** 276
86. Freed, J. H. (1965). *J. Chem. Phys.,* **43,** 1710
87. Nagamiya, T. (1951). *Prog. Theoret. Phys.,* **6,** 702
88. Allen, P. S. (1968). *J. Chem. Phys.,* **48,** 3031
89. Andrew, E. R. and Bersohn, R. (1950). *J. Chem. Phys.,* **18,** 159
90. Apayadin, F. and Clough, S. (1968). *J. Phys. C: Solid State Phys.,* **1,** 932
91. McIntyre, H. M. and Johnson, C. S. Jr. (1971). *J. Chem. Phys.,* **35,** 345
92. Tomita, K. (1953). *Phys. Rev.,* **89,** 429
93. Hennel, J. W. and Lalowicz, Z. T. (1970). *Acta Phys. Polon.,* **A38,** 675
94. Sharp, A. R., Vrscaj, S. and Pintar, M. M. (1970). *Solid State Commun.,* **8,** 1317
95. Lalowicz, Z. T. and Hennel, J. W. (1971). Report No. 766/PL Institute of Nuclear Physics, Cracow
96. Wong, K. P., Noble, J. D., Bloom, M. and Alexander, S. (1969). *J. Mag. Res.,* **1,** 55
97. Jones, E. P. and Montgomery, L. P. (1971). *Phys. Lett.,* **35A,** 229
98. Haupt, J. and Muller-Warmuth, W. (1968). *Z. Naturforsch.,* **23a,** 208
99. Haupt, J. and Muller-Warmuth, W. (1969). *Z. Naturforsch.,* **24,** 1066
100. Allen, P. S. and Howard, C. J. (1969). *Mol. Phys.,* **16,** 311
101. Allen, P. S. and Clough, S. (1969). *Phys. Rev. Lett.,* **32,** 1351
102. Clough, S. (1971). *J. Phys. C: Solid State Phys.,* **4,** 1075
103. Clough, S. (1971). *J. Phys. C: Solid State Phys.,* **4,** 2180
104. Wallach, D. and Steele, W. A. (1970). *J. Chem. Phys.,* **52,** 2534
105. Wallach, D. (1971). *J. Chem. Phys.,* **54,** 4044
106. Haupt, J. Z. (1971). *Naturforsch.,* **26a,** 1578
107. De Witt, G. A. and Bloom M. (1969). *Can. J. Phys.,* **47,** 1195
108. Powles, J. G. (1959). *Arch. Sci. Geneve,* **12,** 87
109. Andrew, E. R. (1961). *J. Phys. Chem. Solids,* **18,** 9
110. Powles, J. G. (1961). *J. Phys. Chem. Solids,* **18,** 17
111. Andrew, E. R. and Allen, P. S. (1966). *J. chim. Phys.,* **63,** 85
112. Andrew, E. R. (1969). *Proc. 2nd Mat. Res. Symp. 'Mol. Dynamics and Struct. of Solids' N. B. S. 1967,* 413. (Washington, D.C.: U.S. Dept. of Commerce)
113. Slichter, W. P. (1969). *Mol. Cryst. Liquid Cryst.,* **9,** 81
114. Smith, G. W. (1965). *J. Chem. Phys.,* **42,** 3341
115. Baughman, R. H. (1970). *J. Chem. Phys.,* **53,** 3781

116. Segel, S. L. and Mansingh, A. (1969). *J. Chem. Phys.*, **51**, 4578
117. Roeder, S. B. W. and Douglass, D. C. (1970). *J. Chem. Phys.*, **52**, 5525
118. Virlet, J. and Rigny, P. (1970). *Chem. Phys. Lett.*, **6**, 377
119. Cvikl, B. and McGrath, J. W. (1970). *J. Chem. Phys.*, **52**, 1560
120. Pederson, B. and Clark, W. G. (1970). *J. Chem. Phys.*, **53**, 1024
121. Barnaal, D. E. and Lowe, I. J. (1968). *J. Chem. Phys.*, **48**, 4614
122. Genin, D. J., O'Reilly, D. E., Peterson, E. M. and Tsang, T. (1968). *J. Chem. Phys.*, **48**, 4525
123. O'Reilly, D. E. (1970). *J. Chem. Phys.*, **52**, 2396
124. Okuma, H., Nakamura, N. and Chihara, H. (1968). *J. Phys. Soc. Jap.*, **24**, 452
125. Kadaba, P. K. and O'Reilly, D. E. (1970). *J. Chem. Phys.*, **52**, 2403
126. Brooks Harris, A. (1970). *Phys. Rev.*, **B1**, 1881
127. Brooks Harris, A. (1970). *Phys. Rev.*, **B2**, 3495
128. De Reggi, A. S., Canepa, P. C. and Scott, T. A. (1969). *J. Mag. Res.*, **1**, 144
129. O'Reilly, D. E., Peterson, E. M., Hogenboorm, D. L. and Scheie, C. E. (1971). *J. Chem. Phys.* **54**, 4194
130. O'Reilly, D. E., Peterson, E. M. and Lammert, S. R. (1970). *J. Chem. Phys.*, **52**, 1700
131. Carolan, J. L. and Scott, T. A. (1970). *J. Mag. Res.*, **2**, 243
132. Van Putte, K. (1970). *J. Mag. Res.*, **2**, 216
133. Van Putte, K. (1969). *Trans. Faraday. Soc.*, **65**, 1709
134. Van Putte, K. (1970). *Trans. Faraday. Soc.*, **66**, 523
135. Haigh, P. J., Canepa, P. C., Matzkanin, G. A. and Scott, T. A. (1968). *J. Chem. Phys.*, **48**, 4234
136. Fyfe, C. A. and Ripmeester, J. (1970). *Can. J. Chem.*, **48**, 2283
137. Yim, C. T. and Gilson, D. F. R. (1970). *Can. J. Chem.*, **48**, 575
138. Dunell, B. A., Fyfe, C. A., McDowell, C. A. and Ripmeester, J. (1969). *Trans. Faraday. Soc.*, **65**, 1153
139. Osredkar, R. (1970). *J. Chem. Phys.*, **52**, 1616
140. McElroy, R. G. C., Dong, R. V., Pintar, M. M. and Forbes, W. F. (1971). *J. Mag. Res.*, **5**, 262
141. Hennel, J. W. and Pollak-Stachura, M. (1969). *Acta Phys. Polon.*, **35**, 239
142. Kydon, D. W., Pintar, M. and Petch, H. E. (1968). *J. Chem. Phys.*, **48**, 5348
143. Woessner, D. E. and Snowden, B. S. (1967). *J. Phys. Chem.*, **71**, 952
144. Woessner, D. E. and Snowden, B. S. (1967). *J. Chem. Phys.*, **47**, 328
145. Sharp, A. R. and Pintar, M. (1971). *Proceedings of XVI Colloque Ampere*, 341 (Bucharest: Academy of the Socialist Republic of Rumania)
146. Blinc, R. and Lahajnar, G. (1967). *J. Chem. Phys.*, **47**, 4146
147. Woessner, D. E. and Snowden, B. S. (1967). *J. Chem. Phys.*, **47**, 2361
148. Tsang, T., Farrar, T. C. and Rush, J. J. (1968). *J. Chem. Phys.*, **49**, 4403
149. Pyykko, P. (1968). *Chem. Phys. Lett.*, **2**, 559
150. Tsang, T. and Farrar, T. C. (1969). *J. Chem. Phys.*, **50**, 3498
151. Niemela, L. and Ylinen, E. (1970). *Phys. Lett.*, **31A**, 369
152. Caron, A. P., Huettner, D. J., Ragle, J. L., Sherk, L. and Stengle, T. R. (1967). *J. Chem. Phys.*, **47**, 2577
153. Huettner, D. J., Ragle, J. L., Stengle, T. R. and Yeh, H. J. C. (1968). *J. Chem. Phys.*, **48**, 1739
154. Rangarajan, G. and Ramakrishna, J. (1969). *J. Chem. Phys.*, **51**, 5290
155. Fukushima, E. and Mastin, S. H. (1969). *J. Mag. Res.*, **1**, 648
156. Fukushima, E. and Hecht, H. G. (1971). *J. Chem. Phys.*, **54**, 4341
157. Farrar, T. C., Tsang, T. and Johannesen, R. B. (1969). *J. Chem. Phys.*, **51**, 3595
158. Blinc, R. (1968). *Advances in Magnetic Resonance*, Vol. 3, 141. (J. S. Waugh, editor). (New York: Academic Press)
159. Blinc, R. Stepisnik, J., Jamsek-Vilfan, M. and Zumer, S. (1971). *J. Chem. Phys.*, **54**, 187
160. Blinc, R. and Pirs, J. (1971). *J. Chem. Phys.*, **54**, 1535
161. Genin, D. J., O'Reilly, D. E. and Tsang, T. (1968). *Phys. Rev.*, **167**, 445
162. Kydon, D. W., Petch, H. E. and Pintar, M. (1969). *J. Chem. Phys.*, **51**, 487
163. Blinc, R. Jamsek-Vilfan, M., Lahajnar, G. and Hajdukovic, G. (1970). *J. Chem. Phys.* **52**, 6407
164. O'Reilly, D. E., Peterson, E. M. and El Saffar, Z. M. (1971). *J. Chem. Phys.*, **54**, 1304
165. Halstead, T. K. (1970). *J. Chem. Phys.*, **53**, 3427

166. El Saffar, Z. M., Eades, R. G. and Llewellyn, J. P. (1969). *J. Chem. Phys.*, **50,** 3462
167. Powles, J. G., Begum, A. and Norris, N. O. (1969). *Mol. Phys.*, **17,** 489
168. Strange, J. H. and Terenzi, M. (1970). *Mol. Phys.*, **19,** 275
169. Van Putte, K. (1970). *J. Mag. Res.*, **2,** 23
170. Davidson, D. W. (1971). *Can. J. Chem.*, **49,** 1224
171. McDowell, C. A. and Ragunathan, P. (1969). *Proc. 2nd Mat. Res. Symp. 'Mol. Dynamics and Struct. of Solids'. N. B. S., 1967,* 261. (Washington, D.C.: U.S. Dept. of Commerce)
172. McDowell, C. A. and Ragunathan, P. (1968). *Mol. Phys.*, **15,** 259
173. McDowell, C. A. and Ragunathan, P. (1968). *J. Mol. Struct.*, **2,** 359
174. Majid, Y. A., Garg, S. K. and Davidson, D. W. (1968). *Can. J. Chem.*, **46,** 1683
175. Majid, Y. A., Garg, S. K. and Davidson, D. W. (1969). *Can. J. Chem.*, **47,** 4697
176. Dianoux, A. J., Sykora, S. and Gutowsky, A. S. (1971). *J. Chem. Phys.*, **55,** 4768
177. Andrew, E. R. and Lipofsky, J. To be published.
178. O'Reilly, D. E. and Eraker, J. H. (1970). *J. Chem. Phys.*, **52,** 2407
179. Brookeman, J. R., McEnnan, M. M. and Scott, T. A. (1971). *Phys. Rev.*, **B4,** 3661

3
Some Applications of Mössbauer Spectroscopy in Chemistry and Chemical Physics

J. R. SAMS
University of British Columbia, Vancouver

3.1 INTRODUCTION

During the past three years well over 1000 papers have been published in the areas covered by this article. To make a selection from this body of work is to face the problems of the anthologist in any large and rich field: the choice is difficult and a truly balanced view improbable. I have therefore limited this article to one quite specific area of Mössbauer research, and I should explain my reasons for the particular choice.

Until recently, perhaps the majority of chemists using Mössbauer spectroscopy have tended to view it as a technique-oriented, rather than a problem-oriented field. One reason for this is that while the chemical isomer shift is reasonably well understood and has been rather widely employed as a diagnostic tool, quadrupole-splitting data for some time remained largely without a firm interpretive basis. (This is not true of inorganic compounds of iron, which attracted much early attention, but it is only quite recently that tin compounds and organic derivatives of iron have received the same sort of careful analysis.) Since the quadrupole splitting contains explicit angular-dependent factors, symmetry arguments play an important role in the study of this parameter. This is not true of the isomer shift, which involves no explicitly angular-dependent terms. The strongest aspects of any bonding theory tend to be those based on symmetry, so that more fundamental information on chemical bonding can in principle be obtained from quadrupole splitting data. A very large amount of effort has been expended recently in attempting to explain the origins of electric field gradients in iron and tin compounds, how one can sort out differences in valence and lattice contributions, σ- and π-bonding effects, and so on. This has gone a long way towards converting the field to a problem-oriented one, and it is to this area that the present article is addressed.

My plan is to review the properties of the electric field gradient (e.f.g.) tensor and how these may be determined experimentally and then to consider recent data and the chemical and structural information they have provided. Only iron and tin compounds are considered for two reasons. First, even now, more than three-quarters of the Mössbauer papers of chemical interest deal with these two elements. Secondly, in both nuclides the sign of the e.f.g. (and in large measure the chemical significance of the quadrupole coupling) cannot in general be deducted from an ordinary Mössbauer spectrum. Methods of extracting this important information, together with some quite simple theoretical considerations, form the backbone upon which a large part of the experimental results discussed depend for support.

3.2 ELECTRIC FIELD GRADIENT TENSOR

A typical Mössbauer spectrum readily yields values for the two parameters of most chemical interest, the chemical isomer shift δ, which measures the effective s-electron density at the Mössbauer nucleus, and the quadrupole splitting ΔE_Q, which measures the distortion of the electron distribution from spherical symmetry. To relate ΔE_Q to details of chemical structure and bonding, it is necessary to evaluate the elements of the electric field gradient (e.f.g.) tensor. For randomly orientated powder samples (the most common absorber type) these parameters cannot be determined from ΔE_Q itself, unless one of the nuclear states involved has spin $I > \frac{3}{2}$. Nuclides with $I > \frac{3}{2}$ usually yield fairly complex spectra which give sufficient information to evaluate the e.f.g. tensor elements; these cases are not considered further here. However, for such important nuclides as ^{57}Fe, ^{119}Sn and ^{197}Au, where the excited and ground state spins are $\frac{3}{2}$ and $\frac{1}{2}$, a quadrupole-split spectrum consists of a simple doublet and other means must be employed to determine the sign of the quadrupole coupling constant and magnitude of the asymmetry parameter.

The various methods available for this purpose are discussed later. Since we are mainly interested in the chemical interpretations of such results, we first outline briefly the relevent properties of the e.f.g. tensor and a theoretical model on which structural and bonding conclusions can be based. We shall adopt here the point-charge formalism[1-3]. Because of its artificiality this model is not totally satisfying chemically, but it has the advantage of simplicity and yields the same results as a McClure-type molecular orbital treatment[4]. Clark[5,6] has presented detailed symmetry considerations of the e.f.g. and has shown that both point-charge and molecular orbital models obtain as particular manifestations of the underlying symmetry features. All three models assume the e.f.g. to be an additive quantity. The point-charge and molecular orbital models further require that the bonds between the central atom and the ligands have axial symmetry, so that the contribution of each ligand to the total e.f.g. at the central atom can be represented by a single parameter.

3.2.1 Point-charge model

The e.f.g. tensor can be written as[7]

$$\text{e.f.g.} = -\begin{bmatrix} V_{xx} & V_{xy} & V_{xz} \\ V_{yx} & V_{yy} & V_{yz} \\ V_{zx} & V_{zy} & V_{zz} \end{bmatrix} \tag{3.1}$$

where V is the electrical potential and $V_{ij} = \partial^2 V / \partial i \partial j$. The tensor is symmetric traceless, and hence can be diagonalised by a proper choice of axes. By convention the principal axis system is labelled such that $|V_{zz}| \geqslant |V_{yy}| \geqslant |V_{xx}|$, and the two independent parameters of the e.f.g. are chosen to be V_{zz} and the asymmetry parameter η,

$$\eta = |(V_{xx} - V_{yy})| / V_{zz} \tag{3.2}$$

so that $0 \leqslant \eta \leqslant 1$. The quadrupole splitting is then given by

$$\Delta E_Q = \tfrac{1}{2} e^2 qQ \left(1 + \frac{\eta^2}{3}\right)^{\frac{1}{2}} \tag{3.3}$$

where $eq = V_{zz} = -(\text{e.f.g.})$, eQ is the nuclear quadrupole moment, and e the positron charge.

The e.f.g. may contain contributions from four sources: (i) an asymmetric distribution of non-bonding electrons at the central atom; (ii) the electrons in the bonds between the central atom and its ligands; (iii) charges on the ligands; (iv) charges on surrounding ions and molecules in the lattice. In the treatment which follows we shall usually assume (i) is absent. This requires spherical symmetry for the distributions of all non-bonding electrons in the central atom. Thus, for example, the model is applicable to low-spin Fe^{II} where the t_{2g} subset of atomic d orbitals is completely filled, and Fe^{-II} in which both the t_2 and e levels are filled, but not to Fe^0 complexes where there are two holes in the 3d shell. Contribution (ii) is generally termed the valence contribution q_{val}, and (iii) and (iv) are usually lumped together[8] as a lattice contribution q_{lat}. Thus,

$$eq = e[(1-R)q_{val}+(1-\gamma_\infty)q_{lat}] \tag{3.4}$$

where R and γ_∞ are the Sternheimer factors[9] accounting for shielding and anti-shielding effects due to polarisation of the inner electrons,

$$q_{val} = -\sum_i p_i <(3\cos^2\theta_i - 1)/r_i^3 > \tag{3.5}$$

$$q_{lat} = \sum_L q_L(3\cos^2\theta_L - 1)/r_L^3 \tag{3.6}$$

Here q_L is the charge on ligand L with polar coordinates θ_L, r_L; p_i is the population of the ith valence-shell orbital, and the angular brackets denote an expectation value for the ith orbital taken over the electron coordinates θ_i, r_i. It will be noted that q_{val} and q_{lat} differ in sign. The Sternheimer factors have been estimated[10] to be $0.2 > R > -0.02$ and $-7 > \gamma_\infty > -100$. Thus $(1-R)$ and $(1-\gamma_\infty)$ are both positive.

In the point-charge model as originally employed by Fitzsimmons and co-workers[1, 11], only the q_{lat} term was considered, but q_{val} is also amenable to a point-charge treatment[4]. Parish and Platt[3] have shown how the latter term can be re-written as

$$q_{val} = \sum_L q_1(3\cos^2\theta_L - 1)/r_1^3 \tag{3.7}$$

q_1/r_1^3 representing the effective electron hole density in the appropriate hybrid orbital directed towards ligand L (note the sign change). One can then define a parameter $[L]$ by

$$(L) = e[(1-R)q_1/r_1^3 + (1-\gamma_\infty)q_L/r_L^3] \tag{3.8}$$

whence

$$eq = \sum_L [L](3\cos^2\theta_L - 1) \tag{3.9}$$

The e.f.g. is thus taken to be a sum of contributions from the various ligands, and the $[L]$ values have been termed[4] partial quadrupole splittings (p.q.s.). Note that $[L]$ is assumed to be independent of the structure of the compound and of the nature of the other ligands attached to the Mössbauer atom. In terms of p.q.s. values of the nine elements of the e.f.g. tensor are given by[7]

$$V_{xx} = \sum_L [L](3\sin^2\theta_L\cos^2\phi_L - 1) \tag{3.10}$$

$$V_{yy} = \sum_L [L](3\sin^2\theta_L\sin^2\phi_L - 1) \tag{3.11}$$

$$V_{zz} = \sum_L [L](3\cos^2\theta_L - 1) \tag{3.12}$$

$$V_{xy} = V_{yx} = \sum_L [L](3\sin^2\theta_L\sin\phi_L\cos\phi_L) \tag{3.13}$$

$$V_{xz} = V_{zx} = \sum_L [L](3\sin\theta_L\cos\theta_L\cos\phi_L) \tag{3.14}$$

$$V_{yz} = V_{zy} = \sum_L [L] (3 \sin \theta_L \cos \theta_L \sin \phi_L) \qquad (3.15)$$

To calculate relative ΔE_Q values from these expressions, one proceeds by assuming a set of bond angles for a given molecular type and finding an axis system which renders the e.f.g. tensor diagonal. It should be noted that the z-axis of the e.f.g. does not always correspond to the axis of highest symmetry in the molecule. Furthermore, for cases in which three or more different kinds of ligands are attached to the central atom it is not always possible to diagonalise the e.f.g. tensor *in general*. Thus, for example, in the octahedral $MA_2B_2C_2$ case with ligands A, B, C, if all like ligands are mutually *cis* the e.f.g. can be diagonalised only if numerical $[L]$ values are known for each ligand.

Point-charge calculations have been performed for a number of different structural types[1-4, 11-16], and the results are summarised in Table 3.1 together with several examples which have not been published previously. We have included only those cases for which the e.f.g. tensor can be diagonalised for arbitrary $[L]$ values, and in every case the highest possible symmetry has been assumed. The effects of distortions from regular bond angles will be considered below in discussion of recent experimental results. In using Table 3.1 it should be remembered that in several instances, although the e.f.g. is diagonal with z-axis as indicated, which of the V_{ii} is principal element will

Table 3.1   Diagonal elements of the e.f.g. tensor computed from the point-charge model for various tetrahedral, trigonal-bipyramidal and octahedral structures. [A] represents the contribution of ligand A to the field gradient. The e.f.g. is diagonal with z-axis as indicated, but this choice is not necessarily unique.

Structure z direction		V_{xx}	V_{yy}	V_{zz}
Tetrahedral				
		0	0	0
	↑	$[A]-[B]$	$[A]-[B]$	$2[B]-2[A]$
	↗	0	$2[A]-2[B]$	$2[B]-2[A]$
Trigonal-bypyramidal				
	↑	$-\frac{1}{2}[A]$	$-\frac{1}{2}[A]$	$[A]$

Structure z direction		V_{xx}	V_{yy}	V_{zz}
A–A–B (A above, A below, A opposite)	↑	$-\frac{5}{2}[A]+2[B]$	$\frac{1}{2}[A]-[B]$	$2[A]-[B]$
B above, A, A, A	↑	$\frac{1}{2}[A]-[B]$	$\frac{1}{2}[A]-[B]$	$2[B]-[A]$
B, A, A, B	↑	$\frac{3}{2}[A]-2[B]$	$\frac{3}{2}[A]-2[B]$	$4[B]-3[A]$
A, B, B, A	↑	$-\frac{1}{2}[B]$	$\frac{5}{2}[B]-3[A]$	$3[A]-2[B]$
B, A, A, B	×	$[B]-\frac{3}{2}[A]$	$[B]$	$\frac{3}{2}[A]-2[B]$
B, A, A, C	↑	$\frac{3}{2}[A]-[B]-[C]$	$\frac{3}{2}[A]-[B]-[C]$	$2[B]+2[C]-3[A]$
B, A, A, C	×	$-\frac{3}{2}[A]-[B]+2[C]$	$2[B]-[C]$	$\frac{3}{2}[A]-[B]-[C]$
A, B, B, C	↑	$-2[A]-\frac{1}{2}[B]+2[C]$	$-2[A]+\frac{5}{2}[B]-[C]$	$4[A]-2[B]-[C]$
A, B, B, C	×	$[A]-\frac{1}{2}[B]-[C]$	$[A]-2[B]+2[C]$	$-2[A]+\frac{5}{2}[B]-[C]$
Octahedral				
A, A, A, A, A, A		0	0	0

Structure	z direction	V_{xx}	V_{yy}	V_{zz}
B above, A A / A A, A below	↑	$[A]-[B]$	$[A]-[B]$	$2[B]-2[A]$
B above, A A / A A, B below	↑	$2[A]-2[B]$	$2[A]-2[B]$	$4[B]-4[A]$
A above, A B / A B, A below	↑	$[B]-[A]$	$[B]-[A]$	$2[A]-2[B]$
B above, A B / A B, A below		0	0	0
B above, A B / A A, B below	↑	0	$3[A]-3[B]$	$3[B]-3[A]$
A above, A B / A C, A below	↑	$-[A]-[B]+2[C]$	$-[A]+2[B]-[C]$	$2[A]-[B]-[C]$
B above, A A / A A, C below	↑	$2[A]-[B]-[C]$	$2[A]-[B]-[C]$	$2[B]+2[C]-4[A]$
C above, A B / A B, A below	↑	$[B]-[C]$	$[B]-[C]$	$2[C]-2[B]$
B above, A A / A C, B below	↑	$3[A]-2[B]-[C]$	$2[C]-2[B]$	$4[B]-3[A]-[C]$
A above, B C / C B, A below	↑	$-2[A]+[B]+[C]$	$-2[A]+[B]+[C]$	$4[A]-2[B]-2[C]$

Structure z direction	V_{xx}	V_{yy}	V_{zz}
(structure diagram) ↑	$-2[A]+[B]+[C]$	$-2[A]+[B]+[C]$	$4[A]-2[B]-2[C]$

depend upon the $[L]$ values involved and rotation of the axis system may be necessary.

The importance of the above formalism lies in the predictions it enables one to make about the sign and magnitude of the quadrupole coupling constant, and the conclusions which can be drawn concerning the relative importance of valence and lattice contributions, σ- and π-bonding, distortions from regular geometry, etc. These topics will be discussed more fully later in connection with experimental results but it is appropriate here to consider briefly the question of the sign of V_{zz}.

Since q_{lat} and q_{val} are opposite in sign, then according to equation (3.4) the sign of $eq = V_{zz}$ will be that of the numerically larger of these two terms. Because of the r^{-3} dependence in equations (3.5) and (3.6), we might expect q_{val} always to dominate the quadrupole splitting. However, the large negative values for γ_∞ may serve to offset the effect of the greater distance from the nucleus and make the lattice contribution dominant.

Consider two ligands A and B, bound to a central Mössbauer atom M, and such that B is considerably more electronegative than A. For the axially symmetric species MA_3B, MA_3B_2, trans-MA_2B_4, cis-MA_2B_4 and MAB_5, the results in Table 3.1 lead us to expect V_{zz} values of relative magnitude $+2$, $+(3-4)$, -4, $+2$, -2, respectively, in units of $[B]-[A]$. Thus, molecules in which the more electropositive (A) groups lie along the z-axis should show V_{zz} of opposite sign to those in which these groups are situated in the xy-plane. To take the trans-MA_2B_4 case as an example, if the quadrupole splitting is caused mainly by imbalances in the σ-bonding interactions, we would expect V_{zz} to be negative since q_{val} is then negative (excess electron density near M along the z-axis over that in the xy-plane). On the other hand, if the splitting is controlled largely by the non-cubic crystal field due to the electronegativity difference between A and B, so that q_{lat} dominates, then V_{zz} will be positive. Hence a knowledge of the sign of V_{zz} is of great importance, and we turn now to a discussion of ways in which this information can be obtained.

3.2.2   Sign of the electric field gradient

Before discussing ways in which the sign of the e.f.g. can be determined, it is important to state the convention regarding the physical meaning of such a sign. We will be concerned principally with ^{57}Fe and ^{119}Sn, and confusion has arisen concerning e.f.g. signs in compounds of these nuclides.

This is largely due to the fact that the nuclear quadrupole moment eQ is positive for ^{57}Fe and negative for ^{119}Sn. Moreover, since $V_{zz} = -$(e.f.g.), one must be careful to specify precisely the quantity under discussion. Following Collins[17], the conventions adopted here for cases of axial symmetry are:

^{57}Fe: For an oblate charge distribution at the nucleus, $V_{zz} > 0$, the quadrupole coupling constant $e^2qQ > 0$, and the $\pm\frac{3}{2}$ level lies above the $\pm\frac{1}{2}$ level in energy. For a prolate charge distribution, $V_{zz} < 0$, $e^2qQ < 0$ and the $\pm\frac{1}{2}$ level is higher in energy.

^{119}Sn: For an oblate charge distribution, $V_{zz} > 0$, $e^2qQ < 0$ and the $\pm\frac{1}{2}$ level is higher in energy. For a prolate field, $V_{zz} < 0$, $e^2qQ > 0$ and the $\pm\frac{3}{2}$ level is at higher energy.

Also, following usual practice, we refer to a nuclear transition of the type $(\pm\frac{1}{2}) \rightarrow (\frac{1}{2})$ as a σ transition, and of the type $(\pm\frac{3}{2}) \rightarrow (\frac{1}{2})$ as a π transition.

The sign of e^2qQ for cases involving nuclear spin states $(\frac{3}{2}, \frac{1}{2})$ can be determined in several ways. If a single-crystal absorber is available, the sign can be determined by measuring the variation in relative intensity of the two lines of the quadrupole-split spectrum with changes in the angle of incidence of the γ-ray. In the simplest case, the π and σ transitions exhibit an intensity ratio

$$\frac{I_\pi(\theta)}{I_\sigma(\theta)} = \frac{1 + \cos^2 \theta}{\frac{5}{3} - \cos^2 \theta} \tag{1.16}$$

where θ is the angle between the z-axis of the e.f.g. and the direction of propagation of the γ-ray. Thus, for $\theta = 0$, $I_\pi/I_\sigma = 3$, while for $\theta = \pi/2$, $I_\pi/I_\sigma = 0.6$. This equation is valid only in the case of an axially symmetric field gradient and in the absence of magnetic line broadening. Zory[18] has given a detailed treatment of the intensity problem, and has also shown how to determine the three Euler angles relating the crystal axes to the principal axes of the e.f.g. This method has been used recently to determine the sign of V_{zz} in FeCl$_2$·4H$_2$O[18], FeCl$_2$·2H$_2$O[19], Na$_2$[Fe(CN)$_5$NO]·2H$_2$O[20, 21], FeSO$_4$ · 7H$_2$O[22], Fe(NH$_4$SO$_4$)$_2$ · 6H$_2$O[22,23] and FeSO$_4$ · 4H$_2$O[24]. It should be noted that this procedure can also be applied to orientated polycrystalline samples, although because the orientation of the crystallites will be imperfect, an area ratio somewhat less than 3 is observed for $\theta = 0$. Nevertheless, this method has been successfully applied to such cases as Fe(CO)$_5$[25] and the mineral gillespite, BaFeSi$_4$O$_{10}$[26].

For a randomly orientated powder sample, one must average the ratio in equation (1.16) over θ to obtain

$$\frac{I_\pi}{I_\sigma} = \frac{\int_0^\pi I_\pi(\theta) \sin \theta \, d\theta}{\int_0^\pi I_\sigma(\theta) \sin \theta \, d\theta} = 1 \tag{1.17}$$

However, in the presence of anisotropic lattice vibrations, equal intensities will in general not be observed owing to the fact that the Lamb–Mössbauer ('recoil-free') fraction f' for the absorber will also depend upon θ. Thus one must include the factor $f'(\theta)$ under the integrals in equation (1.17) and

$I_\pi/I_\sigma \neq 1$. Such quadrupolar anisotropy is generally referred to as the Gol'danskii–Karyagin effect[27, 28], and can often be used to infer the sign of the e.f.g. with some knowledge of the crystal or molecular structure. Flinn et al.[29] have shown that the intensity ratio in this case can be written as

$$ R = \frac{I_\pi}{I_\sigma} = \frac{\int_0^1 (1+u^2)e^{-\varepsilon u^2}du}{\int_0^1 (\tfrac{5}{3}-u^2)e^{-\varepsilon u^2}du} \tag{1.18} $$

where $u = \cos\theta$, $\varepsilon = \kappa^2 (\langle z^2 \rangle - \langle x^2 \rangle)$, κ is the wave vector of the γ-ray, and $\langle z^2 \rangle$ and $\langle x^2 \rangle$ the mean square vibrational amplitudes parallel and perpendicular, respectively, to the z-axis of the e.f.g. If $\varepsilon > 0$ then $R < 1$, and if $\varepsilon < 0$ then $R > 1$. Thus, if the sign of ε is known or can be deduced, one can infer the sign of the e.f.g. from the observed area ratio. In some cases[26] the sign of the anisotropy parameter ε can be assumed with considerable confidence, but the method is particularly advantageous when detailed x-ray diffraction data are available (e.g. thermal vibration ellipsoids) and ε can be calculated directly. Herber and co-workers have used the Gol'danskii–Karyagin effect to deduce the sign of V_{zz} for Me_2SnF_2 [30], Me_3SnCN [31], and Me_3SnF [32].

Three factors appear to limit the general usefulness of such line asymmetries for determining the sign of V_{zz}. First, although a great many compounds apparently show Gol'danskii–Karyagin asymmetry, it is often quite small so that the apparent deviations from $R = 1$ are not significantly larger than the standard deviations in the R values themselves. Secondly, intermolecular bonding appears to be more important than intramolecular bonding in determining the anisotropy of f' [33, 34], and in monomolecular complexes any assumption concerning the sign of ε is likely to be dubious. Thirdly, other causes of line asymmetry may also be present (e.g. partial orientation of the sample[30], fluctuating magnetic hyperfine fields[35], etc.). Polymeric tin compounds probably afford the most fruitful area of application for this method, but even here difficulties can be expected if the polymeric association is not very strong. For example, the x-ray structure of $Me_3SnOCOMe$ shows it to be a linear chain-type polymer[36], but no significant G–K asymmetry has been observed over the temperature range 4–240 K[37].

It is also possible to determine the sign of e^2qQ if a sufficiently large internal magnetic field is present. In a pure magnetic hyperfine-split spectrum ($e^2qQ = 0$) six equally spaced lines are observed due to removal of the $\pm m$ degeneracy of the nuclear spin sub-states. However, the superposition of an electric quadrupole interaction alters the spacings of the excited-state levels and hence of the six lines in the Mössbauer spectrum. When the z-axis of the e.f.g. is parallel to the direction of the internal magnetic field, the $\tfrac{3}{2}$ levels of the excited state are raised in energy while the $\tfrac{1}{2}$ levels are lowered. The converse applies when the z e.f.g. axis is perpendicular to the internal field direction. In favourable cases[38] the pattern of lines observed enables one to obtain both the sign of e^2qQ and the magnitude of η. This procedure is restricted, of course, to compounds which can be studied in a magnetically ordered state.

The most useful and general technique is the magnetic perturbation method suggested by Ruby and Flinn[39]. In principle this method is the same as that above, in that it depends on the presence of combined magnetic and

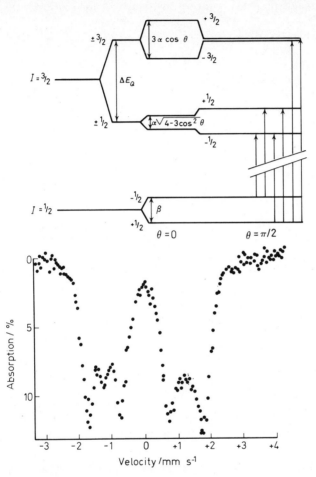

Figure 3.1 Approximate energy level diagram for ^{57}Fe in the presence of combined electric quadrupole and magnetic dipole interactions, assuming $e^2qQ > 0$ and $\eta = 0$. The spectrum illustrated is that of $[Ph_2PCH_2CH_2PPh_2]\cdot Fe_2(CO)_8$ at 4.2 K in a longitudinal magnetic field of 45 kG. $\Delta E_Q = 2.46$ mm s^{-1}, $V_{zz} > 0$, $\eta \simeq 0$ (Data from Clark, M.G. et al.[191])

electric hyperfine interactions, but in this case the magnetic field (usually of the order 15–50 kG) is applied externally. The positions of the resulting levels are dependent upon the quadrupole moment of the excited state, the magnetic moments of both ground and excited states, the magnitude and orientation of the z-component of the e.f.g. with respect to the magnetic field and the degree of mixing of the $\pm\frac{3}{2}$ and $\pm\frac{1}{2}$ levels. In the usual per-turbation treatment[17], this mixing of states of different m_I is ignored. Since the energy level schemes and resulting spectra are rather different for ^{57}Fe and ^{119}Sn, it is worth discussing both cases briefly.

The energy level diagram for an ^{57}Fe nucleus in the presence of both electric and magnetic fields is shown in Figure 3.1 for $e^2qQ>0$ and $\eta = 0$.

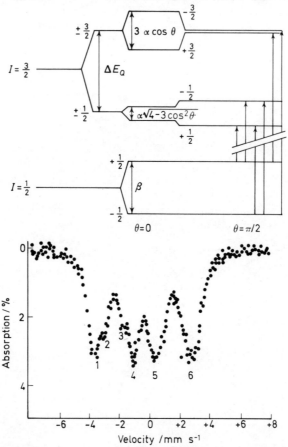

Figure 3.2 Approximate energy level diagram for ^{119}Sn in the presence of combined electric quadrupole and magnetic dipole interactions, assuming $e^2qQ>0$ and $\eta = 0$. The spectrum illustrated is that of $Me_2SnCl_2(C_5H_5NO)_2$ at 4.2 K in a transverse magnetic field of 24 kG. $\Delta E_Q = 4.10$ mm s^{-1}, $V_{zz}<0$, $\eta\simeq0$
(Data from Parish and Johnson[41], by courtesy of North-Holland Publishing Co., and from Fitzsimmons, B. W.[44], by courtesy of the Chemical Society)

Here, $\beta = g_0\beta_nH$, $\alpha = g_1\beta_nH$, where β_n is the nuclear magneton, H the magnetic field, and g_0 and g_1 the ground and excited state gyromagnetic ratios, respectively. θ is the angle between the axes of the magnetic field and the e.f.g. The ratio $g_0/g_1 = -1.715^{40}$ for ^{57}Fe. Averaging over the relative orientations of the electric and magnetic fields produces a typical spectrum of the sort shown in Figure 3.1. The π transitions appear as a broadened doublet and the σ transitions give rise to an apparent triplet. This is because $\beta \sim 2|\alpha|$, so that lines 2 and 3 in the spectrum (reading from left to right) overlap. When $e^2qQ < 0$, the $\pm\frac{1}{2}$ and $\pm\frac{3}{2}$ excited states are reversed, and the doublet appears at lower energy than the triplet.

The presence of a non-vanishing η complicates matters somewhat. As η increases from zero to unity, the spectral pattern progresses smoothly from the doublet–triplet illustrated in Figure 3.1 to a symmetrical triplet–triplet. This is because for $\eta = 1$, $V_{xx} = 0$ and $V_{yy} = -V_{zz}$, so that the sign of V_{zz} is ambiguous in this case. Small values of ΔE_Q and experimental line broadening also create problems because of overlap of the inner branches of the spectrum. Visual inspection is then often insufficient to obtain the sign of e^2qQ and the experimental results must be fitted by theoretical spectra. For ^{57}Fe, compounds having $\Delta E_Q \geq 0.6$ mm s^{-1} are most suitable to the magnetic perturbation technique, although Mazak and Collins[2] have determined the signs of V_{zz} for $Fe(CO)_2(NO)_2$ and $KFe(CO)_3NO$, both of which show splittings of only ~ 0.35 mm s^{-1}.

In the case of ^{119}Sn, the approximate energy level diagram for $e^2qQ > 0$ and $\eta = 0$ is as shown in Figure 3.2. (It should be noted that the energy level scheme given by Goodman and Greenwood[42] is incorrect[8, 41, 43], the m_I substrates being labelled wrongly). The principal differences between the ^{57}Fe and ^{119}Sn cases are the sign reversals for all the m_I substrates, and the fact that for ^{119}Sn the ratio $g_0/g_1 = -4.55$[44], so that spectral lines 2 and 3 do not overlap in this case and the σ transitions appear as a quartet rather than a triplet. For a powdered sample, although all values of θ are possible, the probability that the magnetic field is at an angle θ to the z-axis is proportional to $\sin\theta$, so that the spectrum approximates to that expected for $\theta = 90$ degrees, in which the $\pm\frac{3}{2}$ levels remain degenerate[41, 44]. For a large quadrupole splitting (~ 4 mm s^{-1}) and moderate field (~ 20–30 kG), both doublet and quartet will be reasonably well resolved, as shown in Figure 3.2, although for smaller splittings the inner branches of the σ and π transitions will overlap making resolution of the spectrum more difficult. This problem is exaggerated here because the natural Heisenberg line width for ^{119}Sn (0.63 mm s^{-1}) is more than three times that for ^{57}Fe (0.19 mm s^{-1}). Nevertheless, with machine-fitting programs and good quality spectra, it appears that one can determine the sign of e^2qQ in tin compounds with ΔE_Q as small as ~ 0.6 mm s^{-1} [45]. The above remarks about the effect of a non-zero asymmetry parameter on the magnetically perturbed spectra hold here as well[43, 46], the spectra becoming symmetrical about the centre of the two original quadrupole lines when $\eta = 1$.

3.3 QUADRUPOLE SPLITTINGS IN ^{119}Sn COMPOUNDS

Several different interpretations of the origins of ^{119}Sn quadrupole splittings have appeared and this has generated considerable discussion and confusion during the past five years. A number of different factors might reasonably be expected to influence the occurrence and magnitude of quadrupole splittings: σ-bonding and electronegativity differences; π-bonding asymmetry, charges on ligands and/or surrounding ions in the lattice, gross deviations from cubic symmetry, and spatial arrangement of the ligands (e.g. *cis–trans* isomerism). As noted by Donaldson and co-workers[47], to obtain a detailed interpretation of quadrupole splittings one would need to consider the exact nature and energy of every available orbital on tin and its ligands for each compound, which is clearly impossible. However, with the recent determination of the sign of the quadrupole coupling constant in a number of key compounds, the relative importance of valence and lattice contributions and of σ- and π-bonding interactions has been clarified, and a reasonably lucid picture is now emerging. It is convenient to divide the discussion of tin(IV) compounds according to the coordination number of tin, and to consider derivatives of tin(II) separately. Compounds with tin–metal bonds are discussed in Section 3.3.5.

3.3.1 Tetracoordinate tin(IV) derivatives

In tetracoordinate organotin compounds of the type R_nSnX_{3-n} ($n = 1, 2, 3$), the results of Table 3.1 lead us to expect quadrupole-split spectra, but in several instances (e.g. $X = H$ [48, 49], C_6H_5 [49] and $CH{=}CH_2$ [50]) splitting has not been observed. From the evidence available at the time, Gibb and Greenwood[51] postulated that quadrupole splitting would be observed in such compounds if, and only if, group X possessed a non-bonding electron pair capable of π-bonding to the tin atom, even though there were then one or two known exceptions (e.g. $Ph_3SnC_6F_5$ [52]). However, in the period 1968–1971 a large body of evidence has appeared demonstrating the importance of σ-bonding effects on the e.f.g. in such compounds.

Stöckler and Sano[53] reported ΔE_Q values ranging from 0.92–1.48 mm s^{-1} in several pentafluorophenyltin derivatives. Although these results were interpreted[53] in terms of polarisation effects via $p_\pi{-}p_\pi$ and $p_\pi{-}d_\pi$ bonding, this interpretation was questioned[54, 55] on the grounds that ^{19}F n.m.r. data indicate very little π-electron transfer from pentafluorophenyl to tin in $Me_3SnC_6F_5$ [56–58]. Parish and Platt[54, 59] studied several derivatives of the type R_3SnR' ($R' = CF_3$, C_6F_5, C_6Cl_5, $C{\equiv}CPh$), and concluded that the observed quadrupole splittings ($\Delta E_Q = 0.84{-}1.38$ mm s^{-1}) were largely controlled by imbalances in the polarity of the tin–ligand σ-bonds due to electronegativity differences between R and R', and that any π-bonding effects were of secondary importance. These conclusions were reinforced by results for a variety of compounds of the types Me_3SnR^x, $(Me_3Sn)_2R^x$ and $Me_2Sn(R^x)_2$, where R^x is a polyhalogenoaryl group[55, 60], the two triethyltin derivatives $Et_3SnC_6H_4Cl{-}m$, and $Et_3SnC_6H_4Cl{-}p$[61], some pentachlorophenyltin compounds[62], and aminostannanes[63]. Cullen et al.[64] studied a series of fluorocarbon derivatives of the type Me_3SnR^f. Interpreting the splittings as arising primarily from σ-bonding effects, they suggested that the electronegativity of the R_f group increases in the order: $CH_2F < CHF_2 < CF{=}CF_2 < CF_3 \approx CH(CF_3)_2 < CF_2CF_3 < C{\equiv}CCF_3 < CF(CF_3)_2$.

The sign of e^2qQ has recently been reported for a few tetrahedral tin species[8, 65]. When the more electronegative group lies on the z-axis, as in

$R_3SnC_6F_5$ (R = Me, Ph) and $(Bu_3^nSn)_2O$, e^2qQ is negative ($V_{zz} > 0$)[8], showing clearly that q_{val} makes the major contribution to V_{zz}. It should also be noted that if the e.f.g. in $(Bu_3^nSn)_2O$ were due mainly to π-donation from oxygen to the tin d-orbitals, e^2qQ would be expected to be positive. Thus it appears that σ-bonding effects are predominant here. A positive e^2qQ is observed for Me_2SnO[8] and Bu_2SnO[65], and a large asymmetry parameter ($\eta \sim 0.8$) is reported for the former. If the tin atom is tetracoordinate in these compounds, the positive sign requires a distortion from regular tetrahedral geometry such that the C—Sn—C bond angle is greater than 109.5 degrees. However, these results do not rule out a distorted *trans*-octahedral polymeric structure for the oxides, since again a positive e^2qQ and η close to unity would be expected. The structures of these compounds are therefore still open to question.

The compounds R_nSnX_{3-n} are frequently not simple tetracoordinate species when X is halogen or pseudohalogen (e.g. CN, NCS, NCO), or the anion of a fluoroacid (e.g. AsF_6, SbF_6) or oxyacid (e.g. ClO_4, SO_4, SO_3F, O_2CR). In many of these cases X acts as a bridging group, leading to penta- or hexa-coordination about tin. While the magnitude of the quadrupole splitting is often used to distinguish between differing coordination numbers, in many instances such a distinction is by no means straightforward. This had led to several disagreements in the literature.

Particularly contentious has been the case of Ph_3SnCl, where the quadrupole splitting (2.55 mm s^{-1}) lies between the values many workers associate with tetracoordinate tin compounds on the one hand and pentacoordinate ones on the other. Thus, Maddock and Platt[66, 67] and Fitzsimmons et al.[68] have reached opposite conclusions concerning the structure of Ph_3SnCl, inter alia, using very similar arguments. In both papers a compound of known crystal structure was chosen as a 'standard' tetrahedral molecule and related molecules were assigned tetracoordinate or pentacoordinate structures by comparing their quadrupole splittings with that of the 'standard'. Since the two 'standards' chosen had significantly different ΔE_Q values, the two groups arrived at significantly different conclusions. The wide range of ΔE_Q values reported for compounds known to contain tetracoordinate tin ($0 \leqslant \Delta E_Q \lesssim 3$ mm s^{-1}), and the fact that the ranges for tetra- and penta-coordinate tin compounds show significant overlap should caution one against such a procedure. Unfortunately, knowledge of the sign of the coupling constant is not always helpful either. Goodman and Greenwood[65] have found e^2qQ to be negative for both Ph_3SnF and Ph_3SnCl, and state that this confirms the presence of trigonal-bipyramidal structures since a tetracoordinate structure 'would require unrealistically large distortions from tetrahedral bond angles to account for the sign and magnitude of e^2qQ'[65]. However, it is clear from Table 3.1 and the results of Parish and Johnson[8] that the sign of e^2qQ would be negative in either case (even without distortions), so that Goodman and Greenwood's[65] conclusions are not justified. The x-ray structure of Ph_3SnCl was reported very recently by Russian workers[69], who found discrete tetracoordinate molecules with only small distortions ($\sim \pm 3$ degrees) from tetrahedral angles, and with Sn—Cl and Sn—C bond lengths essentially the same as those found in $SnCl_4$ and Ph_4Sn, respectively. However, this structure was solved at room temperature and there is evidence from ^{35}Cl nuclear quadrupole resonance[70, 71] suggesting the possibility of a phase transition somewhere between 77 and 300 K. Thus, the structure of the species being observed in the Mössbauer spectrum at 77 K is still open to doubt.

In many cases where bridged structures might be envisaged, it appears that the presence of bulky ligands can effectively prevent such association and lead to distorted tetrahedral configurations. Good examples are found in the triorganotin carboxylates, $R_3SnOCOR'$. In general these compounds are linear polymers in the solid state[36, 72-74], but if either R or R' is sufficiently bulky tetracoordination results. This is the case for $R' = Me$, $R = Pr^i$ [72], cyclo-C_6H_{11} [75] and $(Me)PhC(Me)CH_2$ [76], and when $R = Ph$, $R' = C(Me){=}CH_2$, $CH(Et)Bu^n$, CMe_3 and CCl_3 [77-79]. These derivatives usually show quadrupole splittings $\lesssim 2.5$ mm s^{-1}, compared to typical values of ~ 3.3–4.2 mm s^{-1} for polymeric carboxylates[76-80]. Quadrupole splitting data have also been used to infer tetracoordinate structures in a number of di- and tri-organotin halides with bulky organic groups[67], and in R_3SnNCO ($R = Ph$, $PhCH_2$)[81]. Leung and Herber[81] also favour tetracoordination for several dialkyltin cyanates, even though splittings as large as 3.5 mm s^{-1} are observed. However, the situation in these cases may not be unlike that in Me_2SnCl_2 ($\Delta E_Q = 3.60$ mm s^{-1})[82], where weak bridging has been suggested leading to a structure intermediate between tetrahedral and trans-octahedral[83]. This is not unreasonable since NCO appears to act as a bridging group in Me_3SnNCO and other trialkyltin cyanates[81].

3.3.2   Pentacoordinate tin(IV) derivatives

For pentacoordinate derivatives the absence of cubic symmetry dictates a non-vanishing e.f.g. even when all five ligands are identical (e.g. $SnCl_5^-$, $\Delta E_Q = 0.63$ mm s^{-1})[3], although Cunningham et al.[84] report narrow unsplit lines for $SnX_4 \cdot NMe_3$ ($X = Cl$, Br). This fortuitous collapse of the expected doublet structure was explained in terms of apical substitution and p.q.s. values such that $[NMe_3] \sim \frac{1}{2}[X]$ (see Table 3.1). $SnCl_4 \cdot OPPh_3$ also shows a single narrow absorption line[85], presumably for the same reasons, although splitting is observed for $SnCl_4 \cdot OPBu_3^n$ ($\Delta E_Q = 0.52$ mm s^{-1})[85]. This could be due either to different effective electronegativities of the oxygen atoms in the two phosphine oxides or to equatorial substitution in the latter compound. Another, but less likely possibility is that the Lewis base is acting as a bidentate bridging group in these few compounds.

Studies on Lewis base adducts of the type $Me_3SnCl \cdot L$ [86] and $Ph_3SnCl \cdot L$ [87] suggested that σ-bonding effects were more important than π interactions in determining the magnitude of ΔE_Q in pentacoordinate compounds, but did not firmly establish the fact. From a comparison of data for trialkyltin cyanides, thiocyanates and halides it was concluded[88] that the major contributions to the e.f.g. were from the C_{3v} symmetry and electronegativity differences between axial and equatorial ligands, and that any π-bonding effects were of minor importance. For the polymeric halogenoacetate derivatives $Me_3SnOCOCH_nX_{3-n}$ ($X = F$, Cl, Br, I), ΔE_Q shows a linear dependence upon both the pK_a of the parent acid and the Taft σ^* factor[80, 89], clearly indicating the importance of imbalances in the σ-bonding framework. A similar correlation has since been reported for triphenyltin halogenoace-tates[79], but rather surprisingly this trend is apparently not observed for the tributyltin derivatives[89].

Magnetic perturbation studies have recently established negative quad-rupole coupling constants for the ions $R_3SnCl_2^-$ (R = Me, Et, Ph)[8, 41], the polymeric compounds Me_3SnOH [65], Me_3SnNCS[65] and Et_3SnCN[8], and the pentacoordinate adducts of Ph_3SnCl with piperidine and β-picoline[90]. Negative signs for e^2qQ have also been found for Me_3SnCN [31] and Me_3SnF [32] by analysis of lattice dynamic anisotropies. These results definitely establish that the sign of the e.f.g. in pentacoordinate organotin compounds is con-trolled by σ-bonding effects via the q_{val} term, although the actual numerical value may be influenced slightly by π interactions and lattice effects.

Table 3.2 Observed quadrupole splittings and point-charge predictions for some pentacoordinate methyltin(IV) derivatives*

Compound	$\|\Delta E_Q\|$(obs)†/ mm s^{-1}	ΔE_Q(pred)‡/ mm s^{-1}	η(pred)
$Me_3SnSO_3CF_3$	4.57	−4.58	0
Me_3SnSO_3Me	4.21	−4.12	0
$Me_2ClSnSO_3F$	4.69	−4.05	0.64
$MeCl_2SnSO_3F$	3.25	−3.30	0.81
$(Me_3Sn)_2SO_4$	4.06	−4.07	0
Me_3SnF	3.90	−4.10	0
Me_2ClSnF	3.80	−3.58	0.74
$MeCl_2SnF$	2.69	−2.84	0.98

*Data from Reference 16
†For ^{119}Sn, 1 mm s^{-1} = 19.27 MHz
‡The sign given is the sign of the quadrupole coupling constant, e^2qQ.

A particularly interesting result is that e^2qQ is positive for $Me_2SnBr_3^-$ [8] where the methyl groups are expected to be in equatorial positions. This is the sign predicted by the point-charge calculations in Table 3.1 and the large η value observed[8] is also consistent with point-charge predictions. The sign change between $R_3SnX_2^-$ and $R_2SnX_3^-$ suggests a novel potential application of the magnetic perturbation technique to certain structural problems. In the recently reported[91, 92] complexes $R_2SnCl_2 \cdot L$ (R = Me, Ph; L = monodentate Lewis base) neither vibrational nor zero-field Mössbauer spectra are capable of firmly establishing whether the ligand L occupies an apical or equatorial position. However, if the contributions of Cl and L to the e.f.g. are sufficiently different, the sign of e^2qQ will differ for the two cases[37] and thus decide the geometry.

The largest quadrupole splittings yet reported for trigonal-bipyramidal tin compounds are for organotin derivatives of oxyacids and fluoroacids, where generally $\Delta E_Q \gtrsim 4$ mm s^{-1}. In almost every case such compounds are thought to be polymeric[93, 94], although considerable discussion has arisen as to the probable structure of $(Bu_3^nSn)_2SO_4$, which may be an exception[13, 95, 96]. The unusually large isomer shifts and quadrupole splittings for the tri-methyltin sulphonates Me_3SnSO_3X (X = F, CF$_3$, Me) have led Yeats et al.[16] to suggest essentially ionic structures for these derivatives, in which

Me_3Sn^+ cations interact covalently with SO_3X^- anions via polar four-electron–three-centre bonds. Using a self-consistent set of p.q.s. values, Yeats et al.[16] find generally good agreement between point-charge calculations and observed splittings for these sulphonates and for penta-coordinate methyltin fluorides, as shown in Table 3.2. The two cases where agreement is not satisfactory are Me_2ClSnX (X = F, SO_3F), where the vibrational spectra suggest the possibility of bridging Cl as well as bridging X.

3.3.3   Hexacoordinate tin(IV) derivatives

As with tetracoordinated tin compounds, quadrupole splitting in hexa-coordinated derivatives were initially attributed to π-bonding asymmetry. In 1967 the second of 'Greenwood's rules' was proposed[97]: namely, that for hexacoordinate tin compounds quadrupole splittings would be observed only if some, but not all, the ligands attached to tin were capable of d_π–d_π or d_π–p_π interactions. For compounds of the sort R_2SnX_4 (R = alkyl, aryl; X = F, Cl, Br, I, O, P, S, N, etc.), the fact that substantial ΔE_Q values were always observed could be rationalised in terms of this rule. However, on this basis one would not expect to find quadrupole-split spectra for such complexes as SnX_4L_2 (X = halogen, L = N-, O-, P- or S-donor ligand); yet in a large number of such cases small splittings have been observed[85, 98–102]. Poller et al.[103] have also reported quadrupole splittings in cis and trans isomers of SnY_4L_2, where group Y is bonded to tin via sulphur, and L is a nitrogen- or oxygen-donor.

From a systematic study of 30 complexes of $SnCl_4$ with oxygen-donor ligands, Yeats et al.[85] concluded that the primary reason for a non-vanishing e.f.g. in these cases was weak donor–acceptor interaction, leading to an imbalance of charge in the tin σ-bonding orbitals, although in a few cases steric hindrance caused by bulky ligand groups also appeared to play a significant role. There was no evidence that possible π-bonding influenced the splittings. Carty et al.[102] also found that weak interactions between $SnCl_4$ and several tertiary phosphines led to easily resolved splittings, but noted that a quantitative relation between ΔE_Q and ligand donor strength is not likely to be found.

Quadrupole splittings are not expected for simple hexahalogenostannates, SnX_6^{2-}, but have actually been observed in a few SnF_6^{2-} compounds involving complex cations. Splittings between 0.8 and 1.5 mm s^{-1} are reported[104, 105] for SnF_6^{2-} when the cations are BrF_2^+, BrF_4^+, IF_4^+ and ClF_2^+ and Carter et al.[106] found splittings of similar magnitude for $(NO_2)_2SnF_6$ and $(ClO_2)_2$ SnF_6. In all these cases strong cation–anion interaction via fluorine bridges has been suggested, to give structures such as that shown for $(BrF_4)_2SnF_6$.

$$\begin{array}{ccccccc} F & & F & F & & F & \\ | & & \backslash & / & & | & \\ F-Br-F\cdots & Sn & \cdots F-Br-F & \\ / & \backslash & / & \backslash & / & \backslash \\ F & F & F & F & F & F \end{array}$$

This is reminiscent of the situation in SnF_4, where the tin atom is octahedrally coordinated to two terminal and four bridging fluorines. Since e^2qQ is positive in SnF_4 [107], it should be negative in these hexafluorostannates, although no sign determinations have been reported for these compounds.

The SnX_6^{2-} derivatives are also interesting in that they often give fairly strong room-temperature Mössbauer absorption. Kagan[108, 109] has shown that such an effect can arise in non-polymeric tin compounds if there is a low-lying optical branch which dominates the phonon spectrum. The results of Carter et al.[106] are in qualitative agreement with Kagan's model[110].

A number of mixed hexahalogenostannates $SnX_4Y_2^{2-}$ (X = Cl, Br, I; Y = F, Cl, Br, I) have been studied[111–113], and in no case has quadrupole splitting been resolved. However, significantly broader lines were reported for these species of C_{2v} symmetry than for the octahedral SnX_6^{2-} cases[113].

The hexacoordinate tin(IV) compounds which have perhaps received the most attention from Mössbauer spectroscopists are diorganotin derivatives, where octahedral coordination can arise through bridging groups as in Me_2SnF_2 [114] and $Me_2Sn(SO_3F)_2$ [115], via Lewis acid–base interactions as in $Me_2SnCl_2 \cdot 2C_5H_5NO$ [116] and $Me_2SnCl_2 \cdot 2Me_2SO$ [117], or through complexation of diorganotin cations with neutral or anionic ligands as in $Me_2Sn(ox)_2$ [118] (ox = 8-hydroxyquinolinate). Interest has largely centred on the stereochemistries of such compounds, especially the occurrence of cis or trans organic groups. As we have seen, the point-charge model predicts a $-2:1$ trans:cis ratio for quadrupole splittings in R_2SnL_4 species, but it should be kept in mind that ΔE_Q is fairly insensitive to the nature of L since the major contribution to the e.f.g. comes from the most covalent bonds. Thus, $R_2SnX_2Y_2$ complexes show quite similar splittings to R_2SnX_4 or R_2SnY_4. Moreover, for $R_2SnX_2Y_2$ with trans-R groups, cis and trans arrangements of the X and Y ligands are degenerate as regards the e.f.g. tensor[6], and the two cases should give the same ΔE_Q.

For a wide range of cis- and trans-R_2SnL_4 complexes, numerical values of $|\Delta E_Q|$ are in good agreement with the point-charge predictions[1, 3, 47, 112, 119–126]. In the case of trans derivatives, $|\Delta E_Q|$ is typically ~ 4 mm s^{-1}, and about half this for the cis compounds. For every trans-R_2SnL_4 compound where it is known, the sign of e^2qQ is positive[8, 30, 41–44, 65, 90, 107], as predicted, but surprisingly, e^2qQ is also positive for the six cis-R_2SnL_4 cases to which the magnetic perturbation technique has been applied[8, 90]. This apparent anomaly is presumably a result of deviations from octahedral symmetry. In deriving the point-charge expression it was assumed that the C—Sn—C angle was 90 degrees, so that V_{zz} is normal to the C_2Sn plane. It can be shown, however, that V_{zz} is in this direction only when C—Sn—C is less than the tetrahedral angle but greater than its supplement[8]. Outside these limits V_{zz} lies in the C_2Sn plane and the sign of the e.f.g. is opposite to that for the 90 degrees case, although these distortions have relatively little effect on the magnitude of ΔE_Q in the range 60 degrees $<$ C—Sn—C $<$ 120 degrees.

The point-charge model has been applied to a number of hexacoordinated R_2SnX_4 derivatives of strong oxyacids[16, 127]. These compounds show exceptionally large field gradients[93, 127–129], the ΔE_Q values being almost invariably higher than those for the corresponding fluorides and reaching a value of 5.54 mm s^{-1} for $Me_2Sn(SO_3F)_2$ [128], more than 20% greater than that for Me_2SnF_2 [107]. The four series $R_2Sn(SO_3X)_2$ (X = F, CF_3), $R_2Sn(PO_2F_2)_2$ and R_2SnF_2, where R = Me, Et, Pr^n, Bu^n, Oct^n, form three parallel sets of compounds, both isomer shift and quadrupole splitting decrease from one series to the next in the order given (SO_3F and SO_3CF_3 are essentially identical), but remain nearly constant as the R group within a series is varied[127]. This is precisely the type of behaviour required by any additive model of quadrupole interactions, and it is not surprising that the point-charge formalism is very successful here.

More interesting, perhaps, is the smooth transition in both ΔE_Q and δ in the apparent isostructural series $RYSn(SO_3F)_2$ and $RYSnF_2$ as the axial ligands R and Y are varied[16, 107, 130]. In both series ΔE_Q is a linear function of the sum of the Taft inductive factors of the axial ligands (Figure 3.3), showing clearly the importance of σ-bonding asymmetry in these compounds.

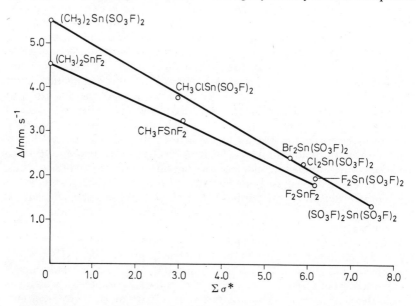

Figure 3.3 Linear relation between [119]Sn Mössbauer quadrupole splittings and the sum of the Taft inductive factors of the axial ligands X for a number of *trans*-octahedral tin(IV) fluorides and bis(fluorosulphates)
(Data from Yeats, P. A. *et al.*[16])

This linear correlation has been used to deduce that e^2qQ is positive in all these derivatives[107]. For F_2SnF_2 and $(SO_3F)_2Sn(SO_3F)_2$ the positive coupling constant indicates greater electron density near tin in the bonds to the two terminal groups than in the bonds to the bridging ligands. At first glance this might seem to require greater electron withdrawal from tin by the bridging groups, but a more likely explanation is a difference in $Sn-X_{terminal}$ and $Sn-X_{bridging}$ bond lengths. In SnF_4, the $Sn-F_t$ bond is 10% shorter than the $Sn-F_b$ bond[131], and since $V_{zz} \propto \langle qr^{-3} \rangle$ this could lead to a 33% difference in the contributions from the two types of fluorines if the q values were the same. For both the fluorides and fluorosulphates the additivity condition appears to be satisfied. Thus, ΔE_Q for $MeFSnF_2$ is very nearly the arithmetic mean of the values for Me_2SnF_2 and F_2SnF_2 [107], and the observed splittings for

$MeClSn(SO_3X)_2$ (X = F, CF_3) are in good agreement with the mean ΔE_Q values for the corresponding dimethyl- and dichloro-tin compounds[16].

3.3.4 Tin(II) compounds

Quadrupole splitting data have been of only limited usefulness in the area of tin(II) chemistry because of difficulties in the interpretation of the ΔE_Q values[132-137a]. Most discussions of ΔE_Q in tin(II) compounds have assumed a linear relation between ΔE_Q and δ[138-141]. Such a correlation implies that the major contribution to the e.f.g. is the p-character of the non-bonding orbital. On this basis it was predicted[139] that SnF_2 should have more electron density in the p_z orbital than in p_x and p_y, whereas SnS should have more $p_{x,y}$ than p_z density. Thus, e^2qQ should be positive for SnF_2 and negative for SnS.

Donaldson and Senior[142] have recently argued that there is no general relationship between the lone-pair p-electron density (estimated from δ) and ΔE_Q, and suggested that the magnitude of ΔE_Q in a number of Sn^{II} derivatives could be explained in terms of an asymmetric distribution of electrons in the tin–ligand bonds if the lone-pair p-electron density is ignored. According to their calculations, SnF_2 should have mainly $p_{x,y}$ and SnS mainly p_z character, leading to a negative e^2qQ for SnF_2 and a positive sign for SnS. Both signs are opposite to those predicted by Lees and Flinn[139], although on both models a *change* in sign between SnF_2 and SnS is predicted.

The first determination of the sign of the quadrupole interaction in Sn^{II} materials was made recently by Gibb et al.[143] for SnO, SnS, SnF_2, SnC_2O_4 and $Sn_3(PO_4)_2$. In all cases e^2qQ was positive. Subsequently, Donaldson et al.[144] reported signs of coupling constants in $SnSO_4$, $NaSnF_3$, $NaSn_2F_5$, $Sn(HCO_2)_2$, $Sn(MeCO_2)_2$ and $K_2Sn(C_2O_4)_2 \cdot H_2O$, and again all were positive. These results make it clear that the p_z-character of the lone pair dominates the e.f.g. yet the expected relationship between δ and ΔE_Q is absent. Donaldson[144] has also found that while the point-charge model yields the correct signs, it cannot predict the magnitude of ΔE_Q in Sn^{II} compounds. The present situation with respect to quadrupole splittings in these materials is still confused and a good deal more understanding is clearly needed.

3.3.5 Compounds with tin–metal bonds

A large number of Mössbauer studies have dealt with compounds (particularly organotin derivatives)[145] in which the tin atom is bonded to another metal atom. Of the investigations prior to 1968, most were concerned with situations in which the organotin moiety was bonded either to another Group IVB atom (Si, Ge, Sn, Pb) or to an alkali metal. In none of these cases was resolvable quadrupole splitting observed, although since SnO_2 was often used as a source there was the possibility that the broadness of the absorption line might be obscuring the splitting[63]. Lappert and co-workers[146] have recently studied the compounds $Me_3Sn—MMe_3$ and $Me_3Sn—MPh_3$ (M = C, Si, Ge, Sn), and $Ph_3Sn—M'Ph_3$ (M' = Si, Ge, Sn), using a narrow-line $BaSnO_3$ source. Although no splitting was observed, in each series the distannane gave the widest line. This is consistent with the fact that the electronegativity difference between tin and carbon is greater than that between silicon or germanium and carbon and suggests a small but finite e.f.g. in such compounds.

The nature of the tin–transition metal bond has received considerable attention[146–159]. In many instances the Mössbauer measurements have been allied with n.m.r. and i.r. studies and more interest has focused on isomer shifts than on quadrupole splittings. From a systematic study of materials with tin–transition metal bonds, Fenton and Zuckerman[151] concluded that Sn—M π-bonding was of little or no importance. Cullen et al.[153] investigated a number of derivatives $(\pi\text{-}C_5H_5)Fe(CO)LSnR_3$ and $(\pi\text{-}C_5H_5)FeL_2 \cdot SnR_3$ (R = Cl, Me, Ph; L = phosphine, arsine, stibine) and suggested that the Fe—Sn bond was essentially pure σ in character. Donaldson[154], Greenwood[156] and their respective co-workers have recently reported the signs of e^2qQ for both Sn and Fe in $(\pi\text{-}C_5H_5)Fe(CO)_2SnCl_3$ (results for several other compounds are also given in Reference 156). For Sn, e^2qQ is positive, indicating an excess of p_z electron density on the tin atom, from which Donaldson[154] concludes that the Sn—Fe bond must have predominately σ_s character. Sano and co-workers[155] have reached similar conclusions concerning the nature of the Sn—Mn bond in derivatives containing the —Mn(CO)$_5$ group, using a combination of ^{119}Sn Mössbauer, ^{1}H and ^{55}Mn n.m.r. data.

Some unusual trends have recently been observed for tin ΔE_Q values in compounds of the type $(CO)_5MnSnR_{3-n}X_n$ (R = Me, Ph; X = Cl, Br) and $(\pi\text{-}C_5H_5)Fe(CO)_2SnPh_{3-n}Cl_n$ [155, 160–162]. In all five series of compounds,

Table 3.3   Bond lengths and bond angles for $(\pi\text{-}C_5H_5)Fe(CO)_2SnCl_3$* and $[(\pi\text{-}C_5H_5)Fe(CO)_2]_2SnCl_2$†.

	$(\pi\text{-}C_5H_5)Fe(CO)_2SnCl_3$	$[(\pi\text{-}C_5H_5)Fe(CO)_2]_2SnCl_2$
Fe—Sn/Å	2.467	2.492
Sn—Cl/Å	2.390	2.43
Fe—Sn—Cl/degrees	119.2	107.2
Cl—Sn—Cl/degrees	98.3	94.1
Fe—Sn—Fe/degrees	—	128.6

*Data from Reference 164
†Data from Reference 163

$|\Delta E_Q|$ values are larger for $n = 1$ and 2 than for $n = 0$ or 3. Independently, Bancroft et al.[160] and Sams and co-workers[161, 162] have been able to rationalise these results on the basis of point-charge calculations. Each group derived a self-consistent set of p.q.s. values (the two sets are very similar) from which calculated values of ΔE_Q and η were obtained after diagonalising the e.f.g. tensor. For each series of compounds it is predicted that e^2qQ is negative for $n = 0$, 1 and positive for $n = 2$, 3. For $MSnCl_3$ [M = $(CO)_5Mn$, $(\pi\text{-}C_5H_5)Fe(CO)_2$] and $(CO)_5MnSnCl_2Me$, the observed[156] signs are as predicted and the negative signs suggested for the compounds in which tin is bonded to two or three organic groups are consistent with the negative e^2qQ found[156] for $(\pi\text{-}C_5H_5)Fe(CO)_2SnBu_3^n$. In a case such as $[(\pi\text{-}C_5H_5)Fe$

$(CO)_2]_2SnCl_2$, if tetrahedral bond angles are assumed $V_{zz} = -V_{yy}$ and $V_{xx} = 0$, so the sign of e^2qQ is ambiguous. However, if the bond angles about tin are in the order Fe—Sn—Fe < Fe—Sn—Cl < Cl—Sn—Cl, one predicts[161] $e^2qQ < 0$, while if this order is reversed one finds $e^2qQ > 0$. The observed angles[163] are Fe—Sn—Fe = 128.6 degrees, Fe—Sn—Cl = 107.2 degrees and Cl—Sn—Cl = 94.1 degrees, and e^2qQ has been shown to be positive[156].

Distortions from regular tetrahedral bond angles are often appreciable in compounds with tin–transition metal bonds, and it is important to take account of these distortions (where possible) in performing point-charge calculations on such systems[160]. It is instructive to examine these effects in some detail for two molecules of known structure: $MSnCl_3$ [164] and M_2SnCl_2 [163] [M = $(\pi-C_5H_5)Fe(CO)_2$]. Pertinent bond distances and angles are given in Table 3.3. For these two molecular types, the point-charge expressions for V_{zz} can be written as [156]

$$V_{zz}(MSnCl_3) = 2[M] + 3(3\cos^2\theta' - 1)[Cl]$$

$$V_{zz}(M_2SnCl_2) = 2\{3\sin^2(\theta/2) - 1\}[M] - 2[Cl]$$

where θ and θ' are the M—Sn—M and M—Sn—Cl bond angles, respectively. In the $MSnCl_3$ case, the z-axis is in the Sn—M bond direction, while in the M_2SnCl_2 case, if $\theta' < 109.47$ degrees $< \theta$, the z-axis is taken normal to the Cl—Sn—Cl plane. If we ignore the small differences in Fe—Sn and Cl—Sn bond lengths between the two compounds (i.e. assume that [M] and [Cl] are the same for both cases) and consider only the angular factors in Table 3.3, we find:

$$V_{zz}(MSnCl_3) = 2[M] - 0.87[Cl], \eta = 0$$

and

$$V_{zz}^\circ(M_2SnCl_2) = 2.87[M] - 2[Cl], \eta = \frac{1.13[M] - 0.43[Cl]}{2.87[M] - 2[Cl]}$$

These results should be compared with those for tetrahedral bond angles:

$$V_{zz}(MSnCl_3) = 2[M] - 2[Cl], \eta = 0$$

$$V_{zz}^\circ(M_2SnCl_2) = |2[M] - 2[Cl]|, \eta = 1$$

Thus, for tetrahedral angles, $\Delta E_Q^\circ(M_2SnCl_2) = 1.15\ \Delta E_Q^\circ(MSnCl_3)$, *independent* of the particular values assumed for [M] and [Cl]. This is not true, however, when distortions are present. The coefficients of [M] and [Cl] are now different for the two cases, giving two linearly independent equations and hence unique values of [M] and [Cl] which satisfy these equations. If we use the measured ΔE_Q values[156] for $MSnCl_3$ (+1.77 mm s^{-1}) and M_2SnCl_2 (+2.35 mm s^{-1}), and ignore the small asymmetry parameter for the latter, we obtain the p.q.s. values [Cl] = −0.253 and [$(\pi-C_5H_5)Fe(CO)_2$] = −0.995. This makes it abundantly clear that the overriding contribution to the e.f.g. in these compounds is from the electron density in the Fe—Sn bonds.

The possibility of a converse application of the above equations should also be noted. For example, if one knows the bond angles in a compound $SnAB_3$ and reliable values of [A] and [B], then from measured ΔE_Q values (including signs) for $SnAB_3$ and SnA_2B_2 one could estimate both θ and θ' to get at least approximate values for the A—Sn—A and A—Sn—B bond angles. No calculations of this type have yet been reported for tin compounds, but we shall discuss below an application of this idea to some iron nitrosyl derivatives.

3.4 QUADRUPOLE SPLITTINGS IN ^{57}Fe COMPOUNDS

The direct application of quadrupole splittings to problems of structure and stereochemistry requires an additive model of the e.f.g. tensor[5-7]. This in turn requires that the non-bonding electrons on the central atom do not contribute to the e.f.g. at the nucleus, i.e. have a spherically symmetric charge distribution. In the case of ^{119}Sn this is not a serious restriction, but for ^{57}Fe an additive model of the e.f.g. is appropriate only for high-spin Fe^{III} and low-spin Fe^{II} and Fe^{-II} derivatives and it is in these cases where quadrupole splitting data will yield the most significant *structural* information. On the other hand, a wealth of *bonding* information (e.g. spin states, nature of orbital ground states, magnetic dimerisation, spin–spin relaxation, etc.) can often be extracted from quadrupole splittings in other types of iron compounds. We therefore divide our discussion along these lines; we consider first those cases where an additive model of the e.f.g. is applicable and then we give some recent examples of the bonding information obtainable in a variety of other situations. For reasons which will be apparent, the high-spin Fe^{III} case is more conveniently discussed amongst the latter examples.

3.4.1 Low-spin iron(II) and iron(−II) derivatives

Most recent studies of diamagnetic Fe^{II} derivatives have dealt with octahedral complexes, although results for some square-pyramidal bis-dithiocarbamatonitrosyliron(II) compounds[165] and a few other structural types have also been reported. An octahedral field splits the iron 3d orbitals into a triply degenerate t_{2g} set (d_{xy}, d_{xz}, d_{yz}) lying below a doubly degenerate e_g set (d_{z^2}, $d_{x^2-y^2}$). Strong ligands lead to a large splitting, such that for Fe^{II} the t_{2g} levels are filled and the e_g levels empty.

The relative contributions of the various metal orbitals to V_{zz} are easily calculated from

$$V_{zz} = eq = -e<3\cos^2\theta - 1><r^{-3}>(1 - R)$$

For hydrogenic orbitals, evaluation of the angular part (which controls the sign of the field gradient) leads to the following contributions to V_{zz}: s = 0, $p_x = p_y = \frac{2}{5}\langle r^{-3}\rangle$, $p_z = -\frac{4}{5}\langle r^{-3}\rangle$, $d_{xz} = d_{yz} = -\frac{2}{7}\langle r^{-3}\rangle$, $d_{xy} = d_{x^2-y^2} = \frac{4}{7}\langle r^{-3}\rangle$, $d_{z^2} = -\frac{4}{7}\langle r^{-3}\rangle$, where the Sternheimer factor has been omitted for brevity but is implicit in these results. Clearly, any filled or spin-unpaired half-filled shell leads to zero ΔE_Q. Likewise, the t_{2g} and e_g subsets contribute nothing to the e.f.g. if they are filled or half-filled.

In the strong ligand field octahedral case then, assuming q_{lat} to be small, the splitting should arise principally from differences in σ-donor and π-

acceptor properties of the axial and equatorial ligands via the metal 3d orbitals. (The greater radial extent of the 4p orbitals diminishes the contributions of these terms, although it is not always possible to ignore them.) Note that the presence of a strong σ-donor and/or weak π-acceptor along the z-axis of the e.f.g. will give a negative contribution to V_{zz}, while if such a ligand lies in the xy-plane the contribution will be positive. Conversely, a weak σ-donor and/or strong π-acceptor ligand yields an overall contribution to V_{zz} which is positive if the ligand is on the z-axis and negative if it is in the xy-plane. For example, Dale, et al.[166] have shown that V_{zz} is positive for several iron(II) compounds in which the strong σ-donor ligands phthalocyanine[167–169] or 1,2-cyclohexanedione dioxime[167] provide the equatorial substituents, with various weaker ligands in axial positions. Imbalances in the σ-bonding lead to more 'metal-orbital character' in $d_{x^2-y^2}$ than in d_{z^2}, and hence to a positive V_{zz}. In the phthalocyanine complexes, replacement of weak axial ligands (e.g. pyridine) by the stronger ligand CN^- lowers ΔE_Q from 1.97 to 0.56 mm s^{-1} [167].

It is interesting that in some cases, changing only two of the six ligands causes a sufficiently large difference in the d-orbital splitting to change the spin state of iron. Thus, $Fe(phen)_2(CN)_2$ (phen = 1,10-phenathroline) is diamagnetic with $\Delta E_Q = 0.60$ mm s^{-1}, while in $Fe(phen)_2Cl_2$ the $t_{2g}-e_g$ splitting is small, orbital filling follows Hund's rule $[(t_{2g})^4(e_g)^2]$, and a large ΔE_Q is observed (3.15 mm s^{-1})[170].

Most of the work on the applicability of an additive model of the e.f.g. to low-spin iron(II) derivatives has been due to Bancroft and co-workers[4,171,172], following the initial use of the point-charge model in such systems by Berrett and Fitzsimmons[11]. The latter authors found that for several $Fe(CN)_2L_4$ complexes, where L is an isocyanide group, $|\Delta E_Q(trans)/\Delta E_Q(cis)|$ was c. 2:1. A discussion of these results was given in terms of a point-charge approach which ignored all valence contributions and attributed the total e.f.g. to lattice effects. Bancroft et al.[171] showed that the $|2:1|$ ratio also held for ligands other than CN^- and that $[FeCl(ArNC)_5]^+$ gave essentially the same splitting as cis-$FeCl_2(ArNC)_4$ (ArNC = p-methoxyphenylisocyanide), as predicted by the model (see Table 3.1). Moreover, ΔE_Q for cis-$FeCl(SnCl_3)$ $(ArNC)_4$ was very close to the arithmetic mean of the values for the corresponding $(Cl)_2$ and $(SnCl_3)_2$ derivatives. It was also noted in this work[171] that ΔE_Q was relatively insensitive to the nature of the neutral ligand, which seemed to imply that the predominant contribution to the field gradient came from the anionic groups via a q_{lat} term.

Subsequently, it was shown[4] that the point-charge predictions also followed from a molecular orbital approach, and p.q.s. values were derived for some 16 ligands. To do this, a 'reference value' was obtained for Cl^- by calculating a hypothetical ΔE_Q (c. -1.2 mm s^{-1}) for the trans-$FeCl_2$ group assuming ionic Fe—Cl bonds 2.3 Å in length. To the extent that the e.f.g. is additive, the relative p.q.s. values and calculated splittings are independent of the particular value taken for the chloride lattice contribution and the p.q.s. values can be obtained from the appropriate expressions in Table 3.1 after getting the valence contribution of the neutral ligands from

$$\Delta E_Q(\text{total}) = -1.2 \, (\text{mm s}^{-1}) + \Delta E_Q(\text{valence})$$

Bancroft *et al.*[4] assumed a negative quadrupole coupling constant for the *trans*-FeX_2B_4 derivatives (X = halide, pseudohalide). This was rationalised by the argument that ΔE_Q values were not strongly dependent upon the nature of B (although B was limited to two isocyanides and three di(tertiary phosphines)). This choice of sign implies that the negative q_{lat} term from the chlorides dominates the splitting, and is opposed by a smaller and positive q_{val} due primarily to σ-donation into $d_{x^2-y^2}$ by the neutral ligands. For the *cis*-complexes, of course, a positive e^2qQ was assumed.

In general, the magnitudes of splittings predicted from this treatment were in adequate agreement with observed values, but it remained to check the as-

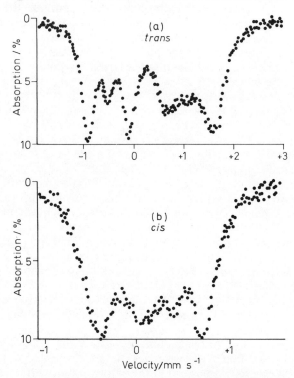

Figure 3.4 Mössbauer spectra of: (a) *trans*-$FeCl_2(p$-MeO·C_6H_4·NC)$_4$ at 4 K in a longitudinal magnetic field of 36 kG ($V_{zz}>0$); (b) *cis*-$FeCl_2(p$-MeO·C_6H_4·NC)$_4$ at 4 K in a longitudinal magnetic field of 28 kG ($V_{zz}<0$).
(Data from Bancroft, G. M. *et al.*[173], by courtesy of the Chemical Society)

sumed signs of V_{zz}. Magnetic perturbation studies on *cis*- and *trans*-$FeCl_2(p$-MeO·C_6H_4·NC)$_4$, showed that the observed[173] signs were opposite in both cases to those previously assumed (see Figure 3.4), the *trans*-compound having $V_{zz}>0$ and the *cis*-compound $V_{zz}<0$. This is a clear demonstration that $|q_{val}| > |q_{lat}|$ in these compounds, and hence controls the sign of the quadrupole coupling, as it does in compounds of ^{119}Sn (*vide supra*) and in ^{129}I derivatives[174]. Presumably, this will be true in general for low-spin iron(II) cases, although with the exception of the above two compounds and sodium nitroprusside[20, 21] (in which the positive V_{zz} is due to the very strong π-acceptor ability of the nitrosyl group which expands the metal d_{xz}, d_{yz} orbitals), signs of the quadrupole interaction do not appear to have been published for such compounds. The small splittings generally observed make the necessary experiments more difficult and may complicate interpretation of the magnetically perturbed spectra. However, I understand[175] that the signs of V_{zz} for several other diamagnetic iron(II) materials are to be reported shortly[176]. It is particularly interesting that replacement of the NO^+ group in the nitroprusside ion by the much weaker field ligand NH_3 to give $[Fe(CN)_5NH_3]^{3-}$ lowers the quadrupole splitting from 1.71 to 0.67 mm s^{-1}, but does not change the sign of V_{zz} [175]. On the other hand, in going from trans-$FeCl_2(p$-MeO $\cdot$ C_6H_4 $\cdot$ NC$)_4$ to trans-$Fe(CN)_2(EtNC)_4$ a sign change is observed; V_{zz} is negative for the latter[175], showing that CN^- is a better σ-donor and/or poorer π-acceptor than the EtNC group.

A very novel application of p.q.s. values has recently been made by Bancroft[177]. Using values derived from Mössbauer spectra of low-spin Fe^{II} compounds, signs of quadrupole coupling constants are predicted for isoelectronic low-spin Co^{III} compounds. Then by comparing the e^2qQ values for the Co^{III} complexes (obtained from ^{59}Co nuclear quadrupole resonance) with those for the corresponding hypothetical Fe^{II} derivatives (calculated from p.q.s. values), Bancroft deduces a value of $+0.16 \pm 0.03$b for the nuclear quadrupole moment of ^{57}Fe. This value compares favourably with a number of recent estimates[178-183] which range from $+0.17$ to $+0.216$.

Comparatively few Mössbauer measurements are available for Fe^{-II} compounds. These studies[2, 184, 185] have been confined to tetrahedral derivatives of the general types $L_2Fe(NO)_2$ and $[Fe(NO)(CO)_3]^-$. The tetrahedral crystal field splits the iron 3d shell into a doubly degenerate e level lying below a triply degenerate t_2 level. Both sets of orbitals are filled to give a d^{10} configuration and the observed diamagnetism.

For the $[Fe(CO)_3(NO)]^-$ ion, the point-charge expression in Table 3.1 predicts $V_{zz} = 2[NO] - 2[CO]$, and the sign should be governed by whichever ligand has the largest $|[L]|$ value. Mazak and Collins[2] found $V_{zz} > 0$ (and $\eta \approx 0$) for the potassium salt, which implies $|[CO]| > |[NO]|$, although the small splitting (0.36 mm s^{-1}) shows that the difference in $[L]$ values is not large. Mazak and Collins commented[2] that the observed sign was somewhat surprising since NO is generally considered a stronger ligand than CO, and hence should have a larger (negative) p.q.s. value. We have argued[184] that since the nitrosyl group bonds in this case as NO^+, one can assume that σ-bonding occurs only through sp^3 hybrids on the metal atom. If the contribution of the 4p orbitals to the e.f.g. is ignored, the effective populations of the iron 3d orbitals (hence V_{zz}) will depend almost entirely upon the extent of metal-to-ligand π back-bonding, and there will be little or no effect on V_{zz} from imbalances in the σ-bonding framework. Since a strong π-acceptor will both reduce the effective iron d-orbital population and tend to delocalise the d-orbital charge density, the best π-acceptor should show the smallest $|[L]|$ value. On this basis, the sign of V_{zz} in $KFe(CO)_3(NO)$ is expected to be positive, as observed.

For $Fe(CO)_2(NO)_2$ and $Fe(Ph_3P)_2(NO)_2$, $V_{zz} < 0$ and large asymmetry parameters ($\eta \approx 0.85$ and 0.76, respectively) are found in both cases[2]. For tetrahedral bond angles, one expects $\eta = 1$, but with distortion of the bond angles $\eta < 1$ and the sign of V_{zz} is determinate. It is found[2] that V_{zz} is negative if the ligands with the larger $|[L]|$ values move apart, and positive if this bond angle decreases. Thus, the negative V_{zz} for $Fe(CO)_2(NO)_2$ implies a

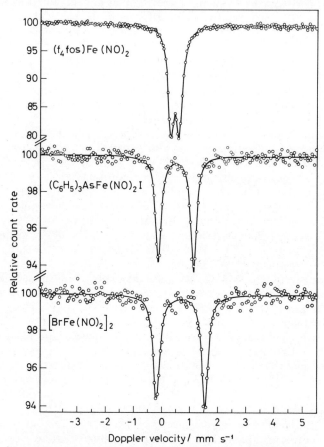

Figure 3.5 Mössbauer spectra at 80 K showing the effect of holes in the iron 3d shell on ΔE_Q: (a) $(f_4fos)Fe(NO)_2$ [(f_4fos) = 1,2-bis(diphenyl-phosphino)tetrafluorocyclobut-1-ene], a d^{10} configuration; (b) PhAsFe $(NO)_2I$, a d^9 configuration; (c) $[BrFe(NO)_2]_2$, a d^8 configuration (Data from Crow, J. P. *et al.*[184] (1971). *Inorg. Chem.*, **10**, 1616. Copyright 1971 by the American Chemical Society. Reprinted by permission of the copyright owner)

C—Fe—C angle slightly greater than 109.5 degrees. From the magnitude of η and the assumption that only the distortion of the ligand with larger $|[L]|$ is important, Mazak and Collins[2] infer that the P—Fe—P angle in $Fe(Ph_3P)_2(NO)_2$ is ~ 116 degrees.

Crow *et al.*[184] have examined a number of phosphine and arsine derivatives of $Fe(CO)_2(NO)_2$, and have used the observed $|\Delta E_Q|$ values to rank the

π-acceptor strengths of the ligands. In the chelate complexes $|\Delta E_Q|$ is only about half the value observed for derivatives containing two mono(tertiary phosphines).

$$
\begin{array}{c}
(Ph)_2 \\
\text{(CF}_2)_n \underset{\big|}{\overset{\big|}{\begin{array}{c} -C \\ -C \end{array}}} \overset{\displaystyle P}{\underset{\displaystyle P}{\Big\langle}} \,Fe \overset{\displaystyle NO}{\underset{\displaystyle NO}{\Big\langle}} \\
(Ph)_2
\end{array}
\qquad
\begin{array}{l}
n = 2,\ f_4fos \\
n = 3,\ f_6fos \\
n = 4,\ f_8fos
\end{array}
$$

From the fact that in $(f_6fos)Fe(NO)_2$ the P—Fe—P angle is only 87 degrees while the N—Fe—N angle is opened out to 125 degrees[186], it was predicted that V_{zz} should be positive in these three chelate complexes[184]. This prediction may be difficult to verify because of the small splittings (~ 0.30 mm s^{-1}).

Finally, it is interesting to note the effect on ΔE_Q caused by the presence of one or two holes in the iron 3d shell (Figure 3.5)[184]. Thus, $(f_4fos)Fe(NO)_2$ with d^{10} configuration shows $\Delta E_Q = 0.29$ mm s^{-1}. In $Ph_3AsFe(NO)_2I$, paramagnetism arises through a single hole in the t_2 level (d^9), and the splitting increases to 1.28 mm s^{-1}. The bromide dimer is again diamagnetic, presumably with d^8 configuration at iron[187], and the double hole in t_2 further increases ΔE_Q to 1.76 mm s^{-1}. An additive field gradient model is not appropriate for the two latter compounds.

3.4.2 Iron(0) carbonyl derivatives

Zero-field Mössbauer spectra have been reported for a very large number of Fe^0 (d^8) carbonyl complexes, and no attempt is made here to review this area. Rather, we consider only those few cases in which signs of the e.f.g. have been determined. In compounds of this type the nature of the non-bonding orbitals is extremely important and theoretical analysis difficult. For example, butadiene tricarbonyl iron and cyclobutadiene tricarbonyl iron might be expected to show very similar field gradients, and indeed their $|\Delta E_Q|$ values are nearly the same. However, application of a magnetic field of 26 kG reveals that V_{zz} is negative in the butadiene compound, but positive in the cyclobutadiene derivative[7]. A satisfactory explanation of this sign reversal does not appear to have been given.

Magnetic perturbation studies on several derivatives in which the (π-C_5H_5)Fe(CO)$_2$ moiety is bonded to a tin atom have shown positive signs for e^2qQ at Fe in every case[154, 156]. However, no attempt has been made to interpret these results, since the nature of the bonding and non-bonding orbitals is not well understood.

For Fe(CO)$_5$, orientation studies on polycrystalline samples have shown that V_{zz} is positive[25]. Under D_{3h} symmetry the metal orbitals transform as shown:

$3d_{z^2}, 4s$	$4_{p_x}, 4_{p_y}$	4_{p_z}	$3d_{x^2-y^2}, 3d_{xy}$	$3d_{xy}3d_{yz}$
a_1'	e'	a_2''	e'	e''

$$\sigma \qquad\qquad\qquad\qquad \pi$$

If one ignores q_{lat} and makes the simplifying assumption that both σ- and π-bonding contributions arise only from the metal d orbitals, one obtains[188]

$$V_{zz} = \tfrac{4}{7}\langle r^{-3}\rangle_d(\pi_c - \bar{\sigma}), \ \eta = 0$$

where $\bar{\sigma}$ and π_c are essentially empirical parameters corresponding, respectively, to the 'effective electron population' and 'effective hole population' in the $3d_{z^2}$ orbital (both parameters being between 0 and 2). A case for which $\pi_c < \bar{\sigma}$ would correspond to a situation in which it would be difficult to justify the usual d^8 description of the iron electronic configuration. (Note that $\bar{\sigma}$ increases with increasing σ-donor character of the ligands, whereas π_c decreases with increasing π-acceptor character.) The fact that $V_{zz} > 0$ for $Fe(CO)_5$ shows that $\pi_c > \bar{\sigma}$.

For both mono- and bis-phosphine and -arsine derivatives of iron pentacarbonyl, e.g. $LFe(CO)_4$ and $L_2Fe(CO)_3$, in which the substituents occupy apical positions, quadrupole splittings are quite insensitive to the number and nature of the substituents and are the same to within $\sim \pm 10\%$ as that for $Fe(CO)_5$ [189]. These C_{3v} and D_{3h} symmetry cases can be treated[188] in a manner analogous to that above. Since $V_{zz} > 0$ for $Ph_3PFe(CO)_4$ [190], $(Ph_3P)_2$ $Fe(CO)_3$ [190, 191], and $(diphos)Fe_2(CO)_8$ [191] (in which diphos($= 1,2$-bis-(diphenylphosphino)ethane) bridges two $Fe(CO)_4$ groups), it is clear that $\pi_c - \bar{\sigma}$ is but little changed by apical substitution of $Fe(CO)_5$.

In cases involving (chelate)$Fe(CO)_3$ complexes[189], where the formal point group symmetry is either C_{2v} (1) or C_s (2), theoretical analysis becomes considerably more difficult. Under C_{2v} both d_{z^2} and $d_{x^2-y^2}$ span a_1, while for symmetry group C_s, d_{yz}, $d_{x^2-y^2}$ and d_{xy} each spans a'.

(1)   (2)

Thus, to calculate the π-bonding contribution in these cases one must consider positive holes in orbitals of a more general nature than a pure d_{z^2} orbital. An approximate analysis of the problem[188] indicates that both η and the sign of V_{zz} are much more sensitive than $|\Delta E_Q|$ to perturbation of the electronic structure of iron for either structure (1) or (2) and that only quite small changes are necessary to change the sign of V_{zz}. Thus, the C_s and C_{2v} cases should show $|\Delta E_Q|$ values similar to those for the C_{3v} and D_{3h} cases, but V_{zz} might have either sign and η any value between 0 and 1, so that the quadrupole interaction is not capable of distinguishing between structures (1) and (2). For $(diphos)Fe(CO)_3$ in an applied field of 45 kG at 4 K, $V_{zz} < 0$, and $\eta \simeq 0.8$[191]. In view of the above discussion, however, interpretation of this result must await an x-ray determination of the molecular structure.

3.4.3 Iron(III) derivatives

Much recent interest has been shown in Mössbauer studies of a number of iron(III)–Schiff-base complexes, particularly the binuclear complexes. There are several reasons for this interest. Firstly, magnetic susceptibility measurements[192, 193] show that the effective moment is strongly temperature dependent and decreases with decreasing temperature. At high temperatures the ferric spin state is clearly $S = \frac{5}{2}$ and the two ferric ions are antiferromagnetically coupled so that the ground state is diamagnetic. The strength of this coupling can show considerable variation. Secondly, as the temperature is raised from 4.2 K the two lines of the quadrupole doublet become very asymmetric, one peak being much broader than the other, an effect which has been attributed to fluctuating hyperfine magnetic fields with a correlation time comparable with the nuclear precession time[194]. Ways of distinguishing this case from other causes of temperature-dependent line asymmetry are obviously important[195]. Thirdly, the dimeric Schiff-base complexes are useful model compounds for certain biologically important systems such as the non-haeme iron protein hemerythrin which is responsible for oxygen transport in certain invertebrates[196]. Here the oxygen is believed to be bound in a bridging manner between two iron atoms in a similar way to the binding in the Schiff-base derivatives.

Reiff *et al.*[197] isolated and characterised three iron(III) complexes of 2,2′,2″-terpyridyl (terpy). Of these, $Fe(terpy)Cl_3$ was found to be high-spin, $Fe(terpy)_2(ClO_4)_3$ was low-spin, and $[Fe_2(terpy)_2O](NO_3)_4 \cdot H_2O$ a binuclear complex. From Mössbauer, magnetic, electronic and i.r. data it was suggested that the latter might be an example of a quartet spin state ($S = \frac{3}{2}$) for iron, as might the binuclear iron(III) complexes of 2,2′-bipyridyl (bipy) and 1,10-phenanthroline (phen). However, additional studies[193] showed that the data for these complexes were most consistent with the presence of two interacting $S = \frac{5}{2}$ iron ions. The compounds all give ΔE_Q values which are exceptionally large ($\gtrsim 1.5$ mm s^{-1}) for high-spin iron(III), although the isomer shifts are normal.

Several similar complexes of phen, bipy and salen (1,2-di(salicylideneamino)ethane), including three binuclear compounds, were examined by Berrett *et al.*[198] and the Mössbauer spectra were interpreted as showing the binuclear materials to contain $S = \frac{5}{2}$ iron.

In the octahedrally coordinated dimeric compound $[Fe(salen)Cl]_2$, susceptibility measurements[192, 193] show that both iron ions are in ^6S states and are antiferromagnetically coupled by an exchange interaction of the form $\mathscr{H} = -JS_1 \cdot S_2$ with an exchange integral $J = -7.5$ cm^{-1} ($|J|/k = 21.5$ K). The system has states with total quantum numbers $S' = 0, 1, ..., 5$ and energies $E(S') = \frac{1}{2}J[2S(S+1) - S'(S'+1)]$. This leads to a series of multiplets with energies spread over $30 J$. However, the splitting of these multiplets by the crystal field is much smaller than $|J|$, and at 4.2 K only the non-degenerate $S' = 0$ ground state is populated. Under these conditions, no fluctuations of the electron spins are possible and the Mössbauer spectrum[194] is a symmetric quadrupole doublet. As the temperature is raised, higher S' states are populated and fluctuations become possible leading to asymmetric line broadening. For $[Fe(salen)Cl]_2$ it is the higher velocity peak which broadens first[194]. (It should be noted that the nature of the asymmetry here is fundamentally different from that associated with the Gol'danskii–Karyagin effect. In the latter case the widths of the two lines are not affected but the area ratio changes with temperature, while the converse is true for the magnetic dimer situation. Hence the two effects can be readily distinguished.) It has been shown[196, 199] by magnetic perturbation studies that $V_{zz} < 0$ in $[Fe(salen)Cl]_2$ so the σ-transitions lie to higher energy. According to Buckley et al.[199] this implies that the fluctuating hyperfine fields are perpendicular to the principal e.f.g. axis.

$[Fe(salen)]_2O$ shows quite different behaviour from that of the chloride. In this case the Fe^{III} ions are pentacoordinated and the susceptibility data suggest a large exchange integral $J \sim -95$ cm^{-1}, so that the coupling is much stronger. However, an $S = \frac{3}{2}$ assignment could not be excluded from susceptibility measurements. Several groups have reported Mössbauer data for this compound[193, 196–202]. Magnetic perturbation of the Mössbauer spectrum[196, 199, 202] is particularly useful here since ΔE_Q is not inconsistent with either $S = \frac{5}{2}$ or $S = \frac{3}{2}$ configuration and spin state assignments based solely on isomer shifts can also be ambiguous[202]. The spectra observed in applied magnetic fields of up to 90 kG show the absence of any substantial hyperfine field at the ferric nuclei, demonstrating conclusively a zero-spin ground state with strong antiferromagnetic coupling[199].

Several Schiff-base–iron(III) adducts have also been investigated[200, 203], primarily with a view to determining if the resultant complex is monomeric or dimeric. Bancroft et al.[200] studied a series of adducts of the type Fe(salen) X·nL (X = Cl, Br; L = MeNO$_2$, MeCN, MeOH, C$_5$H$_5$N) and assigned mononuclear or binuclear structures on the basis of differences in Mössbauer parameters and i.r. spectra. They questioned the ability of magnetic susceptibility measurements to decide the molecularity of such compounds. Buckley et al.[203] have made a detailed Mössbauer and magnetic susceptibility study of Fe(salen)Cl·$\frac{1}{2}$MeNO$_2$, which they conclude is dimeric, i.e. [Fe(salen) Cl]$_2$ · MeNO$_2$. As for the parent dimer, V_{zz} is negative, η is small and the hyperfine fields are perpendicular to the principal axis of the e.f.g.

Many high-spin ferric compounds having lower than cubic symmetry exhibit quadrupole doublets which are symmetrical at 4.2 K but unsymmetrically broadened at higher temperatures. However, this behaviour can be caused by an effect somewhat different from the antiferromagnetic coupling between pairs of iron ions of the type we have been discussing. This, for example, is the situation in haemin, where the magnetic hyperfine broadening arises from a temperature-dependent electron spin–spin relaxation rate[204]. In this case a strong ligand field of the form $\mathcal{H} = DS_z^2$ (with $D/k > 4$ K) splits the ^{6}S state into three Kramers doublets with $S_z = \pm\frac{1}{2}$, $\pm\frac{3}{2}$ and $\pm\frac{5}{2}$. Only the $\pm\frac{1}{2}$ state is occupied at low temperatures and relaxation is fast. At higher temperatures where all the states are populated the relaxation slows down because of the effects of the selection rules on the possible transitions, and line broadening is observed. Magnetic measurements cannot distinguish between these two situations unless they are made at temperatures well below D/k or $|J|/k$, nor can an ordinary Mössbauer spectrum. However, Fitzsimmons and Johnson[195] have pointed out that the effect of an applied magnetic field on the Mössbauer spectra is quite different in the two cases, enabling one to distinguish between them. For the $S_z = \pm\frac{1}{2}$ ground state the applied field H produces a large magnetic hyperfine interaction given by

$$H_{eff} = H_n B_S(\mu H/kT) - H$$

where $B_S(\mu H/kT)$ is the Brillouin function for spin S, and H_n the saturation hyperfine field. For the magnetic dimer $(S' = 0)$ case there is no hyperfine field and only a small splitting due to the direct effect of the applied field.

The compound $Fe(acac)_2Cl$ [205] (acac = acetylacetonato) is particularly interesting in that both types of magnetic broadening appear to operate. The magnetic susceptibility is temperature independent in the range 80–300 K and the magnetic moment has the spin-only value of 5.9 μ_B. Mössbauer spectra show that the application of an external magnetic field produces a hyperfine field which is linear in the applied field but independent of temperature in the range 1.7–4.2 K [195]. This behaviour seems to require both a strong ligand field acting on the ferric ions and an antiferromagnetic exchange interaction between pairs of ions. V_{zz} is positive, consistent with the square-pyramidal coordination about iron, and the arrangement of the molecules in pairs in the unit cell provides a possible path for the magnetic coupling.

Whereas $Fe(acac)_3$ is a high-spin complex, tris(monothio-β-diketonato)-iron(III) chelates typically exhibit a thermal equilibrium between high-spin and low-spin states. Mössbauer measurements appear very well suited to monitor such changes in spin state, since ΔE_Q for the $^6A_1$ state is typically in the range 0.3–0.6 mm s^{-1} and for the $^2T_2$ state typically 1.5–2.0 mm s^{-1}. Cox et al.[206] have examined the four derivatives of the type

$$Fe\left[\begin{array}{c} S\!-\!\!R \\ \langle\quad\rangle \\ O\!-\!\!R' \end{array}\right]_3 \qquad R, R' = Me, Ph$$

and observed that details of the $^6A_1$–$^2T_2$ equilibrium are quite sensitive to the R and R' substituents. Not surprisingly, $Fe(sacsac)_3$ (sacsac = dithio-acetylacetonato) is definitely a low-spin complex at room temperature[207].

Fe^{III} complexes with sulphur-containing ligands have attracted much recent interest[208–216], stimulated by the realisation that in the non-haem iron proteins the iron atoms have sulphur ligands. Special emphasis has focused on the N,N-dialkyldithiocarbamates[208–210, 213, 214], where anomalous magnetic behaviour arising from a high-spin–low-spin equilibrium has provided additional impetus. In the monothio-β-diketone complexes above the spin-state lifetimes are quite long and in some cases four-line Mössbauer spectra due to the presence of a mixture of spin states are observed[206]. The dithiocarbamates, however, have transition times short compared to the ^{57}Fe excited state lifetime (145 ns), so that one sees only an average of the e.f.g.s of the two states and a simple doublet spectrum[210]. ΔE_Q is, however, strongly dependent upon temperature, increasing rapidly with decreasing T as the equilibrium shifts towards the $^2T_2$ state. For the dithiocarbamates studied thus far, it appears that the separation of the two spin states is not more than a few hundred wave numbers. In tris(dithiocarbamate) derivatives of the type $Fe(S_2CNRR')_3$, where the iron is octahedrally surrounded by sulphur, ΔE_Q for the high-spin configuration arises primarily from distortions in the S—Fe—S angles and is typically $\lesssim 0.25$ mm s^{-1}. At low temperatures the dominant contribution to ΔE_Q is from the q_{val} term and the splittings are much larger at 80 K (~ 0.6–1.0 mm s^{-1}) and are quite sensitive to the nature of the nitrogen substituents[210]. Pentacoordinate complexes such as $(R_2NCS_2)_2FeX$ where $X = I^-$, Br^-, Cl^-, SCN^-, $C_6F_5CO_2^-$, show very large ΔE_Q values (~ 2.4–3.0 mm s^{-1})[209], which decrease with change in X in the order given. However, in frozen solutions of such species in dimethyl formamide, ΔE_Q is reduced to ~ 0.7 mm s^{-1}. This has been attributed to the binding of a solvent molecule to yield a hexacoordinated adduct in solution[214]. Rickards *et al.*[208] have studied $(Et_2NCS_2)_2FeCl$ at low temperatures and in external magnetic fields, and have confirmed that the compound is ferromagnetic at 1.6 K.

Johnson and co-workers have recently reported some elegant work on the determination of the orbital ground states in $(Me_2NCS_2)_3Fe$[213], $(Ph_4P)_3$ $[Fe\{S_2C_2(CN)_2\}_3]$[215], $Fe(ttd)_2(dtt)$, $Fe(ttd)(dtt)_2$ and $Fe(dtt)_3 \cdot CHCl_3$ (ttd = trithioperoxy-*p*-toluate, dtt = dithio-*p*-toluate)[216], using a combination of magnetically perturbed Mössbauer spectra and e.s.r. measurements. For the dithiolene complex[215] $\Delta E_Q = -1.85$ mm s^{-1} at 77 K showing a strong distortion from cubic symmetry. The two lines were asymmetrically broadened and application of a small (5 kG) magnetic field was found to decrease the spin–lattice relaxation time. The use of a 30 kG field at 4.2 and 1.6 K showed that V_{zz} was negative and aligned parallel to the major axis of the magnetisation tensor. From the values of the effective hyperfine field produced at the iron nucleus by the applied field it was deduced that the orbital ground state must be a d_{xy} hole well separated from higher states.

The three complexes of dtt[216] all showed similar Mössbauer spectra with shifts and splittings quite characteristic of low-spin iron[III] in a distorted octahedral environment. However, magnetic perturbation of the spectra indicated that V_{zz} was negative for $Fe(ttd)_2(dtt)$ and $Fe(ttd)(dtt)_2$ but positive for $Fe(dtt)_3 \cdot CHCl_3$. E.S.R. spectra were obtained for the two former compounds at 100 K in frozen chloroform solution, and from a combined analysis of the observed g-values and quadrupole splittings it was shown that the ground state in both compounds is an almost pure d_{xy} hole. The opposite sign of V_{zz} for $Fe(dtt)_3 \cdot CHCl_3$ was attributed to an orbital ground state containing essentially equal amounts of d_{xz} and d_{yz} holes.

3.4.4 High-spin–low-spin equilibrium in iron(II) complexes

In this final section we discuss some of the recent applications of Mössbauer spectroscopy (usually in conjunction with magnetic measurements) to the study of the $^5T_2$–$^1A_1$ cross-over in octahedral ferrous complexes where the iron is bonded to six nitrogen atoms. In many cases relatively small changes in the ligands can cause marked changes in the magnetic properties. The value of the Mössbauer effect here lies in the ease with which the two spin states can be distinguished. For the $S = 0$ case, the isomer shift (with respect to sodium nitroprusside) lies in the range 0.3–0.7 mm s^{-1} and $|\Delta E_Q|$ is small ($\lesssim 0.6$ mm s^{-1}). For $S = 2$, the isomer shift is ~ 1.0 mm s^{-1} or greater and $|\Delta E_Q|$ is typically greater than ~ 2 mm s^{-1}. Also, since one of the states involved is diamagnetic, the transition rate is often slow on the Mössbauer time scale and in the cross-over region both spin species may be seen in the spectrum.

König and co-workers[217-222] have investigated a variety of complexes of the types $Fe(phen)_2X_2$ and $Fe(bipy)_2X_2$. The phenanthroline derivatives[217] are high spin if $X = Cl^-$, Br^-, I^-, N_3^-, OCN^- or HCO_2^-, and low-spin if $X = CN^-$ or CNO^-.

For $X = NCS^-$ and $NCSe^-$, the complexes are high spin at room temperature $(\Delta E_Q \simeq 2.6 \text{ mm s}^{-1})$ and low-spin at 80 K $(\Delta E_Q \simeq 0.3 \text{ mm s}^{-1})$, with the cross-over temperatures estimated to be ~ 174 and ~ 232 K, respectively[217, 218]. $Fe(bipy)_2(NCS)_2$ shows similar behaviour with sharp changes in the magnetic moment and the Mössbauer spectrum at ~ 215 K [219].

Octahedral ferrous chelates based on hydrotris(1-pyrazolyl)borate[223] can be fully high spin, fully low spin or exhibit spin equilibrium with surprisingly small changes in the ligand structure. In the spin-equilibrium case studied[223], the complex was fully in the $^5T_2$ state at 269 K and fully in the $^1A_1$ state at 147 K. Between these temperatures the spectra showed thermal mixtures of the two states, so that the transition is spread over a much wider temperature range than observed in the (phen) and (bipy) compounds. Neither of the other two pyrazolyl complexes studied gave any evidence of spin equilibrium between 4.2 and 290 K.

The compounds $[Fe(pyim)_3]X_2$, where pyim = 2-(2-pyridyl)imidazole[224,225], behave similarly to the $Fe(phen)_2X_2$ derivatives[217]; the magnetic behaviour depends on the anion present. For $X = Cl^-$, NCS^- and $\frac{1}{2}SO_4^{2-}$ [225], as well as for $X = ClO_4^-$ [224], four lines are observed in the Mössbauer spectra at room temperature, the two stronger lines attributable to $S = 0$ and the weaker ones to $S = 2$. For ClO_4^- the $S = 2$ lines are absent at 80 K [224], but are still apparent in the spectra of the other three salts[225]. At 4.2 K the $S = 2$ lines have vanished for Cl^- and NCS^- and the application of a 5 kG external field at 1.3 K showed no internal magnetism so that iron has a zero moment in the compounds at this low temperature[225]. The spectrum of the sulphate derivative shows the presence of weak lines from the $S = 2$ state even at 1.3 K. In this case a small applied field greatly broadened the lines indicating a non-zero moment for iron. These results indicate that either part of the $^5T_1$ manifold is still accessible at 1.3 K, or there is a small amount of sulphate compound with iron in a high-spin ground state[225].

Several groups[222, 223, 225] have noted that the energy separation ε between the centres of the $^5T_2$ and $^1A_1$ terms is a function of temperature. Recent magnetic and Mössbauer measurements[222] on the 2-methyl-1,10-phenanthroline (mephen) complexes $[Fe(mephen)_3]X_2$ ($X = ClO_4^-$, BF_4^-) indicate that ε can vary by more than a factor of three between 100 and 300 K. Such spin transitions are often accompanied by a sizeable change in molar volume[223], and an x-ray study of $Fe(bipy)_2(NCS)_2$ at 295 and 100 K has shown that the Fe—N bond lengths and angles are significantly different at the two temperatures[221]. This has led König[222] to conclude that a temperature dependent ε should be expected in any cross-over situation. Variable temperature Mössbauer measurements will undoubtedly be of value in further studies of this problem.

Acknowledgements

I should like to thank Dr. J. N. R. Ruddick and Mr. J. C. Scott for reading and commenting on parts of the manuscript, Mrs. B. Krizan and Mrs. A.

Sallos for technical assistance, and Mrs. B. Stefanek for typing the manuscript.

References

1. Fitzsimmons, B. W., Seeley, R. J. and Smith, A. W. (1969). *J. Chem. Soc. A*, 143
2. Mazak, R. A. and Collins, R. L. (1969). *J. Chem. Phys.*, **51**, 3220
3. Parish, R. V. and Platt, R. H. (1970). *Inorg. Chim. Acta*, **4**, 65
4. Bancroft, G. M., Mays, M. J. and Prater, B. E. (1970). *J. Chem. Soc. A*, 956
5. Clark. M. G. (1969). *Discuss. Faraday Soc.*, **47**, 144
6. Clark, M. G. (1970). *Mol. Phys.*, **20**, 257
7. Collins, R. L. and Travis, J. C. (1967). *Mössbauer Effect Methodology*, Vol. 3. 123. (New York: Plenum Press)
8. Parish, R. V. and Johnson, C. E. (1971). *J. Chem. Soc. A*, 1906
9. Sternheimer, R. (1956). *Phys. Rev.*, **102**, 73
10. Watson, R. E. and Freeman, A. J. (1967). *Hyperfine Interactions*, 71 (Ed by Freeman, A. J., and Frankel, R. B.) (New York: Academic Press).
11. Berrett, R. R. and Fitzsimmons, B. W. (1967). *J. Chem. Soc. A*, 525
12. Debye, N. W. G. and Zuckerman, J. J. (1970). *Devel. Appl. Spectrosc.*, **8**, 267
13. Garrod, R. E. B., Platt, R. H. and Sams, J. R. (1971). *Inorg. Chem.*, **10**, 424
14. Herber, R. H. and Barbieri, R. (1971). *J. Organometallic Chem.*
15. Bancroft, G. M., Butler, K. D. and Rake, A. T. (1971). *J. Organometallic Chem.*, in the press
16. Yeats, P. A., Sams, J. R. and Aubke, F., to be published
17. Collins, R. L. (1965). *J. Chem. Phys.*, **42**, 1072
18. Zory, P. (1965). *Phys. Rev.*, **140**, A1401
19. Johnson, C. E. (1966). *Proc. Phys. Soc.*, **88**, 943
20. Danon, J. and Iannarella, L. (1967). *J. Chem. Phys.*, **47**, 382
21. Grant, R. W., Housley, R. M. and Gonser, U. (1969). *Phys. Rev.*, **178**, 523
22. Chandra, K. and Puri, S. P. (1968). *Phys. Rev.*, **169**, 272
23. Ingalls, R., Ôno, K. and Chandler, L. (1968). *Phys. Rev.*, **172**, 295
24. Garg, V. K. and Puri, S. P. (1971). *J. Chem. Phys.*, **54**, 209
25. Kienle, P. (1963). *Phys. Verhandl*, **3**, 33
26. Clark, M. G., Bancroft, G. M. and Stone, A. J. (1967). *J. Chem. Phys.*, **47**, 4250
27. Gol'danskii, V. I., Makarov, E. F. and Khrapov, V. V. (1963). *Phys Lett.*, **3**, 344
28. Karyagin, S. V. (1963). *Doklady Akad. Nauk SSSR*, **148**, 1102
29. Flinn, P. A., Ruby, S. L. and Kehl, W. L. (1964). *Science*, **143**, 1434
30. Herber, R. H. and Chandra, S. (1970). *J. Chem. Phys.*, **52**, 6045
31. Herber, R. H., Chandra, S. and Hazony, Y. (1970). *J. Chem. Phys.*, **53**, 3330
32. Herber, R. H. and Chandra, S. (1971). *J. Chem. Phys.*, **54**, 1847
33. Stöckler, H. A. and Sano, H. (1968). *Phys. Rev.*, **165**, 406
34. Stöckler, H. A. and Sano, H. (1968). *Chem. Phys. Lett.*, **2**, 448
35. Buckley, A. N., Rumbold, B. C., Wilson, G. V. H. and Murray K. S. (1970). *J. Chem. Soc. A*, 2298
36. Chih, H. and Penfold, B. R., to be published.
37. Ruddick, J. N. R., Sams, J. R. and Scott, J. C., unpublished results.
38. Ôno, K and Ito, A. (1964). *J. Phys. Soc. Japan*, **19**, 899
39. Ruby, S. L. and Flinn, P. A. (1964). *Rev. Mod. Phys.*, **36**, 361
40. Preston, R. S., Hanna, S. S. and Herberle, J. (1962). *Phys. Rev.*, **128**, 2207
41. Parish, R. V. and Johnson, C. E. (1970). *Chem. Phys. Lett.*, **6**, 239
42. Goodman, B. A. and Greenwood, N. N. (1969). *Chem. Commun.*, 1106
43. Erickson, N. E. (1970). *Chem. Commun.*, 1349
44. Fitzsimmons, B. W. (1970). *J. Chem. Soc. A*, 3235
45. Goodman, B. A., Greatrex, R. and Greenwood, N. N. (1971). *J. Chem. Soc. A*, 1868
46. Gibb, T. C. (1970). *J. Chem. Soc. A*, 2503
47. Ali, K. M., Cunningham, D., Frazer, M. J., Donaldson, J. C. and Senior, B. J. (1969). *J. Chem. Soc. A*, 2836
48. Herber, R. H. and Parisi, G. O. (1966). *Inorg. Chem.*, **5**, 769
49. Cordey-Hayes, M., Peacock, R. D. and Vučelić, M. (1967). *J. Inorg. Nucl. Chem.*, **29**, 1177

50. Aleksandrov, A. Yu., Okhlobystin, O. Yu., Polak, L. S. and Shpinel, V. S. (1964). *Doklady Akad. Nauk SSSR*, **157,** 934
51. Gibb, T. C. and Greenwood, N. N. (1966). *J. Chem. Soc. A,* 43
52. Cordey-Hayes, M. (1964). *J. Inorg. Nucl. Chem.,* **26,** 2306
53. Stöckler, H. A. and Sano, H. (1968). *Trans. Faraday. Soc.,* **64,** 577
54. Parish, R. V. and Platt, R. H. (1968). *Chem. Commun.,* 1118
55. Chivers, T. and Sams, J. R. (1970). *J. Chem. Soc. A,* 928
56. Chambers, R. D. and Chivers, T. (1964). *J. Chem. Soc.,* 4782
57. Hogben, M. G., Gay, R. S. and Graham, W. A. G. (1966). *J. Amer. Chem. Soc.,* **88,** 3457
58. Hogben, M. G., Gay, R. S., Oliver, A. J., Thompson, J. A. J. and Graham, W. A. G. (1969). *J. Amer. Chem. Soc.,* **91,** 291
59. Parish, R. V. and Platt, R. H. (1969). *J. Chem. Soc. A,* 2145
60. Chivers, T. and Sams, J. R. (1969). *Chem. Commun.,* 249
61. Watanabe, N. and Niki, E. (1970). *Bull. Chem. Soc. Japan,* **43,** 3034
62. Cordey-Hayes, M., Kemmitt, R. D. W., Peacock, R. D. and Rimmer, G. D. (1969). *J. Inorg. Nucl. Chem.,* **31,** 1515
63. Dalton, R. F. and Jones, K. (1969). *Inorg. Nucl. Chem. Lett.,* **5,** 785
64. Cullen, W. R., Sams, J. R. and Waldman, M. C. (1970). *Inorg. Chem.,* **9,** 1682
65. Goodman, B. A. and Greenwood, N. N. (1971). *J. Chem. Soc. A,* 1862
66. Platt, R. H. (1970). *J. Organometallic Chem.,* **24,** C23
67. Maddock, A. G. and Platt, R. H. (1971). *J. Chem. Soc. A,* 1190
68. Ensling, J., Gütlich, Ph., Hasselbach, K. M. and Fitzsimmons, B. W. (1971). *J. Chem. Soc. A,* 1940
69. Bokii, N. G., Zakharova, G. N. and Struchkov, Yu. T. (1970). *Zh. Struk. Khim.,* **11,** 895
70. Green, P. and Graybeal, J. (1967). *J. Amer. Chem. Soc.,* **89,** 4305
71. Strivastava, T. (1967). *J. Organometallic Chem.,* **10,** 373
72. Janssen, M. J., Luijten, J. G. A. and van der Kerk, G. J. M. (1963). *Rec. Trav. Chim.,* **82,** 90
73. Okawara, R. and Ohara, M. (1963/64). *J. Organometallic Chem.,* **1,** 360
74. Alcock, N. W. and Timms, R. E. (1968). *J. Chem. Soc. A,* 1873
75. Alcock, N. W. and Timms, R. E. (1968). *J. Chem. Soc. A,* 1876
76. Herber, R. H., Stöckler, H. A. and Reichle, W. T. (1965). *J. Chem. Phys.,* **42,** 2447
77. Ford, B. F. E., Liengme, B. V. and Sams, J. R. (1969). *J. Organometallic Chem.,* **19,** 53
78. Ford, B. F. E. and Sams, J. R. (1970). *J. Organometallic Chem.,* **21,** 345
79. Ford, B. F. E. and Sams, J. R. (1971). *J. Organometallic Chem.,* **31,** 47
80. Poder, C. and Sams J. R. (1969). *J. Organometallic Chem.,* **19,** 67
81. Leung, K. L. and Herber, R. H. (1971). *Inorg. Chem.,* **10,** 1020
82. Liengme, B. V., Randall, R. S. and Sams, J. R. (1972). *Can. J. Chem.,* in the press.
83. Davies, A. G., Milledge, H. J., Puxley, D. C. and Smith, P. J. (1970). *J. Chem. Soc. A,* 2862
84. Cunningham, D., Frazer, M. J. and Donaldson, J. D. (1971). *J. Chem. Soc. A,* 2049
85. Yeats, P. A., Sams, J. R. and Aubke, F. (1970). *Inorg. Chem.,* **9,** 740
86. Hill, J. C., Drago, R. S. and Herber, R. H. (1969). *J. Amer. Chem. Soc.,* **91,** 1644
87. Wedd, R. W. J. and Sams, J. R. (1970). *Can. J. Chem.,* **48,** 71
88. Gassenheimer, B. and Herber, R. H. (1969). *Inorg. Chem.,* **8,** 1120
89. Debye, N. W. G., Fenton, D. E., Ulrich, S. E. and Zuckerman, J. J. (1971). *J. Organometallic Chem.,* **28,** 339
90. Goodman, B. A., Greenwood, N. N., Jaura, K. L. and Sharma, K. K. (1971). *J. Chem. Soc. A,* 1865
91. Randall, R. S., Wedd, R. W. J. and Sams, J. R. (1971). *J. Organometallic Chem.,* **30,** C19
92. Liengme, B. V., Randall, R. S. and Sams, J. R., unpublished observations.
93. Ford. B. F. E., Sams, J. R., Goel, R. G. and Ridley, D. R. (1971). *J. Inorg. Nucl. Chem.,* **33,** 23
94. Yeats, P. A., Sams, J. R. and Aubke, F. (1971). *Inorg. Chem.,* **10,** 1881
95. Stapfer, C. H., Leung, K. L. and Herber, R. H. (1970). *Inorg. Chem.,* **9,** 970
96. Herber, R. H. (1971). *J. Chem. Phys.,* **54,** 3755
97. Greenwood, N. N. and Ruddick, J. N. R. (1967). *J. Chem. Soc. A,* 1679
98. Hristov, D., Bonchev, T. and Bourin, K. (1966). *Compt. Rend. Acad. Bulgare Sci.,* **19,** 293
99. Philip, J., Mullins, M. A. and Curran, C. (1968). *Inorg. Chem.,* **7,** 1895
100. Ichiba, S., Mishima, M., Sakai, H. and Negita, H. (1968). *Bull. Chem. Soc. Japan,* **41,** 49
101. Ichiba, S., Mishima, M. and Negita, H. (1969). *Bull. Chem. Soc. Japan,* **42,** 1486

102. Carty, A. J., Hinsperger, T., Mihichuk, L. and Sharma, H. D. (1970). *Inorg. Chem.* **9** 2573
103. Poller, R. C., Ruddick, J. N. R. and Spillman, J. A. (1970). *Chem. Commun.*, 680
104. Sukhovrekhov, V. F. and Dzevitskii, B. E. (1966). *Doklady Akad. Nauk SSSR,* **170,** 1089
105. Sukhovrekhov, V. F. and Dzevitskii, B. E. (1967). *Doklady Akad. Nauk SSSR,* **177,** 611
106. Carter, H. A., Qureshi, A. M., Sams, J. R. and Aubke, F. (1970). *Can. J. Chem.*, **48,** 2853
107. Levchuk, L. E., Sams, J. R. and Aubke, F. (1972). *Inorg. Chem.*, **11,** 43
108. Kegan, Yu. (1962). *Soviet Phys. JETP,* **14,** 472
109. Kagan, Yu. and Maslov, V. A. (1962). *Soviet Phys. JETP,* **14,** 922
110. Sams, J. R., unpublished results.
111. Herber, R. H. and Cheng, H-S. (1969). *Inorg. Chem.* **8,** 2145
112. Davies, A. G., Smith, L. and Smith, P. J. (1970). *J. Organometallic Chem.*, **23,** 135
113. Clausen, C. A. and Good, M. L. (1970). *Inorg. Chem.*, **9,** 817
114. Schlemper, E. O. and Hamilton, W. C. (1966). *Inorg. Chem.*, **5,** 995
115. Allen, F. A., Lerbscher, J. and Trotter, J. (1971). *J. Chem. Soc. A*, 2807
116. Blom, E. A., Penfold, B. R. and Robinson, W. T. (1969). *J. Chem. Soc. A*, 913
117. Isaacs, N. W. and Kennard, C. H. L. (1970). *J. Chem. Soc. A*, 1257
118. Schlemper, E. O. (1967). *Inorg. Chem.*, **6,** 2012
119. Mullins, M. A. and Curran, C . (1967). *Inorg. Chem.*, **6,** 2017
120. Mullins, M. A. and Curran, C. (1968). *Inorg. Chem.*, **7,** 2584
121. Fitzsimmons, B. W., Seeley, N. J. and Smith, A. W. (1968). *Chem. Commun.*, 390
122. Fitzsimmons, B. W. (1968). *Chem. Commun.*, 1485
123. Poller, R. C. and Ruddick, J. N. R. (1969). *J. Chem. Soc. A*, 2273
124. Petridis, D., Mullins, F. P. and Curran, C. (1970). *Inorg. Chem.*, **9,** 1270
125. Fitzsimmons, B. W., Owusu, A. A., Seeley, N. J. and Smith, A. W. (1970). *J. Chem. Soc. A*, 935
126. Naik, D. V. and Curran, C. (1971). *Inorg. Chem.*, **10,** 1017
127. Tan, T. H., Dalziel, J. R., Yeats, P. A., Sams, J. R., Thompson, R. C. and Aubke, F. *Can. J. Chem.,* in the press
128. Yeats, P. A., Ford, B. F. E., Sams, J. R. and Aubke, F. (1969). *Chem. Commun.*, 791
129. Vučelić, M. (1968). *Croat. Chem. Acta,* **40,** 255
130. Yeats, P. A., Poh, B. L., Ford, B. F. E., Sams, J. R. and Aubke, F. (1970). *J. Chem. Soc. A*, 2188
131. Hoppe, R. and Dähne, W. (1962). *Naturwissenschaften,* **49,** 254
132. Donaldson, J. D. and Senior, B. J. (1966). *J. Chem. Soc. A*, 1796, 1798
133. Bearden, A. J., Marsh, H. S. and Zuckerman, J. J. (1966). *Inorg. Chem.*, **5,** 1260
134. Davies, C. G. and Donaldson, J. D. (1968). *J. Chem. Soc. A*, 946
135. Donaldson, J. D., Nicholson, D. G. and Senior, B. J. (1968). *J. Chem. Soc. A*, 2928
136. Donaldson, J. D. and Nicholson, D. G. (1970). *Inorg. Nucl. Chem. Lett.*, **6,** 151
137. Birchell, T., Dean, P. A. W. and Gillespie, R. J. (1971). *J. Chem. Soc. A*, 1777
137a.Greenwood, N. N. and Timmick, A. (1971). *J. Chem. Soc. A*, 676
138. Cordey-Hayes, M. (1964). *J. Inorg. Nucl. Chem.*, **26,** 915
139. Lees, J. K. and Flinn, P. A. (1965). *Phys. Lett.*, **19,** 186
140. Lees, J. K. and Flinn, P. A. (1968). *J. Chem. Phys.*, **48,** 882
141. Gol'danskii, V. I., Khrapov, V. V., Rochev, V. Ya., Sumarokova, T. N. and Surpina, D. E. (1968). *Doklady Akad. Nauk SSSR,* **183,** 364
142. Donaldson, J. D. and Senior, B. J. (1969). *J. Inorg. Nucl. Chem.*, **31,** 881
143. Gibb, T. C., Goodman, B. A. and Greenwood, N. N. (1970). *Chem. Commun.*, 774
144. Donaldson, J. D., Filmore, E. J. and Tricker, M. J. (1971). *J. Chem. Soc. A,* 1109
145. Smith, P. J. (1970). *Organometallic Chem. Rev.*, **A5,** 373
146. Bird, S. R. A., Donaldson, J. D., Keppie, S. A. and Lappert, M. F. (1971). *J. Chem. Soc. A*, 1311
147. Karasev, A. N., Kolobova, N. E., Polak, L. S., Shpinel, V. S. and Anisimov, K. A. (1966). *Teor. Eksp. Khim.*, **2,** 126
148. Gol'danskii, V. I., Borshagovskii, B. V., Makarov, E. F., Stukan, R. A., Anisimov, K. A., Kolobova, N. E. and Skripkin, V. V. (1967). *Teor. Eksp. Khim.*, **3,** 478
149. Herber, R. H. and Goscinny, Y. (1968). *Inorg. Chem.*, **7,** 1293
150. Fenton, D. E. and Zuckerman, J. J. (1968). *J. Amer. Chem. Soc.*, **90,** 6226
151. Fenton, D. E. and Zuckerman, J. J. (1969). *Inorg. Chem.*, **8,** 1771
152. Wynter, C. and Chandler, L. (1970). *Bull. Chem. Soc. Japan,* **43,** 2115

153. Cullen, W. R., Sams, J. R. and Thompson, J. A. J. (1971). *Inorg. Chem.*, **10**, 843
154. Bird, S. R. A., Donaldson, J. D., Holding, A. F. LeC., Senior, B. J. and Tricker, M. J. (1971). *J. Chem. Soc. A*, 1616
155. Onaka, S., Sasaki, Y. and Sano, H. (1971). *Bull. Chem. Soc. Japan.*, **44**, 726
156. Goodman, B. A., Greatrex, R. and Greenwood, N. N. (1971). *J. Chem. Soc. A*, 1868
157. Baranovskii, V. I., Sergeev, V. P. and Dzevitskii, B. E. (1969). *Doklady Akad. Nauk SSSR.*, **184**, 632
158. Jones, M. T. (1967). *Inorg. Chem.*, **6**, 1249
159. Herber, R. H. (1967). *Progr. Inorg. Chem.*, **8**, 1
160. Bancroft, G. M., Butler, K. D. and Rake, A. T. (1972). *J. Organometallic Chem.*, **7**, 1223
161. Liengme, B. V., Newlands, M. J. and Sams, J. R. *Inorg. Nucl. Chem. Lett.*, in the press
162. Liengme, B. V., Sams, J. R. and Scott, J. C. *Bull. Chem. Soc. Japan*, in the press
163. O'Connor, J. E. and Corey, E. R. (1967). *Inorg. Chem.*, **6**, 968
164. Green, P. T. and Bryan, R. F. (1970). *J. Chem. Soc. A*, 1696
165. Frank, E. and Abeledo, C. R. (1969). *J. Inorg. Nucl. Chem.*, **31**, 989
166. Dale, B. W., Williams, R. J. P., Edwards, P. R. and Johnson, C. E. (1968). *Trans. Faraday Soc.*, **64**, 3011
167. Dale, B. W., Williams, R. J. P., Edwards, P. R. and Johnson, C. E. (1968). *Trans. Faraday Soc.*, **64**, 620
168. Hudson, A. and Whitfield, H. J. (1966). *Chem. Commun.*, 606
169. Hudson, A. and Whitfield, H. J. (1967). *Inorg. Chem.*, **6**, 1120
170. Collins, R. L., Pettit, R. and Baker, W. A. (1966). *J. Inorg. Nucl. Chem.*, **28**, 1001
171. Bancroft, G. M., Mays, M. J. and Prater, B. E. (1968). *Chem. Commun.*, 1374
172. Bancroft, G. M., Mays, M. J. and Prater, B. E. (1969). *Discuss. Faraday Soc.*, **47**, 136
173. Bancroft, G. M., Garrod, R. E. B., Maddock, A. G., Mays, M. J. and Prater, B. E. (1970). *Chem. Commun.*, 200
174. Pasternak, M. and Sonino, T. (1968). *J. Chem. Phys.*, **48**, 1997
175. Bancroft, G. M., personal communication.
176. Bancroft, G. M., Garrod, R. E. B. and Maddock, A. G., to be published
177. Bancroft, G. M. (1971). *Chem. Phys. Lett.*, **10**, 449
178. Nozik, A. J. and Kaplan, M. (1967). *Phys. Rev.*, **159**, 273
179. Ham, F. S. (1967). *Phys. Rev.*, **160**, 328
180. Harris, C. B. (1968). *J. Chem. Phys.*, **49**, 1648
181. Lappert, J., Frankel, R. B., Misetich, A. and Blum, N. A. (1969). *Phys. Lett.*, **28B**, 406
182. Rosenberg, M., Mandache, S., Niculescu-Majewska, H., Filotti, G. and Gomolea, V. (1970). *Phys. Lett.*, **31A**, 84
183. Friedt, J. M. (1970). *J. Inorg. Nucl. Chem.*, **32**, 431
184. Crow, J. P., Cullen, W. R., Herring, F. G., Sams, J. R. and Tapping, R. L. (1971). *Inorg. Chem.*, **10**, 1616
185. King, R. B., Epstein, L. M. and Gowling, E. W. (1970). *J. Inorg. Nucl. Chem.*, **32**, 441
186. Harrison, W. and Trotter, J. (1971). *J. Chem. Soc. A*, 1542
187. Dahl, L. F., de Gil, E. R. and Feltham, R. D. (1969). *J. Amer. Chem. Soc.*, **91**, 1653
188. Clark, M. G., personal communication.
189. Cullen, W. R., Harbourne, D. A., Liengme, B. V. and Sams, J. R. (1969). *Inorg. Chem.*, **8**, 1464
190. Fitzsimmons, B. W., personal communication
191. Clark, M. G., Cullen, W. R., Garrod, R. E. B., Maddock, A. G. and Sams, J. R., to be published.
192. Gerloch, M., Lewis, J., Mabbs, F. E. and Richards, A. (1968). *J. Chem. Soc. A*, 112
193. Reiff, W. M., Long, G. J. and Baker, W. A. (1968). *J. Amer. Chem. Soc.*, **90**, 6347
194. Buckley, A. N., Wilson, G. V. H. and Murray, K. S. (1969). *Solid State Commun.*, **7**, 471
195. Fitzsimmons, B. W. and Johnson, C. E. (1970). *Chem. Phys. Lett.*, **6**, 267
196. Okamura, M. Y., Klotz, I. M., Johnson, C. E., Winter, M. R. C. and Williams, R. J. P. (1969). *Biochemistry*, **8**, 1951
197. Reiff, W. M., Baker, W. A. and Erickson, N. W. (1968). *J. Amer. Chem. Soc.*, **90**, 4794
198. Berrett, R. R., Fitzsimmons, B. W. and Owusu, A. A. (1968). *J. Chem. Soc. A*, 1575
199. Buckley, A. N., Herbert, I. R., Rumbold, B. D., Wilson, G. V. H. and Murray, K. S. (1970). *J. Phys. Chem. Solids*, **31**, 1423
200. Bancroft, G. M., Maddock, A. G. and Randl, R. P. (1968). *J. Chem. Soc. A*, 2936
201. Buckley, A. N., Wilson, G. V. H. and Murray, K. S. (1969). *Chem. Commun.*, 718

202. Reiff, W. M. (1971). *J. Chem. Phys.,* **54,** 4718
203. Buckley, A. N., Rumbold, B. D., Wilson, G. V. H. and Murray, K. S. (1970). *J. Chem. Soc. A,* 2298
204. Blume, M. (1967). *Phys. Rev. Lett.,* **18,** 305
205. Cox, M., Fitzsimmons, B. W., Smith, A. W., Larkworthy, L. F. and Rogers, K. A. (1969). *Chem. Commun.,* 183
206. Cox, M., Darken, J., Fitzsimmons, B. W., Smith, A. W., Larkworthy, L. F. and Rogers, K. A. (1970). *Chem. Commun.,* 105
207. Beckett, R., Heath, G. A., Hoskins, B. F., Kelly, B. P., Martin, R. L., Roos, I. A. G. and Weickhardt, P. L. (1970). *Inorg. Nucl. Chem. Lett.,* **6,** 257
208. Rickards, R., Johnson, C. E. and Hill, H. A. O. (1969). *Trans. Faraday Soc.,* **65,** 2847
209. Epstein, L. M. and Staub, D. K. (1969). *Inorg. Chem.,* **8,** 560
210. Epstein, L. M. and Staub, D. K. (1969). *Inorg. Chem.,* **8,** 784
211. Birchall, T. (1969). *Can. J. Chem.,* **47,** 4563
212. Birchall, T. and Greenwood, N. N. (1969). *J. Chem. Soc. A,* 286
213. Rickards, R., Johnson, C. E. and Hill, H. A. O. (1970). *J. Chem. Phys.,* **53,** 3118
214. de Vries, J. L. K. F., Trooster, J. M. and de Boer, E. (1971). *Inorg. Chem.,* **10,** 81
215. Rickards, R., Johnson, C. E. and Hill, H. A. O. (1971). *J. Chem. Soc. A,* 797
216. Rickards, R., Johnson, C. E. and Hill, H. A. O. (1971). *J. Chem. Soc. A,* 1755
217. König, E. and Madeja, K. (1966). *Chem. Commun.,* 61
218. König, E. and Madeja, K. (1967). *Inorg. Chem.,* **6,** 48
219. König, E., Madeja, K. and Watson, K. J. (1968). *J. Amer. Chem. Soc.,* **90,** 1146
220. König, E., Madeja, K. and Böhmer, W. H. (1969). *J. Amer. Chem. Soc.,* **91,** 4582
221. König, E. and Watson, K. J. (1970). *Chem. Phys. Lett.,* **6,** 457
222. König, E. and Kremer, S. (1971). *Chem. Phys. Lett.,* **8,** 312
223. Jesson, J. P., Weiher, J. F. and Trofimenko, S. (1968). *J. Chem. Phys.,* **48,** 2058
224. Goodgame, D. M. L. and Machado, A. A. S. C. (1969). *Inorg. Chem.,* **8,** 2031
225. Dosser, R. J., Eilbeck, W. J., Underhill, A. E., Edwards, P. R. and Johnson, C. E. (1969). *J. Chem. Soc. A,* 810

4
Nuclear Quadrupole Resonance Spectroscopy

HIDEAKI CHIHARA
and
NOBUO NAKAMURA

Osaka University, Toyonaka, Japan

4.1 INTRODUCTION

During the past 22 years since the discovery of nuclear quadrupole resonance (n.q.r.)[1], the technique has found widespread applications in the study of physics and chemistry of the solid state and of chemical bonds. Except for nuclei with $I = \frac{3}{2}$, the measurement of the resonance frequencies gives values of the quadrupole coupling constant e^2Qq and the asymmetry parameter η. In the case of $I = \frac{3}{2}$, these two quantities can only be determined separately by use of the Zeeman effect on a single crystal. In general, deductions that can be made from the n.q.r. spectrum are, inevitably at the present stage of theory, based on a number of assumptions if one is to relate the coupling constant and the asymmetry parameter to the conventional concepts of the chemical bond, e.g. ionicity and double-bond nature. Recent progress on such aspects has been through the use of the molecular orbital theory for calculating the electric field gradient (e.f.g.) *eq*.

Temperature dependence of the resonance frequency is semi-quantitatively understood by the Bayer–Kushida theory. However, usually a single average libration frequency in the harmonic approximation is the best one can hope to deduce even in cases where anisotropic librations are evident from structural considerations. The motion of molecules and ionic groups in solids may be proved by the temperature variation of various relaxation times, T_1, T_2, and T_2^*. Here also n.q.r. is in need of a theory to render the method more useful[2].

Pressure is another auxiliary variable. To assist in the interpretation of the temperature dependence, hopefully by eliminating the effect of thermal expansion, the measurements of the pressure dependence of the resonance frequencies should receive more attention[3].

Unlike nuclear magnetic resonance, the question of line shape and/or line width has attracted little attention from theorists. This is probably not surprising since in many instances small amounts of impurities, or lattice imperfections, completely smear out the resonance which by nature calls for highly sensitive detection. However, the presence of some impurities in $SbCl_5$ was found to improve the crystallinity of the otherwise pure specimen and to enhance its resonances[4].

The number of papers dealing with n.q.r. is growing rapidly, particularly since commercial spectrometers have been made available through the development of the automatic coherence control technique with a super-regenerative spectrometer by Peterson and by Smith. The growth is especially noticeable in the field of coordination compounds. The amount of information being added is still manageable in size, compared to high-resolution n.m.r., but will soon outgrow a single review article. To avoid unnecessary duplication of research work, an international co-operation for storage and retrieval of n.q.r. data is to be greatly desired.

In the present chapter, a brief review will be given of the recent achievements of the n.q.r. method in molecular and crystal studies. The bibliography is not intended to be exhaustive but the reviewers hope that they covered all the important progress that has been made during the past 5 years. The literature survey was concluded in July, 1971.

Because of the limited space, elementary discussions on the principles of nuclear quadrupole resonance and relaxation will be left to the textbooks. The readers are referred to the standard textbooks by Das and Hahn[5] and Lucken[6], the review articles by Weiss[7], Kubo and Nakamura[8], Smith[9], Biryukov[10], Mairinger[11], Brame[12], Schultz and Karr[13], Brown[14], Wendling[15],

Bronswyk[16], Grechishkin and Ainbinder[17], Penkov and Safin[18], Bizot[19], Bose[20], and Maksyutin et al.[21]. Schempp and Bray[22] have given an excellent introduction to n.q.r. A compilation of the n.q.r. frequencies was published by Fedin and Semin[23], Segel and Barnes[24], and Biryukov et al.[25] in Russian and translated into English by Schmorak. Eigenvalue tables for $I = \frac{5}{2}$ have been given by Livingston and Zeldes[26] and for $I = \frac{7}{2}$ by Chihara and Nakamura[27].

4.2 INSTRUMENTATION AND TECHNIQUES

In many aspects, the n.q.r. spectrometers are similar to the n.m.r. analogues. Here we shall confine ourselves to those techniques that are more or less directly relevant to the n.q.r.

The most frequently used n.q.r. spectrometers are the so-called super-regenerative spectrometers either internally (self-) quenched or externally quenched. Borsutskii et al.[28] reported a single spectrometer that covers the frequency between 10 and 350 MHz, whereas Gill et al.[29] showed that by using the grounded-cathode Colpitts-type oscillator, a regenerative spectrometer can be constructed which, compared with Wang's spectro-meter[30], is less sensitive to details of construction and to external disturbances.

The use of field-effect transistors (FET) has become more and more popular. They have the advantages of low microphonic effects, simplification of power supplies, and improved frequency stability besides a high input impedance and low junction noise. The use of FET was discussed by O'Konski and Cartwright[31] for super-regenerative spectrometers and by Viswanathan et al.[32] for marginal oscillators.

To eliminate the need for troublesome adjustment of the quench frequency to obtain the maximum signal-to-noise ratio, an automatic coherence control was devised by Peterson and Bridenbaugh[33] which was improved to cover the 5–1000 MHz range[34]. This constitutes the basis for one of the commercial spectrometers. Graybeal and Croston[35] described a visible, rather than audible, noise monitoring circuit to incorporate with Peterson's spectrometer. Smith and Tong[36] devised a wide-range spectrometer by seeking a simultaneous solution to automatic gain control and side-band suppression; this is the basis for another model of commercial spectrometer. Tong[37] also described a method for reduction of line-shape distortion and Ashley and Tong[38] described elimination of the centre frequency shift for precise frequency measurements. The base-line slope can be eliminated by use of a small magnetic field and subtracting the field-independent portion (base line) of the signal[39]. Muha[40], and Graybeal and Croston[41], described automatic coherence control circuits using operational amplifiers. The effects of the frequency and the wave form of quenching signal on the n.q.r. intensity have been examined both theoretically and experimentally by Clarkson and Sullivan[42], Doolan and Hacobian[43], and Caldwell and Hacobian[44]. Negita[45] has examined the dependence of the line shape and frequency of ^{14}N resonance on the Zeeman modulation field.

N.Q.R. spectrometers for the low frequency region and n.m.r. spectrometers have many features in common. Barton[46] described a dual purpose spectrometer which uses a coherent, self-detecting, self-quenching superregenerative Colpitts oscillator. Modified Robinson or marginal oscillators were reported by Won Choi and Czae-Myungzoon[47], Blum[48], Colligiani[49], Faulkner and Holman[50], Feng[51], and Idoine and Brandenberger[52].

A pulsed oscillator for relaxation measurements was described by Grechishkin and Shishkin[53] who used the double-resonance method to determine the individual relaxation times in a multilevel system. A spin-echo apparatus working between 150 and 300 MHz was reported by Gushchin et al.[54]. Abe[55] determined the relaxation times of the six n.q.r. lines of hydrazine separately by observing the saturation recovery with a continuous-wave spectrometer after a pumping pulse had been applied. Of the components of the pulse apparatus, a high power (>900 V p-p), untuned radio frequency (r.f.) transmitter is described in detail by Lowe and Tarr[56] and an r.f. gate with 140 dB carrier suppression is described by Burnett and Harmon[57]. Ainbinder et al.[58] discussed the slow beats in the quadrupole spin-echo that occur when there is a non-zero asymmetry parameter of the electric field gradient. An automatically stabilised bridge for n.m.r. described by Pryer[59] may probably find a use in n.q.r. A coaxial bridge which was primarily designed for n.m.r. and e.s.r. measurements was successfully used to detect the weak signal from ^{127}I in KIO_3 ($\pm\frac{3}{2} \leftrightarrow \frac{1}{2}$) at 145.876 MHz by Kesselring and Gautschi[60]. To frequency- and sample-minded people, it is perhaps interesting to note that Utton determined the ^{35}Cl frequency in $KClO_3$ to be $28\,213\,372 \pm 2$ Hz at 0 °C and atmospheric pressure[61].

Among accessories, a goniometer assembly for Zeeman studies was described by Peterson and Bridenbaugh[62], Zeeman modulators by Buijs et al.[63], Tong[64], Muha[65], and Carpenter and Forman[66], and an automatic sample-temperature recording device which works at certain intervals, by Tong[67].

Observation of n.q.r. by the nuclear-induction technique was examined both theoretically and experimentally by Smith[68]. The theory was developed for $I = 1$, $\eta = 0$. For the method to work, it is necessary to lift the degeneracy of the $m = \pm 1$ levels by a small magnetic field. Experiments with ^{14}N in hexamethylenetetramine agreed with the predictions of theory in all aspects.

Leppelmeier and Hahn[69] developed the spin-temperature theory for the n.q.r. $I = \frac{3}{2}$ case and discussed the relaxation in the rotating frame. They compared the theory with the experimental results on ^{35}Cl in $KClO_3$ and $Ba(ClO_3)_2 \cdot 2D_2O$.

A double-resonance technique is suggested by Slusher and Hahn[70] which is capable of detecting n.q.r. for molar concentrations as low as 1 in 10^7. They demonstrated the technique by application to Na and Cl nuclei in the vicinity of impurities (K^+ and Br^-) and imperfections in NaCl.

Application of n.q.r. to precision thermometry has been discussed for some time. The precision of ± 0.001 K (50–297 K) was obtained with $KClO_3$ as the sensor by Utton[71]. An apparatus of comparable precision was described by Brodskii and Vorob'ev[72] and Labrie et al.[73]. A locked n.q.r. spectrometer for use in pressure measurements was described by Frisch and VanderHart[74], which could also be used as a thermometer.

4.3 ELECTRONIC STRUCTURE OF MOLECULES[75]

4.3.1 Organic compounds

4.3.1.1 Zeeman study

p-Bromophenol was studied by Rama Rao and Murty[76] who reported the existence of two physically inequivalent bromine sites; this was later refined by Bucci et al.[77] who were able to resolve two pairs of e.f.g.s which make an angle of 2° 52′ to give altogether four inequivalent sites. Bucci et al. also made precise re-measurements on p-chlorophenol[78] and described their IBM 7090 computer program to treat the Zeeman data. The ratio η_{Cl}/η_{Br} was found to be 1.56. Peterson et al.[79] found that in 2,4-dichlorophenoxyacetic acid (the herbicide 2,4-D) the chlorine at the 4 position absorbs at 34.63 MHz with $\eta = 0.0629$ whereas that at the 2 position absorbs at 35.66 MHz with $\eta = 0.1010$. The angle between the two C—Cl bonds was 118.1 ± 0.25 degrees showing a slight distortion of the molecule. Ambrosetti et al.[80] found that in 2,4-dibromoaniline the x-axis of one of the Br atoms (226.08 MHz) makes an unusually large angle (15° 41′) with the normal to the benzene ring, compared with 1 degree which the other ^{81}Br (227.58 MHz) makes. Kantimati[81] assigned the Cl which is ortho to the amino group as the higher frequency (34.854 MHz) and the Cl which is para to the amino group as the lower frequency (34.734 MHz) in 2,4-dichloroaniline. Kamishina[82] reported that the z-axis of N in hydrazine makes an angle of 33 degrees with the direction of the lone-pair orbital if sp^3 hybridisation is assumed. Chihara et al. reported that the CCl$_3$ group in p-chlorobenzotrichloride is tilted towards one side of the plane of the benzene ring[83]. Other Zeeman studies will be found in the references[84-94].

4.3.1.2 Nitrogen quadrupole coupling constant

A knowledge of e^2Qq per p electron of a ^{14}N nucleus is required to interpret the molecular e^2Qq values obtained by experiment in terms of bond properties. No single reliable value of $(e^2Qq)_N$ has been established in spite of many experimental and theoretical attempts. Attempts that have been made so far were to choose a simple molecule or molecular ion and to carry out a cal-

Table 4.1   Values of the quadrupole coupling constant (MHz) of ^{14}N per p electron in the free atomic state

Compound	$(e^2Qq)_N$/MHz	Reference
NH$_3$	−11.12	Kato et al.[95]
Isotopic NH$_3$	−11.7	Lehrer and O'Konski[96]
Diazines	−8.2	Schempp and Bray[97]
NH$_2$D	−2.6 ~ −6.4	Kern[98]
NaNO$_2$	−9.5 ~ −12	Marino and Bray[99]
Pyridine derivatives	$\begin{cases} -8.0 \sim -9.0 \\ -12 \end{cases}$	Guibé and Lucken[100] Schempp and Bray[101]
Nitriles	−9.3	Colligiani et al.[102]
Hydrazine derivatives	−11.3	Ikeda et al.[103]
N-chloropiperidine and N-chlorodimethylamine	−9.8	Schempp[104]

culation of the field gradient $(e^2Qq)_{mol}$ which is then used to estimate $(e^2Qq)_N$. Table 4.1 lists the $(e^2Qq)_N$ values which were calculated from the experimental $(e^2Qq)_{mol}$.

A different approach by Bonaccorsi et al.[105] was to calculate the field gradient eq at the N nucleus in HCN, HCCCN, ClCN, and FCN by ab initio calculation using SCF wave functions and then plotting these against the

Table 4.2   Values of the nuclear quadrupole moment of [14]N

Compound	$10^{26}Q/cm^2$	Reference
HCN	0.71	Bassompièrre[106]
NO	1.6	Lin[107]
NH_3	0.94	Kato[108]
DCN, N_3^-	1.47	Kern and Karplus[109]
Nitriles	1.66	Bonaccorsi et al.[105]
HCN	1.56	O'Konski and Ha[110]

observed $(e^2Qq)_{mol}$. The plot gave a straight line passing through the origin; the slope gave a value of Q as $1.66 \times 10^{-26}cm^2$. A similar procedure was adopted by other investigators who obtained different values of Q as shown in Table 4.2.

4.3.1.3   Nitrogen-containing compounds

Although there is uncertainty in the $(e^2Qq)_N$ value, the fact that nitrogen is a chemically and biologically important element constitutes a good motive force for studying n.q.r. of N-containing compounds. Since its spin is 1, one usually observes two resonance lines from which the asymmetry parameter η may be determined. In thiocyanates such as CH_3SCN, C_2H_5SCN and $NCSC_2H_4SCN$, η values as large as 0.47 were obtained and Ikeda et al.[111] attributed them to the polarisation of π bonds in the C$\equiv$N bond. Tetracyanoethylene and 7,7',8,8'-tetracyanoquinodimethane[112] were also studied. Guibé and Lucken[113] measured $e^2Qq = 2.274$ MHz and $\eta = 0.378$ on formamide and concluded that the values represent the effect of resonance and the increased ionicity of the NH bond due to formation of an intermolecular hydrogen bond if $(e^2Qq)_N$ is taken as 7 MHz. Ikeda et al.[114] treated chloropyridines, cyanopyridines, aminopyridines, aniline, and methyl pyridines by the Pariser–Parr–Pople method and calculated the σ- and π-bond ionicities of ring N—C bonds to compare with their values estimated from experiments. The degree of bond ionicity was derived by assuming 25% s hybridisation and $(e^2Qq)_N = 12$ MHz. Their results show that the donation of π electrons from different substituents to the ring nitrogen decreases in the order $NH_2 > Cl > CH_3 \approx H > CN$, whereas that of σ electrons decreases in the order $NH_2 > CH_3 \approx H > Cl \approx CN$. Guibé et al[115] discussed the inductive effect of a NO_2 group substituted on pyridine. Studies were reported for diazines[97] (including pyridazine, pyrimidine and pyrazine) and for pyrrole[116] for which the coupling constant is less than half that of pyridine. Guibé and Lucken[117] also examined pyrrole and other five-membered heterocyclic compounds. Schempp and Bray[118] made a critical examination of molecular wave functions by comparing the *ab initio* calculations by Clementi[119] with their n.q.r. results for pyridine, pyrazine and pyrrole and found that the agreement was not satisfactory. Finally, Marino, Guibé and Bray[120] demonstrated that in aminopyridines, (a) the amino group is a very strong π electron donor and a weak σ electron donor, (b) there is a marked dependence of the ring-nitrogen spectrum on the amino position, and (c) the amino nitrogen spectrum is very nearly independent of substituent position on the ring. These results were interpreted as suggesting that the amino group delivers approximately a constant amount of electric charge to the ring regardless of its position, while the ring nitrogen accepts the charge only when it is placed *ortho* or *para* to the amino group.

A Townes–Dailey type of analysis was made on a number of saturated cyclic amines by Colligiani *et al.*[121] who showed that in azetidine (1) the e^2Qq and η values may be satisfactorily explained if the CNC angle is taken as 108 degrees thus indicating a tetrahedrally hybridised nitrogen atom. In the case of aziridine (2) the CNC angle was 104 degrees.

$$\begin{array}{c} \text{CH}_2 \\ \diagup \quad \diagdown \\ \text{H}_2\text{C} \quad \text{N–H} \\ \diagdown \quad \diagup \\ \text{CH}_2 \end{array}$$

(1)

$$\begin{array}{c} \text{H}_2\text{C} \\ \quad \diagdown \\ \quad \quad \text{N–H} \\ \quad \diagup \\ \text{H}_2\text{C} \end{array}$$

(2)

They discussed these bent bonds in relation to the molecular orbital calculations[105].

Ikeda *et al.*[103] estimated the ionic character of N—H (42%) and N—C (23–5%) bonds in hydrazine and its derivatives. Very large η values were obtained for hydrazinium bromide (93.37%) and iodide (94.17%) which may be explained in terms of polarisation of the N—N bond.

Marino and Oja[122] rationalised their results of measurements on 4-coordinated nitrogen in terms of the Townes–Dailey approach. For glycine, the pure n.q.r. was difficult to detect and therefore the n.m.r. technique was applied by Andersson *et al.*[123] to determine the principal axis directions, $(e^2Qq) = 1.20$ MHz, and $\eta = 0.5$. On the other hand, Blinc *et al.*[124] made a re-investigation at 140 °C by a double-resonance method and obtained a much lower e^2Qq value of only 745 kHz with $\eta = 0.61$. The direction of the largest principal axis of the coupling tensor did not coincide with the C—N axis but makes an angle of 60 degrees. They offered an explanation for it by considering modes of NH_3^+ motion.

Schempp[104] recently reported his experiments on two compounds, *N*-chloropiperidine and *N*-chlorodimethylamine, in which the N—Cl bond may be studied by n.q.r. of both ^{14}N and ^{35}Cl. Contrary to the Pauling scale of electronegativity, the bond is polarised in the direction of N^+—Cl^-. A consistent set of bond parameters was found which yielded a more reliable value 9.8 MHz for $(e^2Qq)_N$ as given in Table 4.1. Other nitrogen-containing compounds that have been studied include organic nitriles[125, 126], dipyridyls[127], heterocycles[128], ethylenimine[129], benzenesulphonamide[130], aliphatic amines[131], and others[132-134].

4.3.1.4 *Other organic compounds*

Linscheid and Lucken[135] found a correlation between the ^{35}Cl frequency and the difference in the C—Cl bond length between the axial and the equatorial positions in chlorodioxans. Stidham and Farrell[136] studied the relation between the ^{35}Cl n.q.r. frequency and Hammett's σ in chloropyridazines and chloropyrazines. Zeil and Haas[137] found that the π bond character of the C—Cl bond remains constant among $(CH_3)_3X$—C≡C—Cl where X = C, Si, Ge and Sn. Hart and Whitehead[138] observed that in N-chloro molecules the ^{35}Cl frequency at 77 K was about 55 MHz with the N atom in the $tr^1 tr^1 tr^1 \pi^2$ hybridisation (as in N-chlorophthalimide), 46 MHz with N in the $tr^1 tr^1 tr^2 \pi^1$ state (as in N-chloramine B), and 44 MHz with N in the $te^1 te^1 te^1 te^2$ (as in N-chloropiperidine). Considine[139] found a linear relation between the $e^2 Qq$ and Jaffe's group orbital electronegativity in various halogenomethanes. Lucas and Guibé[140] studied the ^{14}N and ^{35}Cl resonances in charge-transfer complexes of chloroform as the acceptor with some amino and heterocyclic compounds as the donors. The amount of electric charge transfered was estimated to be a few per cent of the electronic charge. Maksyutin et al.[141, 142] made a similar study of picryl chloride complexes with aromatic donors and obtained an approximately linear decrease in the frequency shift with the increase in the ionisation potential of the donor molecules.

Many other organic compounds are discussed elsewhere in this review. For others, only references are given at the end of the chapter[143-157].

4.3.2 Molecular orbital calculation of quadrupole coupling constant

The Townes–Dailey type of approach involves a number of conceptual parameters. To overcome this, Cotton and Harris[158] developed a general formulation of $(e^2 Qq)_{mol}$ in the LCAO—MO framework. The Kth MO ψ^K is expressed in terms of mutually orthogonal atomic orbitals

$$\psi^K = \sum_{i=1}^{n} C_i^K \phi_i \qquad (4.1)$$

and the field gradient q_α^K at a particular atom α arising from an electron in ψ is given by

$$q_\alpha = \sum_K \sum_i \sum_j N^K C_i^K C_j^K q_\alpha^{ij} + q_\alpha^{nucl} \qquad (4.2)$$

where

$$q_\alpha^{ij} = \int \phi_i^* \left(\frac{3\cos^2\theta - 1}{r_\alpha^3} \right) \phi_j d\tau \qquad (4.3)$$

We now make some simplifying assumptions, in the atomic-core approximation, that all three-centre integrals are ignored and that a Mulliken type relationship holds, namely

$$q_\alpha^{ij} = S_{ij} q_\alpha^{ii} \tag{4.4}$$

The coupling constant is then given by

$$e^2 Q_\alpha q_\alpha = \sum_i f_i \{ e^2 q_\alpha^{ii} Q_\alpha + (e^2 Q_\alpha P / \Sigma f_i) \} \tag{4.5}$$

with

$$f_i = \sum N^K \{ |C_i^K|^2 + C_i^K \sum_{j>i} C_j^K S_{ij} \} \tag{4.6}$$

P is related to the Sternheimer correction. Equation (4.5) enables one to calculate $e^2 Qq$ without recourse to assumptions as to the s-p hybridisation or the extent of π-bonding. Also the theory can yield the value of η. Cotton and Harris computed the ^{35}Cl coupling constants in a series of hexachloro-metallate(IV) ions; these are compared with the observed values in parentheses; $ReCl_6^{2-}$ 32 MHz (27.8 MHz), $OsCl_6^{2-}$ 38 (33.8), $IrCl_6^{2-}$ 44 (41.6), $PtCl_6^{2-}$ 51 (52.0), $PtCl_4^{2-}$ 34 (36.1).

Self-consistent MO calculations including σ electrons were then developed, some by the semi-empirical method and others by *ab initio* treatment.

Brown and Peel[159] calculated various properties including $e^2 Qq$ of molecules containing atoms from the second row by VESCF–MO method. O'Konski and Ha[110], using LCAO–MO–SCF wave functions constructed with Gaussian-lobe basis sets, calculated eq in HCN, N_2, NH_3, CH_3NH_2, NH_2OH and NH_2NH_2 to derive a value of Q (see Table 4.2). They examined the validity of equation (4.4) in the case of NH_3, HCN and CH_3NH_2 and claimed that it is too crude for $e^2 Qq$ calculations. They[160] also carried out an *ab initio* calculation on the systems NH_3 and $H_3N...H^+$ to see the effect of the crystal field and showed that the solid-state shift (0.61 MHz) of NH_3 is dominated by the indirect effect of electron redistribution. Eletr et al.[161] used one-centre expansion functions generated by a linear combination of the Slater-type orbitals in order to calculate $e^2 Qq$ in NH_3. The calculated values (ranging from -0.802 to 1.123 a.u.) are comparable to that of a multi-centre SCF wave function.

The results of *ab initio* SCF calculation were re-interpreted by Bonaccorsi et al.[105] to check the Townes–Dailey theory which employs a localised-orbital scheme. The results appear to be rather pessimistic in that it seems necessary to introduce in the Townes–Dailey theory some new parameters not easily determinable by means of such semi-empirical considerations.

Avgeropoulos and Ebbing[162] calculated the $e^2 Qq$ at the D nucleus in LiD in the CNDO approximation by separating the wave function into a core and a portion surrounding the D nucleus, the latter being determined by means of a single-centre expansion. They obtained 28.2 kHz for $e^2 Qq$ compared with the observed value of 33 ± 1 kHz.

Eletr[163] compared the extended Hückel, the iterative extended Hückel, and the CNDO/2 semi-empirical treatments in calculation of the coupling constants in pyrrole, pyridine and pyrazine and concluded that the second method gave the best value, based on $(e^2Qq)_N = 8.1$ MHz.

For other MO calculations, the reader is referred to references 164–177.

4.3.3 Inorganic compounds

One of the interesting applications of the n.q.r. technique lies in the study of the behaviour of bridge halogens in a dimerised structure. In the $GaCl_3$ dimer[178], the Zeeman study[179] revealed that the Ga atom has $\eta = 0.867$ whereas the bridge chlorine has $\eta = 0.473$. There is a possibility of bent bonds since the principal axes are not consistent with the x-ray diffraction results. Srivastava[180] studied the Cl and Ga resonances in $Et_2O \cdot GaCl_3$. Tong[181] found a linear relationship between the ^{35}Cl frequency in donor–acceptor complexes between $GaCl_3$ and C_5H_5N, Et_2S, Me_2S, PhCN, $POCl_3$, $PhNO_2$, etc. In many of the complexes the temperature coefficients of the Ga frequency were positive. Okuda et al.[182] determined the geometry of the bridged $AlBr_3$ dimer. Maksyutin et al.[183] measured the ^{79}Br and ^{81}Br resonances in complexes of $AlBr_3$ with such π donors as Et_2O, EtOPh, Ph_2O, Me_2S and C_5H_5N and attributed the frequency variation to a charge-transfer interaction. Chihara et al.[4] suggested that the low-temperature phase of $SbCl_5$ had a bridge-type dimer structure. PCl_5, on the other hand, forms either an ionic crystal $(PCl_4)^+(PCl_6)^-$ or a metastable PCl_5 solid, n.q.r. lines of which were assigned[184]. Halides of Group V elements were also studied by Hart and Whitehead[185], Kaplansky, et al.[186], Brill and Long[187], and DiLorenzo and Schneider[188–190].

The ^{27}Al resonances in $Al_2(CH_3)_6$ reported by Dewar and Patterson[191] show that $e^2Qq = 23.546$ MHz and $\eta = 0.784$. The high η value supports the Longuet-Higgins structure in which Al . . . C . . . Al bridges exist rather than the hydrogen-bonded structure in which there are Al—C—H · · · Al bridges. Because of the large η, the $\frac{1}{2} \leftrightarrow \frac{5}{2}$ transition was observed. The adduct between ICl_3 and $AlCl_3$ was shown to have the structure $(ICl_2^+)(AlCl_4^-)$ by Evans and Lo, in agreement with the x-ray diffraction study[192].

Much attention has been given to compounds containing boron–halogen bonds. Smith and Tong[193] applied the Morino–Toyama method for determining η by the Zeeman study on a powdered specimen, to BCl_3 and obtained $\eta = 0.54 \pm 0.02$ (for Cl) consistent with a MO calculation. Casabella and Oja[194] determined e^2Qq of ^{11}B in BF_3 (2.64 MHz), BCl_3(2.54), BBr_3 (2.4), and BI_3 (2.40) by the n.m.r. technique, which did not agree with the values calculated from e^2Qq of the halogens and the B—X properties derived by the Townes–Dailey theory. Ardjomand and Lucken[195] made a systematic study of the ^{35}Cl resonance in complexes between BCl_3 and various organic donors and classified them into three categories according to the bonding of B and the ^{35}Cl frequency. Wrubel and Voitlander[196] compared the Townes–Dailey treatment, the Cotton–Harris treatment, and their own SCF calculation in the interpretation of the ^{35}Cl data in $Cl_2BN(CH_3)_2$ and suggested that the approximation of equation (4.4) gave too low a frequency.

Single-crystal[197, 198] and powder[199] Zeeman techniques have been applied to $HgCl_2$ to determine the η values at ^{35}Cl nuclei but the values obtained varied between 0.08 and 0.8. Theoretical calculation of e^2Qq and η by Brill et al.[200] was consistent with Negita's result[197]. Creel et al.[201] discussed the reasons why the powder Zeeman method frequently gives unreliable results. Complexes which $HgCl_2$ forms have been studied by Brill and Hugus[202, 203]. Brill[204] studied HgX_2 dioxanates and observed the Hg–O interaction decreases in the order $HgCl_2 > HgBr_2 > HgI_2$. Various organo-mercuric halides were studied by Bregadze et al.[205] and Bryukhova et al.[206–208].

Breneman and Willett[209] measured the Br frequencies of the tribromide ions in PBr_7, $CsBr_3$, and NH_4Br_3. For example, the end atom of the ion absorbs at 251.7 MHz in $Cs^{79}Br_3$ at $0 \, ^\circ C$ whereas the central atom absorbs at 407.0 MHz at $10 \, ^\circ C$. The frequency was found to be very sensitive to the short inter-bromine bond length in the tribromide ion. Sasane et al.[210] measured the e^2Qq of I in RbI_3 at 77 K and assigned the values to the three I atoms as follows: I_a 774.2 MHz ($\eta = 0.056$), I_c 2465.7 MHz, I_b 1449.8 MHz ($\eta = 0.058$) in the I_a–I_c–I_b ion. The positive temperature coefficient in NH_4I_3 was explained by weakening of the hydrogen bond due to thermal motion. Bonding in polyhalogens was discussed using a modified Hückel theory by Wiebenga and Kracht[211]. Evans and Lo[192] discussed the bridge structure in $Cl_2ICl_2ICl_2$. NH_4I_3, CsI_3, Cs_2I_8 and NH_4IBr_2 and tetra-alkylammonium trihalides were studied by Bowmaker and Hacobian[212] and they compared the e^2Qq with a simple LCAO calculation. The Cotton–Harris type of calculation was made to deduce the ground state charge distribution in charge-transfer complexes of amines with halogens and polyhalogens, from the n.q.r. measurements[214]. Charge-transfer complexes of ICl as the acceptor were also examined by Maksyutin et al.[213]. Gupta[215] examined ClF_3 and BrF_3 by the Townes–Dailey theory. Semin et al.[216] ascribed the difference in the I frequency (315.625 and 308.834 MHz) between $I(CF_2)_2I$ and $I(CF_2)_2I \cdot NHEt_2$ to charge-transfer complex formation.

The structure of hydrogen dihalide ion, particularly dichloride ion, has attracted interest since the symmetrical structure was substantiated in the difluoride ion. Evans and Lo[217] obtained a single absorption at 20.22 MHz at 77 K c. 50 kHz wide in the case of $(CH_3)_4NHCl_2$, which was interpreted as suggesting an asymmetrical anion structure. A search for absorption in $(C_2H_5)_4NHCl_2$ [218] provided them with a weak 11.89 ± 0.03 MHz line at $26 \, ^\circ C$, indicating that the HCl_2 anion was symmetrical. The Townes–Dailey treatment led to a net charge of $+0.48e$ at H and $-0.74e$ at Cl in the symmetrical case and $+0.43e$ at H and $-0.57e$ or $-0.86e$ at either Cl in the asymmetrical case. However, Chang and Westrum concluded from the zero-point entropy that $(CH_3)_4NHCl_2$ has the symmetrical anion[219]. Haas and Welsh[220] observed a 21.12 MHz line in $(CH_3)_4NDCl_2$ and no other lines down to 5 MHz. They carried out the Cotton–Harris type of analysis to calculate e^2Qq as a function of the Cl—H distance with a fixed $Cl \cdots Cl$ distance (3.2 Å). To account for the observed $e^2Qq = 40.42$ MHz, the Cl—H distance must be taken to be 1.36 Å, pushing the other distance to 1.84 Å and the other resonance to well below 5 MHz. Ludman et al.[221] measured the ^{35}Cl resonance in ClHCl or ClDCl anion in eight compounds and found that all the compounds fell into two classes according to whether the frequency lies near to 12 or 20 MHz.

Evans and Lo[222] obtained an average $e^2Qq = 46.7$ MHz for Cl of iodobenzene dichloride in which iodine has a T-shaped coordination; the net charge at a Cl atom was estimated as $-0.5e$ by the Townes–Dailey treatment.

Tetrahalides of Group IV elements form a variety of intermolecular compounds. In pure chlorides, $CCl_4(<0.10)$, $SiCl_4(0.45)$, $GeCl_4(0.35)$, and $SnCl_4(0.25)$, Graybeal and Green[223] obtained relatively large values of η as given in the parentheses by the method of the powder Zeeman effect. The η values were probably due in part to d_π–p_π bonding except for CCl_4. Gilson and O'Konski[224] studied the charge-transfer complexes of CBr_4 with methyl-substituted benzenes and pyridines. These Br resonances faded at c. $50°C$ below the melting points of the complexes. Bennett and Hooper[225] attempted to find evidence for charge transfer in chloroform complexes with hydrocarbons or amines but the crystal-field effect prevented it being detected. Graybeal et al.[226, 227] investigated a systematic variation of the Cl frequency in organotin chlorides on progressive substitution. Negita, et al.[228–230] determined the geometry of the $2SbCl_3$–benzene and the $2SbBr_3$–benzene complexes by the Zeeman effect and discussed the Sb—Cl and the Sb—Br bond nature. Biedenkapp and Weiss[231] constructed phase diagrams which clearly show the formation of complexes of $AsCl_3$ with aromatic hydrocarbons. Both As and Cl frequencies are reported. Arsenic resonances and their T_1 values were measured at 77, 195, and 300 K by Pen'kov and Safin[232].

Other compounds of Group V elements include P and S compounds by Clipsham et al.[233], As, Sb, and Bi compounds by Brill and Long[1ᶜ7], $(C_6H_5)_4$ $NAsCl_6$ by DiLorenzo and Schneider[190], P, Sb, As, and Pb compounds by DiLorenzo and Schneider[189], $POCl_3$ complexes by Rogers and Ryan[234], $SCl_3·AlCl_4$ by Doorenbos et al.[235], PCl_5, $SbCl_3$ and $SbCl_5$ by Chihara et al.[184, 236, 4], etc.

Several papers appeared which attempted to account for the e^2Qq at Al sites in corundum (α-Al_2O_3). Thus Hafner and Raymond[237] made a calculation by the multipole expansion technique of electrostatic potentials, field and field gradients by adjusting the ionic dipole and quadrupole moments of the oxygen ions using the e^2Qq data. They also showed[238] that the effect of oxygen octapoles must be taken into account unless the Sternheimer factor of Al is taken to be about twice as large as the theoretical value. Taylor[239] showed that the overlap contribution to e^2Qq is quite sizeable, using Clementi's Al functions and Watson's oxygen functions. Sawatzky and Hupkes[240] calculated the e^2Qq by considering the point charges, electric dipoles, and overlap distortion of cation p orbitals, the last named contribution being important to make the calculated Q close to the atomic value. Sharma[241], on the other hand, considered an Al—O_6 complex taking into account the metal ion–ligand overlaps; he obtained $Q = 0.148$ barns compared with $Q = 0.149$ barns from an atomic beam experiment.

Sholl and Walter[242] suggested that the covalent model could be more favourable than the ionic model to interpret the small temperature gradient of the ^9Be frequency in BeO.

Segel et al.[243] measured the e^2Qq (7.91 ± 0.08 MHz) of ^{127}I in hexagonal AgI, which was favourably compared with Sholl's prediction[244] after a minor modification.

Graybeal and McKown[245] were able to detect the ^{137}Ba and the ^{135}Ba resonances in $BaCl_2 \cdot 2H_2O$ at their natural abundances; the e^2Qq value was 32.6 MHz (if $\eta = 0$) or 29.8 MHz (if $\eta = 0.78$ as in $BaBr_2 \cdot 2H_2O$). The latter leads to a charge distribution $Ba^{1\cdot 8+}$ and $Cl^{0\cdot 9-}$.

4.3.4 Transition metal complexes

The bonding between transition metal ions and chlorine is known to be largely ionic. Bersohn and Shulman[246] found a small covalency from the ^{35}Cl frequency by the Townes–Dailey approach and compared that with the value obtained from the study of the transferred magnetic hyperfine interaction when the central ion is paramagnetic. In $CuCl_2 \cdot 2H_2O$, the fractional spin density in the chlorine 3p orbitals was 9.8% and 15.8%, by theory and experiment, respectively, the latter including the effect of point charges. The π bonding using the metal d orbitals contributes significantly to the covalency. Cotton and Harris[247] made a detailed calculation of the electronic ground state of $PtCl_4^{2-}$ by a semi-empirical MO theory, varying the Pt ionisation potentials and wave functions. The calculated Pt—Cl bond order was 0.403 and is almost entirely σ in character (the ionicity was 0.597) in good agreement with deductions from the n.q.r. frequency of ^{35}Cl. They[248] extended the calculation to $ReCl_6^{2-}$, $OsCl_6^{2-}$, $IrCl_6^{2-}$ and $PtCl_6^{2-}$. Participation of d orbitals in the π-bonding between the metal ion and the ligands was also suggested by Brown et al.[249] who correlated the infrared and the n.q.r. spectra of MCl_6^-. Brill and Hugus[250] observed that the ^{59}Co frequency increased by 15% in going from trans-$[Co(en)_2Cl_2]Cl$ to trans-$[Co(en)_2Cl_2]Cl \cdot HCl \cdot 2H_2O$ and by 25% in the corresponding Br compounds. They also found a 10 MHz difference in e^2Qq of ^{59}Co between trans-$[Co(tn)_2Cl_2]Cl \cdot HCl \cdot 2H_2O$ and the corresponding ethylenediamine complex, and here, they attributed the difference to the difference in Co$\cdots$C interaction[251]. From the variation of the ^{59}Co frequency in tetracarbonylcobalt–tin compounds Spencer et al.[252] supported the hypothesis that the populations of the cobalt $3d_{z^2}$, d_{xz}, d_{yz} orbitals are lower than for the $d_{x^2-y^2}$, d_{xy} orbitals. Brown et al.[253] studied an interesting trigonal coordination of Co in $MX_3Co(CO)_4$ where σ and π bonding of MX_3 to Co is responsible for the variation of the e^2Qq from compound to compound. They obtained, from the Cl resonance, strong evidence for the existence of Si—Cl d_π–p_π bonding in the MX_3 portion. Nesmeyanov et al.[254] found an empirical correlation between the ^{55}Mn e^2Qq in cyclopentadienylmanganese carbonyls and the induction σ_I and the conjugation σ_R constants,

$$e^2Qq_{zz} = (63.72 + 2.89\,\sigma_I - 14.18\,\sigma_R) \pm 0.42 \text{ MHz}$$

Brill and Long[255], on the other hand, attempted to find correlations between the e^2Qq and the infrared, n.m.r. and ultraviolet spectra of the same class of compounds.

Williams and Kocher[256] discussed the magnitude of π back-bonding (Sn sp^3 is more important than d orbitals therefore) from the combined results of n.q.r. and Mössbauer spectra.

Fryer and Smith[257] found that the ^{35}Cl frequencies in trans-L_2PtCl_2 complexes are higher than in the cis-complex, where L is $Bu_3^n P$, etc.

Scaife[258] obtained the metal–chlorine distance in MCl_4^- type complexes as a function of the covalency derived from the ^{35}Cl frequency.

Yesinowski and Brown[259] found a *cis–trans* effect in Pt–olefin halide complexes which may be explained by the directional effect of the σ bond and the net charge donation. The ^{35}Cl frequency in solid solutions K_2IrCl_6–K_2PtCl_6 is linearly dependent on the composition[260].

Sasane *et al.*[261] explained the positive temperature coefficient of a ^{35}Cl frequency in $NaAuCl_4 \cdot 2H_2O$ by weakening of the hydrogen bond due to thermal motion. A similar explanation applies to R_2ReX_6 [262]. Schreiner and Brill[263] proposed an interpretation for the *trans* effect of C_2H_4 in Zeise's anion, which was primarily a σ-effect; Pt atoms in *trans*-$PtCl_2XY$ with X and/or Y capable of π-bonding will suffer loss of Pt electron charge by $Pt(d_\pi) \rightarrow X$, Y (π) bonding. This leaves Pt more positive and in turn causes the formation of more covalent Pt—Cl bonds through σ-bonding. In the case of cyanide complexes of Zn, Cd, Cu, Pt, Co, and Hg, Ikeda *et al.*[264] showed that the d_π—p_π bond character is stronger in Pt or Co than in Zn, Cd, Hg or Cu.

4.4 DYNAMICAL PROPERTIES OF CRYSTALS

4.4.1 Temperature and pressure dependence of the resonance frequency

The n.q.r. frequency $v(T)$ tends to decrease, in general, as the temperature rises unless a phase transition discontinues such a gradual change. Wang[30] and Das and Hahn[5] gave general discussion on this subject. The temperature dependence of $v(T)$ was first treated by Bayer[265] who derived an expression for $v(T)$ in terms of an average frequency of libration v_l of the molecule about the principal axes of the e.f.g. tensor. The theory was later extended to include all vibrations by Kushida[266]. Their equations, based on the harmonic approximation, can be fitted to experimental results over relatively small temperature ranges but a detailed examination disclosed their inadequacy in that the value of the moment of inertia chosen to give the best fit differed from those determined from molecular spectroscopic techniques by several orders of magnitude[267–269]. The failure over a wide temperature range was considered to arise from neglect of anharmonicity in the lattice vibrations including the effect of thermal expansion. Brown[270] used a linear temperature dependence of v_l in the case of *p*-dichlorobenzene. While this modification does extend the applicability of the Bayer theory and though it was employed by a number of authors[3, 271–274], the linear temperature dependence of v_e has not been substantiated by other techniques, particularly Raman spectroscopy[275–277], which showed that v_l decreased more rapidly as the temperature increased. Nakamura and Chihara[269] made precise measurements of $v(T)$ of ^{35}Cl and ^{37}Cl in solid chlorine between 20 K and the m.p. and derived 'quasi-harmonic' v_l values as a function of temperature which can reproduce the experimental $v(T)$ through the Bayer formula. The temperature dependent v_l, was later verified by direct observation by Raman spectroscopy[278]. An attempt was made by Stidham[279] to derive the dependence of v_l on k, the wave-number vector, in a one-dimensional crystal by taking account of nearest neighbour interaction. He was able to derive the expression of $\langle \Theta^2 \rangle$, the mean square angular amplitude and show that it was larger than in the assembly of the Einstein oscillators. The calculation was made for p-dichloro-benzene and compared with the spectroscopic results. However, to the reviewers' knowledge, there has been only one experimental investigation dealing with the dispersion relation of v_l; i.e. Teh et al.[280] constructed a full dispersion diagram for all the lattice modes from inelastic neutron scattering of a single crystal of ND_4Cl, which showed that the Einstein oscillator is a good approximation to the librational motion of the ND_4^+ ions.

Usually it is $v(T)$ rather than e^2Qq that is used in fitting the Bayer formula with the assumption that the asymmetry parameter changes little with temperature. Chihara et al.[236] determined e^2Qq and η of ^{121}Sb and ^{123}Sb in $SbCl_3$ both as a function of temperature, showing that part of the temperature dependence of $v(T)$ is attributed to that of η. Wang's theory[30] was then used to derive the anisotropy in the libration frequencies but one of the v_l became imaginary at high temperatures possibly owing to the effect of intermolecular charge transfer. An earlier similar attempt by Lee et al.[281] in the case of ^{139}La in LaF_3 was not very successful in that the moment of inertia values derived were too small by two orders of magnitude although they were able to determine e^2Qq and η separately at different temperatures.

Kushida et al.[282] discussed the pressure dependence $v(P)$ and showed that in Cu_2O q_0 varies with V^{-1} and in p-dichlorobenzene it is proportional to V^n with $0 < n < 0.04$. The value of the exponent n was obtained in the case of $NaBrO_3$. From the ^{23}Na frequencies $v(T)$ and $v(P)$ together with the experimental values of the compressibility and expansivity, Whidden et al.[283] calculated $v^{-1}(\partial v/\partial T)_V$ which was positive and decreased with the temperature. They interpreted this as suggesting that any model based on a single lattice-vibrational mode is not satisfactory. They observed that the ^{23}Na frequency varied with $V^{-2.5}$ whereas Early et al.[284] reported that the ^{79}Br frequency varied with $V^{-0.225}$ in the same compound. This can be understood if the effect of external pressure is to change primarily the distance between the ions rather than the dimensions of the anions. Other studies relating to the temperature dependence of the n.q.r. frequency are listed in the references[285-299].

Barton attempted to derive the height V_0 of the potential barrier hindering the molecular reorientation by treating $dv(T)/dT$ in the harmonic approximation and obtained the V_0 values of 11 kcal mol^{-1} for 1,4-dichlorobut-2-yne[300], 10.5 kcal mol^{-1} for 3-chloroprop-1-yne[301], and 8 kcal mol^{-1} for 1,1,1-trichloro-2,2,2-trifluoroethane[302]. For the last-named compound, the V_0 value was in agreement with that derived from the temperature dependence of T_2^*.

Extreme sensitivity of the n.q.r. to subtle changes in the crystal field was used to study hydrogen-bond systems, particularly their isotope effect. Thus Blinc et al.[303] studied the ^{35}Cl and ^{37}Cl resonances in $KH(CCl_3COO)_2$, $NH_4H(CCl_3COO)_2$, and their deuterated analogues and suggested that the proton or the deuteron rests at the centre of the hydrogen bond (i.e. symmetrical hydrogen bond) if it is not jumping between the double minima at a faster rate than v_Q. They also concluded that the hydrogen-bond potential function is not affected by deuteration. Absence of change upon D substitution in the temperature coefficients of the three n.q.r. lines was taken to indicate that there is no hydrogen bonding in chloral hydrate CCl_3CH $(OH)_2$ [304]. A particular example of a symmetrical hydrogen bond in NH_4H $(CH_2ClCOO)_2$ will be mentioned in a later subsection[305, 306].

4.4.2 Relaxation times

Further insight into molecular motion may be obtained from measurements of the relaxation times. The general theory of quadrupolar spin–lattice relaxation in ionic crystals was developed in the harmonic approximation by Van Kranendonk who showed that two-phonon (Raman) processes are the dominant mechanism of the relaxation if the Debye model is assumed for the lattice frequency spectrum[2]. In the limit of low temperatures, the transition probability W is approximately proportional to T^7 whereas at high temperatures it has a T^2 dependence. The theory was extended to include anharmonic effects both by ordinary perturbation theory and by the method of phonon Green functions[307]. The anharmonic Raman process is a second-order process due to a combination of the linear spin–lattice coupling and the cubic anharmonic forces. An order-of-magnitude estimate showed that the anharmonic Raman mechanism is 100 times more effective than the first-order (harmonic) Raman process. The theory was later applied by Armstrong and Jeffrey in their study of K_2PtCl_6, Rb_2PtCl_6 and Cs_2 $PtCl_6$ [308]. Bridges[309] modified the Van Kranendonk theory to apply to the diamond lattice by employing a realistic density of phonon states obtained by expanding the Hamiltonian in terms of normal coordinates and dividing Phillips' acoustical phonon spectrum into the transverse and the longitudinal phonons. The calculation agrees with the experimental data[310] for ^{123}Sb in InSb over five orders of magnitude in $1/T_1$ (up to 60 K). The relaxation mechanism arising from molecular librations was first discussed by Bayer[265] and further developed by Woessner and Gutowsky[311], who examined the ratio W_2/W_1 of the transition probabilities in p-dichlorobenzene to assess the relaxation mechanisms. Estimation of W_1 and W_2 is very difficult but experimental methods of determining W_1 and W_2 were discussed in the cases of $I = 5/2$ and $7/2$ [312–314]. The effect of slow reorientation or hindered rotation of molecules and complex ions on the quadrupole relaxation in solids was extensively discussed by Alexander and Tzalmona[315] and also has recently been examined by Russian workers[316–319] who sought 'spectral parameters' which relate the modes of molecular motion and relaxation characteristics.

Hexahalide complexes of the type A_2BX_6 have been mentioned in Section 4.3.4, in relation to the study of the bond nature. The dynamic properties of these salts were extensively studied by Armstrong and co-workers. Thus, from the T_1 and T_2 measurements on K_2PtCl_6, Jeffrey and Armstrong[320] were able to show that the relaxation was dominantly quadrupolar from the ratio of T_1 values for ^{35}Cl and ^{37}Cl and that T_1 was accounted for by the Van Kranendonk theory with a single Einstein frequency of 33 cm^{-1} (which agreed with 38 cm^{-1} derived by use of the Bayer formula between 30 and 90 K). T_1 decreased very rapidly above 320 K, indicating that the reorientation was the dominant mechanism with the barrier of 9.45×10^3 K or 18.8

kcal mol^{-1} hindering the rotation around the fourfold axis of a $PtCl_6^-$ anion. The T_2 data are dominated by a magnetic dipolar spin–spin interaction as verified by the observation of the magnetic field dependence of the spin-echo beats. They also studied the pressure dependence of the resonance frequency $v(T,P)$ and $T_1(T,P)$ at different temperatures[321], from which the lower limit of the activation energy of 12.6 kcal mol^{-1} was derived. The treatment of molecular rotation by Anderson and Slichter[322] in terms of a simple classical model of an activated process was applied to derive the activation volume ΔV^* from the pressure dependence of T_1. The plot of $\log(1/T_1)$ v. P up to 4000 kg cm^{-2} gave $\Delta V^* = 25.6$ cm^3 mol^{-1} independent of temperature for the hindered rotation of a $[PtCl_6]^{2-}$ group in the cage of K^+ cations. The effect of pressure was considered to be such as to change the dimensions of the cages formed by the K^+ ions. Similar analysis was made for the K_2PdCl_6 data in which case the F_{1g} rotary lattice mode of frequency (~41 cm^{-1}) is responsible for the temperature dependence of v and T_1 [323]. O'Leary and Wheeler[324, 325] developed a soft librational-mode theory to account for the measurements of $v(T)$ of K_2ReCl_6. It has a f.c.c. structure at room temperatures and is transformed into a tetragonal phase at 110.9 K where the single n.q.r. line splits into two lines having an intensity ratio of 2:1. In the high-temperature phase, $v(T)$ continues to increase with temperature between 110.9 and 300 K. O'Leary first calculated the contribution of internal modes of vibration to $v(T)$. Subtracting their effect from the observed $v(T)$ above the transition point gave the remainder which is regarded as the part Δv_l that arises from the librational mode which becomes soft at the transition point. The square of the soft mode frequency ω^{-2} was then calculated from $\omega^{-2} = 3kTv(0)/I\Delta v_l$, where $v(0)$ is the resonance frequency at 0 K and I is the moment of inertia. ω^{-2} is very nearly linear in T in agreement with the classical predictions due to Landau[326]. It was estimated that, if account is taken of the small density of states available near the zero wave-number vector, the average frequency $\overline{\omega}$ translates into a librational-mode frequency which changes from ~25 cm^{-1} at 300 K to ~1 cm^{-1} at 112 K. Investigations along similar lines (the multiple-mode analysis) were reported by Armstrong et al.[327] for K_2PtCl_6 and K_2PdCl_6 for which libration frequencies were derived as 58 and 64 cm^{-1}, respectively in good agreement with the normal-mode calculations[328]. This indicates the usefulness of the n.q.r. technique in the study of low-lying librational levels which are rarely accessible by optical spectroscopy because of symmetry restrictions. Armstrong[329] extended his argument to the pressure dependence of the infrared-active vibrations Q_3, Q_4, Q_5, and Q_6 of K_2PtCl_6 and was able to show that the pressure coefficients $\delta\omega^j/\delta P$ derived from the n.q.r. measurements were similar to those derived directly from optical spectroscopy.

To interpret the temperature dependence of v of K_2IrCl_6 a multiple-mode analysis was adapted to include the effect of the destruction of π bonding by the lattice vibrations[330]. The combined pressure and temperature dependences of K_2IrCl_6 were analysed in the light of Kushida–Benedek–Bloembergen theory[282], showing the importance of the volume effect. The ^{35}Cl and ^{37}Cl relaxation times were also measured[331]. At temperatures below 40 K, T_1 is determined by the hyperfine interaction between the Cl nuclear spins and the electron spins of the Ir^{4+} ions. Above 40 K, quadrupolar relaxation mechanisms which result from libration or hindered rotational motions of the $[IrCl_6]^{2-}$ ions dominate. The activation energy of hindered rotation was determined from T_1 data above 350 K as 18.8 ± 0.5 kcal mol^{-1}, identical with the K_2PtCl_6 value. From the measurement of the temperature (130–350 K) and the pressure (1–5000 kg cm^{-2}) dependence of $v(T,P)$ of K_2ReCl_6, Armstrong et al.[332] suggested that a self-consistent picture of the n.q.r. frequency can be obtained by invoking the destruction of the π bonding by lattice vibration; the contribution of the soft librational mode was rather small. From thermodynamic arguments Armstrong and Cooke[333] derived $(\partial v/\partial T)_V$ from $(\partial v/\partial T)_P$ and $(\partial v/\partial P)_T$. This was then used to obtain the average librational frequency by the multiple-mode analysis and compared with theoretical calculations[328]. The result was consistent with the relaxation time results[334]. A number of other ABX_6 or A_2BX_6 complexes were studied by Brown and Kent[335], where A = K$^+$, Rb$^+$, Cs$^+$, B = W, Re, Os, Pt, Sn, Te, Mo, Nb, Ta, and X = ^{35}Cl, ^{81}Br. They compared (dv/dT) at 300 K and found that it was linearly related to d^n, the occupancy of the metal T_{2g} orbital.

^{63}Cu and ^{65}Cu resonances in Cu_2O[336] were also explained by relaxation due to two-phonon anharmonic Raman processes above 20.4 K. The T_1 data under pressure can be explained either by harmonic or anharmonic Raman processes if the Grüneisen constant is taken to be negative[337].

Nitrogen resonances and relaxations in some organic crystals were studied by Alexander, Tzalmona, Zussman and others. Triethylenediamine[338], which has a transition point at 351 K into a plastic crystal phase, shows $1/T_1$ linear in T^2 below 140 K, consistent with the relaxation mechanism by indirect two-phonon processes. Between 140 and 320 K, T_1 shows a minimum at c. 264 K; the dominant mechanism here is the modulation of the nitrogen–hydrogen dipolar interaction by molecular reorientation about the figure axis of the molecule. Since this is the threefold axis, the correlation function has the form, $G(t) = \exp(-3t/\tau_j)$, $1/\tau_j$ being the jump rate between the three equivalent minima, which then leads to $(T_1^{min})^{-1} = 3W_{min} = (81/8)\,(\gamma_N^2\gamma_H^2 \hbar^2/\omega_Q r_{NH}^6)$. Upon substitution of $\omega_Q = 3.665 \times 10^6$ and $r_{NH} = 2.09$ Å, T_1^{min} is obtained as 62×10^{-3} s to be compared with the experimental value of 48×10^{-3} s. The discrepancy may be attributed to the large value of the N—H distance employed. From the temperature-dependence of T_1, Zussman and Alexander derived the activation energy $\Delta E = 8.17 \pm 0.3$ kcal mol^{-1} with $\tau_0 = 7.4 \times 10^{-15}$ s. This can be favourably compared with $\tau_0 = 1.6 \times 10^{-13}$ s and $\Delta E = 7.2 \pm 0.7$ kcal mol^{-1} obtained from the proton n.m.r. line narrowing at 190 K by Smith[339]. This is an interesting example of n.q.r. used to study dipolar interactions. The line width of triethylenediamine defined by $\Delta v = (\pi T_2)^{-1}$ increased from 400 Hz at 80 K to 800 Hz at 290 K, primary contributions to which come from dipole–dipole interactions. A rough estimation gives 45 Hz for N—H, 210 Hz for H—H, and 85 Hz for N—N dipolar interactions. The sum accounts for c. 330 Hz of the width but its increase with temperature remains to be explained. In acetonitrile, CH_3CN[340], two n.q.r. transitions were observed which disappear at 219 K and reappear as a single line at the m.p. (227 K). The relaxation time T_1 has a minimum of 0.1 s at 88.5 K and is explained by the rotation of the CH_3 group about an axis tilted to the C—C≡N axis. Its activation energy was

2.2 kcal mol^{-1}. The effective charge that rotates with the CH_3 group was estimated to be 0.044 e from the T_1 minimum and 0.041 e from the separation of the two n.q.r. frequencies (8.4 kHz). Hydrazine, NH_2NH_2 [341], is a complex crystal in that there are two inequivalent nitrogen sites, each with three energy levels which give rise to six (three pairs) n.q.r. frequencies. The free decay times T_2 are the same for all n.q.r. transitions, being essentially constant below 230 K and decreasing very rapidly above 240 K. In the latter region the decay signal is exponential and gives an activation energy of 16.5 ± 2 kcal mol^{-1}. Saturation-type measurements using repetition of identical 90 degree pulses gave T_1 values of the six n.q.r. lines, of which two pairs of the highest and the lowest frequencies have the same T_1 whereas one pair of the middle frequency have T_1 about twice as long. All show a common temperature dependence leading to the activation energy of 14.3 ± 0.4 kcal mol^{-1} above 210 K. Cross-relaxation measurements were carried out in which one of the three lines at a site is observed after the corresponding line at the other site is saturated. The results were all symmetric between the two sites and an enhancement (Overhauser effect) was observed in the middle frequency pair. All the ratios of the signal amplitudes were independent of temperature. These experimental observations were interpreted as indicating that the relaxation is dominated by the molecular motion by which N nuclei interchange between the two sites. The cross-relaxation and T_1 anisotropy results permitted the determination of the relative orientation of the principal axis systems of the electric field gradient tensor at the two sites.

Using the example of piperazine, for which Barton[342] developed a theory to account for both $v(T)$ and $T_1(T)$ in terms of molecular librations only, Tzalmona[343] demonstrated the usefulness of double-resonance techniques and relaxation measurements in the study of anisotropic molecular rotation. He presented a formulation which relates the transition probabilities with the geometrical parameters representing the two configurations of the principal axes of the field gradient tensor at the N nuclei. The activation energy for the motion is 7 kcal mol^{-1}.

In the case of p-chlorobenzotrichloride[83], $ClC_6H_4 \cdot CCl_3$, $1/T_1$ of ^{35}Cl in the CCl_3 group behaves differently from $1/T_1$ of the Cl atom at the $para$ position in that the former shows a drastic shortening of T_1 above 200 K in contrast to the behaviour of the latter which shows the usual $1/T_1 \propto T^2$ dependence. The abnormal decrease in T_1 is interpreted by the hindered rotation of the CCl_3 group with respect to the benzene nucleus with an activation energy of 10.9 ± 0.3 kcal mol^{-1}, which may be compared with 9.2 kcal mol^{-1} deduced from the temperature dependence of the n.q.r. frequency.

4.5 PHASE TRANSITIONS

The extreme sensitivity of the n.q.r. to the nuclear environment has been utilised in detecting and studying phase transitions in the solid state. Thus a change in the number of resonance lines at a transition temperature T_c simply indicates occurrence of a phase transition and the way in which the frequencies change often helps one to decide the order of the transition. The relaxation measurements give clues to the mechanism by which the molecular motion takes part in the co-operative phenomenon.

Zussman and Alexander[344] studied the complicated phase transitions in malononitrile by the ^{14}N n.q.r. and relaxation measurements. Below 140 K and above 294.55 K, there are four n.q.r. lines, each of which is split into two between these temperatures. The magnitude of the splitting Δv near 294 K was found to obey the relation

$$\Delta v = D(T_c - T)^\beta$$

and $\beta = 0.5 \pm 0.02$ in accordance with the Landau theory[326]. The line width has a critical coefficient $\gamma = 0.3 \pm 0.07$. The activation energy above 250 K was $c.$ 800 K from the T_1 measurements. The available evidence suggests that these transitions are of displacive type. There was a fourth metastable phase reached by a higher-order transition from the middle phase.

The phase transition of chloranil at 95 K was interpreted by the soft librational-mode theory by Chihara et al.[345].

Clément et al.[346] found a phase transition at $c.$ 131 K in a cyclohexane–thiourea inclusion compound by ^{14}N resonance and differential thermal analysis. It is interesting to note that the change seems to occur in the cyclohexane moiety rather gradually (by n.m.r.) and in the thiourea moiety rather abruptly (by n.q.r.).

Ferroelectric HCl was studied by O'Reilly[347] by the ^{35}Cl T_1 measurements which showed an anomalous temperature dependence and was interpreted by the attenuated 90 degree flip process of molecular axes with an activation energy of 2.3 kcal mol^{-1} both for HCl and DCl. The internal local electric field in the ferroelectric phase was calculated by the Fourier transformation summation method. The direction of the field deviates from the molecular axis by as much as 30 degrees. The transition mechanism involves the 90 and 180 degree flip of the molecules. Okuma et al.[348] made detailed measurements of the temperature dependence of the Cl frequencies in HCl and DCl as well as ^{1}H n.m.r. and found that the Bayer theory fails to explain the temperature dependence. They observed the motional narrowing of the Cl n.m.r. in the high-temperature phase. Kadaba and O'Reilly[349] ascribed the rapid increase in the line width above 45 K for HBr and above 65 K for DBr to the 90 degree flip process in the ferroelectric phase as supported by the ^{2}H n.m.r. The activation energy for the process (1.2 kcal mol^{-1}) agrees with what was obtained by the dielectric study. A similar observation was also reported for HI and DI[350].

A theoretical paper by Stankowski[351] showed that in ferroelectric crystals the temperature dependence of the n.q.r. frequency must have a term that takes account of the linear deformation of the field gradient, proportional to the spontaneous polarisation. The theory was fitted nicely to the experimental results of Oja et al.[352] for the ^{14}N resonance in NaNO$_2$. Bonera et al.[353] derived expressions for T_1 near a ferroelectric phase transition which show that T_1^{-1} should behave as $(T - T_c)^{-n}$ for the case of undamped soft-phonon modes, where n depends on the shape of the dispersion curve of the phonon, whereas T_1^{-1} should vary as $(T - T_c)^{-\frac{1}{2}}$ or as $\ln(T - T_c)$ for the case of damped oscillatory modes or diffusive modes. They also measured T_1

of ^{23}Na in $NaNO_2$ single crystals and proposed the existence of either a damped generalised soft mode or critical diffusive mode. The shape of the relaxation rate peak tends to favour the logarithmic singularity suggesting an anisotropic dipolar interaction. Ikeda et al.[354] determined the three components of the e^2Qq tensor of ^{14}N in $NaNO_2$ and found that the mean square amplitude of the libration is greater about the y-axis (lying in the ionic plane and perpendicular to the bisecting z-axis) than about the other axes. The N atom shares $-0.37e$ and O atom has $-0.32e$ of the electronic charge. However, Marino and Bray[99] have reported, from the ^{14}N resonance, that the charge on the N atom was $0 \pm 0.2e$. Marino et al. studied other nitriles in a similar way[355]. The line-width measurements[356] of Kadaba et al. for the ^{14}N resonance in $NaNO_2$ was explained by the order–disorder mechanism of the flipping of the ion between the two equilibrium sites. Betsuyaku[357] calculated the three components of the e^2Qq of ^{14}N by the CNDO approximation which agreed reasonably with experiment[352, 354]. He also calculated the ^{23}Na field gradient by the point charge multipole model[358].

Zhukov et al.[359, 360] found an interesting isotope effect in the ^{75}As n.q.r. of MH_2AsO_4 (M = Li, K, Na, Rb, Cs, Ag, NH_4 and $(CH_3)_4N$): on deuteration the difference in frequencies of K, Rb, Cs and NH_4 salts which are ferroelectric at 77 K vanished completely. T_1 and T_2 were also measured at 77 K.

Oja[361] observed the fading-out of the ^{14}N resonance of some ferroelectric guanidines at about 130 K where the ^2H n.m.r. indicated the onset of hindered rotation of the guanidinium ion $C(NH_2)_3^+$.

Papon and Theveneau[362] derived an expression for T_1 in ferroelectrics assuming that there is a soft-mode coupled to the acoustical phonons through a cubic anharmonic potential. This gives a reasonable agreement for T_1 of $NaNO_2$ though it leads to a different power law.

Yamamoto et al.[305, 306] measured the ^{35}Cl frequency between 20 and 300 K of $NH_4H(ClCH_2COO)_2$ which is ferroelectric below 120 K. The separation between the two frequencies below 120 K decreases with temperature vanishing at T_c as does the spontaneous polarisation.

The Sb, Br and I resonances in SbSI and/or SbSBr were studied by Popov et al.[363–365]. Helg and Graenicher[366] found a peculiar ferroelectricity in KIO_3 (non-invertible but tiltable polarisation). Interesting but complicated multiple transitions were found by Tovborg-Jensen[367] in $CsPbCl_3$ by the ^{35}Cl n.q.r. For the orange modification of $CsPbBr_3$, Volkov et al.[368] discovered a phase transition at 167 K; another at 403 K was suggested to be the first-order transition. In $TlAsSe_2$ a transition occurs near 153 K if the crystal is grown from the glass in the absence of air[369].

Other phase transitions were detected in superconducting and normal La metal[370], trichlorophosphosulphuryl chloride[371], C_6Cl_5F and C_6Br_5F [372], $HGeCl_3$ and $EtSiCl_3$ [373], p-halogenophenols[374], $KAuBr_4 \cdot 2H_2O$ [375], AuI [376] and $(NH_4)_2PtI_6$ and Rb_2TeI_6 [377].

4.6 MISCELLANEOUS STUDIES

A number of problems relevant to n.q.r. spectroscopy have hitherto been left unmentioned. These will be enumerated and only some of the references will be given. The Sternheimer shielding and anti-shielding factors have been discussed by Lyons et al.[378], Ghatikar[379], Sternheimer[380-382], Rao and Murty[383], Gupta et al.[384] and Sternheimer and Peierls[385]. Experimental determinations of the nuclear quadrupole moments were reported for ^{23}Na[386], ^{39}K and ^{41}K[387], ^{51}V[388, 389], ^{55}Mn[390], ^{59}Co[391, 392], ^{57}Fe[393], ^{61}Ni[394], ^{79}Br and ^{81}Br[395, 396], ^{135}Ba[397], ^{139}La[398], ^{155}Gd[399], ^{159}Tb[400-402], ^{185}Re and ^{187}Re[403-405], ^{189}Os[406, 407] and ^{209}Bi[408]. Study of magnetic materials at low temperatures may be aided by n.q.r. in certain cases. Because space is limited and a separate chapter is prepared for electron spin resonance, we would omit description of them. The type of information the n.q.r. method can provide will be seen in the paper by Rinneberg et al.[409]. Solid solutions such as CCl_4 in CBr_4 were studied over a range of concentration by Alderdice and Iredate[410].

Acknowledgement

The reviewers express their gratitude to many authors in the field of nuclear quadrupole resonance for their willingness in sending them reprints and preprints to be included in this chapter. They apologise for not having mentioned all of the papers owing to limited space. Thanks are due to Miss Atsuko Ogura for her help in preparing the manuscript.

References

1. Dehmelt, H. G. and Kruger, H. (1950). Naturwissenschaften, 37, 111
2. Van Kranendonk, J. (1954). Physica, 20, 781
3. Matzkanin, G. A., O'Neal, T. N., Scott, T. A. and Haigh, P. J. (1966). J. Chem. Phys., 44, 4171
4. Chihara, H., Nakamura, N., Okuma, H. and Seki, S. (1968). Bull. Chem. Soc. Japan, 41, 1809
5. Das, T. P. and Hahn, E. L. (1958). Nuclear Quadrupole Resonance Spectroscopy. (New York and London: Academic Press)
6. Lucken, E. A. C. (1969). Nuclear Quadrupole Coupling Constants. (New York and London: Academic Press)
7. Weiss, A. (1967). Proc. Colloq. AMPERE, 14th, Ljubljana, 1076. (Amsterdam: North-Holland Publ.)
8. Kubo, M. and Nakamura, D. (1966). Advan. Inorg. Chem., 8, 257
9. Smith, J. A. S. (1971). J. Chem. Educ., 48, 39, A77, A147, A243
10. Biryukov, I. P. (1966). Latv. PSR. Zinat. Acad. Vestis, Fiz. Teh. Zinat Ser, 22
11. Mairinger, F. (1966). Oesterr. Chem. Zt., 67, 279; Idem. (1967). Allg. Prakt. Chem., 18, 71
12. Brame, E. G., Jr. (1967). Anal. Chem., 39, 918
13. Shultz, H. D. and Karr, C. Jr. (1969). Anal. Chem., 41, 661
14. Brown, E. (1968). Ann. Chim., 3, 323
15. Wendling, E. (1968). Bull. Soc. Chim. Fr., 181
16. Bronswyk, W. Van (1970). Structure and Bonding, 7, 87. (Berlin: Springer-Verlag)
17. Grechishkin, V. S. and Ainbinder, N. E. (1965). Usp. Fiz. Nauk, 91, 645
18. Penkov, I. N. and Safin, I. A. (1966). Isv. Acad. Nauk SSSR, Ser. Geol., 31, 41
19. Bizot, D. (1969). Rev. Chim. Miner., 6, 1007
20. Bose, M. (1969). Progr. Nucl. Mag. Resonance Spectrosc., 4, 335
21. Maksyutin, Yu. K., Gur'yanova, E. N. and Semin, G. K. (1970). Usp. Khim., 39, 727

22. Schempp, E. and Bray, P. J. (1970). *Physical Chemistry, An Advanced Treatise,* D. Henderson, ed., **4,** Chap. 11. (New York and London: Academic Press)
23. Fedin, E. I. and Semin, G. K. (1964). In *The Mössbauer Effect and its Applications to Chemistry.* V. I. Gol'danskii, ed. (New York: Consultants Bureau)
24. Segel, S. L. and Barnes, R. G. (1965). *Catalog of Nuclear Quadrupole Interactions and Resonance Frequencies in Solids,* USAEC Rept. IS-520, Part I and II
25. Biryukov, I. P., Voronkov, M. G. and Safin, I. A. (1969). *Tables of Nuclear Quadrupole Resonance Frequencies.* J. Schmorak, transld. (Jerusalem: Israel Program for Scientific Translations)
26. Livingston, R. and Zeldes, H. (1955). *Table of Eigenvalues for Pure Quadrupole Spectra, Spin 5/2,* ORNL Report 1913. (Oak Ridge, Tenn.: Oak Ridge National Laboratory)
27. Chihara, H. and Nakamura, N. (1968). *Table of Eigenvalues for Pure Quadrupole Spectra, Spin 7/2.* Unpublished, see Appendix in Ref. 236.
28. Borsutskii, Z. R., Grechishkin, V. S., Zakharov, E. A., Izmestev, I. V. and Shishkin, E. M. (1966). *Tr. Estestv. Nauch. Inst. Perm. Univ.,* **11,** 57
29. Gill, D., Hayek, M., Alon, Y. and Simievic, A. (1967). *Rev. Sci. Instr.,* **38,** 1588
30. Wang, T. C. (1955). *Phys. Rev.,* **99,** 566
31. O'Konski, C. T. and Cartwright, B. G. (1968). *Rev. Sci. Instr.,* **39,** 123
32. Viswanathan, T. L. Viswanathan, T. R. and Sane, K. V. (1968). *Rev. Sci. Instr.,* **39,** 472
33. Peterson, G. E. and Bridenbaugh, P. M. (1964). *Rev. Sci. Instr.,* **35,** 698
34. Bridenbaugh, P. M. and Peterson, G. E. (1965). *Rev. Sci. Instr.,* **36,** 702
35. Graybeal, J. D. and Croston, R. P. (1966). *Rev. Sci. Instr.,* **37,** 376
36. Smith, J. A. S. and Tong, D. A. (1968). *J. Sci. Instr., Ser. 2,* **1,** 8
37. Tong, D. A. (1968). *J. Sci. Instr., Ser. 2,* **1,** 1153
38. Ashley, A. and Tong, D. A. (1970). *Proc. 16th Colloq. AMPERE, Bucharest,* Sept., 1
39. Tong, D. A. (1967). *J. Sci. Instr.,* **44,** 875
40. Muha, G. M. (1968). *Rev. Sci. Instr.,* **39,** 416
41. Graybeal, J. D. and Croston, R. P. (1967). *Rev. Sci. Instr.,* **38,** 122
42. Clarkson, T. S. and Sullivan, E. P. A. (1968). *Aust. J. Chem.,* **21,** 2141
43. Doolan, K. R. and Hacobian, S. (1970). *Aust. J. Chem.,* **23,** 653
44. Caldwell, R. A. and Hacobian, S. (1970). *Aust. J. Chem.,* **23,** 1321 (1970)
45. Negita, H. (1966). *J. Chem. Phys.,* **44,** 1734
46. Barton, B. L. (1966). *Rev. Sci. Instr.,* **37,** 605
47. Won Choi, Q. and Czae-Myungzoon (1966). *Daehan Hwahak Hwoejee,* **10,** 143
48. Blum, H. (1966). *Rev. Sci. Instr.,* **37,** 1412
49. Colligiani, A. (1967). *Rev. Sci. Instr.,* **38,** 1331
50. Faulkner, E. A. and Holman, A. (1967). *J. Sci. Instr.,* **44,** 391
51. Feng, S.-Yu (1969). *Rev. Sci. Instr.,* **40,** 963
52. Idoine, J. D., Jr. and Brandenberger, J. R. (1971). *Rev. Sci. Instr.,* **42,** 715
53. Grechishkin, V. S. and Shishkin, E. M. (1969). *Fiz. Tver. Tela,* **11,** 893
54. Gushchin, S. I., Shishkin, V. A. and Derendyaev, B. G. (1967). *Prib. Tekh. Eksp.,* **5,** 195
55. Abe, Y. (1967). *J. Phys. Soc. Japan,* **23,** 51
56. Lowe, I. J. and Tarr, C. E. (1968). *J. Sci. Instr., Ser. 2,* **1,** 604
57. Burnett, L. J. and Harmon, J. F. (1968). *Rev. Sci. Instr.,* **39,** 1226
58. Ainbinder, N. E., Grechishkin, V. S., Gordeev, A. D. and Osipenko, A. N. (1968). *Fiz. Tver. Tela,* **10,** 2026
59. Pryer, C. W. (1969). *J. Sci. Instr., Ser. 2,* **2,** 230
60. Kesselring, P. and Gautschi, M. (1967). *J. Sci. Instr.,* **44,** 911
61. Utton, D. B. (1967). *J. Res. Nat. Bur. Stand.,* **A71,** 125
62. Peterson, G. E. and Bridenbaugh, P. M. (1967). *Rev. Sci. Instr.,* **38,** 387
63. Buijs, B., De Wildt, J. L., Vega, A. J. and De Wijn, H. W. (1966). *J. Sci. Instr.,* **43,** 330
64. Tong, D. A. (1968). *J. Sci. Instr., Ser. 2,* **1,** 1162
65. Muha, G. M. (1968). *Rev. Sci. Instr.,* **39,** 916
66. Carpenter, B. J. and Forman, R. A. (1970). *J. Phys., E., Sci. Instr.,* **3,** 922
67. Tong, D. A. (1969). *J. Phys., E., Ser. 2,* **2,** 906
68. Smith, G. W. (1966). *Phys. Rev.,* **149,** 346
69. Leppelmeier, G. W. and Hahn, E. L. (1966). *Phys. Rev.,* **142,** 179
70. Slusher, R. E. and Hahn, E. L. (1968). *Phys. Rev.,* **166,** 332
71. Utton, D. B. (1969). *Metrologia,* **3,** 98
72. Brodskii, A. D. and Vorob'ev, I. V. (1967). *Ismer. Iekh.,* 39

73. Labrie, R., Infantes, M. and Vanier, J. (1971). *Rev. Sci. Instr.*, **42,** 26
74. Frisch, R. C. and VanderHart, D. L. (1970). *J. Res. Natl. Bur. Stand.*, **74C,** 3
75. Townes, C. H. and Dailey, B. P. (1949). *J. Chem. Phys.*, **17,** 782. See also ref. 5, p.119
76. Rama Rao, K. V. S. and Murty, C. R. K. (1968). *J. Phys. Soc. Japan*, **25,** 1424
77. Bucci, P., Cecchi, P. and Colligiani, A. (1969). *J. Chem. Phys.*, **50,** 530
78. Peterson, G. E. and Bridenbaugh, P. M. (1967). *J. Chem. Phys.*, **46,** 2644
79. Peterson, G. E., Steed, N. and Bridenbaugh, P. M. (1967). *J. Chem. Phys.*, **47,** 2262
80. Ambrosetti, R., Bucci, P., Cecchi, P. and Colligiani, A. (1969). *J. Chem. Phys.*, **51,** 852
81. Kantimati, B. (1968). *Curr. Sci.*, **37,** 371
82. Kamishina, Y. (1971). *J. Phys. Soc. Japan*, **31,** 242
83. Kiichi, T., Nakamura, N. and Chihara, H. (1972). *J. Mag. Res.* To be published.
84. Rama Rao, K. V. S. and Murty, C. R. K. (1966). *Curr. Sci.*, **35,** 252
85. Rama Rao, K. V. S. and Murty, C. R. K. (1966). *J. Phys. Soc. Japan*, **21,** 1627
86. Peneau, A. and Guibé, L. (1968). *C. R. Acad. Sci. Paris, Ser. A, B.*, **226B,** 1321
87. Kantimati, B. (1969). *Curr. Sci.*, **38,** 335
88. Angelone, R., Cecchi, P. and Colligiani, A. (1970). *J. Chem. Phys.*, **53,** 4096
89. Ambrosetti, R., Angelone, R., Cecchi, P. and Colligiani, A. (1971). *J. Chem. Phys.*, **54,** 2915
90. Kantimati, B. (1967). *Curr. Sci.*, **36,** 40
91. Rao, K. K. and Murty, C. R. K. (1966). *J. Phys. Soc. Japan*, **21,** 2725
92. Rama Rao, K. V. S. and Murty, C. R. K. (1966). *Phys. Lett.*, **23,** 404
93. Sasikala, D. and Murty, C. R. K. (1967). *J. Phys. Soc. Japan*, **23,** 139
94. Lucken, E. A. C. and Mazeline, C. (1968). *J. Chem. Soc. A*, 153
95. Kato, Y., Furukane, U. and Takeyama, H. (1959). *Bull. Chem. Soc. Japan*, **32,** 527
96. Lehrer, S. S. and O'Konski, C. T. (1965). *J. Chem. Phys.*, **43,** 1941
97. Schempp, E. and Bray, P. J. (1967). *J. Chem. Phys.*, **46,** 1186
98. Kern, C. W. (1967). *J. Chem. Phys.*, **46,** 4543
99. Marino, R. A. and Bray, P. J. (1968). *J. Chem. Phys.*, **48,** 4833
100. Guibé, L. and Lucken, E. A. C. (1968). *Molec. Phys.*, **14,** 79
101. Schempp, E. and Bray, P. J. (1968). *J. Chem. Phys.*, **49,** 3450
102. Colligiani, A., Guibé, L., Haigh, P. J. and Lucken, E. A. C. (1968). *Molec. Phys.*, **14,** 89
103. Ikeda, R., Noda, S., Nakamura, D. and Kubo, M. (1971). *J. Mag. Res.*, **5,** 54
104. Schempp, E. (1971). *Chem. Phys. Lett.*, **8,** 562
105. Bonaccorsi, R., Scrocco, E. and Tomasi, J. (1969). *J. Chem. Phys.*, **50,** 2940
106. Bassompièrre, A. (1955). *Discuss. Faraday Soc.*, **19,** 260; *C. R. Acad. Sci. Paris*, **240,** 285
107. Lin, C. C. (1960). *Phys. Rev.*, **119,** 1027
108. Kato, Y. (1961). *J. Phys. Soc. Japan*, **16,** 122; *J. Chem. Phys.*, **34,** 619
109. Kern, C. W. and Karplus, M. (1965). *J. Chem. Phys.*, **42,** 1062
110. O'Konski, C. T. and Ha, T.-K. (1968). *J. Chem. Phys.*, **49,** 5354
111. Ikeda, R., Nakamura, D. and Kubo, M. (1966). *J. Phys. Chem.*, **70,** 3626
112. Onda, S., Ikeda, R., Nakamura, D. and Kubo, M. (1969). *Bull. Chem. Soc. Japan*, **42,** 2740
113. Guibé, L. and Lucken, E. A. C. (1966). *C. R. Acad. Sci. Paris*, **263,** *B*185
114. Ikeda, R., Onda, S., Nakamura, D. and Kubo, M. (1968). *J. Phys. Chem.*, **72,** 2501
115. Guibé, L., Linscheid, P. and Lucken, E. A. C. (1970). *Molec. Phys.*, **19,** 317
116. Schempp, E. and Bray, P. J. (1968). *J. Chem. Phys.*, **48,** 2380
117. Guibé, L. and Lucken, E. A. C. (1968). *Molec. Phys.*, **14,** 73
118. Schempp, E. and Bray, P. J. (1968). *J. Chem. Phys.*, **48,** 2381
119a.Clementi, E., Clementi, H. and Davis, D. R. (1967). *J. Chem. Phys.*, **46,** 4725
119b.Clementi, E. (1967). *J. Chem. Phys.*, **46,** 4731
119c.Clementi, E. (1967). *J. Chem. Phys.*, **46,** 4737
120. Marino, R. A., Guibé, L. and Bray, P. J. (1968). *J. Chem. Phys.*, **49,** 5104
121. Colligiani, A., Ambrosetti, R. and Angelone, R. (1970). *J. Chem. Phys.*, **52,** 5022
122. Marino, R. A. and Oja, T. (1970). *Chem. Phys. Lett.*, **4,** 489
123. Andersson, L.-O., Gourdji, M., Guibé, L. and Proctor, W. G. (1968). *C. R. Acad. Sci. Paris*, **267,** 803
124. Blinc, R., Mali, M., Osredkar, R., Prelesnik, A., Zupančič, I. and Ehrenberg, L. (1971). *Chem. Phys. Lett.*, **9,** 1971
125. Onda, S., Ikeda, R., Nakamura, D. and Kubo, M. (1969). *Bull. Chem. Soc. Japan*, **42,** 1771

126. Colligiani, A. and Ambrosetti, R. (1971). *J. Chem. Phys.*, **54**, 2105
127. Negita, H., Hayashi, M., Hirakawa, T. and Kuwata, H. (1970). *Bull. Chem. Soc. Japan*, **43**, 2262
128. Osokin, D. Ya., Safin, I. N. and Nuretdinov, I. A. (1970). *Dokl. Akad. Nauk SSSR*, **190**, 357
129. Ha, T.-K. and O'Konski, C. T. (1970). *Z. Naturforsch.*, **A25**, 1155
130. Singh, K. and Singh, S. (1967). *Indian J. Phys.*, **41**, 862
131. Osokin, D. Ya., Safin, I. A. and Nuretdinov, I. A. (1969). *Dokl. Akad. Nauk SSSR*, **186**, 1128
132. Abe, Y., Kamishina, Y. and Kojima, S. (1966). *J. Phys. Soc. Japan*, **21**, 2083
133. Ha, T.-K. and O'Konski, C. T. (1970). *Z. Naturforsch.*, **A25**, 1509
134. Krause, L. and Whitehead, M. A. (1970). *J. Chem. Phys.*, **52**, 2787
135. Linscheid, P. and Lucken, E. A. C. (1970). *Chem. Commun.*, 425
136. Stidham, H. D. and Farrell, H. H. (1968). *J. Chem. Phys.*, **49**, 2463
137. Zeil, W. and Haas, B. (1967). *Z. Naturforsch.*, **A22**, 2011
138. Hart, R. M. and Whitehead, M. A. (1971). *Can. J. Chem.*, **49**, 2508
139. Considine, W. J. (1966). *J. Chem. Phys.*, **44**, 4036
140. Lucas, J. P. and Guibé, L. (1970). *Molec. Phys.*, **19**, 85
141. Maksyutin, Yu. K., Babushkina, T. A., Gur'yanova, E. N. and Semin, G. K. (1969). *Zh. Strukt. Khim.*, **10**, 1025
142. Maksyutin, Yu. K., Babushkina, T. A., Gur'yanova, Ye. N. and Semin, G. K. (1969). *Theor. Chim. Acta*, **14**, 48
143. Watts, V. S. and Goldstein, J. H. (1967). *Theor. Chim. Acta*, **7**, 181
144. Koltenbah, D. E. and Silvidi, A. A. (1970). *J. Chem. Phys.*, **52**, 1270
145. Neimysheva, A. A., Semin, G. K., Babushkina, T. A. and Knunyants, I. L. (1967). *Dokl. Akad. Nauk SSSR*, **173**, 585
146. Bryukhova, E. V., Stanko, V. I., Klimova, A. I., Titova, N. S. and Semin, G. K. (1968). *Zh. Strukt. Khim.*, **9**, 39
147. Grechishkin, V. S., Izmest'ev, I. V. and Soifer, G. B. (1969). *Zh. Fiz. Khim.*, **43**, 757
148. Agranat, I., Gill, D., Hayek, M. and Loewenstein, R. M. J. (1969). *J. Chem. Phys.*, **51**, 2756
149. Brevard, C. and Lehn, J. M. (1968). *J. Chim. Phys. Physicochim. Biol.*, **65**, 727
150. Grechishkin, V. S., Gordeev, A. D. and Galishevskii, Yu. A. (1969). *Zh. Strukt. Khim.*, **10**, 743
151. Semin, G. K., Babushkina, T. A. and Robas, V. I. (1966). *Zh. Fiz. Khim.*, **40**, 2564
152. Nesmeyanov, A. N., Kravtsov, D. N., Zhukov, A. P., Kochergin, P. M. and Semin, G. K. (1968). *Dokl. Akad. Nauk SSSR*, **179**, 102
153. Venkatacharyulu, P. and Premaswarup, D. (1971). *Curr. Sci.*, **40**, 154
154. Narayana, K. L. and Nagarajan, V. (1968). *J. Shivaji Univ.*, **1**, 109
155. Kinastowski, S. and Peitrzak, J. (1968). *Bull. Acad. Pol. Sci., Ser. Sci. Chim.*, **16**, 155
156. Rao, K. K. and Murty, C. R. K. (1967). *Chem. Phys. Lett.*, **1**, 323
157. Hart, R. M. and Whitehead, M. A. (1971). *Can. J. Chem.*, **49**, 2508
158. Cotton, F. A. and Harris, C. B. (1966). *Proc. Nat. Acad. Sci. U.S.A.*, **56**, 12
159. Brown, R. D. and Peel, J. B. (1968). *Aust. J. Chem.*, **21**, 2589, 2605
160. Ha, T.-K. and O'Konski, C. T. (1969). *J. Chem. Phys.*, **51**, 460
161. Eletr, S., Ha, T.-K. and O'Konski, C. T. (1969). *J. Chem. Phys.*, **51**, 1430
162. Avgeropoulos, G. N. and Ebbing, D. D. (1969). *J. Chem. Phys.*, **50**, 3493
163. Eletr, S. (1970). *Molec. Phys.*, **18**, 119
164. Harrison, J. F. (1967). *J. Chem. Phys.*, **47**, 2990
165. Harrison, J. F. (1968). *J. Chem. Phys.*, **48**, 2379
166. Unland, M. L., Letcher, J. H. and Van Wazer, J. R. (1969). *J. Chem. Phys.*, **50**, 3214
167. Davies, D. W. (1967). *Chem. Commun.*, 1226
168. Davies, D. W. and Mackrott, W. C. (1967). *Chem. Commun.*, 3456
169. Sichel, J. M. and Whitehead, M. A. (1968). *Theoret. Chim. Acta*, **11**, 263
170. Kaplansky, M. and Whitehead, M. A. (1969). *Trans. Faraday Soc.*, **65**, 641
171. Kaplansky, M. and Whitehead, M. A. (1968). *Molec. Phys.*, **15**, 149
172. Kaplansky, M. and Whitehead, M. A. (1969). *Molec. Phys.*, **16**, 481
173. Purcell, K. F. (1967). *J. Chem. Phys.*, **47**, 1198
174. Bader, R. F. W. and Bandrauk, A. D. (1968). *J. Chem. Phys.*, **49**, 1653
175. McLean, A. D. and Yoshimine, M. (1967). *J. Chem. Phys.*, **47**, 3256

176. White, W. D. and Drago, R. S. (1970). *J. Chem. Phys.*, **52,** 4717
177. Dewar, M. J. S., Lo, D. H., Patterson, D. B., Trinajstic, N. and Peterson, G. E. (1971). *Chem. Commun.*, 238
178. Wallwork, S. C. and Worrall, I. J. (1965). *J. Chem. Soc.*, 1816
179. Peterson, G. E. and Bridenbaugh, P. M. (1969). *J. Chem. Phys.*, **51,** 238
180. Srivastava, T. S. (1968). *Curr. Sci.*, **37,** 253
181. Tong, D. A. (1969). *Chem. Commun.*, 790
182. Okuda, T., Terao, H., Ege, O. and Negita, H. (1970). *J. Chem. Phys.*, **52,** 5489
183. Maksyutin, Yu. K., Bryukhova, E. V., Semin, G. K. and Gur'yanova, E. N. (1968). *Isv. Akad. Nauk SSSR. Ser. Khim.*, 2658
184. Chihara, H., Nakamura, N. and Seki, S. (1967). *Bull. Chem. Soc. Japan*, **40,** 50
185. Hart, R. M. and Whitehead, M. A. (1971). *J. Chem. Soc. A*, 1738
186. Kaplansky, M., Clipsham, R. and Whitehead, M. A. (1969). *J. Chem. Soc. A*, 584
187. Brill, T. B. and Long, G. G. (1970). *Inorg. Chem.*, **9,** 1980
188. Schneider, R. F. and DiLorenzo, J. V. (1967). *J. Chem. Phys.*, **47,** 2434
189. DiLorenzo, J. V. and Schneider, R. F. (1967). *Inorg. Chem.*, **6,** 766
190. DiLorenzo, J. V. and Schneider, R. F. (1968). *J. Phys. Chem.*, **72,** 761
191. Dewar, M. J. S. and Patterson, D. B. (1970). *Chem. Commun.*, 544
192. Evans, J. C. and Lo, G. Y.-S. (1967). *Inorg. Chem.*, **6,** 836
193. Smith, J. A. S. and Tong, D. A. (1971). *J. Chem. Soc. A*, 173, 178
194. Casabella, P. A. and Oja, T. (1969). *J. Chem. Phys.*, **50,** 4814
195. Ardjomand, S. and Lucken, E. A. C. (1971). *Helv. Chim. Acta*, **54,** 176
196. Wrubel, H. and Voitlander, J. (1969). *Z. Naturforsch.*, **24A,** 282
197. Negita, H., Tanaka, T., Okuda, T. and Shimada, H. (1966). *Inorg. Chem.*, **5,** 2126
198. Ramakrishna, J. (1966). *Phil. Mag.*, **14,** 589
199. Dinesh and Narasimhan, P. T. (1966). *J. Chem. Phys.*, **45,** 2170
200. Brill, T. B., Hugus, Z. Z., Jr. and Schreiner, A. F. (1970). *J. Phys. Chem.*, **74,** 469
201. Creel, R. B., Segel, S. L. and Anderson, L. A. (1969). *J. Chem. Phys.*, **50,** 4908
202. Brill, T. B. and Hugus, Z. Z., Jr. (1970). *Inorg. Chem.*, **9,** 984
203. Brill, T. B. and Hugus, Z. Z., Jr. (1970). *J. Inorg. Nucl. Chem.*, **33,** 371
204. Brill, T. B. (1970). *J. Inorg. Nucl. Chem.*, **32,** 1869
205. Bregadze, V. I., Babushkina, T. A., Okhlobystin, O. Yu. and Semin, G. K. (1967). *Teor. Eksp. Khim.*, **3,** 547
206. Bryukhova, E. V., Velichko, F. K. and Semin, G. K. (1969). *Isv. Akad. Nauk SSSR, Ser. Khim.*, 960
207. Bryukhova, E. V., Babushkina, T. A., Kashutina, M. V., Okhlobystin, O. Yu. and Semin, G. K. (1968). *Dokl. Akad. Nauk SSSR*, **183,** 827
208. Nesmeyanov, A. N., Okhlobystin, O. Yu., Bryukhova, E. V., Bregadze, V. I., Kravtsov, D. N., Faingor, B. A., Golovchenko, L. S. and Semin, G. K. (1969). *Isv. Akad. Nauk SSSR, Ser. Khim.*, 1928
209. Breneman, G. L. and Willett, R. D. (1967). *J. Phys. Chem.*, **71,** 3684
210. Sasane, A., Nakamura, D. and Kubo, M. (1969). *J. Phys. Chem.*, **71,** 3249
211. Wiebenga, E. H. and Kracht, D. (1969). *Inorg. Chem.*, **8,** 738
212. Bowmaker, G. A. and Hacobian, S. (1968). *Aust. J. Chem.*, **21,** 551
213. Maksyutin, Yu. K., Gur'yanova, E. N. and Semin, G. K. (1968). *Zh. Strukt. Khim.*, **9,** 701
214. Bowmaker, G. A. and Hacobian, S. (1969). *Aust. J. Chem.*, **22,** 2047
215. Gupta, L. C. (1967). *Ind. J. Pure Appl. Phys.*, **5,** 437
216. Semin, G. K., Babushkina, T. A., Khrlakyan, S. P., Pervova, E. Ya., Shokina, V. V. and Knunyants, I. L. (1968). *Teor. Eksp. Khim.*, **4,** 275
217. Evans, J. C. and Lo, G. Y.-S. (1966). *J. Phys. Chem.*, **70,** 2702
218. Evans, J. C. and Lo, G. Y.-S. (1967). *J. Phys. Chem.*, **71,** 3697
219. Chang, S. and Westrum, E. F. (1962). *J. Chem. Phys.*, **36,** 2571
220. Haas, T. E. and Welsh, S. M. (1967). *J. Phys. Chem.*, **71,** 3363
221. Ludman, C. J., Waddington, T. C. and Salthouse, J. A. (1970). *Chem. Commun.*, 405
222. Evans, J. C. and Lo, G. Y.-S. (1967). *J. Phys. Chem.*, **71,** 2730
223. Graybeal, J. D. and Green, P. J. (1969). *J. Phys. Chem.*, **73,** 2948
224. Gilson, D. F. R. and O'Konski, C. T. (1968). *J. Chem. Phys.*, **48,** 2767
225. Bennett, R. A. and Hooper, H. O. (1967). *J. Chem. Phys.*, **47,** 4855
226. Swiger, E. D. and Graybeal, J. D. (1965). *J. Amer. Chem. Soc.*, **87,** 1464

227. Green, P. J. and Graybeal, J. D. (1967). *J. Amer. Chem. Soc.*, **89**, 4305
228. Negita, H., Okuda, T. and Kashima, M. (1966). *J. Chem. Phys.*, **45**, 1076; (1967). *J. Chem. Phys.*, **46**, 2450
229. Okuda, T., Nakao, A., Shiroyama, M. and Negita, H. (1968). *Bull. Chem. Soc. Japan*, **41**, 61
230. Okuda, T., Terao, H., Ege, O. and Negita, H. (1970). *Bull. Chem. Soc. Japan*, **43**, 2398
231. Biedenkapp, D. and Weiss, A. (1968). *Z. Naturforsch.*, **23B**, 174
232. Pen'kov, I. N. and Safin, I. A. (1968). *Kristallografiya*, **13**, 330
233. Clipsham, R., Hart, R. M. and Whitehead, M. A. (1969). *Inorg. Chem.*, **8**, 2431
234. Rogers, M. T. and Ryan, J. A. (1968). *J. Phys. Chem.*, **72**, 1340
235. Doorenbos, H. E., Evans, J. C. and Kagel, R. O. (1970). *J. Phys. Chem.*, **74**, 3385
236. Chihara, H., Nakamura, N. and Okuma, H. (1968). *J. Phys. Soc. Japan*, **24**, 306
237. Hafner, S. and Raymond, M. (1968). *J. Chem. Phys.*, **49**, 3570
238. Hafner, S. S. and Raymond, M. (1970). *J. Chem. Phys.*, **52**, 279
239. Taylor, R. D. (1968). *J. Chem. Phys.*, **48**, 3570
240. Sawatzky, G. A. and Hupkes, J. (1970). *Phys. Rev. Lett.*, **25**, 100
241. Sharma, R. R. (1970). *Phys. Rev. Lett.*, **25**, 1622
242. Sholl, C. A. and Walter, J. A. (1969). *J. Phys. C*, **2**, 2301
243. Segel, S. L., Walter, J. A. and Troup, G. J. (1969). *Phys. Status Solidi*, **31**, K43
244. Sholl, C. A. (1966). *Proc. Phys. Soc.*, **87**, 897
245. Graybeal, J. D. and McKown, R. J. (1969). *J. Phys. Chem.*, **73**, 3156
246. Bersohn, R. and Shulman, R. G. (1966). *J. Chem. Phys.*, **45**, 2298
247. Cotton, F. A. and Harris, C. B. (1967). *Inorg. Chem.*, **6**, 369
248. Cotton, F. A. and Harris, C. B. (1967). *Inorg. Chem.*, **6**, 376
249. Brown, T. L., McDugle, W. G., Jr. and Kent, L. G. (1970). *J. Amer. Chem. Soc.*, **92**, 3645
250. Brill, T. B. and Hugus, Z. Z., Jr. (1970). *J. Phys. Chem.*, **74**, 3022
251. Brill, T. B. and Hugus, Z. Z., Jr. (1970). *Inorg. Nucl. Chem. Lett.*, **6**, 753
252. Spencer, D. D., Kirsch, J. L. and Brown, T. L. (1970). *Inorg. Chem.*, **9**, 235
253. Brown, T. L., Edwards, P. A., Harris, C. B. and Kirsch, J. L. (1969). *Inorg. Chem.*, **8**, 763
254. Nesmeyanov, A. N., Semin, G. K., Bryuchova, E. V., Babushkina, T. A., Anisimov, K. N., Amisimov, K. N., Kolobova, N. E. and Makarov, Yu. V. (1968). *Tetrahedron Lett.*, **37**, 3987
255. Brill, T. B. and Long, G. G. (1971). *Inorg. Chem.*, **10**, 74
256. Williams, D. E. and Kocher, C. W. (1970). *J. Chem. Phys.*, **52**, 1480
257. Fryer, C. W. and Smith, J. A. S. (1970). *J. Chem. Soc. A*, 1029
258. Scaife, D. E. (1970). *Aust. J. Chem.*, **23**, 2205
259. Yesinowski, J. P. and Brown, T. L. (1971). *Inorg. Chem.*, **10**, 1097
260. Fergusson, J. E. and Scaife, D. E. (1971). *Aust. J. Chem.*, **24**, 1325
261. Sasane, A., Matuo, T., Nakamura, D. and Kubo, M. (1970). *Bull. Chem. Soc. Japan*, **43**, 1908
262. Ikeda, R., Sasane, A., Nakamura, D. and Kubo, M. (1966). *J. Phys. Chem.*, **70**, 2926
263. Schreiner, A. F. and Brill, T. B. (1970). *Theoret. Chim. Acta*, **17**, 323
264. Ikeda, R., Nakamura, D. and Kubo, M. (1968). *J. Phys. Chem.*, **72**, 2982
265. Bayer, H. (1951). *Z. Phys.*, **130**, 227
266. Kushida, T. (1955). *J. Sci. Hiroshima Univ., Ser. A-II*, **19**, 327
267. Gutowsky, H. S. and McCall, D. W. (1960). *J. Chem. Phys.*, **32**, 548
268. Scott, T. A. (1962). *J. Chem. Phys.*, **36**, 1459
269. Nakamura, N. and Chihara, H. (1967). *J. Phys. Soc. Japan*, **22**, 201
270. Brown, R. J. C. (1960). *J. Chem. Phys.*, **32**, 116
271. Moross, G. G. and Story, H. S. (1966). *J. Chem. Phys.*, **45**, 3370
272. Utton, D. B. (1967). *J. Chem. Phys.*, **47**, 371
273. Tipsword, R. F., Allender, J. T., Stahl, E. A., Jr. and Williams, C. D. (1968). *J. Chem. Phys.*, **49**, 2464
274. Barton, B. L. (1971). *J. Chem. Phys.*, **54**, 814
275. Bazhulin, P. A. and Bakhimov, A. A. (1965). *Fiz. Tver. Tela*, **7**, 2088
276. Ito, M. and Shigeoka, T. (1966). *Spectrochim. Acta*, **22**, 1029
277. Cahill, J. E. and Leroi, G. E. (1969). *J. Chem. Phys.*, **51**, 1324
278. Cahill, J. E. and Leroi, G. E. (1969). *J. Chem. Phys.*, **51**, 4514
279. Stidham, H. D. (1968). *J. Chem. Phys.*, **49**, 2041
280. Ten, H. C., Brockhouse, B. N. and Dewit, G. A. (1969). *Phys. Lett.*, **29A**, 694

281. Lee, K., Sher, A., Andersson, L. O. and Proctor, W. G. (1966). *Phys. Rev.,* **150,** 168
282. Kushida, T., Benedek, G. B. and Bloembergen, N. (1956). *Phys. Rev.,* **104,** 1364
283. Whidden, C. J., Williams, C. D. and Tipsword, R. F. (1969). *J. Chem. Phys.,* **50,** 507
284. Early, D. D., Tipsword, R. F. and Williams, C. D. (1971). *J. Chem. Phys.,* **55,** 460
285. Utton, D. B. (1971). *J. Chem. Phys.,* **54,** 5441
286. Tipsword, R. F. and Moulton, W. G. (1963). *J. Chem. Phys.,* **39,** 2730
287. Ward, R. W., Williams, C. D. and Tipsword, R. F. (1969). *J. Chem. Phys.,* **51,** 823
288. Schempp, E., Peterson, G. E. and Carruthers, J. R. (1970). *J. Chem. Phys.,* **53,** 306
289. Vijaya, M. S. and Ramakrishna, J. (1970). *Molec. Phys.,* **19,** 131
290. Vijaya, M. S. and Ramakrishna, J. (1970). *J. Chem. Phys.,* **53,** 4714
291. Kaplansky, M. and Whitehead, M. A. (1967). *Can. J. Chem.,* **45,** 1669
292. Dixon, M. and Jenkins, H. D. B. (1967). *Trans. Faraday Soc.,* **63,** 2852
293. Kumar, U. V. and Rao, B. D. N. (1971). *Phys. Status Solidi,* **B44,** 203
294. Yim, C. T., Whitehead, M. A. and Lo, D. H. (1968). *Can. J. Chem.,* **46,** 3595
295. Gilson, D. F. R. and Hart, R. M. (1970). *Can. J. Chem.,* **48,** 1976
296. Kaplansky, M. and Whitehead, M. A. (1970). *Can. J. Chem.,* **48,** 697
297. De Wijn, H. W. and de Wildt, J. L. (1966). *Phys. Rev.,* **150,** 200
298. Kantimati, B. (1966). *Indian J. Pure Appl. Phys.,* **4,** 131
299. Hashimoto, M., Morie, T. and Kato, Y. (1971). *Bull. Chem. Soc. Japan,* **44,** 1455
300. Barton, B. L. (1969). *J. Chem. Phys.,* **51,** 4670
301. Barton, B. L. (1969). *J. Chem. Phys.,* **51,** 5726
302. Barton, B. L. (1969). *J. Chem. Phys.,* **51,** 4672
303. Blinc, R., Mali, M. and Trontelj, Z. (1967). *Phys. Lett.,* **25A,** 289
304. Biedenkapp, D. and Weiss, A. (1967). *Z. Naturforsch.,* **22A,** 1124
305. Yamamoto, T., Nakamura, N. and Chihara, H. (1968). *J. Phys. Soc. Japan,* **25,** 291
306. Yamamoto, T., Nakamura, N. and Chihara, H. (1970). *J. Phys. Soc. Japan,* **28,** *Suppl.,* 112
307. Van Kranendonk, J. and Walker, M. B. (1967). *Phys. Rev. Lett.,* **18,** 701; (1968). *Can. J. Phys.,* **46,** 2441
308. Armstrong, R. L. and Jeffrey, K. R. (1971). *Can. J. Phys.,* **49,** 49
309. Bridges, F. (1967). *Phys. Rev.,* **164,** 299
310. Bridges, F. and Clark, W. G. (1967). *Phys. Rev.,* **164,** 288
311. Woessner, D. E. and Gutowsky, H. S. (1963). *J. Chem. Phys.,* **39,** 440
312. Daniel, A. C. and Moulton, W. G. (1964). *J. Chem. Phys.,* **41,** 1833
313. Grechishkin, V. S., Gordeev, A. D. and Ainbinder, E. E. (1966). *Izv. Vysshikh Uchebn. Zavedenii Radiofiz.* No. 3
314. Grechishkin, V. S. (1966). *Opt. Spektrosk.,* **21,** 517
315. Alexander, S. and Tzalmona, A. (1965). *Phys. Rev.,* **138,** A845
316. Lotfullin, R. Sh. and Semin, G. K. (1969). *Phys. Status Solidi,* **35,** 133
317. Lotfullin, R. Sh. and Semin, G. K. (1969). *Kristallografiya,* **14,** 809
318. Lotfullin, R. Sh. and Semin, G. K. (1970). *Dokl. Akad. Nauk, SSSR,* **193,** 1044
319. Ainbinder, N. E., Amirkhanov, B. F., Izmestev, I. V., Osipenko, A. N. and Soifer, G. B. (1971). *Fiz. Tver. Tela,* **13,** 424
320. Jeffrey, K. R. and Armstrong, R. L. (1968). *Phys. Rev.,* **174,** 359
321. Armstrong, R. L. and Jeffrey, K. R. (1969). *Can. J. Phys.,* **47,** 1095
322. Anderson, J. E. and Slichter, W. P. (1966). *J. Chem. Phys.,* **44,** 309
323. Armstrong, R. L. and Cooke, D. F. (1969). *Can. J. Phys.,* **47,** 2165
324. O'Leary, G. P. (1969). *Phys. Rev. Lett.,* **23,** 782
325. O'Leary, G. P. and Wheeler, R. G. (1970). *Phys. Rev.,* **B1,** 4409
326. Landau, L. D. and Lifshitz, E. (1958). *Statistical Physics.* (Cambridge, Mass.: Addison-Wesley)
327. Armstrong, R. L., Baker, G. L. and Jeffrey, K. R. (1970). *Phys. Rev.,* **B1,** 2847
328. Debeau, M. and Poulet, H. (1969). *Spectrochim. Acta,* **25A,** 1553
329. Armstrong, R. L. (1971). *J. Chem. Phys.,* **54,** 813
330. Baker, G. L. and Armstrong, R. L. (1970). *Can. J. Phys.,* **48,** 1649
331. Jeffrey, K. R., Armstrong, R. L. and Kisman, K. E. (1970). *Phys. Rev.,* **B1,** 3770
332. Armstrong, R. L., Baker, G. L. and Van Driel, H. M. (1971). *Phys. Rev.,* **B3,** 3072
333. Armstrong, R. L. and Cooke, D. F. (1971). *Can. J. Phys.,* **49,** 2381
334. Cooke, D. F. and Armstrong, R. L. (1971). *Can. J. Phys.,* **49,** 2389
335. Brown, T. L. and Kent, L. G. (1970). *J. Phys. Chem.,* **74,** 3572

336. Jeffrey, K. R. and Armstrong, R. L. (1966). *Can. J. Phys.*, **44**, 2315
337. Armstrong, R. L. and Jeffrey, K. R. (1969). *Can. J. Phys.*, **47**, 309
338. Zussman, A. and Alexander, S. (1968). *J. Chem. Phys.*, **48**, 3534
339. Smith, G. W. (1965). *J. Chem. Phys.*, **43**, 4325
340. Tzalmona, A. (1971). *Phys. Lett.*, **34A**, 289
341. Zussman, A. and Alexander, S. (1968). *J. Chem. Phys.*, **49**, 5179
342. Barton, B. L. (1967). *J. Chem. Phys.*, **46**, 1553
343. Tzalmona, A. (1969). *J. Chem. Phys.*, **50**, 366
344. Zussman, A. and Alexander, S. (1967). *Solid State Commun.*, **5**, 259; (1968). *J. Chem. Phys.*, **49**, 3792
345. Chihara, H., Nakamura, N. and Tachiki, M. (1971). *J. Chem. Phys.*, **54**, 3640
346. Clément, R. Gourdji, M. and Guibé, L. (1971). *Molec. Phys.*, **21**, 247
347. O'Reilly, D. E. (1970). *J. Chem. Phys.*, **52**, 2396
348. Okuma, H., Nakamura, N. and Chihara. H. (1968). *J. Phys. Soc. Japan*, **24**, 452
349. Kadaba, P. K. and O'Reilly, D. E. (1970). *J. Chem. Phys.*, **52**, 2403
350. O'Reilly, D. E., Peterson, E. M. and Kadaba, P. K. (1970). *J. Chem. Phys.*, **52**, 6444
351. Stankowski, J. (1969). *Phys. Status Solidi*, **34**, K173
352. Oja, T., Marino, R. A. and Bray, P. J. (1967). *Phys. Lett.*, **26A**, 11
353. Bonera, G., Borsa, F. and Rigamonti, A. (1970). *Phys. Rev.*, **B2**, 2784
354. Ikeda, R., Mikami, M., Nakamura, D. and Kubo, M. (1969). *J. Mag. Res.*, **1**, 211
355. Marino, R. A., Oja, T. and Bray, P. J. (1968). *Phys. Lett.*, **27A**, 263
356. Kadaba, P. K., O'Reilly, D. E. and Blinc, R. (1970). *Phys. Status Solidi*, **42**, 855
357. Betsuyaku, H. (1969). *J. Chem. Phys.*, **50**, 3117; *J. Chem. Phys.*, **50**, 3118
358. Betsuyaku, H. (1969). *J. Chem. Phys.*, **51**, 2546
359. Zhukov, A. P., Golovchenko, L. S. and Semin, G. K. (1968). *Chem. Commun.*, 854
360. Zhukov, A. P., Rez, I. S., Pakhomov, V. I. and Semin, G. K. (1968). *Phys. Status Solidi*, **27**, K129
361. Oja, T. (1969). *Phys. Lett.*, **30A**, 343
362. Papon, P. and Theveneau, H. (1969). *Phys. Lett.*, **30A**, 362
363. Krainik, N. N., Popov, S. N. and Myl'nikova, I. E. (1966). *Fiz. Tverd. Tela*, **8**, 3664
364. Popov, S. N., Krainik, N. N. and Myl'nikova, I. E. (1969). *Izv. Akad. Nauk SSSR, Ser. Fiz.*, **33**, 271
365. Popov, S. N., Krainik, N. N. and Myl'nikova, I. E. (1970). *J. Phys. Soc. Japan*, **28**, *Suppl.*, 120
366. Helg, U. and Graenicher, H. (1970). *J. Phys. Soc. Japan*, **28**, *Suppl.*, 169
367. Tovborg-Jensen, N. (1969). *J. Chem. Phys.*, **50**, 559
368. Volkov, A. F., Venevtsev, Yu. N. and Semin, G. K. (1969). *Phys. Status Solidi*, **35**, K167
369. Kravchenko, E. A., Dembovskii, S. A., Chernov, A. P. and Semin, G. K. (1969). *Phys. Status Solidi*, **31**, K19
370. Poteet, W. M., Tipsword, R. F. and Williams, C. D. (1970). *Phys. Rev.*, **B1**, 1265
371. Hart, R. M. and Whitehead, M. A. (1970). *Molec. Phys.*, **19**, 383
372. Babushkina, T. A., Kozhin, V. M., Robas, V. I., Safin, I. A. and Semin, G. K. (1967). *Kristallographiya*, **12**, 143
373. Babushkina, T. A. and Semin, G. K. (1968). *Kristallographiya*, **13**, 527
374. Babushkin, A. A., Babushkina, T. A., Orlova, E. A., Sperantova, I. B. and Semin, G. K. (1969). *Zh. Fiz. Khim.*, **43**, 1999
375. Sasane, A., Matuo, T., Nakamura, D. and Kubo, M. (1971). *J. Mag. Res.*, **4**, 257
376. Machmer, P. (1968). *J. Inorg. Nucl. Chem.*, **30**, 2627
377. Nakamura, D. and Kubo, M. (1964). *J. Phys. Chem.*, **68**, 2986
378. Lyons, J. D., Langhoff, P. W. and Hurst, R. P. (1966). *Phys. Rev.*, **151**, 60
379. Ghatikar, M. N. (1966). *Proc. Phys. Soc.*, **88**, 536
380. Sternheimer, R. M. (1967). *Phys. Rev.*, **164**, 10
381. Sternheimer, R. M. (1967). *Phys. Rev.*, **195**, 266
382. Sternheimer, R. M. (1967). *Int. J. Quantum Chem., Symp.*, No. 1, 67
383. Rao, K. K. and Murty, C. R. K. (1969). *Indian J. Pure Appl. Phys.*, **7**, 320
384. Gupta, R. P., Rao, B. K. and Sen, S. K. (171). *Phys. Rev.*, **A3**, 545
385. Sternheimer, R. M. and Peierls, R. F. (1971). *Phys. Rev.*, **A3**, 837
386. Schoenberner, D. and Zimmermann, D. (1968). *Z. Phys.*, **216**, 172
387. Ney, J. (1969). *Z. Phys.*, **223**, 126
388. Talmi, I. (1967). *Phys. Lett.*, **B25**, 313

389. Bennett, R. A. and Hooper, H. O. (1970). *J. Chem. Phys.*, **52**, 5485
390. Handrich, E., Steudel, A. and Walther, H. (1969). *Phys. Lett.*, **A29**, 486
391. Murakawa, K. (1969). *J. Phys. Soc. Japan*, **27**, 1690
392. Childs, W. J. and Goodman, L. S. (1968). *Phys. Rev.*, **170**, 50
393. Rosenberg, M., Mandache, S., Niculescu-Majewska, H., Filotti, G. and Gomolea, V. (1970). *Phys. Lett.*, **A31**, 84
394. Childs, W. J. and Goodman, L. S. (1968). *Phys. Rev.*, **170**, 136
395. Hebert, A. J. and Street, K., Jr. (1969). *Phys. Rev.*, **178**, 205
396. Brown, H. H. and King, J. G. (1966). *Phys. Rev.*, **142**, 53
397. Becker, W., Fischer, W. and Huehnermann, H. (1968). *Z. Phys.*, **216**, 142
398. Childs, W. J. and Goodman, L. S. (1971). *Phys. Rev.*, **A3**, 25
399. Unsworth, P. J. (1969). *Proc. Phys. Soc., London, At. Mol. Phys.*, **2**, 122
400. Klinkenberg, P. F. A. and Dekker, J. W. M. (1969). *Physica*, **43**, 92
401. Childs, W. J. (1970). *Phys. Rev.*, **A2**, 316
402. Klinkenberg, P. F. A. (1970). *Physica*, **46**, 119
403. Kuhl, J., Steudel, A. and Walther, H. (1966). *Z. Phys.*, **196**, 365
404. Krause, H., Krebs, K., Winkler, R. and Zschimmer, M. (1969). *Naturwissenschaften*, **56**, 84
405. Semin, G. K. and Bryukhova, E. V. (1968). *Yad. Fiz.*, **7**, 1346
406. Himmel, G. (1968). *Z. Phys.*, **211**, 68
407. Guthoehrlein, G. and Himmel, G. (1969). *J. Phys. (Paris), Colloq.*, **30**, 66
408. Eisele, G., Koniordos, I., Mueller, G. and Winkler, R. (1968). *Phys. Lett.*, **B28**, 256
409. Rinneberg, H., Haas, H. and Hartmann, H. (1969). *J. Chem. Phys.*, **50**, 3064
410. Alderdice, D. S. and Iredate, T. (1966). *Trans. Faraday Soc.*, **62**, 1370

5
Carbon-13 Nuclear Spin Relaxation

J. R. LYERLA, JR.
University of Toronto, Ontario
and

D. M. GRANT
University of Utah

5.1 INTRODUCTION

5.1.1 General relaxation considerations

An attractive feature of nuclear magnetic resonance (n.m.r.) spectroscopy in the study of chemical and physical phenomena is its amenability to the investigation of both molecular structure and molecular dynamics. The structural aspect of magnetic resonance is generally associated with the chemical shielding and spin–spin coupling constants, while the motional aspect is manifest primarily in the spin–lattice (T_1) and spin–spin (T_2) nuclear relaxation times. Although the primary focus of the majority of n.m.r. experiments has been the measurement of chemical shifts and coupling constants, recent advances in instrumentation have provided for a significant improvement in the facility with which nuclear relaxation rates may be determined. The fundamental purpose of these relaxation studies has been to characterise better the molecular dynamics of liquid systems.

To develop the relationship between liquid dynamics and relaxation parameters, it is necessary to describe the interactions[1] which affect the T_1 and T_2 time constants. The T_1 interaction returns the longitudinal component, M_z, of the magnetisation along the applied field, H_0, to its equilibrium value after any perturbation. Such a process requires a flow of energy between the spin system and the lattice. The spin–spin relaxation process arises from interactions within the nuclear-spin system that return any magnetisation, $M_{x,y}$, transverse to H_0 to its equilibrium value of zero. As this is an adiabatic process involving phase relationships between nuclear spins, the energy of the spin system is conserved. Under certain conditions both processes may be treated as first-order decays and T_1 and T_2 become inverse exponential rate constants. Unlike T_1, the T_2 time is affected significantly by inhomogenities in the applied magnetic field. This limitation often prevents a detailed interpretation of T_2 data in terms of molecular motion. Thus, extensive work on carbon-13 T_2 relaxation processes has not appeared in print and this review will focus primarily on spin–lattice processes and T_1 values.

Spin–lattice coupling, requisite for nuclear relaxation, results from the interaction of individual nuclear moments with fluctuating magnetic fields produced by the lattice at the nuclear site. The time dependence of these fields results directly from the motion of the molecular system thereby providing the link between molecular dynamics and spin–lattice relaxation[2].

It is necessary for these various oscillating fields to have a frequency component at the nuclear Larmor frequency in order to induce the transitions between nuclear spin levels giving rise to relaxation. Unlike most molecular relaxation processes in liquids, the time scale for the spin–lattice relaxation process is relatively long because the nuclear spin system is only weakly coupled to the motions of the system[3]. Experimental values of T_1 are correspondingly of the order of $10^{-4} - 10^4$ s [4] and usually in the range of several seconds for diamagnetic liquids.

Assuming the absence of paramagnetic impurities, the significant mechanisms for spin–lattice relaxation in diamagnetic liquids are inter- and intramolecular nuclear dipole–dipole, electric quadrupole, spin–rotation, chemical-shift anisotropy, and scalar-coupling interactions[2]. In formalistic terms each of these contributions to T_1 can be treated in terms of the magnitude of the oscillating magnetic field arising from the lattice and of a correlation time characterising the motion. In any particular liquid any combination of the above-mentioned relaxation mechanisms may be operative and may contribute to T_1. In order to extract motional information, however, the relaxation rate of each individual mechanism must be separated from the overall rate in order to characterise the motional details. For nuclei possessing a spin quantum number greater than $\frac{1}{2}$, electric quadrupole interactions usually dominate the spin–lattice process[5], and other mechanisms need not be considered. However, for spin-$\frac{1}{2}$ nuclei, several mechanisms can be important simultaneously, and the separation of the overall relaxation rates into the individual components is usually required.

The preponderance of existing relaxation data on liquid systems is for either the hydrogen or fluorine nucleus. This situation primarily reflects the widespread availability of ^1H and ^{19}F instruments and the relative ease of detection for these nuclei rather than any particular simplicity or benefit in interpretating ^1H and ^{19}F relaxation data. In fact, delineating the relaxation contributions of the various mechanisms for ^1H and ^{19}F generally proceeds with considerable difficulty because the proton and fluorine nucleus are almost exclusively located on the periphery of molecules and thus subject to extensive relaxation by intermolecular dipolar interactions. To separate inter- and intra-molecular dipolar relaxation rates necessitates time-consuming dilution studies in a medium devoid of magnetic nuclei. Furthermore, multiplet structure generally found in ^1H and ^{19}F spectra complicates the interpretation of relaxation data on the several lines as the process is no longer governed by a *single* exponential time constant. Finally, even in ^{19}F relaxation where the T_1 process has been shown to be dominated by the spin–rotation interaction[6], the paucity of measured spin–rotation interaction constants usually limits the quantitative interpretation of these results.

Carbon-13 spin–lattice relaxation rates, compared with those for ^1H and ^{19}F, are usually more susceptible to ready interpretation because the carbon nucleus is usually not found at the periphery of the molecule where intermolecular dipole–dipole interactions can be significant. The inverse sixth power dependence of the dipole–dipole interaction upon spin–spin separations requires the dominance of the intramolecular dipole–dipole mechanism over intermolecular interactions for all carbons with directly bonded protons[7-9]. Intermolecular dipole–dipole relaxation is limited, therefore, to non-hydrogen-bearing carbons. This particular facet of ^{13}C relaxation often makes the separation of individual C—H dipolar relaxation times from T_1 quite straightforward[9-12]. The simple formalism of the intramolecular dipolar process in terms of correlated molecular reorientation then permits ready dynamical interpretation of C—H dipolar relaxation rates. As a further advantage, ^{13}C spectra are generally observed under random-noise proton-decoupled[13] conditions where, in the absence of magnetic nuclei other than hydrogen, the relaxation rate of each ^{13}C singlet is governed by a single exponential time constant[8]. The low isotopic abundance of ^{13}C eliminates any need for consideration of C—C dipolar couplings. One final advantage over the proton is claimed for the carbon-13 isotope, and this is the relatively larger chemical shift range found for ^{13}C spectra. Such spectral dispersion makes it possible to focus on the relaxation of several individual resonance lines found in the same molecule, whereas with hydrogen serious over-lapping of lines is often encountered. Multiple relaxation times for a single molecule can be used to characterise internal molecular motion and devia-tions from isotropy in overall molecular reorientation. These features point out the advantage of employing ^{13}C spin–lattice relaxation data in the study of molecular dynamics and provides the primary justification for this review.

5.1.2   Background for carbon-13 relaxation studies

Until recently the availability of relaxation data on the ^{13}C nucleus has been very limited for two reasons. First, ^{13}C magnetic resonance is difficult to observe due to its low natural abundance and low relative sensitivity of detection (0.016 based on 1.0 for the ^{1}H nucleus). Secondly, early reports implied that ^{13}C relaxation times were unusually long and not amenable to study. These conclusions were based on T_1 values[14, 15] for ^{13}C nuclei isolated from hydrogen or on deuterium-substituted materials where the efficient proton dipolar term is absent. Recent instrumental advances, however, have removed many of the experimental difficulties and greatly increased the opportunity of routinely measuring ^{13}C T_1 constants. With random-noise proton-decoupling techniques, time-averaging capabilities, and ^{13}C spectro-meters operating at 14.1 and 23.5 kG, Grant and co-workers[8-12, 16, 17] have demonstrated the feasibility of determining ^{13}C spin–lattice relaxation times on small molecules and have discussed the motional information available from such results.

Of paramount importance to the future exploitation of ^{13}C relaxation data in molecular dynamics is the application of Fourier transform (FT) n.m.r. procedures[18, 19]. This method decreases by one or two orders of magnitude the time required to obtain a given signal-to-noise ratio over conventional averaging techniques. As the FT–n.m.r. method is based on pulsed spectrometer techniques, it provides the necessary capabilities for investigation of very short ^{13}C relaxation times in macromolecular systems[20]. As carbon atoms form the structural framework of all organic and biologic-ally important molecules, ^{13}C relaxation data can be expected to provide significant information on segmental motion of the carbon backbone in macromolecular systems.

5.1.3   Scope of the review

The facility with which spin–lattice relaxation data can be acquired by
FT–n.m.r. methods and the potential wealth of information available in
such relaxation results now necessitate a general understanding and appreciation of nuclear relaxation theory and processes by the practicing chemist.
It is the purpose of this review to provide a conceptual basis, both of a
theoretical and empirical nature, of ^{13}C spin–lattice relaxation. The approach
consists of (a) an outline of the experimental procedures available to measure
^{13}C T_1 values, (b) a discussion in general terminology of the theoretical
considerations necessary to understand relaxation mechanisms and correlated liquid and molecular motion, (c) a comparison of the relative
importance of mechanisms contributing to ^{13}C relaxation, and the manner
in which these individual processes may be separated one from another in
the overall rate, and (d) a collection of specific examples which demonstrate
the extraction of motional parameters from ^{13}C relaxation data.

5.2   EXPERIMENTAL

In this section the experimental aspects of measuring spin–lattice relaxation
times and nuclear Overhauser enhancement (NOE) factors are reviewed with
emphasis placed on both continuous-wave (CW) and pulse methods.

5.2.1   Continuous-wave determination of relaxation parameters

5.2.1.1   Adiabatic rapid-passage (ARP) method

Utilising a continuous-wave, radio-frequency field, H_1, to perturb the sample,
the adiabatic rapid-passage method[21] (ARP) consists of passing the nuclei
rapidly through resonance either by sweeping the external magnetic field,
H_0, or by varying the frequency of the H_1 field. Under proper experimental
conditions[22], the populations of the two spin states ($I = \frac{1}{2}$ nuclei) giving rise
to the resonance line will be inverted and the time dependent return to
equilibrium characterised by the T_1 process may be followed. Monitoring
of the non-equilibrium magnetisation is conveniently achieved by sweeping
in the reverse direction through the signal at a time τ after the first pass
through the resonance. The signal intensity as a function of τ exhibits the
exponential dependence from which T_1 may be obtained.

   The first of two experimental conditions[21] which must be satisfied in the
ARP technique is that the passage through resonance must be fast compared
with both the longitudinal and transverse relaxation rates ($1/T_1$ and $1/T_2$
respectively). Secondly, the H_1 intensity must be of sufficient magnitude to
lock the nuclear magnetisation vector to the effective magnetic field, H_{eff},
which in the rotating reference frame is a vector sum of H_1 and the virtual
magnetic field $(\omega - \omega_0)/\gamma$ represented by the offset from resonance of the
oscillating H_1 field. The constant γ is the magnetogyric ratio associated with
the nuclei resonating at ω_0. The first condition requires that the sweep be

initiated at frequency ω_i and finished at ω_f such that $|\omega_i - \omega_0| \gg \gamma H_1$ and $|\omega_f - \omega_0| \gg \gamma H_1$. Furthermore, the period of the sweep $(\omega_f - \omega_i)/(d\omega/dt)$ must be short compared with both T_1 and T_2. These inequalities combine to give the following requirements:

$$\frac{1}{T_1} \ll \frac{1}{\gamma H_1}\left(\frac{d\omega}{dt}\right) \text{ and } \frac{1}{T_2} \ll \frac{1}{\gamma H_1}\left(\frac{d\omega}{dt}\right) \qquad (5.1)$$

As $T_2 \leqslant T_1$, it is sufficient to invoke only the right-hand expression in equation (5.1). The second general requirement is met when the nuclear magnetisation vector, M, is locked to H_{eff} or precesses in a tight angle about H_{eff} with a frequency, $\gamma H_{eff} = [(\omega - \omega_0)^2 + \gamma^2 H_1^2]^{\frac{1}{2}}$, which is large compared with the rate of sweeping through resonance. Recognising that the minimum value for γH_{eff} is γH_1 at resonance and employing the previously discussed inequalities the second requirement becomes

$$\frac{1}{\gamma H_1}\left(\frac{d\omega}{dt}\right) \ll \gamma H_1 \qquad (5.2)$$

Combination of equations (5.1) and (5.2) yields the following combined requirements for the ARP experiment:

$$\frac{\gamma H_1}{T_2} \ll \left(\frac{d\omega}{dt}\right) \ll \gamma^2 H_1^2 \qquad (5.3)$$

A similar expression for a field-swept spectrometer may be obtained by dividing through by γ and imposing the definition $dH_0/dt = d\omega/\gamma dt$. The

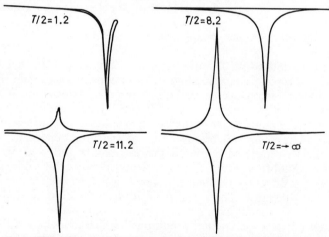

Figure 5.1 Characteristic features of the adiabatic rapid passage experiment for the carboxyl carbon of acetic acid at various times
(Taken from *Ph.D. Thesis* of J. R. Lyerla, Jr., University of Utah, 1971)

right-hand condition of equation (5.3) ensures that the sweep rate is not so rapid as to leave M behind H_{eff}, while the left inequality will guarantee that relaxation processes are negligible during the time of passage through resonance[21]. In classical terms these two conditions lead to a 180 degree rotation of M, or in quantum mechanical concepts a population inversion of the two spin levels is realised. Most high-resolution spectrometers used for ^1H and ^{13}C have r.f. units of sufficient power to meet the right-hand condition providing reasonable sweep rates are employed. However, the technique is limited by the relaxation requirement whenever the value of T_2 becomes too short for effective instrument response in the manual mode such as may be found in the time constant of the recording module. Field inhomogenities should be carefully minimised in order to prevent very short T_2 values arising from this source.

Characteristic spectral features of the ARP experiment are illustrated in Figure 5.1 for the relaxation of the carboxyl carbon of acetic acid. For a relatively short τ (c. 2.4 s) between the two passes through resonance, one still observes the inversion in spin population while at longer values of τ (c. 16.4 s) it is observed that a null condition is achieved. The spectrum at still longer times (c. 22.4 s) depicts the situation in which the population of spin is once again greater in the lower magnetic energy level. Finally at $t = \infty$ (usually taken to be at least five times T_1) equilibrium has once again been restored and the height of the reverse pass is equivalent to the forward sweep. The 180 degree phase shift results from the corresponding phase relationship between M and H_{eff} when the resonance is approached from opposite sides of the resonance frequency. The directional cosine of the H_{eff} vector relative to H_0 (and therefore M_0) is given by $(\omega - \omega_0)/\gamma H$. As the signal S_τ at time τ is proportional to M_z, S_τ expressed in terms of T_1 and τ is given by the expression,

$$S_\tau = S_\infty (1 - 2e^{-\tau/T_1}) \qquad (5.4)$$

where S_∞ represents the signal intensity at $\tau = \infty$.

The ARP method is readily applied to the ^{13}C isotope since the usual range of carbon chemical shifts is sufficiently dispersed that chemically different carbon nuclei in a molecule may be studied individually[23]. Because of the low natural abundance of ^{13}C, accumulation techniques[12] have at times been necessary when the signal cannot be observed in a single sweep. This automatic method consists of storing the two opposing sweeps in two halves of a Varian C-1024 time-averaging device. A triggering pulse at the mid-point in the C-1024 sweep is used to reverse the direction of sweep.

Janzen et al.[24] have described a steady state modification of the ARP experiment in which a periodic saw-tooth or trapezoidal sweep is applied to the sample. The resonance line is placed at the mid-point of the sweep so that a constant time τ separates successive fast passages through the signal. Two equal and opposite signals S_τ are observed corresponding to a steady state value of M_z. The expression for S_τ is

$$S_\tau = S_\infty \frac{1 - e^{-\tau/T_1}}{1 + e^{-\tau/T_1}} \qquad (5.5)$$

By measuring S_τ as a function of τ, T_1 can be obtained from equation (5.5) If it is not convenient to measure S_∞, any two measurements of S_τ can be used to obtain T_1.

As the ARP requirements embodied in equation (5.3) may not be met at all times, it is preferable if a plot of $\ln(S_\infty - S_\tau)$ $v. \tau$ is used to obtain T_1 rather than an absolute calculation using equation (5.4). Janzen et al.[24] have shown that the slope of such a plot is given by $1/T_1$ even when the magnetisation inversion falls short of 180 degrees. The error does not affect the slope even though the anticipated intercept, $\ln(2S_\infty)$ is not achieved. Similar considerations also apply of course to equation (5.5). The use of only a null measurement such as shown in Figure 5.1 is associated with much larger errors. The relative errors in measuring relaxation times with the ARP method, providing optimal instrument adjustments are made, generally do not exceed 10%.

5.2.1.2   Measurement of nuclear Overhauser enhancement factors

The nuclear Overhauser enhancement (NOE) factor is obtained in a straight-forward manner whenever the integrated line intensities of the coupled and proton-decoupled spectra are obtainable. For molecules with simple proton-induced multiplets, the intensities of the lines in the coupled spectrum can usually be determined without too much difficulty, but complicated multiplet structure reduces the relative signal-to-noise of the coupled spectra by an order of magnitude over that of the decoupled singlets. Any minor non-linearity in the response of the spectrometer can lead to serious errors in the NOE ratio, and the direct measurement of NOE values becomes unreliable. One means of avoiding this problem is to measure the relative intensities of a collection of decoupled singlets for which the NOE of one of the lines is determinable by some reliable method. For many molecules one of the carbon peaks will have a simpler multiplet structure in the coupled spectrum and one relatively accurate NOE can be obtained in this manner. The use of a second molecule in a known concentration can also provide a reference NOE for the relative intensity approach. While these methods are admittedly indirect, the errors appear to be somewhat smaller than encountered in intensity measurements of a very complex multiplet structure.

5.2.1.3   Dynamic proton-decoupling method

Whenever the proton–carbon-13 dipolar relaxation time contributes significantly to the carbon relaxation process, a dynamic proton-decoupling technique can be employed to measure not only T_1 but also the NOE factor. This approach described by Kuhlmann and Grant[12] consists of switching on a noise-modulated proton decoupler at high power, thereby immediately saturating all protons, and observing the resulting carbon-13 singlet at a time τ later. While multiplet collapse is essentially instantaneous the development of a steady state NOE is characterised by the relaxation parameters for the system. Thus, the signal S_τ grows according to the appropriate exponential laws with the delay time τ. Extrapolation of this plot to both $\tau = 0$ and ∞ yields a value for S_∞/S_0, the NOE factor. Furthermore, T_1 can be obtained from a plot of $\ln(S_\infty - S_\tau)$ $v. \tau$ in accordance with equation (5.6) which governs the time development of S_τ.

$$\ln(S_\infty - S_\tau) = -\tau/T_1 + \ln(S_\infty - S_0) \tag{5.6}$$

The method clearly will not work when the NOE factor is unity as $S_0 = S_\infty$ and the terms in equation (5.6) are no longer finite.

This technique is most valuable for determining directly the NOE factors when extensive multiplet structure exists in the coupled spectrum as the intensity of only singlet lines need to be observed. Furthermore, the intensities vary by only a small factor and not by an order of magnitude. Thus, this method offers an excellent means of obtaining NOE parameters even though errors in obtaining T_1 values are sufficiently serious that other methods of obtaining T_1 are required to obtain this parameter.

5.2.2   Pulsed method for determination of T_1

5.2.2.1   Inversion-recovery method

If a very intense H_1 field is applied in the form of a square pulse to a nuclear spin system at equilibrium, the magnetisation vector is nutated through 180 degrees in a manner similar to the ARP method but differing in that the M vector rotates 180 degrees in the plane perpendicular to H_{eff}. At the conclusion of the 180 degrees pulse a time delay of τ is imposed, followed by a 90 degree observing pulse which turns the recovering M vector into the x, y-plane where its free induction decay is observed. As the signal intensity at time τ is proportional to the amplitude of the decaying magnetisation, data on the time development of the magnetisation vector may be obtained. The experiment can be repeated at various values of τ providing a time delay of at least $5T_1$ between successive experiments is allowed so that the system can return to equilibrium. Equation (5.4) is again used to obtain T_1. Unlike the ARP experiment, the pulsed inversion-recovery technique can be used to measure simultaneously the T_1 values of all resonances of a multi-line spectrum[25, 26]. A very short 180 degree pulse of high power is non-selective and simultaneously inverts all nuclei of a given type even though they have differing chemical shifts. The resulting free-induction signal is then digitised and stored in a computer. The process may be repeated as desired to improve the signal-to-noise ratio. As the Fourier transform of this free-induction decay yields the normal frequency-domain spectrum, one obtains time-dependent spectra in which the signal intensity or net magnetisation associated with the various peaks relates to both T_1 and τ in accordance with equation (5.4). There also are several experimental requirements that must be met in this mode of operating to ensure that all the nuclear spins are inverted. First, the technique still requires a waiting period of at least $5T_1$ between the 90 degree pulse and the next 180 degree pulse to allow for equilibrium conditions to be re-established between successive independent measurements. Secondly, the difference in resonance frequency and the carrier frequency of the pulse must be relatively small compared with γH_1 so that H_{eff} will be perpendicular to M. This requires that $\gamma H_1 \gg (\omega - \omega_0)$ for all resonance lines. Actually, some of the first commercial spectrometers marketed with Fourier attachments utilise H_1 power levels which are too low to sample simultaneously peaks which are separated by 5000 Hz and still meet the conditions required by this inequality, and care needs to be taken in restricting interpretation to only those regions of the spectrum where inversion is complete.

5.2.2.2 Progressive saturation method

A pulsed method of determining T_1 which avoids the long delays required in re-establishing equilibrium has been suggested by Freeman and Hill[27, 28]. The technique termed 'progressive' saturation, owing to its similarity to the CW progressive saturation experiment, makes use of a repetitive $(90-\tau)_n$ pulse sequence, where τ is the time between pulses. After the first few pulses a steady-state condition for the longitudinal magnetisation is established in which there is a dynamic balance between the effect of the H_1 pulse and the relaxation of the spin system. In this steady-state situation, the amplitude of the free-induction decay is directly proportional to the z-magnetisation which begins to recover during the τ period between pulses. Thus, monitoring the amplitudes of the free-induction signal as a function of τ once again yields values for T_1. Unlike the inversion-recovery technique, however, a long delay period is not required before a repetition of the process can be carried out when accumulation techniques are used. When the value of τ is changed, however, the time necessary to establish steady state (usually only three or four repetitive pulses)[27] does introduce some inefficiencies into the process whenever the T_1 values in the sample become relatively short compared with the acquisition time associated with each pulse.

The progressive saturation technique requires that the magnetisation before the pulse be completely longitudinal and completely transverse after the pulse[27]. Thus, any residual magnetisation in the x,y-plane must be destroyed before the next 90 degree pulse in order that echoes from earlier pulses be avoided and that the decay be purely a T_1 phenomenon. The general use of proton decoupling with incoherent noise modulation to obtain ^{13}C spectra does reduce[27] some of the x,y-magnetisation, but non-identical gradient pulses can be used to achieve this requirement when necessary. It is important to note that saturation in the context of this experiment refers to the elimination of any z-axis magnetisation (i.e. all the magnetisation is nutated 90 degrees by the pulse) and not to the elimination of magnetisation in all directions which is the context in which the term saturation is normally used.

5.2.2.3 Multiple-pulse saturation method

Another saturation technique due to Markley, Horsley and Klein[29] has been used for measuring T_1. This technique makes use of a burst of pulses to saturate the spin system in the normal sense. The pulse train is then followed at time τ with a 90 degree pulse to sample the z-magnetisation which has recovered during the period τ. If time accumulation is required, the system can then be immediately re-cycled with the same pulse-train–τ–90 degree sequence. The accumulated free-induction decays are Fourier transformed and the entire frequency spectrum as a function of τ is once again available for extraction of T_1 data. A field-gradient pulse is employed to suppress any echo formation due to rapid repetition rates of the 90 degree pulses. As total saturation can be achieved by very high H_1 levels of relatively long duration, this latter approach would also be equivalent to the multi-pulse technique.

5.2.2.4 Comparison of pulse methods

While other pulse sequences and variations on the above techniques will probably be developed to measure T_1, the above methods are representative of the power of Fourier pulsed techniques. A discussion of some of the limitations confronting each of the above three methods will illustrate some of the advantages and limitations of each approach. The inversion-recovery method primarily suffers from the long time period required for the magnetisa-

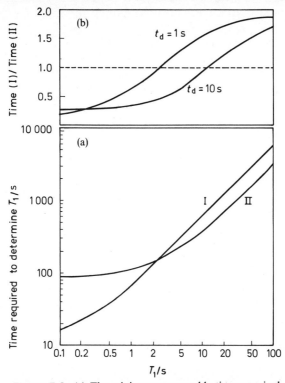

Figure 5.2 (a) The minimum comparable times required to determine T_1 by Method I (spin inversion) and Method II (spin saturation) plotted as a function of T_1. The times are calculated using a saturating time of 1 s and a digitising time t_d of 1 s. (b) The relative advantage of Method II over Method I [time (I)/time (II)] as a function of T_1 for a 1 s saturation time and for digitising times of 1 s (1 Hz resolution) and 10 s (0.1 Hz resolution)
(From Markley, Horsley and Klein[29], reproduced by courtesy of the American Institute of Physics)

tion to be restored to its equilibrium value. For long T_1 values this method becomes inefficient. Furthermore, if the T_1 values are not known, experimental conditions must be set up such that the maximum possible T_1 will be properly accommodated by the delay period after data acquisition. This method, however, will work nicely for very short T_1 values and has the advantage that the maximum $\Delta M_z = 2M_0$ is greater by a factor of two than the M_0

upper limit on the other two methods. If time-accumulation techniques are employed, it becomes necessary to acquire four times as many transients to achieve the same signal-to-noise figure as represented by $\Delta M_z = 2M_0$. This feature does gain back some efficiency for the inversion method.

The progressive saturation method, as pointed out above, suffers from the smaller net ΔM_z value but avoids the long recovery requirement for the short τ determinations. The main problem confronting this approach is the requirement of a short T_2 relative to T_1. Unless the spins can be completely dephased in the x,y-plane between the 90 degrees pulses, one cannot obtain the full efficiency from this method. It is less convenient to study very short T_1 values as the period τ must allow for acquiring the free-induction decay between pulses. Thus, if one wishes to study the region where $\tau \to 0$, the resolution in the FT–n.m.r. mode will seriously degenerate as it is proportional to $1/\tau$.

The multi-pulse saturation technique suffers from the same limitation on ΔM_z as the progressive saturation method. It avoids the resolution problem, however, as no data is accumulated during the period τ between the perturbing pulse train and the analysing 90 degree pulse. This permits τ to be short compared to T_1. Once again the period between the 90 degree pulse and the next sequence of saturating pulses must allow for the transverse relaxation to dephase the spins completely. Markley, Horsley, and Klein[29] have compared the relative benefits of the inversion-recovery and multiple-pulse saturation method, and their results are given in Figure 5.2. For very short T_1 values, such as found for many macromolecules[20] or very viscous media, it can be observed that the inversion-recovery method is superior. However, for long T_1 values, the total saturation method is seen to be more efficient. Similar results would be obtained for the progressive saturation experiment which may be slightly more efficient than the multiple-pulse saturation techniques as the 90 degree pulse takes less time than the pulse train. Even for those molecules which are best treated with the inversion-recovery technique, one of the two saturation methods offers the preferable approach for exploratory work as one may not have *a priori* knowledge of the approximate values of the relaxation times of the system needed to set up the inversion-recovery method.

Regardless of the pulse method used to determine T_1, the greater efficiency in information generation by pulsed over CW methods makes these approaches preferable to the ARP approach. The investment in a pulsed, Fourier attachment to a spectrometer, however, is not trivial and may force the use of the less efficient but adequate CW methods. The ultimate simplicity of the ARP approach also should not be overlooked in any comparison of the two methods. The non-specialist may employ the ARP technique with a minimal investment of time as well as resources.

5.2.3 Data reduction and analysis

The most reliable method to treat the T_1 data is to use the method of plotting $\ln (S_\infty - S_\tau)$ v. τ and extracting T_1 from the slope. However, the ARP and inversion-recovery techniques can be used in a null experiment to derive T_1 from one measurement. If τ is varied so as to give a null S_τ voltage, T_1 is given by

$$T_1 = \tau_{null}/\ln 2 \qquad (5.7)$$

However, this relation assumes that the perturbing H_1 inverts the magnetisation by exactly 180 degrees. Any violation of this requirement can result in sizeable errors in the T_1 value. T_1 is also obtainable from two measurements at two different τ periods. Janzen et al.[24] have shown that the ARP (the argument is also applicable for the 180 degree–τ–90 degree pulse sequence) gives accurate T_1 values even if the initial rotation of M is not quite equal to 180 degrees as the errors in the two values of S_τ for all purposes are mutually cancelling. However, the far more accurate method is the plot technique which allows for better statistical analysis of the data.

Freeman and Hill[27] have demonstrated that determining T_1 from two points using the (90 degree–τ–)$_n$ pulse method gives results within the experimental error inherent in all methods. They have developed criteria in tabular and graphic form[28] for selecting the optimal values of the two τ times for a given T_1. For best accuracy, it has been shown that the two pulse intervals should be set so that the intensity ratio of the two signals lies in the range of 0.2–0.6. Deviations from this condition can lead to large errors in the experimental T_1 values due to the difficulty of measuring the small variations in intensities outside of this range.

5.2.4 Sample preparation

The presence of a paramagnetic species will often cause the relaxation rate of a nucleus to be dominated by the electron–nuclear dipolar interaction due to the much larger dipole moment of the electron relative to nuclear moments. As most liquids will contain dissolved oxygen, a paramagnetic molecule, it is necessary that oxygen be eliminated from the sample before very accurate T_1 values can be determined. This is usually accomplished by purging the sample with nitrogen or argon or by freeze–pump–thaw vacuum techniques. Because of the inverse sixth power dependence of dipolar interactions, arguments have been advanced that the presence of dissolved oxygen will rarely play an important role for carbons with directly attached protons[7]. However, reductions of up to 25% have been found by Alger and Grant[30] in the T_1 determinations in benzene saturated with oxygen as compared with thoroughly degassed samples. Although this effect of O_2 on ^{13}C T_1 values is not as great as for proton relaxation times, where the reduction is almost fourfold, it is still necessary to exclude O_2 from samples when very accurate T_1 data is desired.

5.3 THEORETICAL CONSIDERATIONS

5.3.1 Liquid dynamics and the phenomenological basis of nuclear relaxation

A brief, almost qualitative treatment of the theoretical foundation of relaxation processes is developed in this section to provide a conceptual framework on which the remaining parts of the review can be based. The reader is referred to standard works[2, 31-33] for a rigorous development of the equations used herein. As previously indicated, only interactions that fluctuate at the Larmor frequency can be effective in causing magnetic relaxation. As these time oscillations arise from the thermal motions in liquids, any treatment of relaxation requires a means of characterising this frequency distribution and the corresponding magnetic field intensities. Local magnetic fields, H_{loc}, associated with these interactions can be treated as correlated fluctuations which have a mean value of zero[31]. If these processes are formalised with a Fourier expansion, the resulting Fourier coefficients at each frequency yield the intensity or 'spectral density' of the oscillating H_{loc}. The auto-correlation function[31, 32] $G(\tau)$ given as follows in equation (5.8) provides a statistical measure of the time development of these interactions.

$$G(\tau) = H_{loc}^*(t)\,H_{loc}(t+\tau) \approx \overline{H_{loc}^*(t)\,H_{loc}(t)}\,e^{-|\tau|/\tau_c} \tag{5.8}$$

It is assumed[2] in equation (5.8) that any inherent motional order in the system decays out exponentially with a characteristic constant τ_c, referred to as the correlation time. In an elementary sense $G(\tau)$ is a measure of the persistence of a given fluctuation in H_{loc}. If τ is very short, $G(\tau)$ will assume a value of unity and as τ increases, $G(\tau)$ will fall off to zero (i.e. no correlation or phase coherence between the value of H_{loc} at t and $t+\tau$). The absolute value sign on τ in equation (5.8) arises from the irreversible nature of the liquid motional processes[34] and the bar on H_{loc} represents an ensemble average over time. $G(\tau)$ is both an even and real function of τ; furthermore, $G(\tau)$ as defined in equation (5.8) is independent of the time origin and hence depends only on τ (reference 2). The power density, $J(\omega)$, of H_{loc} is the Fourier inverse of $G(\tau)$ and is given by[2, 31, 32]

$$J(\omega) = \int_{-\infty}^{\infty} G(\tau)\,e^{i\omega\tau}\,d\tau = \int_{-\infty}^{\infty} \overline{H_{loc}^*(t)\,H_{loc}(t)}\,e^{-|\tau|/\tau_c}e^{i\omega\tau}\,d\tau \tag{5.9}$$

Equation (5.9) represents a transformation from the time spectrum to the frequency domain.

Evaluation of the integral in equation (5.9) results in the relation

$$J(\omega) = \overline{H_{loc}^*(t)H_{loc}(t)}\cdot\frac{2\tau_c}{1+\omega^2\tau_c^2} \tag{5.10}$$

It is evident that $J(\omega)$ is a maximum at $\omega = 0$ and begins to fall off with increasing frequency as ω becomes comparable to $1/\tau_c$. A typical plot of $J(\omega)$ versus ω is given in Figure 5.3 for various values of τ_c. Whenever τ_c is very short, then ω must be large before attenuation of $J(\omega)$ is realised and the converse is noted for longer τ_c values. As $\int_0^{\infty} J(\omega)d\omega$ is a constant independent of the values of τ_c, the area under the various curves are all equivalent, and it is apparent that variations in τ_c only change the distribution of the spectral density. When τ_c is long, the lower motional frequencies have a high probability density but $J(\omega)$ falls to zero at a lower frequency value. Conversely, if τ_c is short, the spectral density extends to the higher frequency range with a concomitant decrease in $J(\omega)$ over the lower frequency spectrum. Thus, in both extremes a smaller power density is realised for some intermediate constant frequency such as the nuclear Larmor frequency which is represented by ω_0 in Figure 5.3. Thus, $J(\omega)$ is maximised in the intermediate case where τ_c is comparable to ω_0.

The spin–lattice relaxation process is now related to $J(\omega)$. As the spin–lattice process results in induced transitions between nuclear spin energy

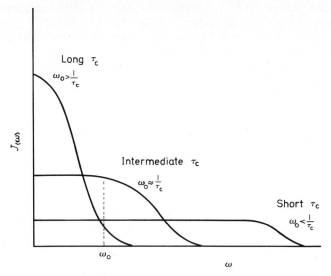

Figure 5.3 Spectral density curves plotted against frequency for various correlation times. The Larmor frequency is represented by ω_0

levels, a quantum mechanical description of the relaxation process can be formulated in which the fluctuating local magnetic fields are treated as time-dependent perturbations ($H_{loc} \ll H_0$) of the magnetic energy levels of the nucleus in the static field, H_0. Time-dependent perturbation theory is formulated in terms of a transition probability per unit time, $W_{\alpha\beta}$, associated with $H_{loc}(t)$ that connects a state α to a state β with frequency separation $\omega_{\alpha\beta}$. As $J(\omega)$ is the amplitude or power density of H_{loc} at ω, the transition probability is given by[2]

$$W_{\alpha\beta} = \gamma^2 J(\omega_{\alpha\beta}) \qquad (5.11)$$

or in terms of the auto-correlation function[2],

$$W_{\alpha\beta} = \gamma^2 \int_{-\infty}^{\infty} G_{\alpha\beta}(\tau) e^{-i\omega_{\alpha\beta}\tau} \, d\tau \qquad (5.12)$$

Generally, the exponential form of $G(\tau)$ embodied in equations (5.9) and (5.10) is used in evaluation of equation (5.12) and the transition probability becomes:

$$W_{\alpha\beta} = \gamma^2 \overline{H_{loc}^*(t) H_{loc}(t)} \frac{2\tau_c}{1 + \omega_{\alpha\beta}^2 \tau_c^2} \qquad (5.13)$$

Thus, it is noted that $W_{\alpha\beta}$ shows the same dependence on τ_c as does $J(\omega)$.

For an isolated single spin system in a magnetic field, only two energy levels are present, and $1/T_1 = W_{\alpha\beta}$. In coupled multi-spin systems, however, the relationship is generally more complex between T_1 and the several different $W_{\alpha\beta}$ which characterise the system. In fact, the concept of a single exponential relaxation time for some coupled systems has no validity and multiple parameters are required to characterise the relaxation process. Use of individual transition probabilities, $W_{\alpha\beta}$, of course, avoids these problems as only one parameter is needed at most to describe the relaxation between any two spin levels in a coupled multi-spin system.

In many small molecules whose relaxation rate is studied at room temperature or above, equation (5.13) can often be simplied. For such systems the correlation times τ_c may be of the order of 10^{-10} s or shorter, and since at the usual n.m.r. field strengths of 14.1 and 23.5 kG, ω is of the order of $10^7 - 10^8$ rad s^{-1}, the condition $\omega^2\tau_c^2 \ll 1$ holds. Applying this condition to equation (5.13) the following expression for $W_{\alpha\beta}$ is obtained.

$$W_{\alpha\beta} = 2\gamma^2\overline{H_{\text{loc}}^*(t)H_{\text{loc}}(t)}\tau_c \tag{5.14}$$

$W_{\alpha\beta}$ is now independent of the frequency separation between the several energy levels and is dependent only on τ_c and $\overline{H_{\text{loc}}^*(t)H_{\text{loc}}(t)}$. This region of the spectral density is termed the 'white spectrum' or region of 'motional narrowing'[2, 31, 32]. So long as the 'motional narrowing' requirement is met, the transition probability, $W_{\alpha\beta}$, and the inverse relaxation time, $1/T_1$, for simple isolated spin systems depend directly on τ_c which is proportional to the temperature of the sample. A plot of T_1 and T_2 versus correlation time is given in Figure 5.4. As the correlation time increases, T_1 decreases until the extreme narrowing condition is violated and a minimum is reached in the range of $\omega_0 \approx 1/\tau_c$. As $\omega^2\tau_c^2$ becomes much greater than unity, the transition probability becomes:

$$W_{\alpha\beta} = 2\gamma^2\overline{H_{\text{loc}}^*(t)H_{\text{loc}}(t)}(1/\omega^2\tau_c) \tag{5.15}$$

and $1/T_1 \approx 1/\tau_c$. In this region, however, $1/T_1 \approx 1/\omega_0^2$ and relaxation measurements as a function of field strength may be used to establish this fact. The plot of T_2 $v.\tau_c$, also given in Figure 5.4, indicates that T_2 continues to decrease to a lower limiting value as τ_c increases. This results from the appearance of frequency independent terms in expressions for T_2. In the region of motional narrowing, when dipolar processes dominate both the T_1 and T_2 relaxation of the spin system, then $T_1 = T_2$, which is the case shown in Figure 5.4.

Figure 5.4 also portrays the effect on T_1 as the static magnetic field is increased to larger ω_0 values. At higher field strengths the onset of the T_1 minimum is encountered at shorter τ_c values. The favourable consequence of $T_2 < T_1$ becomes apparent when it is recalled from the last section that some of the experimental pulse techniques for measuring T_1 required very efficient T_2 processes to avoid echoes from the magnetisation in the x,y-plane. Unfortunately, line width is inversely proportional to T_2 and very high resolution spectra are not obtainable at very short T_2.

As higher-field spectrometers of a superconducting type are employed in the study of macromolecules where correlation times can be relatively long, it is likely that one will be on the right side of the minimum in the T_1

curve of Figure 5.4. This will lead to longer T_1 values at the higher ω_0 value with the consequence that longer delay times between repetitive FT pulses will be required to avoid saturation along the z-axis. This loss in efficiency may, in fact, offset any advantages realised for the higher field

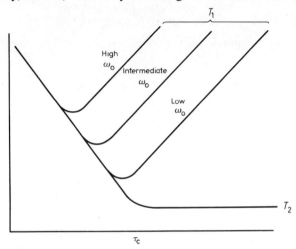

Figure 5.4 Relaxation times v. correlation times are plotted for various resonance frequencies

strengths. Furthermore, the greater importance of the chemical-shielding relaxation mechanism may lead to even shorter T_2 values with the concomitant loss in resolution and signal-to-noise ratio. Thus, the improved resolution of closely-spaced peaks normally expected for superconducting systems may not in every case be obtainable.

5.3.2   Mechanisms of nuclear spin relaxation

It is now appropriate to introduce the individual relaxation mechanisms contributing to T_1. In each case, a brief discussion of the origin of the mechanism, the form of the perturbing Hamiltonian and the specific T_1 formulation is presented. In general, each relaxation process can be described by a bi-linear coupling of the nuclear spin to its lattice via a tensor A which is a function of lattice parameters[2]. In the molecular coordinate system, A has well-defined constant components; however, n.m.r. is concerned with the orientation of nuclear spins in a laboratory frame of reference. A thus must be transformed to the laboratory frame through the usual rotation matrices. As liquid systems are subject to rapid molecular motion, the components of A in the laboratory frame become time-dependent thereby providing a process for spin relaxation.

5.3.2.1   *Magnetic dipole–dipole*

The principle mode of spin–lattice relaxation in most molecular systems arises from the local magnetic fields produced by spin–spin dipolar interactions. Any magnetic nucleus produces an instantaneous local magnetic dipole field proportional to its magnetic moment and modified by distance and geometrical factors[35, 36] as follows:

$$H_{\text{loc}} = \pm \frac{\mu_j}{r_{ij}^3}(3\cos^2\theta_{ij} - 1) \tag{5.16}$$

where H_{loc} is the magnetic field produced at nucleus i by nucleus j, μ_j the dipole moment of j, r_{ij} the nuclear separation of i and j, and θ_{ij} the angle of r_{ij} vector relative to the applied magnetic field H_0. Thus, both inter- and intra-molecular interactions are possible except that the former will tend to be attenuated more readily by the r^{-3} dependence. Local fields from this source may be as large as several dekagauss[37]. In a rigid system of intramolecular dipoles, θ_{ij} assumes a time dependence in liquids due to molecular tumbling while for intermolecular interactions r_{ij} and θ_{ij} can both fluctuate with time owing to translational and rotational diffusion.

For the interaction of two spins I and S, the perturbing Hamiltonian is[2]

$$\mathscr{H}'_{\text{D}}(t) = \hbar I \cdot H_{\text{loc}}(t) = \hbar I \cdot D \cdot S \tag{5.17}$$

where D is the dipolar-coupling tensor and contains the time dependence of the system. When I and S are both spin $\frac{1}{2}$ nuclei and the same nuclear species, the intramolecular contribution[38] to the relaxation of I becomes

$$1/T_{1,\text{D}} = \frac{6}{20}\frac{\hbar^2\gamma^4}{r_{IS}^6}\left[\frac{\tau_{\text{r}}}{1+\omega^2\tau_{\text{r}}^2} + \frac{4\tau_{\text{r}}}{1+4\omega^2\tau_{\text{r}}^2}\right] \tag{5.18}$$

whereas the expression for heteronuclear dipolar relaxation becomes

$$\frac{1}{T_{1,\text{D}}} = \frac{1}{10}\frac{\hbar^2\gamma_I^2\gamma_S^2}{r_{IS}^6}\left[\frac{\tau_{\text{r}}}{1+(\omega_I-\omega_S)^2\tau_{\text{r}}^2} + \frac{3\tau_{\text{r}}}{1+\omega_I^2\,\tau_{\text{r}}^2} + \frac{6\tau_{\text{r}}}{1+(\omega_I+\omega_S)^2\tau_{\text{r}}^2}\right] \tag{5.19}$$

The several terms in the two expressions for $1/T_{1,\text{D}}$ arise from the several $W_{\alpha\beta}$ terms associated with a two spin system, and τ_{r} is a reorientational correlation time. For the purpose of comparison and of later discussion, the formulation for the homonuclear dipolar contribution to T_2 is also presented[38]:

$$\frac{1}{T_{2,\text{D}}} = \frac{3}{20}\frac{\hbar^2\gamma^4}{r_{IS}^2}\left[3\tau_{\text{r}} + \frac{5\tau_{\text{r}}}{1+\omega^2\tau_{\text{r}}^2} + \frac{2\tau_{\text{r}}}{1+4\omega^2\tau_{\text{r}}^2}\right] \tag{5.20}$$

and for the heteronuclear case

$$\frac{1}{T_{2,\text{D}}} = \frac{\hbar^2\gamma_I^2\gamma_S^2}{20\,r_{IS}^6}\left[4\tau_{\text{r}} + \frac{\tau_{\text{r}}}{1+(\omega_I-\omega_S)^2\tau_{\text{r}}^2} + \frac{3\tau_{\text{r}}}{1+\omega_S^2\tau_{\text{r}}^2} + \frac{6\tau_{\text{r}}}{1+\omega_I^2\tau_{\text{r}}^2} + \frac{6\tau_{\text{r}}}{1+(\omega_I+\omega_S)^2\tau_{\text{r}}^2}\right] \tag{5.21}$$

From the correspondence between the bracketed terms of equations (5.18–5.21) and the frequency dependence of equation (5.10), it is obvious that for homonuclear coupled spins $T_{1,\text{D}}$ depends on the spectral density at ω_I and

$2\omega_I$, while $T_{2,D}$ depends on the power available at zero frequency $J(0)$ as well as $J(\omega_I)$ and $J(2\omega_I)$. For heteronuclear spins, the formulae are somewhat more complex due to the presence of the $(\omega_I - \omega_S)$ term linking the $\alpha\beta$ and $\beta\alpha$ spin states. Again, $T_{2,D}$ depends on a $J(0)$ term while $T_{1,D}$ does not. This zero frequency component in $T_{2,D}$ accounts for the different dependence of T_1 and T_2 upon τ_c as exhibited in Figure 5.4. The dependence of $T_{1,D}$ on either $J(2\omega_I)$ or $J(\omega_I + \omega_S)$ is peculiar to the dipole–dipole and quadrupolar mechanisms and its origin as well as the origin of the $J(0)$ dependence in $T_{2,D}$ are discussed in reference 39. Under the extreme narrowing approximation $(\omega^2\tau_r^2 \ll 1$ and $(\omega_I + \omega_S)^2\tau_r^2 \ll 1)$, the following simplified expressions are obtained for $T_{1,D}$

$$\frac{1}{T_{1,D}} = \frac{1}{T_{2,D}} = \frac{3\,\hbar^2\gamma^4}{2\,r_{IS}^6}\tau_r \tag{5.22}$$

for homonuclear relaxation, and for the heteronuclear case the result is

$$\frac{1}{T_{1,D}} = \frac{1}{T_{2,D}} = \frac{\hbar^2\gamma_I^2\gamma_S^2}{r_{IS}^6}\tau_r \tag{5.23}$$

These expressions are the usual ones given for the dipolar relaxation rate for a $I = \frac{1}{2}$ nucleus coupled to another $S = \frac{1}{2}$ nucleus in the motional narrowing approximation and for isotropic rotational diffusion characterised by a single correlation time constant. The absence of frequency terms in equations (5.22) and (5.23) follows directly from the frequency invariance of $J(\omega)$ over the range of Larmor frequencies encountered in a coupled two-spin system.

Equation (5.23) can be modified to handle the case of multi-nuclear spin relaxation as follows[8]:

$$\frac{1}{T_{1,D}} = \sum_S \frac{\hbar^2\gamma_I^2\gamma_S^2}{r_{IS}^6}\tau_{eff} \tag{5.24}$$

providing $S \neq I$ and an effective isotropic correlation time, τ_{eff}, can be employed. This expression has often been utilised as a means of acquiring rough quantitative estimates of the correlated motion even though anisotropic motion may be present. Under such conditions, however, details of the motional features are obscured and some care should be taken not to over interpret these approximate correlation times. Due to the rapid attenuation of the r_{IS}^{-6} term, a simplified form of equation (5.24) can be obtained by summing S only over the directly-bonded nuclei to yield,

$$\frac{1}{T_{1,D}} = \frac{n_S\hbar^2\gamma_I^2\gamma_S^2}{r_{IS}^6}\tau_{eff} \tag{5.25}$$

where n_S is the number of directly bonded S nuclei. This form presumes that all I–S distances are constant and reorientation of all I–S vectors are governed by the same effective correlation time. In addition, the isotropic assumption requires that internal molecular motions do not provide effective means for nuclear relaxation. Very low frequency vibrations and some internal rotations can significantly invalidate the assumption of an overall constant, isotropic correlation time.

It is worth noting that the C—H dipole–dipole interaction is the most efficient dipolar process due to the relatively larger γ_H and to the small C—H bond distance. Using unity for the proton–carbon-13 coupling, the relative values of $S(S+1)\gamma_S^2/r_{IS}^6$ for several directly-bonded nuclei are given as follows: bromine-81, 0.013, iodine-127, 0.0095; chlorine-35, 0.0030; nitrogen-14, 0.0021. Standard single-bond covalent radii were used to calculate r_{IS} values and γ_S was obtained from standard tables. Thus, to the extent that a molecule has the same τ_{eff}, the other dipolar processes will not compete effectively with a proton–carbon-13 process whenever a carbon has a directly-bonded hydrogen. Furthermore, many of the other mechanisms will be competitive with ^{13}C—X dipolar processes even when they are not comparable to the ^{13}C—^1H process.

For non-spherically-symmetrical systems or anisotropic tumblers, it is convenient to use a rotational diffusion tensor in place of correlation times. This second-rank tensor in diagonal form specifies the rotational diffusion about the three principal axes of a molecule. The relationship between each D_i component and the corresponding τ_r for motion about that principal axis is given by[40]

$$D_i = \tfrac{1}{6}\tau_{ri} \tag{5.26}$$

The simplest anisotropic system to treat is that of a symmetric top in which two of the three diagonal elements are equal. Woessner[34] and others[41] have developed an expression for T_1 in terms of these two unique parameters $D_{||}$ and $D_\perp$, which specify, respectively, the rotational diffusion about the C_3 axis and about the two perpendicular axes, as follows:

$$\frac{1}{T_{1,D}} = \frac{n_S\hbar^2\gamma_I^2\gamma_S^2}{r_{IS}^6}\left[\frac{A}{6D_\perp} + \frac{B}{5D_\perp + D_{||}} + \frac{C}{2D_\perp + 4D_{||}}\right] \tag{5.27}$$

where A, B, and C are geometrical constants given as follows:

$$A = \tfrac{1}{4}(3l^2 - 1)^2$$
$$B = 3l^2(1 - l^2) \tag{5.28}$$
$$C = \tfrac{3}{4}(l^2 - 1)^2$$

where l is the direction cosine of the angle between the C—H vector and the C_{3v} symmetry axis.

Thus,

$$\tau_{eff} = D_\perp^{-1}[A/6 + B/(5+\sigma) + C/(2+4\sigma)] \tag{5.29}$$

where $\sigma = D_{||}/D_\perp$. Note, when $D_{||} \gg D_\perp$ then $\sigma \gg 5$, $4\sigma \gg 2$, and the two terms involving B and C vanish giving $\tau_{eff} \approx A/6D_\perp$. Thus, very rapid symmetric-top motions about the parallel axis may become totally ineffective in dipolar–nuclear relaxation. In order to evaluate both $D_{||}$ and $D_\perp$ or conversely $D_\perp$ and σ, the dipolar relaxation times of at least two I nuclei with different geometrical dependence on the IS vectors are required to solve the simultaneous equations which ensue.

For asymmetric tops three rotational diffusion constants are required to characterise the motional details. Likewise, at least three different T_1 values for three nuclei with non-equivalent IS geometrical configurations relative to the principal axes are needed so that three independent simultaneous expressions may be established for evaluating the three components of a diagonalised diffusion tensor. The expression for one of these T_1 values in terms of the various geometrical and diffusion parameters is given by[39]

$$\frac{1}{T_{1,D}} = \frac{n_s \hbar \gamma_I^2 \delta_S^2}{r_{IS}^6} \left[C_+ \tau_+ + C_- \tau_- + C_1 \tau_1 + C_2 \tau_2 + C_3 \tau_3 \right] \tag{5.30}$$

where

$$\frac{1}{\tau_1} = 4R_1 + R_2 + R_3 \qquad \frac{1}{\tau_\pm} = 6[R \pm (R^2 - L^2)^{\frac{1}{2}}]$$

$$\frac{1}{\tau_2} = 4R_2 + R_1 + R_3 \qquad R = \tfrac{1}{3}(R_1 + R_2 + R_3) \tag{5.31}$$

$$\frac{1}{\tau_3} = 4R_3 + R_1 + R_2 \qquad L^2 = \tfrac{1}{3}(R_1 R_2 + R_1 R_3 + R_2 R_3)$$

The various R values in equation (5.31) refer to the reorientation rates about the principal axes, while the C values in equation (5.30) are geometrical constants defined in reference 34. Numerical solution of equations (5.30) and (5.31) in R_1, R_2 and R_3 for at least three different $T_{1,D}$ values presumes that no internal vibrational or rotational modes will contribute to the relaxation of the three I nuclei required to obtain the data.

As in the symmetric top case, very rapid rotational diffusion about any one principle axis will reduce the efficiency of that motion in the relaxation process.

The treatment of internal rotation of a top attached to a rigid isotropic tumbler requires no additional formulation as one may treat the system with equation (5.27) by substituting R for $D_\parallel$ and D for $D_\perp$. Here R is the internal rotational diffusion constant and D is the overall molecular diffusion parameter. Treatment of internal rotation of a methyl top attached to a rigid symmetric-top tumbler, however, involves an expression of considerable more complexity. Woessner[42] has again given formulation for this case as follows:

$$\frac{1}{T_{1,D}} = \frac{n_s \hbar^2 \gamma_I^2 \gamma_S^2}{r_{IS}^6} \left[\frac{A_1}{6D_\perp} + \frac{A_2 + A_3}{6D_\perp + R} + \frac{B_1}{5D_\perp + D_\parallel} + \frac{B_2 + B_3}{5D_\perp + D_\parallel + R} + \right.$$
$$\left. + \frac{C_1}{2D_\perp + 4D_\parallel} + \frac{C_2 + C_3}{2D_\perp + 4D_\parallel + R} \right] \tag{5.32}$$

The reader is referred to reference 42 for definition of the geometrical constants used in equation (5.32) and for the more complete treatment of an internal top attached to a completely asymmetric rigid tumbler. The extent of this latter formulation unfortunately exceeds the scope of this review and further details on the dipolar relaxation mechanism must be obtained from the original literature.

The intermolecular dipolar relaxation rate is more difficult to formulate because of the translational diffusion features important in this interaction. In most cases it is usually assumed[43] that spin I is relaxed by the relative changes in the intermolecular dipolar vector arising from translation while rotational effects are neglected. It is further assumed that molecules have a single effective distance of closest approach to give the following expression[44]:

$$\frac{1}{T_{1,\mathrm{D\,inter}}} = \frac{2\pi\hbar^2\gamma_I^2 N\tau_t}{a^2}\left[4\gamma_I^2 I(I+1)\sum_S\frac{1}{d_{IS}}+\frac{8}{3}\sum_F\gamma_F^2 F(F+1)\frac{1}{d_{IF}}\right] \quad (5.33)$$

where the sum over S includes all nuclei of the same type as I while the sum over F is for all non-identical nuclei on neighbouring molecules. The d_{IS} and d_{IF} parameters represent distances of closest approach, while N represents the number of molecules per unit volume, a is the effective radius of the spherical molecules and τ_t the translational correlation time. As will be seen in Section 5.4, intermolecular dipole relaxation has rarely needed to be considered for ^{13}C spin–lattice relaxation.

5.3.2.2 Spin–rotation

Electron and nuclear currents associated with overall molecular rotation can give rise to correlated fluctuating magnetic fields which can lead to spin relaxation. These fluctuations in the local magnetic fields may result from a modulation of both the magnitude and direction of the angular momentum vector associated with the rotating molecular system[31]. Although currents arising from completely symmetrical negative and positive charge distributions, of course, can be expected to cancel one another, it is to be noted that any angular momentum possessed by electrons on a given nucleus, no matter how symmetrical the charge distribution may be, will on the average lead to a local magnetic field at that nucleus. In liquids the interruption of angular momentum associated with rotation will result from strongly perturbing intermolecular interactions. In general, the period between such encounters is very short giving rise to correspondingly very short spin–rotation correlation times, τ_{SR} which will lead to inefficient power densities at the nuclear Larmor frequencies[46]. Furthermore, only those molecules with relatively small moments of inertia will generate rotational velocities which can give rise to charged currents with sufficiently large associated magnetic fields that relaxation by this mechanism will be significant.

The Hamiltonian for the spin–rotation interaction is given as[31]

$$\mathcal{H}_{SR}(t) = -\boldsymbol{I}\cdot\boldsymbol{C}\cdot\boldsymbol{J}(t) \quad (5.34)$$

where $\boldsymbol{I}$ is the spin angular momentum, $\boldsymbol{C}$ the spin–rotation interaction tensor, and $\boldsymbol{J}(t)$ the time-dependent angular momentum associated with overall molecular rotation. While equation (5.34) is relatively easy to employ in the gas phase where a set of 'good' J quantum numbers can be used to characterise such systems, only an approximate solution for liquids is possible because the states are rendered indistinguishable by lifetime broadening resulting from extensive intermolecular interactions[46]. Thus, relatively simple ensemble averaging of $J(t)$ over all possible angular momenta is employed to estimate the magnitude of $H_{loc}(t)$ due to rotation. For spherically symmetric molecules the effective $T_{1,SR}$ (T_1 due to spin rotation) for a magnetic nucleus at the centre of symmetry is given by[46]

$$\frac{1}{T_{1SR}} = \frac{2kT}{\hbar^2} I_m C^2 \tau_{SR} \tag{5.35}$$

where I_m is the moment of inertia, C the isotropic spin–rotation interaction constant and τ_{SR} is the spin–rotation correlation time. Equation (5.35) presumes that the condition of extreme narrowing holds (i.e. $\omega^2 \tau_r^2 \ll 1$) and requires the assumption that the effective epicentre of rotation is the centre of gravity. If the magnetic nucleus lies away from the centre of gravity but in a cylindrically-symmetric electronic environment, the expression becomes[47]

$$\frac{1}{T_{1,SR}} = \frac{2kT}{3\hbar^2} I_m (C_{\parallel}^2 + 2C_{\perp}^2) \tau_{SR} \tag{5.36}$$

where the parallel direction is given by the principal vector of the rotation axis. The extension of $T_{1,SR}$ to a nucleus in a totally asymmetric electronic environment, but in an overall spherical or isotropic tumbler, involves using the two independent C values for $2C_{\perp}$ in equation (5.36).

While the formulation $T_{1,SR}$ for asymmetric tops has not yet been published, recent results[16, 17] have indicated that $T_{1,SR}$ for symmetric-top molecules may be written as

$$\frac{1}{T_{1,SR}} = \frac{2kT}{3\hbar^2} \left\{ C_{\parallel}^2 I_{m\parallel} \left(\frac{\tau_{SR\parallel}}{1+2D_{\perp}\tau_{SR\parallel}} \right) + 2C_{\perp}^2 I_{m\perp} \left(\frac{\tau_{SR\perp}}{1+(D_{\perp}+D_{\parallel})\tau_{SR\perp}} \right) \right\} \tag{5.37}$$

in the extreme narrowing limit where C, I_m and τ_{SR} have the same definitions as before, while their tensor properties are specified by the parallel, $\parallel$, and perpendicular $\perp$, designations. The parallel axis is taken to be the principal axis of the top. The microscopic rotational diffusional tensor terms are given by $D_{\parallel}$ and $D_{\perp}$, respectively. Equation (5.37) assumes that the duration of correlation between the orientation and angular momentum is short compared either to the reorientational or to the angular momentum correlation times, thereby allowing an easier separation of these terms in the averaging process[48]. In the case of many methyl tops where spin–rotation is often very important, both $D_{\perp}$ and $\tau_{SR\perp}$ are usually relatively small, leading to the condition that $2D_{\perp}\tau_{SR\parallel} \ll 1$ and $(D_{\perp}+D_{\parallel})\tau_{SR,\perp} \ll 1$. Equation (5.37) now becomes,

$$\frac{1}{T_{1,SR}} = \frac{2kT}{3\hbar^2} \{ C_{\parallel}^2 I_{m,\parallel} \tau_{SR,\parallel} + 2C_{\perp}^2 I_{m,\perp} \tau_{SR,\perp} \} \tag{5.38}$$

Note, that this expression does *not* require $D_{\parallel} \tau_{SR\cdot\parallel} \ll 1$, a condition which would not hold for rapidly-rotating methyl tops and one that is fortunately not required to simplify equation (5.37). It is to be noted that the same approximations which lead to equation (5.37) also minimise the importance of the second term and one often obtains the very simple but good approximation

$$\frac{1}{T_{1,SR}} = \frac{2kT}{3\hbar^2} C_\parallel^2 I_{m,\parallel} \tau_{SR,\parallel}$$ (5.39)

$\tau_{SR,\parallel}$ values, therefore, can be determined readily from $T_{1,SR}$ in this instance providing values for $C_\parallel$ can be obtained from other spectroscopic sources.

The paucity of experimental values for C does pose a fairly significant limitation on the present use of spin–rotation formalism. Flygare[49] has presented a reliable way of estimating the magnitudes of C from the paramagnetic shielding constant which fortunately dominates the variations in carbon-13 chemical shifts. This method utilises the known spin–rotation constant[50] and the carbon-13 chemical shift of CO to scale and calibrate the relationships which are needed to estimate C from the shielding constants. To avoid the need for shift anisotropies in the calculation of $C_\parallel$ and $C_\perp$, it is noted that $C_\parallel I_\parallel$ is almost equal to $C_\perp I_\perp$ for many systems and, therefore, approximately equal to the trace of the combination $(CI_m)_t$ tensor. Equation (5.38) now reduces to

$$\frac{1}{T_{1,SR}} \approx \frac{2kT(CI_m)_t^2}{3\hbar^2} \left\{ \frac{\tau_{SR,\parallel}}{I_{m,\parallel}} + \frac{2\,\tau_{SR,\perp}}{I_{m,\perp}} \right\}$$ (5.40)

where $(CI_m)_t = \frac{1}{3} \overset{x,y,z}{\Sigma} C_i I_{mi}$ is the trace of this combination tensor. Now, the negligible magnitude of the second term can be readily seen as $\tau_{SR,\perp} < \tau_{SR,\parallel}$ and usually $I_{m,\perp} > I_{m,\parallel}$ for rotating tops. The importance of a low moment of inertia, stressed earlier, about the axis of prime rotation is now easily observed from equation (5.40). Reference 87 also supports many of the conclusions contained in this section.

5.3.2.3 Chemical-shift anisotropy

The magnetic field at a nucleus due to the external magnetic field H_0 is given by[45]

$$H_{loc} = (1 - \sigma) H_0$$ (5.41)

where σ is the chemical shielding tensor. The σ contribution has its origins in the magnetic screening produced by the surrounding electrons. If the screening is not isotropic, σ will have directional components which vary with time as the molecules tumble relative to the H_0 axis. The nuclei then experience a fluctuating magnetic field which can relax the nuclear spins.

The bilinear coupling between H_0 and spin I is given[31] by the following Hamiltonian:

$$\mathscr{H}'_A(t) = -H_0 \cdot \hbar \sigma \cdot I$$ (5.42)

which yields in the motional narrowing limit the appropriate expression for $T_{1,SA}$, the shift anisotropy relaxation time, as follows:

$$\frac{1}{T_{1,SA}} = \frac{\gamma_I^2 H_0{}^2}{5} (\sigma_{12}^2 + \sigma_{23}^2 + \sigma_{31}^2) \tau_r$$ (5.43)

where the several σ_{ij}'s represent the anisotropic magnitudes, $(\sigma_i - \sigma_j)/3$, of the three principal terms in the diagonalised shielding tensor, $\boldsymbol{\sigma}$. The mechanism is dependent on τ_r, the re-orientational correlation time important in the dipole–dipole mechanism. If $\boldsymbol{\sigma}$ is axially symmetric (C_{3v} or higher symmetry), two of the terms in $\boldsymbol{\sigma}$ are identical and equation (5.43) reduces to

$$\frac{1}{T_{1,\mathrm{SA}}} = \frac{2}{15} \gamma_I^2 H_0^2 (\sigma_\parallel - \sigma_\perp)^2 \tau_r \tag{5.44}$$

where $\sigma_\parallel$ and $\sigma_\perp$ are the components of $\boldsymbol{\sigma}$ parallel and perpendicular to the symmetry axis. Of importance in this mechanism is the quadratic field dependence of $T_{1,\mathrm{SA}}$ in the motion narrowing limit. In the limit, $\omega^2 \tau_r^2 \gg 1$ the relaxation will become field invarient, however.

5.3.2.4 Quadrupole

Although this process occurs only with nuclei that have a spin quantum number $I \geqslant 1$ and thus is not directly applicable to ^{13}C relaxation, it does indirectly enter in the consideration of ^{13}C relaxation rates when a scalar-coupling mechanism of the second kind is important. Nuclei with spins $\geqslant 1$ possess an electric quadrupole moment owing to the non-spherical symmetry of the charge distribution at the nucleus[45]. The quadrupole interacts with the electric field gradient produced at the nuclear position by the surrounding electric charges. Molecular motion of the system imparts a temporal variance to the local field gradient producing a fluctuating electric field at the nucleus which is capable of inducing transitions between the nuclear quadrupolar levels and hence magnetic relaxation.

The perturbation Hamiltonian, $\mathscr{H}'_Q(t)$, for the ineraction between the nuclear quadrupole moment and the local electric field gradient is[31]

$$\mathscr{H}'_Q(t) = I \cdot Q(t) \cdot I \tag{5.45}$$

where $Q(t)$ is a function of the quadrupole coupling tensor. Evaluation of the pertinent transition probabilities gives[2]

$$\frac{1}{T_{1,Q}} = \frac{1}{T_{2,Q}} = \frac{3}{40} \frac{(2I+3)}{I^2(2I-1)} \cdot \left(1 + \frac{\xi^2}{3}\right) \left(\frac{e^2 qQ}{\hbar}\right)^2 \tau_r \tag{5.46}$$

where I is the spin quantum number, $e^2 qQ/\hbar$ the quadrupole coupling constant, and ξ the asymmetry parameter. In the motional narrowing approximation, this mechanism is also proportional to τ_r. The correlation time τ_r is again for molecular re-orientation and, therefore, identical to that used for dipole–dipole relaxation. In most cases, this mechanism dominates the relaxation of nuclear spins $I \geqslant 1$ unless the electric field gradient is very small or zero due to symmetry.

5.3.2.5 Scalar-coupling

The multiplet structure observed in high-resolution n.m.r. spectra results from a bilinear coupling between magnetically non-equivalent nuclei found within the same molecule. This coupling, distinct from dipolar coupling, is a second-order effect originating from the Fermi hyperfine coupling of nuclei through the molecular electrons. The observation of spin–spin coupling requires[2] that the relaxation time of both spins be long compared with the inverse of the coupling constant (J) and that there are no relatively rapid time dependent processes which will either average the J value to zero or scramble the spin designations of one or both of the nuclei. Whenever these criteria are violated, the coupling will collapse and a single resonance line will be observed. Rapid chemical exchange provides one such process of collapsing the multiplet structure providing the frequency of exchange ($1/\tau_e$) is much greater (i.e. $1/\tau_e \gg 2\pi J$) than the frequency of the induced coupling. Whenever the exchange rate is of proper frequency, the process can act as an efficient relaxation mechanism for the I spin. The mechanism, referred to by Abragam[2] as scalar coupling of the 'first kind', requires that the power density be peaked near the Larmor frequency as the local magnetic fields due to scalar coupling are usually two or three orders of magnitude less than the corresponding dipolar couplings.

Alternatively, when spin 'S' has a relaxation mechanism such that $1/T_{1,s} \gg 2\pi J$ (i.e. the local field due to 'S' fluctuates at a rate fast compared to the magnitude of the splitting), then the local magnetic field due to scalar coupling again fluctuates and can once again give rise to significant nuclear relaxation providing the modulating frequency is comparable to the Larmor frequency of I. This mechanism, referred to as scalar coupling of the 'second kind'[2] becomes important for spin $-\frac{1}{2}$ nuclei which are bound to spins with $I > \frac{1}{2}$ that relax via a rapid quadrupole process.

The interaction Hamiltonian is represented by[2]

$$\mathcal{H}'_{SC}(t) = 2\pi\hbar J \, \mathbf{I}\cdot\mathbf{S} \tag{5.47}$$

J being the scalar coupling constant. The $\mathcal{H}$ results in a formulation of the scalar-coupling process as

$$\frac{1}{T_{1,SC}} = \frac{8\pi^2 J^2}{3} S(S+1) \frac{\tau_{SC}}{1+(\omega_I - \omega_S)^2 \tau_{SC}^{\,2}} \tag{5.48a}$$

$$\frac{1}{T_{2,SC}} = \frac{1}{2T_{1,SC}} + \frac{4\pi^2 J^2}{3} S(S+1)\tau_{SC} \tag{5.48b}$$

J being the scalar coupling constant. The $\mathcal{H}$ results in a formulation of the where τ_{SC} is the chemical exchange time (τ_e) for scalar coupling of the first kind and is T_Q the quadrupolar relaxation time of the S spin when the scalar coupling mechanism is of the second kind. The respective resonant frequencies of the two nuclei are ω_I and ω_S. In the usual case quadrupolar relaxation of S is such that $T_Q = T_{1,Q} = T_{2,Q}$. The two modes of relaxation by scalar coupling differ only in the origin of the time dependent process and otherwise the formulae are identical. As scalar-coupling relaxation requires τ_e or T_Q to take values so as to maximise the power densities one is often in the region where $\tau_e^2(\Delta\omega)^2 \approx 1$ or $T_Q^2(\Delta\omega)^2 \approx 1$ in violation of the motional narrowing approximations for these correlation times. Thus, the field dependence of the $(\Delta\omega)^2$ terms may be manifest in the relaxation.

5.3.3 Separation of various mechanisms important in carbon-13 relaxation

In this section the separation of individual relaxation rates for the various mechanisms is outlined. It is assumed for convenience that paramagnetic impurities are not present, and as the carbon-13 nucleus has a spin of $\frac{1}{2}$ the direct quadrupolar relaxation need not be of concern. One or more of the four remaining relaxation mechanisms, however, may simultaneously be important, and the overall T_1 relaxation time can be calculated from the sum of the inverse relaxation times of the individual mechanisms as follows[2]:

$$\frac{1}{T_1} = \sum_j \frac{1}{T_{1,j}} \tag{5.49}$$

This formulation supposes independent contributions from each mechanism as all cross-terms are neglected. In some cases the cross-correlation effects actually vanish due to different transformation properties of the perturbing Hamiltonians under rotation[2].

As many carbons are directly bound to important magnetic nuclei such as ^1H, ^{19}F and ^{31}P, the heteronuclear dipolar process is usually a prime contributor in ^{13}C relaxation[9, 11, 51]. The separation of this process from the overall relaxation rate can be accomplished through a measurement of the $^{13}C-\{X\}$ nuclear Overhauser enhancement (NOE) factor[8, 52]. The Overhauser phenomenon, which may contribute significantly to the increased S/N ratio for a ^{13}C signal upon heteronuclear decoupling, is solely dependent upon the dipolar relaxation rate[38]. Detailed consideration of this inherent relationship for ^{13}C has been given by Kuhlmann, Grant and Harris[8] following the general theory given by Solomon[38].

For a heteronuclear two-spin system of spin $-\frac{1}{2}$ nuclei in the region of motional narrowing, the differential NOE factor for ^{13}C is given by[10]

$$\eta = \frac{1}{2}\frac{\gamma_X}{\gamma_C} \cdot \frac{T_1}{T_{1,D}} \tag{5.50}$$

where γ_X and γ_C are the respective magnetogyric ratios and $T_{1,D}$ is the heteronuclear dipolar relaxation time. The overall signal enhancement due to the NOE is given by $(1+\eta)$. The maximum in η occurs when $T_{1,D} \approx T_1$ and is 1.988 for the proton case[52], while for ^{19}F and ^{31}P the maximum values are 1.871 and 0.805. When other relaxation pathways effectively contribute to T_1, $T_{1,D} > T_1$ and the value of η_{C-H} will drop towards zero. Equation (5.50) is strictly correct only for the situation in which all of the coupled nuclei are saturated and no other spin $-\frac{1}{2}$ nuclei are present[8]. Under these conditions the relaxation of the enhanced ^{13}C lines is governed by a single exponential time constant[51]. It should be noted that this proton-decoupled T_1 value measures the characteristic time to re-establish the non-Boltzmann steady-state[38] situation which arises upon proton decoupling.

Separation of the dipolar mechanism from the other relaxation processes allows attention to then be focused on spin–rotation, shift-anisotropy, and scalar-coupling mechanisms. In the extreme narrowing limit, the shift-anisotropy mechanism is the only one which has a quadratic dependence upon the magnetic field, and becomes more efficient as the applied magnetic field is increased. A study of T_1 as a function of H_0 therefore provides for the separation of the chemical-shift anisotropy relaxation rate from the remaining terms in equation (5.49). Whenever the motional narrowing approximation does not hold, all of the other mechanisms can manifest a field dependence due to the inverse $(1+\omega^2\tau_c^2)$ term which has a field dependence. Thus, increased fields lead to longer T_1 values and not to the more efficient shorter values noted for the shift-anisotropy mechanism.

The scalar-coupling mechanism may exhibit the inverse field dependence even for small molecules due to the great range of characteristic times governing this mechanism. The effect has been observed especially when carbon and bromine are directly bonded to each other. When it is clear from other arguments that the scalar-coupling process is the only important field dependent mechanism, then evaluation of $T_{1,SC}$ is possible from variable H_0 data. The simultaneous importance of two mechanisms both exhibiting field dependence invalidates the use of T_1 $v.$ H_0 data as a means of separating relaxation rates. When the scalar-coupling mechanism is field dependent and dominated by the $(\omega_I - \omega_S)$ term in the denominator of equation (5.48), $T_{1,\rho}$, the spin–lattice relaxation time for the system in the rotating-reference frame, is controlled by the scalar-coupling process and a method for determining $T_{1,SC}$ has been given by Rhodes et al.[53]. Measurement of $T_{1,}o$ is usually accomplished by aligning the spin magnetisation along H_1 and observing its decay to zero. In this manner $T_{1,SC}$ can be isolated from the other terms even when competitive field-dependent processes prevent separation of $T_{1,SC}$ from the T_1 $v.$ H_0 data. The unique dependence which the spin–rotation relaxation rate has on temperature provides an excellent method for separating the effect of spin–rotation from the other processes. Because of the more facile molecular re-orientation at higher temperatures, rotational angular momentum increases with temperature to give a more efficient spin–rotation relaxation process. This temperature effect is clearly observed in equations (5.35–5.40). Conversely, increasing temperature shortens τ_r and all mechanisms depending on the re-orientation correlation process lose efficiency accordingly. Often, T_1 will pass through a maximum in a plot against temperature as mechanisms dependent on τ_r give way to the dominance of τ_{SR} processes at higher temperatures. Careful analysis of this temperature dependence allows $T_{1,SR}$ to be extracted from T_1.

In general, only two different relaxation mechanisms will be significant at the same time under a given set of experimental conditions and thus the problem of separating relaxation rates for three or four mechanisms is only rarely encountered. Usually, NOE data, field dependence of T_1, and temperature effects on T_1 is sufficient to provide the information necessary to unscramble the competing mechanisms.

Interpretation of results is, of course, significantly complicated whenever the motional narrowing approximation does not hold as this may invalidate the simple NOE arguments and make the H_0 dependence of T_1 impossible to characterise. The expression for η in terms of the specific transition probabilities appropriate to a two spin system has been given by Solomon[38] as follows:

$$\eta = \frac{\gamma_x}{\gamma_c} \frac{W_2 - W_0}{W_2 + 2W_1 + W_0} \tag{5.51}$$

The equivalent expression given by equation (5.50) is obtained in the extreme narrowing limit with the neglect of scalar coupling as $1/T_1 = W_2 + 2W_1 + W_0$ and $1/T_{1,D} = 2(W_2 - W_0)$. Whenever the extreme narrowing condition begins to break down in the region $\omega_I \approx 1/\tau_c$, the transition probabilities for the higher frequencies $(\omega_I + \omega_S)$ and ω_I are attenuated before the low frequency $(\omega_I - \omega_S)$ term is affected. Under these conditions η is greatly reduced even though the dipolar term continues to dominate the relaxation process. Thus, some care is needed in the measurement of NOE values on macromolecules when the motional narrowing condition does not hold.

5.3.4 Correlation times of importance in nuclear spin relaxation

Four different correlation times were encountered in the survey of the various relaxation mechanisms. These included the translational diffusion time, τ_t, important in the intermolecular dipole–dipole mechanism, the average chemical exchange period, τ_e, which regulates the scalar coupling mechanism of the first kind, the angular momentum (or velocity) correlation time, τ_{SR}, which is characteristic of the spin–rotation mechanism, and the angular re-orientation time constant, τ_r, which is of importance in several of the mechanisms: the intramolecular dipolar, the shift anisotropy, and indirectly in scalar coupling of the second kind. Simply stated, all of these time constants specify the characteristic period required for a given magnetic property to vary by the factor e. They differ in that they each characterise different dynamic processes which manifest their own interaction periods. Thus, τ_t describes a characteristic period relevant to the diffusion of a magnetic dipole in and out of a given sphere of interaction for the relaxing nucleus. Similarly, τ_e specifies an average period for two chemically-exchanging nuclear species to be coupled together through the scalar-coupling mechanism.

The angular re-orientation time constant, τ_r, and the angular momentum correlation time, τ_{SR}, each measure different features of molecular rotational diffusion important in liquid dynamics. The parameter τ_r specifies a period for molecular rotation through a given angular displacement, while τ_{SR} provides a measure of the persistency of a given angular momentum and does not depend directly on the magnitude of the rotational angle. For instance, a molecule could rotate through the same angle at two different velocities and, therefore, have two very different τ_{SR} values. It should be noted, also, that all angular displacements are not equally effective in relaxing nuclear spins in those mechanisms governed by τ_r. This can be appreciated, for example, in the cases of interacting dipoles where the dipoles assume a specific orientation relative to H_0. Rotations parallel and perpendicular to H_0 will in this example yield vastly different local magnetic fields even when the rotating molecule possesses spherical symmetry.

When all factors are taken into account in the averaging process then the average angular displacement in a period τ_r is approximately 1 radian. The factors governing the value of τ_r are different depending on whether the rotational re-orientation involves on the average large or small angular displacements. For large displacements, angular momentum persists for relatively long periods of time due to fewer intermolecular interactions (crudely referred to as collisions) which interrupt the molecular rotation. Under these conditions angular displacements comparable or larger than 1 radian may be realised and $\tau_{SR} \gtrsim \tau_r$. This results in the moments of inertia of the molecule governing τ_r more than microviscosity factors[54-57]. The time τ_r now becomes bounded at the lower end by the equipartition principle, which for free rotation through one radian is $(I_m/kT)^{\frac{1}{2}}$. Conversely, for small average angular displacements, $\tau_{SR} \ll \tau_r$ and normal diffusion factors control the molecular rotations more than I_m. It is this latter diffusion case which has been treated extensively in the literature[39] and the one which fortunately holds for most molecules. Using equipartition principles and a simple two dimensional random-walk argument in the diffusion limit ($\tau_{SR} \ll \tau_r$) the Hubbard[58] relationship can be obtained as follows:

$$\tau_{SR, i} = I_{m, i}/6kT\tau_r \tag{5.52}$$

The factor of 6 comes from the $l(l+1)$ quantum number associated with the second-order ($l = 2$) Legendre polynomial which characterises the geometrical features important in the special averaging process required to obtain the appropriate τ_r from the components of the rotational diffusion tensor given by equation (5.26) in an earlier section. The simple diffusion model used by Bloembergen, Purcell and Pound[39] in their treatment of nuclear relaxation is due to Debye, Stokes and Einstein[48] and assumes that molecules re-orient in random steps of small angular displacements. The medium is taken to be a continuous hydrodynamic liquid with the usual viscosity properties. As indicated above, this model holds for most relatively large molecules and reasonable agreement is found for these systems between diffusion constants obtained from n.m.r. and hydrodynamic studies. Microviscosity corrections have been made to achieve similar agreement for smaller molecules[59]. These corrections partially account for the low I_m and high symmetry sometimes found in small molecules, especially methyl tops. The large rotational steps encountered in such systems invalidate the diffusion assumptions, and D, which depends on an angular velocity correlation function, can no longer be averaged independently of the re-orientation transformations. Some workers have chosen to use the following emperical expression[60, 61]

$$\tau_j = n_j I_m (6kT\tau_r \tag{5.53}$$

where $n_j = 1$ in the diffusion limit and some number between 1 and 6 is used when the diffusion requirement gives way to the moment of inertia controlled process. Corrections of this type have generally been required only for very small molecules such as methyl top molecules or in some cases methyl moieties attached to larger molecules with no significant barrier restricting the internal rotation about the C_3 axis. However, if τ_r is obtained from either $T_{1,D}$ or $T_{1,SA}$ and τ_{SR} from $T_{1,SR}$, it no longer becomes necessary to employ the

Hubbard relationship to characterise many of the microscopic features of the molecular dynamics, and the data provide a quantitative measure of the extent of diffusion dominance in the rotational processes.

5.4 RESULTS AND DISCUSSION

While data available on ^{13}C relaxation times is quite limited, detailed separation into the contributing relaxation mechanisms is even less available. The majority of such analyses are due to Grant and co-workers[8–11, 16, 17, 51] on small molecules and Allerhand and co-workers[20] on larger systems. The results of these investigations combine to provide a consistent overview of ^{13}C relaxation for molecules of diverse size at ambient temperatures and verify the anticipated result that ^{13}C T_1 values are controlled in the main by the C—H intramolecular dipolar relaxation process. Thus, the dynamical information on liquid systems obtained from relaxation studies has, for the most part, been extracted from the effective re-orientational correlation time associated with the C—H dipolar mechanism. However, in specific instances other mechanisms have been shown to be important and significant motional information derived from the corresponding correlation time of importance[16, 17]. In this section specific examples are used to define the circumstances in which the various mechanisms are operative and to illustrate the type of dynamical results which may be obtained. Such a development is best approached by successive treatment of the individual relaxation mechanisms.

5.4.1 C—H dipolar relaxation

Using ^{13}C—{^{1}H} NOE and T_1 values, Grant and co-workers[8–11, 16, 17, 51, 86] have separated the T_1 data into $T_{1,D}$ times and an effective time, $T_{1,0}$ for all other relaxation processes. Selected representative results of such studies are presented in Table 5.1 to aid in the discussion both in this and later sections. As indicated from these data, the significance and magnitude of the C—H dipolar process can vary greatly from molecule to molecule. However, it is quite apparent from the results on the larger molecules such as benzene, cyclohexane and the methylbenzenes (exclusive of the methyl carbons) that the role of the C—H dipolar process will usually be dominant whenever a carbon is directly bonded to a proton. In general, significant deviations from maximum NOE values are only encountered in large molecules for carbons not directly bound to hydrogens, and in small molecules or in small moieties of larger molecules where rapid re-orientation is possible.

Employing pulsed techniques, Allerhand et al.[20] have determined the T_1 and NOE values for aqueous solutions of cholesteryl chloride and adenosine 5′-monophosphate (AMP). The T_1 values of these systems are presented in Figures 5.5 and 5.6. By determining quantitatively the NOE values of the quaternary carbons, the NOE values of all the carbons were assessed from relative intensity measurements. The maximum NOE parameter (2.988) obtained for all carbons except the C_4 and C_5 ring carbons of AMP demonstrates that the $1/T_1$ relaxation rate for even the non-proton bearing carbons in such intermediate sized molecules is dominated by the C—H dipolar process. The deviations from maximum NOE at C_4 and C_5 in AMP have been attributed to an effective chemical-shift anisotropy mechanism.

Table 5.1   Carbon-13 T_1 and NOE results for some typical systems*

Molecule	$\eta_{C-(H)}$**	T_1/s	$T_{1,D}/s$	$T_{1,0}/s$
Formic acid	2.00	10.2	10.2	>70
Cyclohexane	2.00	17.5	18	>70
Benzene	1.85	22.0	24	>70
Methyl iodide	0.42	11.1	52.5	14.1
Methyl alcohol	0.73	13.4	36.5	21.2
Acetonitrile				
methyl	0.53	13.1	49.1	17.9
cyanide	0.08	46.0	>200	47.8
Acetic acid				
methyl	1.40	10.5	14.9	35.5
carboxyl	1.08	29.1	53.6	63.7
Acetone				
methyl	0.76	17.3	45.3	28.1
carbonyl	0.08	22.0	>100	22.8
o-Xylene				
ring (1,2)	0.74	38	103	60
ring (3,6)	2.00	13.2	13.2	>70
ring (4,5)	2.06	12.4	12.4	>70
methyls	2.12	12.4	12.4	>70
trans-But-2-ene				
methyls (1,4)	0.94	23.0	49	43
vinyl (2,3)	0.31	16.0	103	19
trans-Decalin				
C-1,4,5,8	2.00	6.8	6.8	>70
C-2,3,6,7	2.00	6.2	6.2	>70
C-9,10	2.00	10.0	10.0	>70

*Values taken from References 10, 12, 17, 51, 67, 68. Errors are in the range of 10%–15%.
**$\eta_{c-(H)} = $ NOE – 1

Allerhand[62] has also carried out studies on the ^{13}C relaxation rates in native and acid-denatured ribonuclease A. The ^{13}C T_1 values in this protein, given in Table 5.2, are assumed to be governed by the C—H dipolar process, and it is important to note the increased efficiency of this process with the increased size of the molecule. With regard to macromolecular systems, it should be cautioned that use of the NOE parameter to determine the extent of T_1 domination by the C—H dipolar process is invalid if the motional narrowing condition is violated. In such cases, the NOE parameter becomes frequency dependent and may deviate strongly from the maximum value (in the limit $\omega^2 \tau_c^2 \gg 1$, NOE = 1.153 [8]) even though the dipolar process dominates the relaxation. Conversely, if it can be assumed that the C—H dipolar process dominates the T_1 of large molecules, monitoring of the NOE parameter (in the absence of spectrometers operating at different field strengths) should indicate whether the motional-narrowing condition has been violated. Some caution in using this indirect argument is, of course, needed, especially when the carbon is subject to internal motions.

As the NOE parameter does not distinguish between inter- and intramolecular effects[8], the C—H dipolar relaxation rate derived from T_1 and NOE data cannot separate these two contributions. However, sufficient arguments have been advanced in the preceding section to indicate that the $T_{1,D}$ for proton-bearing carbons can be assessed for dynamical information in terms of only the directly bound C—H interactions. Experimental results for the ^{13}C T_1 values in benzene and cyclohexane and a 50% volume–volume

Figure 5.5 Carbon-13 spin–lattice relaxation times, in seconds, of 1M cholesteryl chloride in CCl_4, determined at 15.08 MHz and 42 °C. T_1 values were not obtained for C-7, C-8, C-10, and C-22, because of chemical-shift degeneracies
(From Allerhand, Doddrell and Komoroski[20] reproduced by courtesy of the American Institute of Physics)

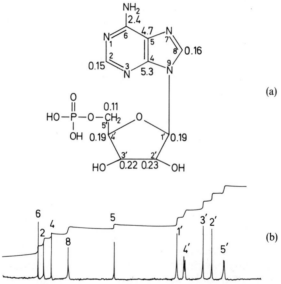

(a)

(b)

Figure 5.6 (a) Structure of adenosine 5′-monophosphate, showing standard numbering system (small digits) and carbon-13 T_1 values of aqueous 1M-AMP at 15.08 MHz and 42 °C (large digits). (b) Proton-decoupled natural-abundance carbon-13 FT–n.m.r. spectrum of 1M AMP in H_2O at pH 5.2, obtained at 15.08 MHz and 42 °C, after 2048 accumulations, with a recycle time of 10.9 s. The range 21.6–146.6 p.p.m. upfield from liquid CS_2 is shown (From Allerhand, Doddrell and Komoroski[20], reproduced by courtesy of the American Institute of Physics)

mixture of the two[9] demonstrate the minor role of the intermolecular term. The T_1 values in the mixture of the two compounds are identical to the values of the two neat liquids despite the differences in proton density of the three different samples. Results[20] on 0.5 mol l^{-1} sucrose solutions in H_2O and in D_2O also reveal no intermolecular effects arising from the solvent protons for all carbons including non proton bearing nuclei. Finally, results[12] on mesitylene also demonstrate that the C—H dipolar rates for proton-bonded carbons is little influenced by intermolecular effects, although calculations do indicate that the intermolecular and long-range intramolecular processes can be of importance at the non proton bonded positions.

Thr τ_{eff} correlation time derivable from $T_{1,D}$ data via equation (5.25) can be considered as quantitative only in the case of isotropic re-orientation of a molecule. In general, however, the quantitative details of the motion are not directly available from τ_{eff} because symmetry features required for isotropic motion do not hold. In this case, an assumed model of re-orientation is required before quantitative information is obtainable. Specifically, this arises because the n.m.r. relaxation depends only on the power density of motional fluctuations at the Larmor frequency, and hence, the explicit form of the correlation function at other motional frequencies (and thereby all of the specific details of the molecular dynamics) is not available. Explicit formulations assuming a diffusion controlled re-orientation process have been presented for the case of anisotropic motion in the preceding section. However, these formulations may have no value for molecules without some symmetry, as the principle axes system of the moment of inertia tensor cannot be expected to diagonalise the diffusion tensor[40]. In fact, for molecules of high asymmetry, anisotropy in the off-diagonal elements of the diffusion tensor may prevent any diagonalisation of this tensor and more than three rotational diffusion parameters will be required to characterise the motion. In this sense, use of τ_{eff} parameters for admittedly non-isotropic motion is partially justified by the inability to obtain the necessary detailed information with which to treat the motion.

Results[20] on the ^{13}C T_1 values of cholesteryl chloride (Figure 5.5) provide an excellent illustration of the information available from τ_{eff} parameters. As the NOE is maximum for all carbons in this compound, $T_1 = T_{1,D}$ and $1/T_{1,D} \propto \tau_{eff}$. The carbons of the ring backbone with one attached proton have T_1 values in the range 0.44–0.51 s while the carbons with two direct C—H interactions have T_1 values ranging from 0.23–0.26 s in accordance with the n_H term in equation (5.25). Hence, it may be concluded that the carbons of the ring backbone all have the same re-orientational correlation time (c. 9×10^{-11} s) and thus the entire ring system re-orients as a rigid structure. In contrast, the methyls attached directly to the rings, although having more C—H interactions than the ring carbons, have relaxation times (1.5 s) which are three times longer than the ring carbons with a single hydrogen. Thus, the effective correlation time for the methyls is roughly nine times that of the structural backbone which, on the basis of theoretical considerations given in the last section, indicates the presence of rapid internal re-orientational motion about the C_3 axis of the methyl group. In addition, the T_1 values of the side chain demonstrate that the τ_{eff} of the carbons decrease along the chain, suggesting increased internal motion in the chain as the free end is approached[63]. Thus, in a semi-quantitative fashion it can be seen that the dynamical features of the re-orientation of cholesteryl chloride have been obtained from the dipolar relaxation data. Values for τ_{eff} have also been used to describe molecular motion in Ribonuclease A (RNase A)[62]. Here, the T_1 process was assumed to be dominated for the majority of carbons by the C—H dipolar process. Calculation of $\tau_{r,\text{eff}}$ values, using 5.19 and 5.21 required computer simulation of the spectrum using T_2 as a variable in order to compensate for a breakdown in the motional narrowing requirement. This method allowed solution of the problem of double-valued dependence of T_1 on τ_{eff} near the minimum in the T_1 $v.$ $\tau_{r'\text{eff}}$ curve. Similar analysis was carried out on acid-denatured RNase A and the results are given

Table 5.2* Some carbon-13 spin–lattice relaxation times and rotational correlation times in aqueous ribonuclease A†

Type of carbon	Native protein‡ (pH 6.51) T_1/ms	$\tau_{r,\text{eff}}$/ns	Denatured protein (pH 1.65) T_1/ms	$\tau_{r,\text{eff}}$/ns
Carbonyls	416		539	
α-Carbons	42	30	120	0.40
β-Carbons, Thr	c. 40	c. 30	99	0.48
Rigid side chains‡	c. 30			
ε-Carbons, Lys	330	0.070	306	0.076

*Table taken from Reference 62, reproduced by courtesy of The American Chemical Society.
†Obtained for 0.019 M protein at 45 °C and 15.08 MHz. The T_1 and τ_R values have an estimated maximum error of ±30%.
‡Difficulties in separating these resonances from the narrow ones make their T_1 only an estimate of an average value.

in Table 5.2. The change in the molecular dynamics upon denaturation is observed in the sixtyfold shortening of $\tau_{r,\text{eff}}$ at the α- and β-carbons of the residues, and in the shortening of $\tau_{r,\text{eff}}$ for the carbonyl carbons. Thus, an appreciable degree of segmental motion is noted in the protein backbone as the structural rigidity is decreased upon denaturation. In contrast, the $\tau_{r,\text{eff}}$ value for the lysine ε-carbon is relatively unchanged in the two forms indicating the side chain motion for this amino acid residue is relatively independent of the state of the molecule.

Derivation of more quantitative features of re-orientational motion is represented by ^{13}C $T_{1,\text{D}}$ data on CH_3CN, CH_3I and $CHCl_3$ [10, 23, 43]. As the ^{14}N relaxation rate in acetonitrile is dependent only upon the motions perpendicular to the molecular symmetry axis, combination of this result[5, 64] with the measured $T_{1,\text{D}}$ value (49.1 s) for the methyl carbon yields a value of $\sigma = 9.2$ from equation (5.27) at 38 °C. Thus, these data predict that acetonitrile re-orients approximately nine times $(D_\parallel = 13.6 \times 10^{11}$ s$^{-1})$ faster about its symmetry axis than about the perpendicular $(D_\perp = 1.5 \times 10^{11}$ s$^{-1})$ axis. The re-orientation of CH_3CN as a prolate top is certainly not unreasonable in view of the low moment of inertia about the symmetry axis and the fact that the $D_\perp$ motion requires re-orientation of an electric dipole moment $(\mu = 3.5$ D). In the case of CH_3I the $T_{1,\text{D}}$ value of 52.5 s and dielectric relaxation data may be used to arrive at a value of $\sigma = 13.8$ from equation (5.27). Hence, the motional details $(D_\parallel = 19.3 \times 10^{11}$ s and $D_\perp = 1.4 \times 10^{11}$ s) for CH_3I mirror those of CH_3CN and in view of the similarities between the systems this result is not surprising. Using temperature dependent relaxation data, Noggle et al.[43] have found activation parameters for $D_\perp$ of 1.9 kcal mol^{-1} and $D_\parallel$ of 0.8 kcal mol^{-1} in CH_3I. These constants are consistent with the relative facility of motion in the parallel mode. In the molecule $CHCl_3$, ^{13}C and ^{35}Cl relaxation data[65] have been combined[23] to give the value of

$\sigma = 1.2$ and $D_\perp = 1.3 \times 10^{11}\,s^{-1}$. Thus, this molecule with no relatively low moment of inertia is near isotropic in its re-orientational motion.

Quantitative calculation of the diffusion tensor for asymmetric top molecules have been carried out on *trans*-decalin, and norbornane[66] where symmetry now requires three diffusion parameters but is still sufficient to diagonalise the diffusion tensor. Thus, ^{13}C $T_{1,D}$ data from three carbons with different geometrical dependencies of the C—H vector relative to the principal axis are required to set up and solve the three simultaneous

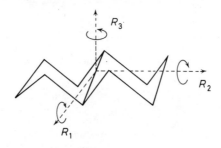

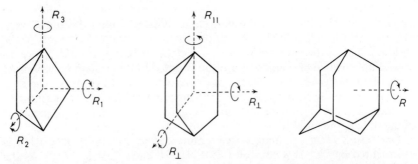

Figure 5.7 The principal axes are given for *trans*-decalin, norbornane, bicyclo-octane and adamantane. The values for the various rotational diffusion constants in units of 10^9 rad s^{-1} are: $R_1 = 50, R_2 = 110, R_3 = 220$ for *trans*-decalin; $R_1 = 130, R_2 = 200, R_3 = 340$ for norbornane; $R_\| = 330$, $R_\perp = 170$ for the symmetrical-top molecule bicyclo-octane; and $R = 160$ in the isotropic adamantane system

equations of the form given by equation (5.30). A diagram of *trans*-decalin and norbornane showing the principal axes and the re-orientation rates about each is given in Figure 5.7. The degree of anisotropy in the motion is noted to be considerable for *trans*-decalin as might be expected from the general planar structure of the molecule, whereas norbornane is more nearly isotropic and has rotational diffusion parameters similar to the related bicyclo-octane and adamantane systems also shown in Figure 5.7 for comparison.

Analysis of internal re-orientational effects upon dipolar relaxation have been examined in the context of equation (5.32). If the rate of internal re-orientation $R \gg D_\perp = D_\|$, $1/T_{1,D}$ in equation (5.32) reduces to $\frac{1}{9}$ the value for the isotropic case with no significant internal rotation in which $D_\perp = D_\| \gg R$. In mesitylene the ring carbons have $T_{1,D}$ values which are three times shorter than the methyls (7.8 s v c. 23 s)[12] even though there are three times as many C—H interactions for the methyls. Combining these two factors of three it is observed that the effectiveness of each C—H interaction has been reduced by $\frac{1}{9}$ indicating very rapid internal rotation. This result is similar to that found in cholesteryl chloride where the methyls attached to the ring had relaxation times three times those of mono-proton-bearing ring carbons. While the two systems are similar, they do differ in one important aspect. Being a small molecule, mesitylene has a τ_{eff} which is much shorter than cholesteryl chloride and insufficient to dominate completely the spin–rotation contribution due to internal rotation. On the other hand, τ_{eff} for cholesteryl chloride is of sufficient magnitude for the dipolar process to overwhelm completely any spin–rotation effects even if they should be as large as in mesitylene. A comparison of the respective methyl $T_{1,D}$ values of 23 and 1.5 s for mesitylene and cholesteryl chloride clearly demonstrate the effect. The 23 s time in mesitylene is only comparable to the $T_{1,0} = 25$ s attributed to the spin–rotation effect. Conversely, even if $T_{1,SR}$ were as low as 25 s in cholesteryl chloride, the 1.5 s dipole term would still dominate the overall relaxation process.

In a more quantitative sense, equation (5.32) has been used to calculate methyl rotational barriers in o-xylene[12] and in acetone, DMSO, methyl acetate, t-butyl chloride and methylchloroform[67]. In the work on o-xylene an isotropic τ_{eff} was calculated from the ring carbons and R, the internal rotation rate determined from $T_{1,0}$ for the methyls from which a barrier of 1.4 kcal was estimated. Using microviscosity calculations and dielectric relaxation

Table 5.3   Methyl internal rotational barriers from ^{13}C dipolar relaxation rates

Molecule	$D \times 10^{-11}$ /s^{-1}	$R \times 10^{-11}$ /s^{-1}	V_0 (eqn. 5.54)/ kcal mol^{-1}	V_0 (lit.)/ kcal mol^{-1}
$(CH_3)_2SO$	0.35	4.0	2.2	3.07†, 2.87*
$(CH_3)_2CO$	1.6	29.9	0.92	0.76‡, 0.78*
$(CH_3)CCl_3$	1.3	1.2	2.9	2.91*
$(CH_3)_3C–Cl$	1.4	0.5	3.5	4.3§*
$(CH_3)COOCH_3$	1.4	49.4	0.67	0.48*
$CH_3COO(CH_3)$	1.4	21.0	1.1	1.19¶*

*Values represent gas phase data and unless otherwise noted derive from Gordy, W. and Cook, R. L. (1970), in *Technique of Organic Chemistry*, Vol. IX, 477. (A. Weissberger, editor) (New York – Interscience).
†Dreizler and Dendl, G. (1965). *Z. Naturforschg.*, **20a**, 1431
‡Swalen, J. D. and Costain, C. C. (1959). *J. Chem. Phys.*, **31**, 1562
§Value for t-butyl fluoride
¶Value for OCH in methyl formate
Details of estimates of D given in *PhD. Thesis* of J. R. Lyerla, Jr., University of Utah (1971). Contents of this table are derived from this service.

data, R was determined for the remaining molecules. The values of R were related to rotational barriers using a conventional expression for an activation process of the form

$$R = R_0 e^{-V_0/RT} \tag{5.54}$$

where R_0 is $\frac{3}{2}$ the rate of re-orientation for zero barrier and V_0 is the potential barrier in cal mol^{-1}. Using the equipartition principle, the rate of rotation of a methyl fragment in the gas phase $(kT/I_{methyl})^{\frac{1}{2}}$ multiplied by the $\frac{3}{2}$ degeneracy factor was taken as a measure of R_0 $(1.3 \times 10^{13}$ s$^{-1})$. The V_0 values derived from equation (5.54) are given in Table 5.3 along with a summary of the other parameters used in the calculations. The good agreement between these barriers and those from microwave data indicate the value of ^{13}C relaxation times and the correlation times derived therefrom in evaluating features of molecular motion. Furthermore, all of the above results also point out the general utility of $T_{1,D}$ and NOE data in studies of motional dynamics in the liquid state.

5.4.2   Spin–rotation relaxation

Since it is only recently that ^{13}C temperature probes have been available, the role of the spin–rotation mechanism has only been investigated to date in an incomplete manner. In the absence of variable-temperature capability, it has been assumed that the $T_{1,0}$ time derived from NOE and T_1 data can be assigned to spin–rotation provided it can be shown that no field-dependent or scalar-coupling mechanism is important[10, 17, 23, 43, 68]. Molecules in which scalar-coupling is possible can also be studied provided quadrupolar data clearly justifies neglect of this term. Table 5.4 contains a collection of compounds where $T_{1,0}$ can be attributed to spin–rotation. This result is confirmed by the very short $T_{1,SR}$ values in CH_3I, CH_3CN, and $(CH_3)_2CO$ where spin-rotation actually dominates the relaxation at room temperatures. Very recent

Table 5.4   Spin–rotation relaxation times at 38 °C for several molecules*

Molecule	$T_{1,0} = T_{1,SR}$
CH_3I	18.0
CH_3CN	17.9
$(CH_3)_2CO$	28.1
$(CH_3)_2CO$	22.8
CH_3COOCH_3	24.6
CH_3COOCH_3	33.1
CH_2Cl_2	37.7
$CHCl_3$	91.9
Mesitylene methyls	25.0
CH_3CCl_3	59.1
trans-But-2-ene	
methyl (1,4)	43.0
vinyl (2,3)	19.0

*Values taken from References 10–12, 17, 68

^{13}C T_1 temperature studies[69] on many of the same small molecules reported in Table 5.4 further verify the assignment of $T_{1,0}$ to the spin–rotation process. These results also reveal that many molecules such as benzene, cyclohexane, and formic acid which are totally dominated by the C—H dipolar process at room temperature exhibit considerable spin–rotational effects at higher temperatures (> 80 °C). However, as far as room temperature studies are concerned, the findings[23, 68] to date indicate that this mechanism will be of importance in small symmetrical molecules with a low moment of inertia or for methyl carbons subject to internal re-orientation[23]. The $T_{1,SR}$ values for methyl carbons decrease as the barrier to methyl rotation decreases for the compounds reported in Table 5.3. This result is consistent with the increased angular velocity expected for methyl groups with lower barriers. However, as indicated previously, the overall re-orientation rate of very large molecules may make the C—H dipolar process so efficient that the

Table 5.5   Relaxation data and molecular parameters important to spin–rotation in CH_3I and CH_3CN
(From Lyerla, Grant and Bertrand[73], reproduced by courtesy of the American Institute of Physics)

	CH_3I	CH_3CN
$I_{m,\perp}$	5.5×10^{-40} gm cm^2	5.5×10^{-40} gm cm^2
$I_{m,\parallel}$	111.9×10^{-40} gm cm^2	91.1×10^{-40} gm cm^2
$(CI_m)_t$	-9.6×10^{-35} rad erg s	-11.6×10^{-35} rad erg s
$T_{1,SR}$	18.0 ± 3 s	17.9 ± 3 s
$\tau_{SR,\parallel}$	0.13 ps	0.09 ps
$\tau_{r,\parallel}$	0.09 ps	0.12 ps
$\tau_{f,\parallel}$	0.11 ps	0.11 ps
$\tau_{SR,\perp}$	0.036 ps	0.032 ps
$\tau_{r,\perp}$	1.2 ps	1.1 ps
$\tau_{f,\perp}$	0.51 ps	0.46 ps

spin–rotation mechanism does not contribute even in the case of zero barrier. In fact, the maximum effectiveness of the spin–rotation interaction at room temperature has not been observed to reduce $T_{1,SR}$ below 15–20 s [23, 68].

As in the case of the dipolar mechanism, more than one spin–rotational correlation time is needed when anisotropic motion is present. Unfortunately, such motional details are not easily obtained from the spin–rotation relaxation times owing to the need for the components of the spin–rotational interaction tensor which are not generally available at the present time. However, equation 5.39, using the trace of this matrix, has been used to analyse spin–rotational data for CH_3CN and CH_3I, and the pertinent data for these systems are given in Table 5.5. The results show that $\tau_{SR,\parallel}/I_{m,\parallel} \gg 2\tau_{SR,\perp}/I_\perp$ as assumed[16], and thus the spin–rotation interaction arises primarily from the fast re-orientation motion of these molecules about their symmetry axis. Indeed the proton and carbon $T_{1,SR}$ relaxation times have a temperature dependence that has an activation energy comparable to that of the correlation time governing the parallel re-orientational motion[43]. It should be noted that while the spin–rotation formulation given by equation 5.39 predicts the correct temperature dependence of $T_{1,SR}$ other formulations[48] in the literature are not successful in doing so.

The Hubbard[58] relation, equation 5.52, cannot be used to estimate $\tau_{SR,\parallel}$ correctly since the fast re-orientational motion in the parallel mode would invalidate the $\tau_{c,\parallel} \gg \tau_{SR,\parallel}$ condition (i.e. the re-orientation does not occur via small step diffusion). The respective τ_{SR} correlation time can be compared to the period for rotation of a molecule through a radian τ_f, given by the equipartition principle as follows:

$$\tau_f = (I/kT)^{\frac{1}{2}} \qquad (5.55)$$

Comparison of $\tau_{SR, \parallel}$ and $\tau_{SR, \perp}$ with the respective τ_f values allows the minimum average angle through which CH_3I and CH_3CN move during a single jump to be estimated. For CH_3I this is 68 degrees in the parallel motion and 4 degrees in the perpendicular, while for acetonitrile, 47 and 4 degrees respectively[17]. The actual angular displacement during a single jump could be longer if the molecule were to librate for any appreciable time in some equilibrium position before undergoing a rotational re-orientation. The results, although semi-quantitative, do indicate that the motion about the parallel axis in CH_3I and CH_3CN involves large angular displacements which are governed by the moment of inertia and are not diffusion-controlled. This result is consistent with recent i.r. band-shape studies[70] on CH_3I that demonstrate that the parallel motion in CH_3I is dominated by inertial effects up to 0.2×10^{-12} s, after which normal diffusional processes characterise the interactions. The ^{13}C relaxation rate of the methyl group in toluene has been investigated by Schmidt and Chan[60]. These authors used a correction factor for the Hubbard relation of the form given by equation 5.53 in their evaluation. A value of n_j equal to c. 3 was found to fit the experimental results. It is interesting to note that the same n_j value of 3 gives the correct result for the spin–rotation relaxation time of the methyl protons of toluene. This result coupled with the results in CH_3CN and CH_3I firmly establish that the effective spin–rotation interaction field originates from the fast re-orientation of methyl groups about their C_3 axis.

Regardless of whether one uses the modified Hubbard relation or ignores it totally, the spin–rotation data derived from ^{13}C relaxation times demonstrates that for many small molecules at room temperature the re-orientational motion is not controlled by small step diffusion. Several models of n.m.r. relaxation based on inertial controlled re-orientation have been proposed[56, 71, 72] to treat this particular situation, but to date they have been found either to be inadequate or else not sufficiently investigated. It does seem clear, however, that an acceptable model of liquid re-orientation would have to be one that incorporates both inertial and diffusion characteristics. As more detailed analyses of ^{13}C relaxation results become available, it is anticipated that a suitable model of general utility will be developed.

5.4.3 Scalar-coupling relaxation

The existence of an effective scalar-coupling process for ^{13}C spin–lattice relaxation has been demonstrated only for carbons with directly-bonded bromines[73]. In the bromomethanes proof for the presence of this mechanism depends upon theoretical estimates of the magnitude of the relaxation time using equation 5.48a and estimates of quadrupolar data on bromine, but more specifically upon the field dependence of the scalar-coupling term which equation 5.48a predicts will be important for the parameters used. As the $(\Delta\omega)^2$ and $T_{1,Q}^2$ terms in the denominator of equation 5.48a are of comparable value, the scalar coupling term will exhibit a field dependence through the $(\Delta\omega)^2 T_{1,Q}^2$, term. Thus, the extreme narrowing approximation for this mechanism does not hold in this case.

The ^{13}C T_1 results at 14.1 and 23.5 kG and the respective NOE factors for seven members of the halomethane family are given in Table 5.6. The significant field dependence displayed by methyl bromide is not as readily observed in the other bromomethanes due in part to the lower T_1 values where errors decrease the accuracy of the method but also in part to the isotopic mixture problem. Only the ^{79}Br isotope can be expected to yield a significant scalar relaxation, and it is limited to 50.5% isotopic concentration in natural

Table 5.6   Carbon-13 relaxation data for some halogenomethanes
(From Lyerla, Grant and Bertrand[73], reproduced by courtesy of the American Chemical Society)

Compound	Low field (14.1 kG)			High field (23.5 kG)		
	T_1/s	η	$T_{1,D}$/s	T_1/s	η	$T_{1,D}$/s
CH_3Br	8.5 ± 0.6	0.30 ± 0.03	56 ± 5	11.6 ± 1.8	0.43 ± 0.06	54 ± 5
CH_3I	11.1 ± 0.4	0.42 ± 0.05	52 ± 5	12.8 ± 0.6	0.51 ± 0.04	50 ± 5
CH_2Cl_2	18.0 ± 1.3	1.07 ± 0.07	32 ± 3	16.6 ± 1.5	1.00 ± 0.10	33 ± 3
CH_2Br_2	4.1 ± 0.4	0.36 ± 0.08	22 ± 4	5.9 ± 0.6	0.50 ± 0.09	24 ± 4
CH_2I_2	3.9 ± 0.5	1.55 ± 0.12	5 ± 2	4.3 ± 0.9	1.60 ± 0.16	5 ± 2
$CHCl_3$	23.8 ± 1.8	1.47 ± 0.11	32 ± 4	24.0 ± 1.5	1.53 ± 0.30	31 ± 6
$CHBr_3$	2.0 ± 0.3	0.30 ± 0.05	13 ± 3	2.2 ± 0.6	0.33 ± 0.08	13 ± 4

abundance. Even so, the very short T_1 values and low NOE observed for CH_2Br_2 and $CHBr_3$ clearly indicate that a mechanism other than the C—H dipolar process is operative. Compared with the other halogenomethanes the bromine series are uniquely low in their T_1 values and NOE values. As indicated earlier in this review, the direct C—Br dipolar term cannot effectively compete with the corresponding proton relaxation process. As the spin–rotation process for the bromine series should be intermediate between the chloro and iodo families, only the scalar-coupling process can provide a reasonable explanation of the results. It should also be noted that the field dependence is of the opposite sense from that of a shift anisotropy process. The estimates of T_{1p} for the chlorine, bromine and iodine nuclei in

Table 5.7   Predicted values of $T1,SC$ at 14.1kG and 23.5 kG for some halogenomethanes
(From Lyerla, Grant and Bertrand[73], reproduced by courtesy of the American Chemical Society)

Compound	$\tau_c \times 10^{12}$/s	$(e^2q\,Q/h)$/MHz	$T_{1,Q} \times 10^7$/s	$T_{1,SC}^{14.1}$/s	$T_{1,SC}^{23.5}$/s
$CH_3^{79}Br$	0.87	553	9.4	8.3	9.7
$CH_3^{81}Br$		462	13.5	476	1326
CH_3I	1.1	1850	2.8	278	753
$CH_2^{35}Cl_2$	0.64	72	707	3.2×10^6	14×10^6
$CH_2^{37}Cl_2$		57	1230	6.5×10^6	27×10^6
$CH_2^{79}Br_2$	0.84	586	8.8	3.1	3.6
$CH_2^{81}Br_2$		490	12.6	157	435
CH_2I_2	3.7	1993	0.72	36	78
$CH^{35}Cl_3$	1.2	79	339	0.9×10^6	2.5×10^6
$CH^{37}Cl_3$		62	560	1.9×10^6	5.0×10^6
$CH^{79}Br_3$	2.8	629	2.3	5.3	5.5
$CH^{81}Br_3$		523	3.3	24	61

the compounds given in Table 5.7 are obtained from equation (5.46) using the C—H dipolar relaxation times to obtain the τ_r used in this expression. Quadrupole coupling constants were taken from studies on gases and solids[74]. The values of J_{Cx} in equation (5.48a) were approximated from corresponding C—H couplings with the expression, $J_{CX} = \gamma_X J_{CH}/\gamma_H$. The estimates of $T_{1,sc}$ given by Lyerla, Grant and Bertrand[73] are given in Table 5.7 for the various isotopic species. The calculations demonstrate that the scalar-coupling process is sufficiently important in carbon–bromine systems to account for the low T_1 values, and the low Overhauser effects in these systems are readily explicable in terms of this efficient mechanism. The minor field dependence of the bromomethanes is also readily accounted for by $\Delta\omega$, the difference in resonance frequency between ^{13}C (15.087/25.144 MHz at 14.1/23.5 kG) and ^{79}Br (15.032/25.054 at 14.1/23.5 kG), and the appropriate $T_{1,Q}$ value for the bromine nucleus. As $1 > (\Delta\omega)^2 T_{1,Q}^2$, $T_{1,sc}$ displays only a slight but measurable field dependence over the range 14.1–23.5 kG.

While theoretical estimates indicate that minor contributions to ^{13}C T_1 values in iodo compounds are possible, other common quadrupolar nuclei such as chlorine and nitrogen will have virtually no effect on T_1 due to the very large $(\Delta\omega)^2 T_{1,Q}^2$ terms in the denominator cutting into the efficiency of the $T_{1,sc}$ process. In this situation one is at the extreme right of the T_1 v. τ_c plot given in Figure 5.4 A shortened T_2 can be expected, however, from Figure 5.4 and equation 5.48b in the region where $(\Delta\omega^2\tau_{sc}^2 \gg 1)$. Thus, significant effects on T_2 by Cl and N can be expected even though T_1 values are not perturbed. Hence, this mechanism accounts for the usual broad lines associated with carbons attached to chlorine or nitrogen nuclei.

Returning again to the bromine situation it is noted that $1/T_{2,sc} \approx 1/T_{1,sc}$ whenever $(\Delta\omega)^2 T_{1,Q}^2 < 1$. Recent results for the T_1 and T_2 values of ^{13}C nuclei in p-bromobenzonitrile[27] illustrate this general equivalence of $T_{1,sc}$ and $T_{2,sc}$ for the carbon attached to ^{79}Br, while the quaternary carbon bound to nitrogen has a markedly more efficient T_2 than T_1. Results on chloroform[19, 23] and o-dichlorobenzene[75] also illustrate the same type of relationship between the ^{13}C T_1 and T_2 values.

Preliminary data[27, 73, 81, 82] thus indicate that scalar coupling of the second kind is a potent relaxation mechanism in 'normal' organic systems only in carbon–bromine systems owing primarily to the close resonance frequencies of these nuclei and the very fast $T_{1,Q}$ relaxation times usually associated with carbon–bromine systems. It is expected other nuclei such as ^{27}Al with resonance frequencies near that of ^{13}C will also display similar behaviour. A final qualification regarding such systems is that the fast $T_{1,Q}$ value of Br depends on the large quadrupole coupling constant which in turn is a function of the asymmetry in the nuclear charge density.

The existence of scalar coupling of the first kind has not been demonstrated as yet, although there could possibly be conditions under which a fast-exchanging proton could provide ^{13}C T_1 relaxation via this route.

5.4.4 Chemical-shielding anisotropy relaxation

This field-dependent mechanism has been shown to contribute to ^{13}C spin–lattice relaxation in specific instances. Several studies have been conducted on CS_2 [73, 76, 77, 85] where the absence of an abundant magnetic isotope for sulphur and the low abundance of the carbon magnetic species renders any dipole–dipole process negligible. Early results[15] at 9.3 and 14.1 kG suggested the dominance of the anisotropy mechanism at room temperature, but recent studies[73, 77] have established the relative unimportance of anisotropy except at very low temperatures. Instead, the relaxation of CS_2 is dominated by the spin–rotation process at room temperature even at superconducting field strengths. At very low temperatures the slower re-orientational motions increase the efficiency of the shift anisotropy mechanism and correspondingly de-emphasise spin–rotation effects. Due to the opposite temperature dependence of $T_{1.SR}$ and $T_{1.SA}$, T_1 passes through a maximum

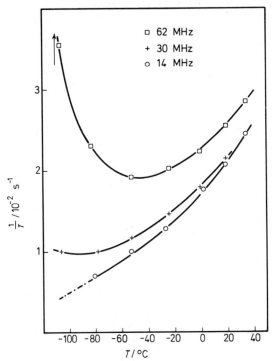

Figure 5.8 Experimental relaxation rates in $^{13}CS_2$ at 14.0369, 30.1026, and 61.9164 MHz
(From Spiess, Schweitzer, Haeberlen, and Hausser[77], reproduced by courtesy of Academic Press)

with temperature if both processes are operative. Figure 5.8 demonstrates the temperature dependent behaviour of the T_1 in CS_2 [77] at several field strengths. The effect is, of course, most pronounced at the higher field: where $T_{1,A}$ processes compete more efficiently. Similar temperature results have been obtained from studies at 23.5 kG [73] where it was shown even at $-90\,°C$, the anisotropy mechanism contributes only about 30% of the total CS_2 relaxation rate at this field.

The other principle example of shift anisotropy relaxation was found in the carbonyl carbon of acetic acid[19]. A plot of $1/T_1$ at room temperature $v.$ ω_0^2 gives a straight line over the range 9–60 MHz. T_1 decreases from 40 s at 9 MHz to 21 s at 63 MHz, which corresponds to a $T_{1,A}$ value of c. 42 s at 63 MHz or a value of c. 265 s at 25 MHz. Thus, at non-superconducting fields strengths it again appears the mechanism will assume a minor role in ^{13}C relaxation at room temperature, but use of superconducting systems can be expected to bring this mechanism into importance. (Recent results by Levy[83, 84] demonstrated the importance of this mechanism at room temperature in the relaxation rate at 57 kG for the methyne carbons of diphenylacetylene.)

Although very few values of $\Delta\sigma$ (the term of prime importance in equation 5.44) are known for carbon nuclei, preliminary liquid crystal results[78, 79] indicate that the highest anisotropies are found for carbons forming multiple bonds of which the carbons of $CS_2(\Delta\sigma = -440)$[80] and $CH_3^{13}COOH$ are good examples. Other carbons not possessing multiple-bond character will generally have relatively small anisotropies and thus these carbons will be little affected by the anisotropy mechanism.

5.5 SUMMARY AND FUTURE CONSIDERATIONS

This review has examined the carbon-13 spin–lattice relaxation process in terms of its basic theoretical formulation and the nature of its relationship to molecular dynamics. The role of the various relaxation mechanisms in causing ^{13}C relaxation has been assessed and experimental results explored with regard to molecular motion.

Although ^{13}C relaxation studies are still in the early stages, the limited data available make it abundantly clear that the C—H dipolar mechanism generally dominates carbon relaxation. Furthermore, as the dipolar relaxation process can be readily separated from the overall relaxation rate, it is possible to focus on this particular mechanism which embodies a wealth of motional information. Only in small symmetrical molecules or molecular moieties has spin–rotation been found to be of significant importance. In these few cases, however, important details of inertial controlled motional processes have been revealed. Scalar coupling effects on T_1 and T_2 have been noted for Br compounds, but only T_2 variations should be expected for N- or Cl-bearing carbons. Chemical-shift anisotropy effects at room temperature appear to be limited to superconducting systems owing to general dominance by the dipolar term.

The facility with which pulsed FT–n.m.r. techniques allow determination of T_1 relaxation data should make it possible to obtain the T_1 parameter in as routine a manner as the chemical shift and spin-coupling constant. In this respect, it is expected that the relaxation area of n.m.r. research will be one of the most exploited during the coming decade. Because of the general applicability of FT–n.m.r. to macromolecules, the biological area should realise some of the greatest benefits. Work on dynamical features of macromolecules is currently proceeding and initial results hold high promise for future studies. It is expected that biomolecules selectively enriched in the ^{13}C isotope at amino acid residues in the active site will allow ^{13}C T_1 studies to be focused on the enzyme binding process. Temperature data will allow for the determination of activation parameters for a variety of motional features in all molecular systems, including such facets as segmental motion in polymeric chains. Finally, the facility of data acquisition in making available a large body of relaxation results will enable theories of liquid reorientation to be evaluated and should provide for a much greater understanding of the microdynamic behaviour of molecules in the liquid state.

References

1. Bloch, F. (1946). *Phys. Rev.,* **70,** 460
2. Abragam, A. (1961). *The Principles of Nuclear Magnetism,* Chapt. 8, (London: Oxford Univ. Press)
3. Pople, J. A. (1967). *Discuss. Faraday Soc.,* **43,** 192
4. Slichter, C. P. (1963). *Principles of Magnetic Resonance,* 45, (New York: Harper and Row)
5. Bopp, T. T. (1967). *J. Chem. Phys.,* **47,** 3621
6. Green, D. K. and Powles, J. G. (1965). *Proc. Phys. Soc. (London),* **85,** 87
7. Jaeckle, H., Haeberlen, U. and Schweitzer, D. (1971). *J. Mag. Res.,* **4,** 198
8. Kuhlmann, K. F., Grant, D. M. and Harris, R. K. (1970). *J. Chem. Phys.,* **52,** 3439
9. Alger, T. D. and Grant, D. M. (1971). *J. Phys. Chem.,* **75,** 2538
10. Lyerla, J. R. Jr., Grant, D. M. and Harris, R. K. (1971). *J. Phys. Chem.,* **75,** 585
11. Alger, T. D., Grant, D. M. and Lyerla, J. R. Jr. (1971). *J. Phys. Chem.,* **75,** 2539
12. Kuhlmann, K. F. and Grant, D. M. (1971). *J. Chem. Phys.,* **55,** 2998
13. Ernst, R. R. (1966). *J. Chem. Phys.,* **45,** 3845
14. McConnell, H. M. and Holm, C. J. (1956). *J. Chem. Phys.,* **25,** 1289
15. Olivson, A., Lippmaa, E. and Past, J. (1967). *Est. NSV Tead. Aka. Toim. Fuus.-Mat.,* **16,** 390
16. Wang, C. H., Grant, D. M. and Lyerla, J. R. Jr. (1971). *J. Chem. Phys.,* **55,** 4674
17. Lyerla, J. R. Jr., Grant, D. M. and Wang, C. H. (1971). *J. Chem. Phys.,* **55,** 4676
18. Ernst, R. R. and Anderson, W. A. (1966). *Rev. Sci. Instrum.,* **37,** 93
19. Farrar, T. C. and Becker, E. D. (1971). *Pulse and Fourier Transform NMR,* (New York: Academic Press, Inc.)
20. Allerhand, A., Doddrell, D. and Komoroski, R. (1971). *J. Chem. Phys.,* **55,** 189
21. Abragam, A. (1961). *The Principles of Nuclear Magnetism,* Chap. 3, (London: Oxford Univ. Press)
22. A. Powles, J. G. (1958). *Proc. Phys. Soc. (London),* **71,** 497
    B. Pake, G. E. (1962). *Paramagnetic Resonance,* Chap. II, (New York: W. A. Benjamin, Inc.)
23. Lyerla, J. R. Jr. (1971). Ph.D. Thesis, University of Utah.
24. Janzen, W. R., Cyr, T. J. R. and Dunell, B. A. (1968). *J. Chem. Phys.,* **48,** 1246
25. Vold, R. L., Waugh, J. S., Klein, M. P. and Phelps, D. E. (1968). *J. Chem. Phys.,* **48,** 3831
26. Freeman, R. and Hill, H. D. W. (1970). *J. Chem. Phys.,* **53,** 4103
27. Freeman, R. and Hill, H. D. W. (1971). *J. Chem. Phys.,* **54,** 3367, and references therein.
28. Freeman, R. and Hill, H. D. W. (1971). Presented 5th Conference on Molecular Spectroscopy, Brighton, England (Sept. 22, 1971).
29. Markley, J. L., Horsley, W. J. and Klein, M. P. (1971). *J. Chem. Phys.,* **55,** 3604, and references therein.
30. Alger, T. D. and Grant, D. M. Unpublished results.
31. Carrington, A. and McLachlan, A. D. (1967). *Introduction to Magnetic Resonance,* Chap. 11, (New York: Harper and Row)
32. Slichter, C. P. (1963). *Principles of Magnetic Resonance,* Chap. 5, (New York: Harper and Row)
33. Morrish, A. H. (1965). *The Physical Principles of Magnetism,* Chap. 4, (New York: John Wiley and Sons, Inc.)
34. Woessner, D. E. (1962). *J. Chem. Phys.,* **37,** 647
35. Carrington, A. and McLachlan, A. D. (1967). *Introduction to Magnetic Resonance,* 187, (New York: Harper and Row)
36. Saupe, A. (1968). *Agnew. Chem. Int. Ed. Engl.,* **7,** 97

37. Slichter, C. P. (1963). *Principles of Magnetic Resonance*, 45, (New York: Harper and Row)
38. Solomon, I. (1955). *Phys. Rev.*, **99**, 559
39. Bloembergen, N., Purcell, E. M. and Pound, R. V. (1948). *Phys. Rev.*, **73**, 679
40. Wallach, D. and Huntress, W. T., Jr. (1969). *J. Chem. Phys.*, **50**, 1219
41. Huntress, W. T., Jr. (1970). *Advan. Mag. Res.*, **4**, 1
42. Woessner, D. E., Snowden, B. S. Jr. and Meyer, G. H. (1969). *J. Chem. Phys.*, **50**, 719
43. Gillen, K. T., Schwartz, M. and Noggle, J. H. (1971). *Mol. Phys.*, **20**, 899
44. Mitchell, R. W. and Eisner, M. (1960). *J. Chem. Phys.*, **33**, 86
45. Becker, E. D. (1969). *High Resolution NMR*, 205, (New York: Academic Press, Inc.)
46. Green, D. K. and Powles, J. G. (1965). *Proc. Phys. Soc. (London)*, **85**, 87
47. Deverell, C. (1970). *Molec. Phys.*, **18**, 319
48. Huntress, W. T. Jr. (1968). *J. Chem. Phys.*, **48**, 3524
49. Flygare, W. H. (1964). *J. Chem. Phys.*, **41**, 793
50. Ozier, I., Crapo, L. M. and Ramsey, N. F. (1968). *J. Chem. Phys.*, **49**, 2314
51. Alger, T. D., Collins, S. W. and Grant, D. M. (1971). *J. Chem. Phys.*, **54**, 2820
52. Kuhlmann, K. F. and Grant, D. M. (1968). *J. Amer. Chem. Soc.*, **90**, 7355
53. Rhodes, M., Aksnes, D. W. and Strange, J. H. (1968). *Molec. Phys.*, **15**, 541
54. Gierer, V. A. and Wirtz, K. (1953). *Z. Naturforsch.*, **8a**, 532
55. Rothschild, W. G. (1970). *J. Chem. Phys.*, **53**, 3265
56. Moniz, W. B., Steele, W. A. and Dixon, J. A. (1963). *J. Chem. Phys.*, **38**, 2418
57. Maryoff, A. A., Farr, T. C. and Malmberg, M. S. (1971). *J. Chem. Phys.*, **54**, 64
58. Hubbard, P. S. (1963). *Phys. Rev.*, **131**, 1155
59. Gillen, K. T. and Noggle, J. H. (1970). *J. Chem. Phys.*, **53**, 801
60. Schmidt, C. F. Jr. and Chan, S. I. (1971). *J. Mag. Res.*, **5**, 151
61. Burke, T. E. and Chan, S. I. (1970). *J. Mag. Res.*, **2**, 120
62. Allerhand, A., Doddrell, D., Glushko, V., Cochran, D., Wenkert, E., Lawson, P. J. and Gurd, F. R. N. (1971). *J. Amer. Chem. Soc.*, **93**, 544
63. Doddrell, D. and Allerhand, A. (1971). *J. Amer. Chem. Soc.*, **93**, 1558
64. Woessner, D. E., Snowden, B. S. Jr. and Meyer, G. H. (1969). *J. Chem. Phys.*, **50**, 719
65. Huntress, W. T. Jr. (1969). *J. Phys. Chem.*, **73**, 103
66. Pugmire, R. J. and Grant, D. M. Unpublished results.
67. Lyerla, J. R. Jr. and Grant, D. M. Unpublished results.
68. Collins, S. W. (1971). Ph.D. Thesis, University of Utah
69. Olivson, A. and Lippmaa, E. (1971). *Chem. Phys. Lett.*, **11**, 241
70. Favelukes, C. E., Clifford, A. A. and Crawford, B. Jr. (1968). *J. Phys. Chem.*, **72**, 962
71. Atkins, P. W. (1969). *Molec. Phys.*, **17**, 321
72. Atkins, P. W. (1969). *Molec. Phys.*, **17**, 329
73. Lyerla, J. R., Jr., Grant, D. M. and Bertrand, R. D. (1972). *J. Phys. Chem.*, **75**, 3967
74. Dailey, B. P. (1960). *J. Chem. Phys.*, **33**, 1641
75. Freeman, R. and Hill, H. D. W. (1971). *J. Chem. Phys.*, **55**, 1985
76. Shoup, R. R. and VanderHart, D. L. (1971). *J. Amer. Chem. Soc.*, **93**, 2053
77. Speiss, H. W., Schweitzer, D., Haeberlen, U. and Hausser, K. H. (1971). *J. Mag. Res.*, **5**, 101
78. Yannoni, C. S. and Whipple, E. B. (1967). *J. Chem. Phys.*, **47**, 2508
79. Caspary, W. J., Millett, F., Reichbach, M. and Dailey, B. P. (1969). *J. Chem. Phys.*, **51**, 623
80. Pines, A., Rhim, W. and Waugh, J. S. (1971). *J. Chem. Phys.*, **54**, 5438
81. Farrar, T. C., Druck, S. J., Shoup, R. R. and Becker, E. D. (1972). *J. Amer. Chem. Soc.*, **94**, 699
82. Levy, G. C. (1972). *Chem. Commun.*, 352
83. Levy, G. C., White, D. M. and Anet, F. A. L. (1972). *J. Mag. Res.*, **6**, 453
84. Levy, G. C. Private communication
85. Schmidt, C. F., Jr. and Chan, S. I. (1971). *J. Chem. Phys.*, **55**, 4670
86. Alger, T. D., Grant, D. M. and Harris, R. K. (1972). *J. Phys. Chem.*, **76**, 281
87. Bender, H. J. and Zeidler, M. D. (1971). *Ber. Bunsenges. Phys. Chem.*, **75**, 236

6
N.M.R. and E.S.R.
in Liquid Crystals

J. BULTHUIS
Scheikundig Laboratorium van de Vrije Universiteit, Amsterdam

C. W. HILBERS
Katholieke Universiteit, Nijmegen, The Netherlands

and

C. MACLEAN
Scheikundig Laboratorium van de Vrije Universiteit, Amsterdam

6.1   INTRODUCTION

Thermotropic liquid crystals are compounds exhibiting a so-called meso-phase which exists between the solid state and the isotropic liquid state. A substance in this mesophase possesses the characteristics of a normal liquid, but it appears that a long-range molecular order exists. Three types of liquid crystal can be distinguished: the nematic, the smectic and the chloresteric (twisted nematic) type. The molecules constituting a liquid crystal generally have a rod-like shape.

Most of the magnetic resonance work in liquid crystalline solvents has been carried out with nematic liquids, although experiments have been performed with cholesteric and smectic compounds. In the nematic phase the molecules tend to orientate with their long axis parallel to each other. This relative orientation extends over macroscopic distances and can be charac-terised by a unit vector N pointing in the direction of the long molecular axis. This vector N, the so-called director, does not need to have the same orientation over a macroscopic sample due to disturbing effects such as temperature gradients, flow, and boundaries, but the application of a modest magnetic or electric field gives a uniform distribution of the vectors N throughout the sample.

From a dynamical point of view the orientational molecular motion has cylindrical symmetry with respect to the optical axis, in contrast to a normal liquid. Because of the anisotropic motion of the molecules, orientation dependent interactions between the spins may manifest themselves in e.s.r. and n.m.r. spectra. As an example the direct magnetic interactions between the nuclear spins may be mentioned. The intermolecular dipole–dipole interactions, however, are averaged to zero because of rapid translational diffusion.

Apart from *thermotropic* liquid crystals, there are *lyotropic* liquid crystals that exist for specific concentration ranges in some multicomponent systems. Examples are certain soap solutions; biological membranes are likewise important representatives of this type of substance. A recent review on liquid crystalline substances has been published by Brown, Doane, and Neff[1].

Research on liquid crystals in which magnetic resonance is involved may be roughly divided into two parts: (a) The properties of molecules dissolved in a liquid crystal are studied. (b) The liquid crystal is studied for its own interest, in most cases via solute molecules. Since the pioneering n.m.r. experiments of Saupe and Englert[2] in 1963, attention has been directed mainly to the first subject. The method has been applied as a new technique for the determination of molecular geometry, and for the study of anisotropies of chemical shifts and of indirect nuclear spin–spin coupling constants. Non-rigid molecules have been studied in the nematic phase and barriers hindering internal motion have been obtained. At present, as we will see, the method suffers from several limitations. Molecules containing more than seven or eight magnetic nuclei give rise to n.m.r. spectra composed of so many lines that extensive overlapping occurs. The interpretation of the results is hampered by the difficulty in accounting properly for molecular vibrations and for pseudo-dipolar couplings.

In e.s.r. spectroscopy in nematic phases the emphasis has been on studies of the molecular distribution function of the liquid crystal rather than on anisotropies of g-factors and of hyperfine couplings.

An excellent and extensive review on n.m.r. in liquid crystals by Diehl and Khetrapal[3] covers the period up to *medio* 1970. Other reviews on the same subject have been published by Luckhurst[4], by Meiboom and Snyder[5, 6] and by Bernheim and Lavery[7]. A more general account of n.m.r. of oriented molecules has been given by Buckingham and McLauchlan[8]. In the present review results of n.m.r. as well as of e.s.r. in liquid crystalline solvents are discussed, up to *medio* 1971.

6.2   THE SPIN HAMILTONIAN

6.2.1   Introduction

As stated in the introduction, n.m.r. and e.s.r. spectra of compounds dissolved in liquid crystalline solvents are dominated by anisotropic interactions, which in normal isotropic liquids either average to zero or remain manifest as scalar interactions. To discuss the influence of these anisotropic interactions we shall first describe the spin Hamiltonian appropriate to liquid crystal spectra.

Let us consider a rigid diamagnetic molecule fixed with respect to a laboratory frame of reference. In the presence of a magnetic field H the following terms contribute to the spin Hamiltonian. The Zeeman interaction of the nuclear spins is given by

$$H_{Z_N} = -\hbar \sum_i \gamma_i \, \boldsymbol{H} . \boldsymbol{I}_i \tag{6.1}$$

The energy associated with chemical shielding is

$$H_{cs} = \hbar \sum_i \gamma_i \, \boldsymbol{H} . \boldsymbol{\sigma}_i . \boldsymbol{I}_i \tag{6.2}$$

The indirect spin–spin interaction between spins i and j in a molecule is represented by

$$H_J = \sum_{i<j} \boldsymbol{I}_i . \boldsymbol{J}_{ij} . \boldsymbol{I}_j \tag{6.3}$$

In normal high-resolution n.m.r. spectra this term simplifies to the well-known scalar J-coupling. We shall see later (Section 6.5) that the anisotropic parts may complicate geometry determinations with liquid crystal n.m.r.

The dominant perturbation for a set of spin $I = \frac{1}{2}$ nuclei is the direct dipole–dipole interaction

$$H_D = \sum_{i<j} I_i . D_{ij} . I_j \tag{6.4}$$

In the case of nuclei with spin $I \geq 1$, however, the quadrupole coupling may completely dominate. This interaction is represented by

$$H_Q = \sum_i I_i . Q_i . I_i \tag{6.5}$$

The components of the dipolar coupling tensor D and the quadrupolar coupling tensor Q are

$$D_{\alpha\beta}^{ij} = \hbar^2 \frac{\gamma_i \gamma_j}{r_{ij}^5} (r_{ij}^2 \delta_{\alpha\beta} - 3 r_{ij\alpha} r_{ij\beta}) \tag{6.6}$$

$$Q_{\alpha\beta} = \frac{eQ}{2I(2I-1)} \frac{\partial^2 V}{\partial\alpha\partial\beta} \tag{6.7}$$

where r_{ij} is the distance between nuclei i and j, δ is the Kronecker symbol, and α, β refer to axes of the reference system. Note that the tensors are traceless and symmetric.

For paramagnetic molecules additional interaction terms, which mainly determine the form of the e.s.r. spectrum, must be introduced. If only one unpaired electron is involved, the Zeeman energy is

$$H_{Z_E} = \beta H . g . S \tag{6.8}$$

The hyperfine coupling between the unpaired electron and the nuclear spins in the system is given by

$$H_{SI} = \sum_i S . A . I_i \tag{6.9}$$

where A is the hyperfine tensor. It consists of the isotropic Fermi interaction and a part which represents the direct dipole–dipole coupling between the nuclei and the unpaired electron. If the radical contains two unpaired electrons, the interaction between the spins also contributes to the energy through

$$H_{ss} = S_1 . D_{el} . S_2 \tag{6.10}$$

where the D_{el} tensor comprises the scalar electron exchange coupling and the dipolar interactions between the electrons. In paramagnetic species containing five or more unpaired electrons, higher than second rank tensor interactions may play a role[9, 10]. We shall, however, not be concerned with the effects of such electron–electron interactions because, as far as we know, they have not been measured in liquid crystal spectra.

6.2.2 Transformation of tensorial properties

In the presence of a magnetic field, which in practice provides a space fixed quantisation axis, it is convenient to define the component form of equations (6.1–6.5) and (6.8–6.10) in the laboratory frame of reference, with the z'-axis parallel to the magnetic field. Tensorial properties, such as the chemical shift, the dipole–dipole coupling, etc., which are molecular properties, are preferably expressed in the molecular frame of reference. To relate the tensor components in the different coordinate systems we shall define them in the spherical coordinate representation[11], because then they have convenient transformation properties under rotation.

According to their definition, spherical (or standard irreducible) tensors transform under rotation as the spherical harmonics, i.e.[11]

$$F^{(k,q')} = \sum_p \mathcal{D}^{(k)}_{pq'}(\alpha\beta\gamma) F^{(k,p)} \tag{6.11}$$

k is the rank of the tensor and p denotes a particular component of the irreducible tensor $(-k \leqslant p \leqslant k)$. The primed elements denote the value of the tensor components in the laboratory frame, whereas the unprimed component is the value in the molecular frame of reference. The $\mathcal{D}^{(k)}_{pq'}(\alpha\beta\gamma)$ are elements of the Wigner rotation matrix[11] and α, β and γ are the Euler angles of rotation which take the unprimed into the primed axes.

After decomposing the tensors in equations (6.1)–(6.5) and (6.8)–(6.10) into an isotropic, an antisymmetric, and a traceless symmetric part, the total Hamiltonian can easily be written as a sum of scalar products of spherical tensors of equal rank k [11]

$$H = \sum_{\lambda, k, q'} (-1)^{q'} F^{(k, q')}_\lambda A^{(k, -q')}_\lambda \tag{6.12}$$

where λ indicates the type of magnetic interaction. $F^{(k,q')}_\lambda$ are elements of the magnetic interaction tensors and $A^{(k, -q')}_\lambda$ represent components of tensor operators acting on the spin functions. Insertion of equation (6.11) in equation (6.12) yields equation (6.13)

$$H = \sum_{\lambda, k, q', p} (-1)^{q'} \mathcal{D}^{(k)}_{pq'}(\alpha\beta\gamma) F^{(k, p)}_\lambda A^{(k, -q')}_\lambda \tag{6.13}$$

In the presence of molecular motions the Hamiltonian is time dependent, because the Euler angles in equation (6.13) depend on the orientation of the molecule with respect to the magnetic field. Moreover, the magnetic interaction terms $F^{(k, p)}_\lambda$ may depend on the orientation of the molecule and on the internal molecular motions. For instance, the nuclear dipole–dipole coupling will be a function of the molecular vibrations and, if present, of intramolecular chemical exchange. In the following we assume that the molecular motions are fast, so that we are allowed to deal with an effective time-independent Hamiltonian. We also assume that the averaging of the $\mathcal{D}^{(k)}_{pq'}$ functions and of the magnetic interaction tensors $F^{(k, q')}_\lambda$ over the molecular motion can be performed independently. The mean values of the function $\mathcal{D}^{(k)}_{pq'}$, which then describe the average orientation of the molecule with respect to the magnetic field, are called motional constants. Their number may be reduced on account of molecular symmetry or because of the symmetrical molecular motion. If the action of the constraint on the molecule is axially symmetric about the laboratory z'-axis (the magnetic field direction), all terms with $q \neq 0$ will vanish after averaging over γ [12]. Thus under these circumstances equation (6.13) transforms into the following expression

$$H = \sum_{\lambda, k, p} \overline{\mathscr{D}_{p0}^{(k)}} \langle F_{\lambda}^{(kp)} \rangle A_{\lambda}^{(k, 0)} \qquad (6.14)$$

The bar and the brackets $\langle \rangle$ denote an average over the appropriate molecular motions. If, however, the axial constraint is not in the direction of the magnetic field, a situation which can be reached by applying an electric field not parallel to the magnetic field, the Hamiltonian (equation (6.13)) reads[13]

$$H = \sum_{\lambda, k, q', p} (-1)^{q'} \overline{\mathscr{D}_{0q'}^{(k)} \mathscr{D}_{p0}^{(k)}} \langle F_{\lambda}^{(k, p)} \rangle A_{\lambda}^{(k, -q')} \qquad (6.15)$$

In n.m.r. the high-field approximation can be invoked, i.e. the Zeeman energy levels are affected by the terms in equations (6.2)–(6.5) only in first order. A truncated Hamiltonian can then be used, which only connects terms with the same $I_z = \Sigma I_{zi}$ value. From the properties of irreducible tensor operators[11] it follows immediately that in equation (6.15) only products containing factors $A_{\lambda}^{(k, 0)}$ survive. The angular dependence of the Hamiltonian (equation (6.15)) on the direction of the axial constraint is then represented by $\mathscr{D}_{00}^{(2)} = 2^{-1}(3\cos^2\varepsilon - 1)$, where ε is the angle between the director and the magnetic field[14]. The elements $\mathscr{D}_{p0}^{(k)}$ are related to the spherical harmonics by[11]

$$\mathscr{D}_{p0}^{(k)}(\alpha\beta) = \left(\frac{4\pi}{2k+1}\right)^{\frac{1}{2}} Y_p^{*(k)}(\beta\alpha) \qquad (6.16)$$

Thus the spin Hamiltonian (6.15) can also be written in terms of average values of spherical harmonics.

In the magnetic resonance experiments under discussion only terms with $k = 0$ and 2 contribute to the spin Hamiltonian. Even if the tensors contain antisymmetric parts, as is the case for the indirect spin–spin interaction and the chemical shift, these do not contribute[15]. Traceless symmetric tensors like the dipole–dipole coupling and the quadrupole coupling tensor also do not contain the isotropic term with $k = 0$.

The general expressions for the operators $A_{\lambda}^{(0, 0)}(M, N)$ and $A_{\lambda}^{(2, 0)}(M, N)$ are given by

$$A_{\lambda}^{(0, 0)}(M, N) = 3^{-\frac{1}{2}}(M_{z'}N_{z'} + \tfrac{1}{2}(M^+N^- + M^-N^+)) \qquad (6.17)$$

and

$$A_{\lambda}^{(2, 0)}(M, N) = 2^{\frac{1}{2}} \cdot 3^{-\frac{1}{2}}(M_{z'}N_{z'} - \tfrac{1}{4}(M^+N^- + M^-N^+)) \qquad (6.18)$$

where the vector operators M and N may correspond to H, S or I. When the magnetic field is along the laboratory z'-axis and/or the magnetogyric ratios of the coupling particles are largely different, the $M^{\pm}N^{\pm}$ terms may be neglected. The magnetic interaction terms connected with equations (6.17) and (6.18) are

$$F_{\lambda}^{(0,0)} = 3^{-\frac{1}{2}} \mathrm{Tr}\, F \tag{6.19}$$

$$F_{\lambda}^{(2,0)} = 2^{\frac{1}{2}} \cdot 3^{-\frac{1}{2}} (F_{z'z'} - 3^{-\frac{1}{2}} \mathrm{Tr}\, F) \tag{6.20}$$

The high-field approximation formulated above, is not always valid in e.s.r. The operators $A_{\lambda}^{(2, \pm 1)}$ may remain in the spin Hamiltonian (equation (6.12)) because the nuclear spins are quantised in an effective field involving the external magnetic field plus the field arising from the unpaired electron via the hyperfine interaction. Below we shall assume that the constraint acting on the molecules is along the magnetic field; then the $A_{\lambda}^{(2, \pm 1)}$ terms are eliminated as discussed above[12]. Under these conditions and as a consequence of the neglect of the non-secular terms, effective hyperfine interactions and g values are measured in e.s.r. spectra;

$$A' = \langle A^{(0,0)} \rangle + \sum_p \overline{\mathscr{D}_{p0}^{(2)}} \langle A^{(2,p)} \rangle \tag{6.21}$$

and

$$g' = \langle g^{(0,0)} \rangle + \sum_p \overline{\mathscr{D}_{p0}^{(2)}} \langle g^{(2,p)} \rangle \tag{6.22}$$

An analogous formula is valid for the chemical-shift term in the n.m.r. Hamiltonian.

As a consequence of the restrictions mentioned above, the following points are noteworthy. The nuclear quadrupole coupling does not influence the line positions in e.s.r., neither in the isotropic nor in the anisotropic case, because the e.s.r. transitions take place between levels with the same m_z value. In n.m.r. the coupling between nuclei with strongly different magnetogyric ratios is measured as the sum $(J^{(0,0)} + J^{(2,0)} + D^{(2,0)})$. $D^{(2,0)}$ is always measured in combination with $J^{(2,0)}$.

6.2.3 Tensor averages in the formalisms of Saupe and Snyder

To relate the tensor components measured in the laboratory frame of reference to the components in the molecular frame, two other methods have been used, which have both found wide application in n.m.r. Saupe[14] has worked out the problem in terms of Cartesian tensors. The alignment of the solute is described in terms of an ordering matrix S, with elements

$$S_{ij} = \tfrac{1}{2} \overline{(3 \cos \theta_{\alpha} \cos \theta_{\beta} - \delta_{\alpha\beta})} \tag{6.23}$$

where α, β denote the coordinates x, y and z in the molecular frame, and θ_x, θ_y and θ_z are the angles between the molecular axes and the direction of the magnetic field. For a constraint exhibiting rotational symmetry it has been shown[14] that the mean tensor components $\langle T_{z'z'}\rangle$, $\langle T_{x'x'}\rangle$, and $\langle T_{y'y'}\rangle$ of a second-rank tensor $\boldsymbol{T}$ are given by the relations

$$\langle T_{z'z'}\rangle = \tfrac{1}{3}\sum_{\alpha}\langle T_{\alpha\alpha}\rangle + \tfrac{2}{3}\sum_{\alpha,\beta} S_{\alpha\beta}\langle T_{\alpha\beta}\rangle \tag{6.24}$$

and

$$\langle T_{x'x'}\rangle = \langle T_{y'y'}\rangle = \tfrac{1}{3}\sum_{\alpha}\langle T_{\alpha\alpha}\rangle - \tfrac{1}{3}\sum_{\alpha,\beta} S_{\alpha\beta}\langle T_{\alpha\beta}\rangle \tag{6.25}$$

The tensor components $T_{\alpha'\beta'}$, with $\alpha'\beta' = x', y', z'$ are defined in the laboratory frame of reference.

Snyder[16] has related the tensor components $T_{\alpha'\beta'}$ to the anisotropic molecular motion via a probability function $P(\theta,\phi)$, which is the probability per unit solid angle that the magnetic field is at polar angles θ and ϕ with respect to the molecular axes. $P(\theta, \phi)$ is expanded in real spherical harmonics

$$P(\theta,\phi) = \frac{1}{4\pi} + c_x P_x + c_y P_y + c_z P_z + c_{3z^2-r^2} D_{3z^2-r^2}$$
$$+ c_{x^2-y^2} D_{x^2-y^2} + c_{xy} D_{xy} + c_{xz} D_{xz} + c_{yz} D_{yz} + \dots. \tag{6.26}$$

When the components of the second-rank tensors, defined in the laboratory frame, are averaged with $P(\theta,\phi)$, their relation with the components in the molecular frame is given by[16]

$$\langle T_{z'z'}\rangle = \tfrac{1}{3}\langle T_{xx}+T_{yy}+T_{zz}\rangle + c_{3z^2-r^2}(2/3)5^{-\frac{1}{2}}\langle T_{zz}-\tfrac{1}{2}(T_{xx}+T_{yy})\rangle$$
$$+ c_{x^2-y^2}15^{-\frac{1}{2}}\langle T_{xx}-T_{yy}\rangle + c_{xz}15^{-\frac{1}{2}}\langle T_{xz}-T_{zx}\rangle + c_{yz}15^{-\frac{1}{2}}\langle T_{yz}+T_{zy}\rangle$$
$$+ c_{xy}15^{-\frac{1}{2}}\langle T_{xy}+T_{yx}\rangle \tag{6.27}$$

The motional constants $c_{3z^2-r^2}$, $c_{x^2-y^2}$, etc. are averages of second-order real spherical harmonics over the molecular tumbling[16]. As a specific example of equation (6.27) we quote the expression for the dipole coupling between spins i and j

$$D_{ij} = \frac{-2}{\sqrt{5}} K_{ij}\left(c_{3z^2-r^2}\langle \frac{(\Delta z_{ij})^2}{r_{ij}^5} - \tfrac{1}{2}\frac{(\Delta x_{ij})^2}{r_{ij}^5} - \tfrac{1}{2}\frac{(\Delta y_{ij})^2}{r_{ij}^5}\rangle \right.$$
$$+ c_{x^2-y^2}\frac{\sqrt{3}}{2}\langle \frac{(\Delta x_{ij})^2}{r_{ij}^5} - \frac{(\Delta y_{ij})^2}{r_{ij}^5}\rangle + c_{xz}\sqrt{3}\langle \frac{\Delta x_{ij}\Delta z_{ij}}{r_{ij}^5}\rangle$$
$$+ c_{yz}\sqrt{3}\langle \frac{\Delta y_{ij}\Delta z_{ij}}{r_{ij}^5}\rangle + c_{xy}\sqrt{3}\langle \frac{\Delta x_{ij}\Delta y_{ij}}{r_{ij}^5}\rangle \tag{6.28}$$

where $\Delta z_{ij} = z_i - z_j$, etc.; z_i is the z-coordinate of nucleus i in the molecular frame; $K_{ij} = (2\pi)^{-1}\gamma_i\gamma_j\hbar$ Hz cm^3.

The motional constants in the different schemes are related:

$$c_{3z^2-r^2} = 5^{\frac{1}{2}}S_{zz} \qquad\qquad = 5^{\frac{1}{2}}\overline{\mathscr{D}_{00}^{(2)}}$$

$$c_{x^2-y^2} = 5^{\frac{1}{2}}3^{-\frac{1}{2}}(S_{xx}-S_{yy}) = 5^{\frac{1}{2}}2^{-\frac{1}{2}}(\overline{\mathscr{D}_{20}^{(2)}}+\overline{\mathscr{D}_{-20}^{(2)}})$$

$$c_{xz} = 5^{\frac{1}{2}}3^{-\frac{1}{2}}S_{xz} \qquad = 5^{\frac{1}{2}}8^{-\frac{1}{2}}(\overline{\mathscr{D}_{10}^{(2)}}-\overline{\mathscr{D}_{-10}^{(2)}}) \tag{6.29}$$

$$c_{yz} = 5^{\frac{1}{2}}3^{-\frac{1}{2}}S_{yz} \qquad = -i5^{\frac{1}{2}}8^{-\frac{1}{2}}(\overline{\mathscr{D}_{-10}^{(2)}}+\overline{\mathscr{D}_{10}^{(2)}})$$

$$c_{xy} = 5^{\frac{1}{2}}3^{-\frac{1}{2}}S_{xy} \qquad = -i5^{\frac{1}{2}}8^{-\frac{1}{2}}(\overline{\mathscr{D}_{-20}^{(2)}}-\overline{\mathscr{D}_{20}^{(2)}})$$

On account of molecular symmetry, motional constants may be zero. For instance, if a molecule possesses a threefold or higher symmetry axis only one motional constant, $c_{3z^2-r^2}$, suffices to describe the molecular orientation. If the molecule has C_{2v} symmetry, two motional constants are sufficient. The relationship between molecular symmetry and the number of independent components of a tensor property of a molecule can be found in a paper by Jahn[17].

6.3 EXPERIMENTAL ASPECTS

6.3.1 Liquid crystals

To be suitable as a solvent in n.m.r. spectroscopy, the n.m.r. spectrum of the liquid crystal should preferably be quasi-continuous. This is in general the case for the ^{1}H n.m.r. spectrum if the number of protons per molecule is large. The line widths of the solute spectrum should be as small as possible. The liquid crystal should therefore preferably have a low viscosity, a requirement easily met by nematic liquids.

The quality of a spectrum is enhanced when good temperature stability and homogeneity over the probe can be maintained. These conditions can most easily be attained with liquid crystalline solutions that are nematic at the magnet temperature. At present, several liquid crystals and mixtures of liquid crystals are known, which are nematic in a wide temperature range, including the magnet temperature. Nematic liquids, which are useful for n.m.r., are listed in reference 3, together with their temperature ranges. The ranges for nematic solutions may shift appreciably to lower temperatures compared with those of the pure compounds.

In e.s.r. the applicability of liquid crystals, e.g. as a solvent for free radicals, is mainly restricted by their chemical reactivity.

So far, we have tacitly assumed that the nematic liquids are of the thermotropic type. *Lyotropic* liquids have also been used successfully as solvents in n.m.r. spectroscopy[18-25].

Only in relatively few instances[26-30], the use of (viscous) smectic solvents in n.m.r. and e.s.r. has been reported. Homogeneously orientated smectic layers are not easily obtained, but once prepared, the angle between the optic axis and the magnetic field can be chosen arbitrarily.

6.3.2 Instrumentation

N.M.R. spectra of molecules dissolved in liquid crystals can be recorded with standard equipment. Frequency-swept spectrometers are more suitable than field-swept spectrometers, because the field homogeneity does not change over the spectrum, the width of which may well amount to a few kHz. In conventional n.m.r. spectroscopy, the effect of inhomogeneities in a plane perpendicular to the cylindrical sample is reduced by spinning. This is not, in general, possible for liquid crystalline samples as the orientation of the molecules may easily be destroyed. This disadvantage does not apply, however, to spectrometers with a superconducting coil.

As liquid crystal spectra are often composed of many relatively broad lines (c. 5–15 Hz), the sensitivity of the n.m.r. spectrometer may be insufficient to show a complete spectrum in a single scan; spectrum accumulation will then be necessary to detect lines of relatively low intensity. ^{13}C spectra of molecules dissolved in liquid crystals, have been obtained using Fourier-transform spectroscopy[32]. The dependence of liquid-crystal n.m.r. spectra on the spinning speed, as well as on the temperature and concentration, has been studied systematically for a number of different solutes in the nematic liquid anisole azophenyl-n-capronate by Diehl and Khetrapal[31].

6.4 N.M.R. SPECTRUM ANALYSIS

Liquid-crystal n.m.r. spectra of spins $I = \frac{1}{2}$ can be interpreted in terms of effective chemical shifts, isotropic indirect couplings, and anisotropic couplings. The latter ones consist of the direct dipolar interactions and possibly of anisotropic contributions from indirect couplings: the pseudo-dipolar interactions.

Diehl and Kellerhals have shown that for a spin system without symmetry, the determination of these parameters is unique, except that the nuclei may be permuted, and that the sign of the coupling constants relative to the Larmor frequencies is arbitrary[33]. In practice, ambiguities arising in the analysis of an anisotropic n.m.r. spectrum can often be removed; for example, a certain assignment of the couplings can be rejected if the direct couplings are incompatible with any acceptable values of the geometrical and ordering parameters. Spin tickling may also prove useful[34].

Spectrum analysis is generally accomplished using one of the modifications[20, 35, 36] of the iterative program LAOCOON III [37]. In a further modification (LEQUOR) the treatment of groups of fully equivalent spins has been introduced[38].

6.4.1 Equivalent spins

The line positions and intensities can be easily written down as analytical functions of the parameters only for simple systems. For larger molecules it is desirable to trace the features in the spectrum that are a consequence of the symmetry of the spin system (reference 3 for examples). The concept of equivalence of spins is useful in this respect. In the isotropic case, two or more spins within a molecule are said to be magnetically equivalent if the nuclei have (a) the same chemical shift, (b) the same coupling constant to any nucleus outside the set. For magnetic equivalence in the anisotropic case— or 'full equivalence'—a third condition is necessary; (c) equivalent nuclei have the same coupling constant within the set[39]. The coupling constant now includes both the scalar and dipolar coupling.

These conditions are equivalent with the condition that $\sum_{i<j} J_{ij} I_i \cdot I_j$, where i and j indicate equivalent spins, commutes with the spin Hamiltonian[35, 40, 41]. The spectrum of a system of n fully-equivalent nuclei consists of a binominal multiplet of n lines, equally spaced by $\frac{3}{2}D_{eq}$, where D_{eq}

denotes the anisotropic coupling between the nuclei[35, 41]. It should be stressed, that the definition of full equivalence only applies to nuclei with $I = \frac{1}{2}$; nuclei with $I \geqslant 1$ cannot be fully equivalent[35, 41, 42].

When one or more sets of equivalent nuclei are present, introduction of the total spin[35, 44] may be advantageous; n.m.r. spectra can often be broken down into a number of sub-spectra, each of which corresponds to a state of the total spin. The classification of a single set of equivalent nuclei according to the total spin is equivalent with the classification according to the irreducible representations of the n.m.r. symmetry group of the molecule[43]. The n.m.r. symmetry group, which is the group of operations that leaves the spin Hamiltonian invariant, is in general a sub-group of the molecular symmetry group.

6.4.2 Non-rigid molecules

The application of group theory is not restricted to rigid molecules, although the symmetry operations of non-rigid molecules are defined differently. Longuet-Higgins has introduced a new system of permutation operations in order to classify the energy levels of non-rigid molecules[45]. All operations have to be feasible, i.e. they can be achieved without passing over an insuperable energy barrier within the time scale of the experiment.

A different approach was followed by Altmann[46], who introduced 'isodynamic' operations: operations that transform conformations, which have identical energy eigenvalues, into one another. Again, the operations have to be feasible. The discrepancy between the two approaches is a consequence of Altmann's neglect of rovibronic interactions in some of the operations defined. If these are taken into account, the two methods coincide[47, 48].

Both schemes have been used to derive the character tables for non-rigid molecules. (For examples see references 49 and 50.) For n.m.r. spectrum analysis of non-rigid molecules, simple rules have been derived[51] which are useful for finding the equivalent nuclei, for identifying the distinct sets of coupling constants and for classifying the spin states.

6.5 STRUCTURAL DATA FROM N.M.R. SPECTRA

The dipolar coupling D_{ij} between two nuclei i and j is related to the internuclear distance r_{ij} by

$$D_{ij} = -K_{ij} \left\langle \frac{3\cos^2 \theta_{ij} - 1}{r_{ij}^3} \right\rangle \tag{6.30}$$

θ_{ij} is the angle between r_{ij} and the magnetic field. If no correlation is present between the solute orientation and the interatomic distances within the solute, the angular term $(3\cos^2\theta_{ij} - 1)$ and the term (r_{ij}^{-3}) can be averaged independently. This assumption underlies all geometry determinations with the liquid crystal n.m.r. method and, although it has been questioned at times[52], no convincing evidence to the contrary has been reported in the literature.

It should be noted that in equation (6.28) a geometrical quantity (such as $\langle \Delta z_{ij}^2/r_{ij}^5 \rangle$) is always multiplied by a motional constant. Hence, a unique set of motional constants and internuclear distances cannot be obtained from the dipolar coupling. Accordingly, one might say that the liquid-crystal n.m.r. method determines the molecular shape rather than the geometry. In a sense one could also say that angles are determined, rather than distances[53].

6.5.1 Limitations of the method

An experimental anisotropic coupling may contain contributions ($J^{(2,0)}$) from indirect nuclear spin–spin interactions, which are not of the familiar Fermi type. Indirect proton–proton couplings are generally assumed to be caused by Fermi interaction; hence they are assumed to be isotropic. On the other hand, indirect fluorine–fluorine couplings may possess significant pseudo-dipolar parts and neglect thereof in the determination of molecular geometries by n.m.r. may lead to erroneous results. Vice versa, liquid crystal n.m.r. may provide valuable data related to pseudo-dipolar spin–spin interactions (Section 6.6).

Another limitation of the liquid-crystal n.m.r. method to determine molecular geometries lies in the limited number of spins, say seven or eight per molecule, that can be handled. The number of energy levels is 2^n and the number of transitions is of the same order of magnitude. If n is too large, a poorly resolved or even quasi-continuous signal remains. Meiboom and Snyder[6] have devised an elegant substitution method to circumvent this problem, which is promising to study geometries of large molecules (see Section 6.5.3).

It may happen that the geometrical information contained in a liquid-crystal n.m.r. spectrum is very sparse[3, 54]. It seems worthwhile to discuss a few examples. Consider the planar molecules (1) and (2), which possess C_{2v} symmetry. The geometry, as far as the protons and fluorine nuclei are concerned, is characterised by two parameters: $\rho = r_{34}/r_{12}$ and $\kappa = r_{14}/r_{12}$. Clearly, $\rho_{(1)} \simeq 1$ and $\rho_{(2)} \simeq 2$

(1)   (2)   (3)

These molecules possess parallel internuclear vectors $r_{12} \| r_{34}$; hence[55]

$$\left(\frac{\gamma_H}{\gamma_F}\right)^2 D_{34} = \rho^{-3} D_{12} \tag{6.31}$$

An experiment performed in the region where D_{12} and D_{34} are small, cannot provide the value of ρ with acceptable precision. For molecule (3) the parameter ρ equals 1 by symmetry. The following relation holds for molecules (1)–(3)

$$\frac{D_{14}}{D_{12}} = \frac{1}{\kappa^{-5}}\left\{\frac{D_{13}}{D_{12}}(\rho+\kappa^2)^{\frac{5}{2}} - \frac{\gamma_3}{\gamma_1}\rho\right\} \qquad (6.32)$$

This relation is visualised for $\rho = 1$ in Figure 6.1. Again no accurate conclusions concerning the molecular geometry can be drawn if the experiment is performed in the neighbourhood of the intersections of the lines.

Other cases where an n.m.r. spectrum, measured in the nematic phase, may not provide detailed geometrical information are, of course, those where the spectrum is of the deceptively simple type[3, 56].

Liquid-crystal n.m.r. spectra of nuclei with electric quadrupole moments are characterised by large line splittings, induced by incomplete averaging of the quadrupole interaction (see Section 6.10). Because of the unfavourable

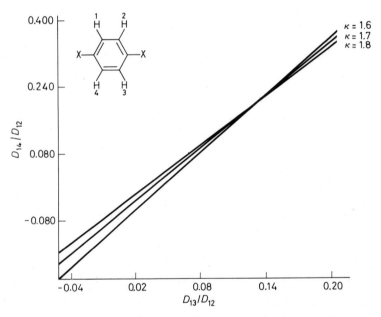

Figure 6.1 D_{14}/D_{12} as a function of D_{13}/D_{12} in a 4-spin system with D_{2h} symmetry, given for various values of $\kappa = r_{14}/r_{12}$ (see reference 100)

line widths, n.m.r. spectroscopy of quadrupolar nuclei has been restricted to studies of quadrupole coupling constants.

Finally, a limitation of geometry studies by n.m.r. in liquid crystals is that the experimental dipolar coupling are averages of all motions of the solute molecule in the solvent, including vibrations. The problem of correcting the latter will be discussed briefly in Section 6.5.4.

6.5.2 Rigid molecules

In spite of the previously mentioned assumptions underlying relative geometry determinations by n.m.r. in liquid crystalline solvents, the comparison
of n.m.r. geometries with those obtained by other techniques remains valuable,
especially when dipolar couplings are involved which may be expected to
contain no significant contributions from indirect couplings. As practically
all structural information obtained from n.m.r. in liquid crystals is contained
in reference 3, we shall only mention recent results.

Barili and Veracini[57] have suggested that changes in the molecular
geometry may be produced by solute–solvent interactions. This was inferred
from the results of n.m.r. measurements of p-dinitrobenzene in a nematic
solvent, by varying the temperature over a range of only 15 °C. The calculated
distance of the *meta*-protons relative to that of the *ortho*-protons varied
gradually from 1.6899 ± 0.005 at 75 °C to 1.7102 ± 0.005 at 90 °C. The
significance of this change, which is attributed by the authors to a charge-
transfer interaction between solvent and solute molecules, is doubtful in
view of the accuracy reported for the smallest dipolar coupling.

The unreliability of geometrical data obtained from fluorine couplings, is
illustrated by a comparison of the results obtained for fluoroethylenes
(Table 6.2, reference 101). If a comparison with results from other techniques
is possible, a large discrepancy of the n.m.r. results may be indicative of
anisotropic indirect couplings.

Recently, data on other molecules, although not restricted to geometry
determinations, have been reported for 1,4-naphthoquinone[58], isoxazole[59],
pyrimidine[60], benzenebicarbonylchromium[61], π-cyclopentadienyltricarbonyl-
manganese[62], tetrafluoro-1,3-dithiethane[63], benzo-, and 7.7-difluorobenzo-
cyclopropene[64] and bicyclobutane[5, 65].

6.5.3 Non-rigid molecules

6.5.3.1 General remarks

In a non-rigid molecule the rate of internal motion is fast on the time scale
of the n.m.r. experiment if the barrier to the internal motion is sufficiently
small. The n.m.r. spectrum can then be interpreted in terms of the molecular
symmetry group, as defined by Longuet-Higgins[45] or by Altmann[46]. However, to relate the direct couplings to the geometry of distinct conformations,
one has to distinguish two situations[66]: (a) If the internal motion is faster
than the reorientational motion, the symmetry operations are 'feasible' in
the time scale of the latter motion, and the number of motional constants is
in accordance with the molecular symmetry group. (b) If the reorientational
motion is faster, the direct couplings are the weighted averages of the corresponding couplings in each conformation. Moreover, there is not a single
set of motional constants; each conformation has its own set, in which the
number of motional constants depends on the symmetry of the conformation.
It is not always possible to discriminate between the two situations. For
example, the dipolar couplings in cyclobutane[67] can be expressed in terms of

a single motional constant, independent of whether the non-planar conformation or the effective planar structure is considered.

The complication arising from the competition between the rates of internal motion and reorientational motion should be borne in mind when studying non-rigid molecules. This aspect has been overlooked in the interpretation of the direct proton–proton couplings in 1,2,2,3-tetrachloro-propane[68], and also in the discussion of the n.m.r. spectra of furan- and thiophene-2,5-dialdehyde in a nematic solvent[69]. If a certain conformation of a molecule is expected to dominate, one may of course try to explain the direct couplings in terms of this conformation, provided that the appropriate symmetry group is used in deriving the number of motional constants[70, 71].

6.5.3.2   Bullvalene

In bullvalene all nuclei permute with each other in the course of an isomerisation process. The n.m.r. spectrum in a nematic solvent was measured by Yannoni[72]. It shows a binomial multiplet of ten lines, corresponding to ten equivalent protons. It is supposed that the equivalence arises because the rate of reorientation is faster than the rate of isomerisation (which is 10^6 s^{-1} at 130 °C). Using the geometry from diffraction data, the value of 0.054 was obtained for the absolute value of the ordering parameter, describing the orientation of the threefold axis of the static molecule. This value is large compared with that of other globular molecules (e.g. reference 73). Another interesting feature of the spectrum is line broadening, which decreases with increasing temperature. Obviously, the isomerisation rate is not sufficiently fast to cause extreme line narrowing.

6.5.3.3   Substituted toluenes

Diehl *et al.* have studied thoroughly a series of orientated substituted toluenes, namely 2,6-dichloro-[74], 3,5-dichloro-[75], *o*-chloro-[76], *o*-bromo- and *o*-iodo-toluene[77] and toluene[38], with the object of obtaining the structure of these molecules and of the rotation barrier of the CH_3 group. The assumptions explicitly made in their analysis and common to all molecules studied are: (a) the molecule has a plane of symmetry; (b) vibrations and distortions are neglected; (c) the plane of the methyl protons is perpendicular to the rotation axis of the methyl group; (d) the rotation of the methyl group is much faster than the molecular reorientation (except in the case of 2,6-dichlorotoluene, when applying a sixfold potential for the methyl group rotation). For each molecule it was necessary to make one or more additional assumptions about the molecular geometry.

In some cases the magnitude of the barrier to the methyl group rotation could be obtained by calculating the degree of orientation and the relative geometry from the experimental direct couplings, which were weighted according to the model considered for the internal motion. In a later stage, the calculations were performed with the aid of an iterative computer program (SHAPE) based on a least-squares method.

Barriers to the methyl group rotation could only be evaluated for *o*-chloro-[76], *o*-bromo-[77], and *o*-iodo-toluene[77]. The calculations involved three types of methyl group rotation: (a) stable rotamers, one with a methyl proton pointing towards the halogen atom (eclipsed), and the other with a staggered conformation; (b) free rotation; (c) a threefold potential: $V = V_3 \frac{1}{2}(1 - \cos 3\phi)$. In the latter case the eigenvalues were computed for the calculation of the rotational angle distribution. In Figure 6.2 the weighted r.m.s. error for the best-fit geometry of *o*-chlorotoluene is plotted against the barrier height for two positions of minimal energy. The minimal r.m.s. value was obtained with a barrier height of *c.* 1.2 kcal mol^{-1}, and with the staggered conformation

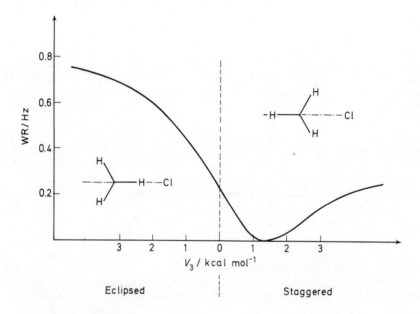

Figure 6.2 Variation of the weighted r.m.s. error (WR) for the best-fit geometry of *o*-chlorotoluene, with various barrier heights (V_3) and two positions of minimal energy. (Reproduced from Diehl *et al.*[76], by courtesy of Taylor & Francis Ltd. and the authors)

as the conformation of minimal energy. The error of the barrier height was rather large; the standard deviation was 0.6 kcal mol^{-1}. For *o*-bromotoluene and *o*-iodotoluene the values are, respectively: 0.89 ± 0.55 and 1.80 ± 0.58 kcal mol^{-1}.

In toluene[38] the assumed sixfold potential affected the direct couplings only very slightly (contrary to a corresponding threefold potential). The height of the hindering potential could therefore not be determined, although the results indicate a potential minimum when a C—H bond is in a plane perpendicular to the benzene plane.

The precision of the derived geometrical data was better than for the previously mentioned molecules. This was attributed to the overdetermination of the system due to the presence of eight protons and to a higher quality of the spectra.

6.5.3.4   Benzaldehyde[58] and salicylaldehyde[78]

These molecules have also been investigated by Diehl *et al.*[66, 78]. In contrast with toluene, these systems are highly underdetermined, consequently several assumptions were to be made to obtain information about their structures and the potential of the aldehyde group rotation.

In case of benzaldehyde[66], the authors concluded that a planar form corresponds to a minimum in the energy. The barrier height could, however, not be determined, nor could it be established whether the molecule is able to reorientate after a transition from one stable planar conformation to the other.

For salicylaldehyde[78], the interesting result was obtained that the data are consistent with a model in which the carbonyl and the hydroxyl groups are twisted outside the plane of the benzene ring, with both groups twisted to the same side.

6.5.3.5   Cyclohexane

In a study of the geometry of cyclohexane, Meiboom and Snyder[6] measured the ^1H n.m.r. spectrum of a deuterated sample (isotopic abundance *c.* 98 %). The ^2D—^1H couplings were erased by irradiation at the centre frequency of the two deuterium lines (which originate from the interaction of the deuterium quadrupole with the electric field gradient in the molecule) to induce double-quantum transitions. The ^1H n.m.r. spectrum of those molecules containing one proton (*c.* 78 %) consists of a single line; that of molecules with two protons (*c.* 20 %) consists of a series of doublets, the line splittings of which are related to the inter-proton distances. More information could be obtained by taking dipolar couplings with a ^{13}C nucleus into account.

Preliminary calculations indicate that agreement can be reached between the experimental dipolar couplings and those calculated from a reasonable geometry of cyclohexane[6].

6.5.4   Vibrational corrections of dipolar couplings

From the experimental direct couplings relative *equilibrium* distances can be obtained if the vibrational motions can be taken into account. If no vibration corrections are made, the molecular structures obtained from various techniques cannot be compared directly. For a diatomic molecule, the experimental observable related to the internuclear distance r, is $\langle r \rangle$ in electron diffraction, $\langle r^{-2} \rangle^{\frac{1}{2}}$ in microwave spectroscopy, and $\langle r^{-2} \rangle^{\frac{1}{3}}$ —assuming that the orientation is known—in liquid crystal n.m.r.

To discuss vibrational corrections of n.m.r. geometries, we write r, the instantaneous separation of the two nuclei, as the sum of the equilibrium distance r_e and δr, the time dependent displacement. $1/r^3$ can be expanded in a series

$$\frac{1}{(r_{\text{n.m.r.}})^3} = \frac{1}{(r+\delta r)^3} = \frac{1}{r_e^3}\left(1 - \frac{3\langle\delta r\rangle}{r_e} + 6\frac{\langle(\delta r)^2\rangle}{r_e^2} - \ldots\right) \quad (6.33)$$

Although the second and third terms of the right-hand side are often comparable in magnitude, the higher terms can safely be neglected. Purely harmonic vibrations contribute only to the third term, while anharmonicity effects determine the second term. For pure stretching vibrations, these two terms may cancel in the series expansion, as for instance in the hydrogen molecule[79]. In methane the linear correction term dominates the quadratic term[79, 80].

In practical calculations of vibrational corrections, the experimental direct couplings are expanded in a Taylor series about the equilibrium geometry

$$D_{ij} = D_{ij}^e + \sum_m \left(\frac{\partial D_{ij}}{\partial q_m}\right)\langle\Delta q_m\rangle + \tfrac{1}{2}\sum_{m,n}\left(\frac{\partial D_{ij}}{\partial q_m}\frac{\partial D_{ij}}{\partial q_n}\right)\langle\Delta q_m\Delta q_n\rangle + \cdots (6.34)$$

q represents a convenient set of coordinates. The mean displacements from equilibrium $\langle\Delta q_m\rangle$, and the mean square amplitudes $\langle\Delta q_m^2\rangle$ can be estimated, using an appropriate force field (see also reference 80). This has been carried out for the methyl halides, applying different force field models[81]. The magnitudes of the calculated correction are relatively large; expressed in distances, they are of the order of 0.01 Å. A different method to account for molecular vibrations was proposed by Lucas[82], and applied to methyl fluoride. He has shown that a function of the internal coordinates may be expressed in terms of effective structural parameters plus a contribution from the harmonic force field. Thus, his method has the important advantage that the elaborate calculation of the anharmonic contribution from a rather hypothetical force field is circumvented (see Note added on proof).

Starting from equation (6.34), the vibrational corrections for benzene and hexafluorobenzene were calculated by Meiboom and Snyder[6], using a very simple model; only the C—H bond stretchings and bendings were considered, while only the stretching vibrations were assumed to be anharmonic. Furthermore, the cross terms in equation (6.34) were assumed to be negligible. The derivatives of the direct couplings were not calculated from the analytical expressions, but evaluated numerically. The mean square amplitudes and the mean displacements were estimated from average stretching and bending frequencies. The correction of D_{CH}, thus calculated, appeared to be much larger than that of D_{HH}; when expressed in distances, the C—H bond length decreased by $c.$ 1.8%, relative to the H—H distance.

Spiesecke[83] estimated the vibration corrections, neglecting anharmonic contributions, for acetylene, in order to obtain a relative equilibrium geometry from the n.m.r. spectra of acetylene and its ^{13}C isomers. The correspondence with an equilibrium geometry inferred from infrared data did not, however, improve by correcting for harmonic vibrational motions only. Significantly better results were obtained when the Bastiansen–Morino shrinkage effect was taken into account[83]. The improvement was attributed to a cancelling of the anharmonic contributions to the shrink-age effect in linear molecules. This result suggests that the discrepancies in the n.m.r. geometries of dimethyldiacetylene[84] and allene[85], relative to geo-metries from electron-diffraction data, may also be explained in terms of the shrinkage effect (see also reference 5).

6.6 PSEUDO-DIPOLAR INTERACTIONS

Electron-coupled nuclear spin–spin couplings may arise from several nuclear–electron interactions: (a) Fermi-type interactions between nuclear spins and electron spins; (b) dipolar interactions between nuclear spins and electron spins, and (c) interactions between the nuclei and the orbital momentum of the electrons. Calculations of the relative contributions of the terms (a)–(c) and of cross terms between them have been reported in the literature[86–93, 161].

Both theoretical calculations and experimental evidence seem to agree that indirect couplings between protons are predominantly caused by a Fermi-type mechanism. Nevertheless, the contributions of the other inter-actions to J_{HH} and not *vanishingly* small. This is suggested by results of Barfield, who calculated the tensor elements due to the scalar–dipolar cross terms in H_2 [87]. If the z-axis is the internuclear axis, then $J_{xx} = J_{yy} = +33.7$ Hz and $J_{zz} = -67.4$ Hz. Calculations by Barfield of the geminal proton–proton coupling of ethylene[87] give $J_{xx} = +16.37$ Hz, $J_{yy} = -4.98$ Hz, $J_{zz} = -11.39$ Hz, where the z-axis is perpendicular to the molecular plane and the x-axis is along the double bond (see also reference 161).

Detailed calculations of the anisotropic parts of indirect couplings of $^{13}CH_3F$ have been reported[88, 89, 92, 93]; it was found that the Fermi term dominates in the H—H, H—F and ^{13}C—H couplings; the Fermi spin–dipolar cross term is negligibly small. The ^{13}C—F coupling, however, was shown to be strongly anisotropic[88, 89].

6.6.1 Experimental data on pseudo-dipolar interactions

One of the few experimental methods to establish contributions from mechanisms (b) and (c) is n.m.r. in liquid crystalline solvents. Consider, for example, hexafluorobenzene. In the absence of pseudo-dipolar interactions the dipolar couplings should, to a good approximation, be in the ratios expected for a regular hexagon, namely 1:0.192:0.125. The experimental ratios were quite different: 1:0.187:0.134 [6, 94]. This has been attributed to pseudo-dipoler couplings between the fluorine nuclei[6, 94]. In accordance with the assumption that the coupling mechanism between protons is of the Fermi type, the ratios for the dipolar couplings in benzene were as expected for a regular hexagon[20, 95]. Molecular vibrations do not provide an alternative explanation[6], since the corrections are probably larger for benzene than for hexafluorobenzene, because of the smaller mass of the protons.

In special cases liquid-crystal n.m.r. can provide the principal values of the indirect spin–spin coupling tensor. The following relation holds for molecules with D_{2h} or C_{2v} symmetry

$$J_{ij}^{(2, 0)} = \frac{2}{3\sqrt{5}} c_{3z^2 - r^2}(J_{zz} - \tfrac{1}{2}(J_{xx} + J_{yy}))_{ij} + \left(\frac{1}{\sqrt{15}}\right) c_{x^2 - y^2}(J_{xx} - J_{yy})_{ij} \qquad (6.35)$$

A plot of $J_{ij}^{(2, 0)}/c_{3z^2 - r^2}$ against $c_{x^2 - y^2}/c_{3z^2 - r^2}$ can provide two combinations of the tensor components. The isotropic coupling, preferably measured in the liquid crystal, gives the third, so that J_{xx}, J_{yy} and J_{zz} can be extracted. Although this approach is simple in principle, the experimental difficulties are considerable owing to the problem of finding a liquid crystal in which $c_{x^2 - y^2}/c_{3z^2 - r^2}$ can be varied over a sufficiently large range. This method has been applied to the indirect fluorine–fluorine coupling of 1,1-difluoro-ethylene[96, 97] with the following results: $J_{xx} = -720$ Hz; $J_{yy} = +339$ Hz; $J_{zz} = +478$ Hz. The axes were as in ethylene. Vibrations were not taken into account in the calculations. The values depend on the precise geometry adopted. In all cases investigated[96, 97] the absolute values of the tensor components were an order of magnitude larger than the isotropic coupling constant.

An illustration of the way in which the presence, though not the magnitude, of pseudo-dipolar couplings may be demonstrated, is the following. In a molecule, an FF vector may be parallel to an HH vector. If $(D+J)_{FF}^{(2, 0)}/(D+J)_{HH}^{(2, 0)}$ varies from one experiment to another, this indicates pseudo-dipolar coupling(s). D_{FF} and D_{HH} may even have opposite signs, a fact not explainable by taking vibrations into account. *Meta* and *para* FF couplings in fluorine substituted benzenes contain detectable pseudo-dipolar contributions[54, 100]. The experimental results for *ortho*-difluorobenzene indicate a very small pseudo-dipolar contribution to the F–F coupling. Moreover, if D_{FF} is predicted from the geometry, calculated from the H–H– and H–F couplings, the agreement with the experimental D_{FF} is within experimental error[98, 99].

For investigations on other molecules the reader is referred to references 52, 101 and 102 (fluoroethylenes) and to reference 63 (tetrafluorodithiethane).

Krugh and Bernheim[103, 104] have inferred values of pseudo-dipolar couplings in $^{13}CH_3F$, by comparing an assumed geometry with measured dipolar couplings. The reliability of this procedure is, however, questionable[81, 82].

6.7 E.S.R. STUDIES OF DISSOLVED PARAMAGNETIC MOLECULES

The first contribution to e.s.r. in liquid crystals was by Carrington and Luckhurst[105], who measured the spectra of diphenylpicrylhydrazyl and the tetracyanoethylene anion, dissolved in the nematic phase of *p*-azoxyanisole. Anisotropies of *g* tensors and of hyperfine couplings of numerous paramagnetic molecules have since been obtained. In an extensive paper, Falle and Luckhurst[10] have discussed these topics for a number of doublet ground-state radicals, namely, perinaphthenyl[106–108], triphenylmethyl[108], perinaphthenyl-3-carboxylic acid, 3,7,9,13-tetramethyldiphenylmethyl, carbazole-*N*-

oxyl, diphenylnitroxide, N-Coppinger's radical[109], 2-amino-4,6-di-t-butyl phenoxyl, N-(trifluoraceto)-2,2,6,6-tetramethylpiperidinazyl and Wurster's blue perchlorate. The previously measured compounds have been indicated by references. The experimentally determined hyperfine components were compared with the values obtained with the theory of McConnell and Strathdee[110]. The calculated tensor elements of nuclei such as ^{14}N and ^{13}C, which have unpaired spin density in their $2p_z$ orbitals, were in satisfactory agreement with experiment. The proton hyperfine couplings, which are determined by π spin densities on adjacent atoms, were not always correctly predicted[10, 107]. This may be due to the relatively small value of the $A^{(2, 0)}$ component.

For the radicals perchlorodiphenylmethyl(PDM), $(C_6Cl_6)_2\dot{C}Cl$ and perchlorotriphenylmethyl(PTM), $(C_6Cl_6)_3\dot{C}$, the chlorine hyperfine coupling was investigated[111].

Liquid-crystal e.s.r. spectra may, in principle, also yield information about molecules containing two unpaired electrons, namely the electron exchange coupling and the dipolar coupling between the electrons. A demonstration of these interactions has been given for the ground state triplet molecules tetramethyl-2,2,6,6-piperidinol-4 glutarate (TPG) and tetramethyl-2,2,6,6-piperidinol-4 terephthalate (TPT)[112, 164]. It was shown that for TPT the exchange coupling was much smaller than the hyperfine interaction, contrary to the case of TPG. Lemaire[113] has studied the biradical (tetramethyl-2,2,6,6-piperidinol-4oxyl-1)$_2$-carbonate, dissolved in p-azoxyanisole and in dimethyl formamide. The exchange coupling turned out to depend on temperature; this was attributed to a dynamical equilibrium between two molecular conformations. The form of the e.s.r. spectra depends on the relative magnitudes of the electron exchange coupling and the nuclear hyperfine splitting. This phenomenon proved useful in distinguishing between the occurrence of mono- or bi-radicals in reactions of binitrones[114] and bianthrone systems[115].

Krebs, Sackmann and Schwarz[116] have studied photoexcited triplet states of the charge transfer complexes of 1,2,4,5-tetracyanobenzene (TCNB) with several perdeuterated donors (pyrene, naphthalene, triphenylene, phenanthrene and anthracene). The $\Delta m = \pm 2$ as well as the $\Delta m = \pm 1$ transitions were measured. For the TCNB–pyrene complex the same orientation was found as for molecular pyrene.

6.8   INVESTIGATIONS OF LIQUID-CRYSTAL PROPERTIES

6.8.1   Introduction

An equilibrium property $\langle X_k \rangle$ of a molecule k of an assembly of molecules is given by

$$\langle X_k \rangle = \int X_k e \quad d\omega_k dr_k / \int e^{-\beta U} d\omega_k dr_k \qquad (6.36)$$

$\beta = 1/kT$; ω_k and r_k represent the orientational and positional coordinates of molecule k, U is the potential of average force acting on molecule k[118]. A manageable expression for U can be obtained if certain approximations are made. For example, by assuming that in a liquid crystal U is only determined by dispersion forces, Maier and Saupe[119] derived the following expression

$$U = \frac{\delta\varepsilon}{V^2} \overline{\mathscr{D}_{00}^{(2)}} \tfrac{1}{2} (3\cos^2\theta - 1) \qquad (6.37)$$

$\delta\varepsilon$ is a parameter determining the nematic–isotropic transition temperature, V is the molecular volume, and θ is the angle between the long molecular axis and the optic axis of the liquid crystal. The form of U can be investigated by magnetic resonance studies of the motional constants, which are mono-molecule equilibrium properties. Because of the low concentration of paramagnetic molecules needed in e.s.r., this method is particularly suited and one then obtains information about the behaviour of the solute in a potential entirely generated by the solvent. For an axially symmetric paramagnetic molecular the motional constant $\overline{\mathscr{D}_{0,0}^{(2)}}$ is related to the hyperfine coupling by (see equation 6.21)

$$\overline{\mathscr{D}_{00}^{(2)}} = \frac{\langle A \rangle - A_{\mathrm{is}}}{A_{\parallel} - A_{\perp}} \qquad (6.38)$$

$\langle A \rangle$ is the experimentally measured hyperfine coupling; A_{is} is the isotropic hyperfine coupling; $A_{\parallel}$ and $A_{\perp}$ are the tensor components defined in the molecular frame.

6.8.2 Experimental results

Using the principle of corresponding states, Chen and Luckhurst[120] found evidence for the assumption of a potential entirely generated by the solvent. Mixtures of p-azoxyanisole with benzene, carbon tetrachloride, ethanol, phenanthrene and o-terphenyl were investigated, using vanadyl acetylacetonate (VAA) as a probe.

In a subsequent study[121] the motional constant of VAA was measured as a function of temperature in eight nematogens. A potential proportional to $\exp(-\,\mathrm{const.}\,\beta\sin^2\theta)$ turned out not to be quantitatively correct. In principle, the anisotropic dipole–dipole, induction and repulsion interactions should also be included in the potential function.

These investigations were restricted to systems in which the average orientation of all solute molecules can be described by one motional constant. In viscous liquid crystals the assumption of fast rotational reorientation is not valid and a spatial—rather than a time average—has to be performed. This has been demonstrated by Fryberg and Gelerinter[122], who investigated VAA in bis (4′-n-acetyloxybenzal)-2-chloro-1,4-phenylenediamine at temperatures at which the solvent still shows liquid crystalline behaviour, but where the spectra are 'glass like' in appearance. Independently, Schwerdtfeger and Diehl[123] discovered analogous phenomena, using the liquid crystal 4-methoxybenzylidene-4-amino-α-methyl-cinnamic acid n-propyl ester. The authors noted that, at high viscosities, the molecular distribution remained anisotropic. By application of an electric field perpendicular to the magnetic field, in order to rotate the optic axis, an increase of the parallel transitions was observed. The heights of the hyperfine lines in the e.s.r. spectrum depend on the angular distribution of the solute molecules and, as was recognised by the authors[124], the angular distribution function of the solvent, $f(\theta)$, can then be deduced (θ is the angle between the optic axis and the VO bond). Schwerdtfeger and Diehl assumed the intensity $I_{\parallel}(\theta)$ of the parallel hyperfine lines to be proportional to $f(\theta)$. After choosing an appropriate expression for $f(\theta)$:

$$f(\theta) \sim \exp\left(-\text{const. } \beta \sin^2 \theta\right), \tag{6.39}$$

excellent agreement with experiment could be obtained. James and Luckhurst[125] have shown that $I_{\parallel}(\theta)$ is *not* proportional to $f(\theta)$. However, this does not invalidate the conclusion concerning the form of equation (6.39).

James and Luckhurst[125] studied the distribution function of a mixture of 60 mole% of 4-methoxy benzylidene-4-amino-α-methyl-cinnamic acid n-propyl ester and 40 mole% of (4-methoxy-azobenzene-4-oxy)capronate, which has a nematic range from 52–96 °C, but may easily be supercooled to -50 °C without losing its ordering. It turned out that the distribution function is proportional to $\exp\left(a \sin^2 \theta + b \sin^4 \theta\right)$. The quartic term proved to be dominant at the higher orientations (see Table 6.1), meaning that the distribution function of the mixture is dominated by repulsion forces. In contrast, the potential function for 4-methoxybenzylidene-4-amino-α-methyl-cinnamic acid n-propyl ester is determined by dispersion forces.

Table 6.1   **Parameters *a* and *b* of the probability distribution function given by James and Luckhurst[125]. *S* is the ordering parameter**

(From James and Luckhurst[125], by courtesy of Taylor & Francis Ltd.)

a	b	S
0.437	0.097	-0.069
0.175	0.983	-0.157
-0.195	2.147	-0.250
-0.303	3.120	-0.319

This change in potential, after addition of a second nematogen, is surprising; the authors suggest that this is a result of the low temperature (-50 °C) at which the experiments were conducted[169].

In a further analysis of the results, the solute pseudo-potential was expanded as a power series. The terms of the expansion are products of Legendre polynomials, representing the solvent order and the angular dependence of the potential acting on the solute. The appropriate coefficients are a measure of the solute–solvent interaction and are determined by the solute–solvent pair distribution function. Only the first two terms, i.e. the quadratic and quartic terms of the series were required to fit the experimental distribution function. Preliminary computations of the coefficients were performed; their values turned out to be relatively independent of temperature. This should be expected if the pair distribution function does not vary strongly with temperature.

It has been demonstrated by magnetic resonance[126–128] that the director is in general parallel to an applied d.c. electric field. This is not to be expected for 4,4′-dimethoxyazoxybenzene for which it has been found that the largest component of the dielectric tensor is perpendicular to the director[129]. In order to get additional information on this problem, an electric field was applied to the sample at different angles with respect to the magnetic field[130]. The angular dependence of the line splittings of VAA obtained in this experiment was analysed by Luckhurst[13] in terms of a model containing cybotactic groups. The mesophases of certain nematic liquid crystals are thought to consist of cybotactic groups in which the ends of the molecules lie in well defined planes. The long axes of the molecules may have some constant angle with respect to the planes. When the angle is zero the mesophase is a classically nematic one[131]. The electron resonance spectrum can be interpreted if it is assumed that the planes of these groups are parallel to the electric field. The analysis further shows that the solute molecules are at an angle of 13 degrees with respect to these planes.

6.8.3   Relaxation phenomena

An interesting relaxation phenomenon in e.s.r. liquid-crystal spectra that is governed by the angular part of the distribution function has been discussed by Glarum and Marshall[132]. Widths of hyperfine lines in e.s.r. spectra can be expressed as a polynomial in m, the nuclear magnetic quantum number:

$$T_2^{-1} = A + Bm + Cm^2 \qquad (6.40)$$

For partially-orientated molecules the coefficients A, B and C involve the averages $\overline{\cos^2 \theta}$ and $\overline{\cos^4 \theta}$, where θ is the angle between the molecular z-axis (symmetry axis) and the magnetic field. The former average can be obtained, in principle, from the measured splittings. The average $\overline{\cos^4 \theta}$ has been calculated using the distribution function, equation (6.36). As a function of the average orientation, B and C exhibit an interesting property. In going from a disordered to a completely ordered system the coefficient C decreases and passes through zero at $\mathscr{D}_{00}^{(2)} = 0.71$. At still higher degrees of order it reverses sign and for $\overline{\mathscr{D}_{00}^{(2)}} = 1$ it vanishes, but not in an asymptotic manner. The linear coefficient B gradually decreases as the ordering increases (see Figure 6.3). It was concluded that the outermost lines of vanadyl complexes should be broad at low ordering, but that for $\overline{\mathscr{D}_{00}^{(2)}} > 0.7$ their widths should be smaller than those in the central part. This prediction has been corroborated by experiment (Figure 6.4).

Careful examination by Brooks et al.[133] of the e.s.r. transitions of vanadyl complexes showed that the lines were asymmetric. The line shapes could be simulated by introducing the distribution function (equation (6.36)) in the formula for the line intensity of the e.s.r. transitions. The asymmetry was attributed to the fluctuations in the orientation of the solute molecules. Using de Gennes'[134] theory on the thermal fluctuations in nematic mesophases, the 'misalignment' of the solute was computed.

6.8.4   Systems of biochemical interest

Interesting examples of lyotropic systems are the biologically important membranes and phospholipid dispersions. Hubbell and McConnell[135, 136],

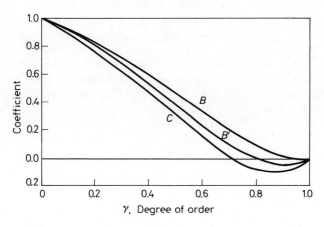

Figure 6.3 Variation of coefficients B and C (equation (6.40)) with degree of order. The curve B' includes second-order contributions to B.
(Reproduced from Glarum and Marshall[32], by courtesy of the American Institute of Physics, and the authors)

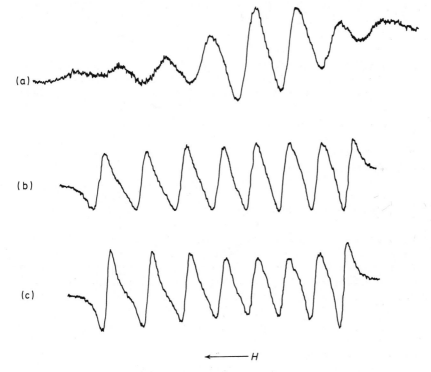

Figure 6.4 E.S.R. spectra of the vanadyl dibenzoylmethane complex at (a) 150 °C, (b) 130 °C and (c) 113 °C
(Reproduced from Glarum and Marshall[132], by courtesy of the American Institute of Physics, and the authors)

have introduced the spin label technique to study the structural and kinetic features of such substances. It was found that labels of the following type

$$CH_3-(CH_2)_m-\overset{\displaystyle O\quad N\to O}{\underset{\displaystyle \qquad}{C}}-(CH_2)_n-COOH$$

(5)

(4)

exhibited rapid anisotropic motion in neuronal membranes and in phospholipid bilayers.

Seelig[29] has studied the behaviour of the spin-labelled fatty acids (5) in the smectic phase of the mixture sodium decanoate–decanol–water. The labels are supposed to behave chemically and physically similar to the amphiphilic components of the liquid crystal, so that the e.s.r. spectrum of the label reflects the dynamical behaviour of the amphiphilic region of the liquid crystal. This is of interest because the smectic phase of this liquid crystal has a known bilayer structure and may serve as a model system for the study of bilayer membranes. It was found by the author that the degree of order of the spin labels decreases when the distance between the carboxyl group and the label is increased. However, the amphiphilic materials of the bilayer were found in an almost extended conformation.

In subsequent experiments, Hubbell and McConnell[137] were led to analogous conclusions concerning their experiments in aqueous dispersions of lecithin–cholesterol and nerve fibres.

6.8.5   Investigations with n.m.r.

Saupe and Nehring have studied the interaction of a solute molecule with its liquid crystalline environment through the preferential orientation the solvent imposes on the solute molecule[138, 139]. Their approach is related to the statistical theory of Maier and Saupe[119] for the nematic state. For the two series of molecules investigated, chlorobenzenes[138] and fluorobenzenes[139], the orientation was supposed to be governed by mainly two factors, namely, dispersion forces and permanent electric dipole moments; the molecular shape was not explicitly taken into account. The experimental results do not indicate significant contributions of electrical dipole–dipole forces[138] to the orientation and furthermore support the assumption that differences in the orientation of the different molecules studied are mainly caused by localised contributions of the substituent bonds to the intermolecular interactions.

The tendency of solute molecules to orientate with their longest dimension parallel to the optic axis of a nematic solvent has been studied by Robertson *et al.*[140] and by Anderson[141] (see also reference 24). The former authors considered only dispersion forces and the molecular shape was introduced in the solute–solvent interaction energy via the polarisability anisotropies. Assuming that the dispersion interactions may be approximated by the dipole–dipole part, it was shown that the ratio

$$R = (a_2 - a_3)/(a_1 - a_3) \qquad (6.41)$$

is approximately equal to the ratio involving molecular dimensions:

$$R = (L_2 - L_3)/(L_1 - L_3) \qquad (6.42)$$

a_1, a_2 and a_3 are the components of the interaction energy along the molecular axes, as defined by Saupe[138]. L_1, L_2 and L_3 are the molecular dimensions along the same axes.

For 27 different solute molecules with C_{2v} or D_{2h} symmetry, the dependence of S_1/S_2, the ratio of the ordering parameters, on the dimension ratio R was calculated[140], and compared with the experimental values. The experimental data follow the theoretical trend, but especially for small values of R considerable deviations occur. This is partly attributed to uncertainties in the determination of R from the molecular geometries. The dependence on specific solute and solvent properties is held responsible for the larger deviations. Evidence for the neglect of solvent dependent terms was obtained from the scatter of the ratio S_1/S_2 for *p*-difluorobenzene in different nematic solvents. Molecules which possess a large dipole moment do not show particular deviations, in agreement with Saupe's conclusion that dipolar interactions are relatively unimportant[138].

In view of the foregoing, it is not surprising that Anderson[141] has found a correlation between the asymmetry parameters of the ordering matrix and the inertia tensor for a series of substituted benzenes.

6.9   ANISOTROPIES OF CHEMICAL SHIFTS

The anisotropy of a chemical shift can be obtained from the shift of a spectrum as the solvent changes from the isotropic to the nematic phase. Discussions of this phenomenon and references up to 1970 can be found in reference 3.

The spectral shift, mentioned above, provides only one relation between the components of the shielding tensor. Hence, the method may be useful for molecules with a threefold, or higher, axis; if the motional constant involved is known from the n.m.r. spectrum, then $(\sigma_{||} - \sigma_{\perp})$ can be inferred. The usefulness is restricted if more than two tensor components are necessary to describe the chemical shift.

A notable source of error, especially for protons, may lie in changes of solute–solvent interactions as the solvent goes nematic[142, 143]. To circumvent referencing with respect to the isotropic phase, Diehl *et al.*[144] have proposed to utilise an external electric field to align the optic axis perpendicular to the magnetic field. As the authors remark, this approach may suffer from different influences of the solvent anisotropy effect on the solute and internal reference molecules. For a discussion of the determination of proton chemical shift anisotropies by liquid crystal n.m.r. see also references 165 and 166.

Fluorine shift anisotropies measured with the liquid–crystal n.m.r. method and with other methods, have been compared by Yannoni et al.[145]. A large spread in the values of $\Delta\sigma$ is found for aromatic as well as for aliphatic compounds. Related molecules, like CH_3F and CHF_3 or 1,3,5- and 1,2,4-trifluorobenzene, are reported to have opposite signs. It should be recalled that the 'experimental' values from various techniques (e.g. liquid-crystal n.m.r.) are obtained by a transformation of the molecular symmetry axis to a bond axis; the assumption that the latter possesses axial symmetry may be dubious. Discrepancies may also be attributable to medium effects and possibly to differences in mobility. Waugh et al.[146] have developed a promising pulse method to measure the principal values of the shielding tensor. This method does not suffer from the above limitations and it is applicable to polycrystalline samples. For a recent account of chemical shift anisotropies of hydrogen and fluorine, see reference 147.

Recent measurements of proton chemical shift anistropies have been published by Hayamizu and Yamamoto[23] (methyl halides) and by Lindon and Dailey[167] (benzene). Anisotropies of ^{19}F shifts of substituted fluoro-benzenes have been correlated with Taft parameters by Yim and Gilson[148]. ^{19}F Chemical shift anisotropies of another series of substituted fluoro-benzenes have been measured by Nehring and Saupe[149]. Their results could not be explained satisfactorily by the theory of ^{19}F chemical shifts for fluorobenzenes due to Karplus and Das because a very high ionicity of the C—F bond had to be assumed.

^{13}C Chemical shift anisotropies have been reported by Millett and Dailey[150] for $H^{13}CN$ ($\Delta\sigma = 280$ p.p.m.) and by Morishima et al.[168] for $^{13}CH_3I$ and $^{13}CH_3CN$.

The ^{14}N shift anisotropy in methylisocyanide (CH_3NC) has been investi-gated by Yannoni[151], using spin-tickling techniques ($\Delta\sigma = +360$ p.p.m.).

Theoretical calculations of chemical shift anisotropies have been carried out by several authors[152].

6.10 NUCLEAR QUADRUPOLE COUPLING CONSTANTS

The study of nuclear quadrupole coupling constants (q.c.c.) has also been the subject of liquid-crystal n.m.r. (reference 3). From equations (6.17)–(6.20) it can be deduced that the induced splitting between neighbouring lines is given by

$$\Delta v = \sqrt{6}\, h^{-1}\, Q^{(2,\,0)}\ \text{Hz} \qquad (6.43)$$

For molecules with D_{2h} or C_{2v} symmetry

$$Q^{(2,\,0)} = \tfrac{3}{4}\frac{e^2 qQ}{h}\cdot\frac{1}{I(2I-1)}\cdot\overline{([3\cos^2\theta-1]}+\eta\,\overline{\sin^2\theta\cos 2}\,\phi) \qquad (6.44)$$

η is the asymmetry parameter: $(V_{xx}-V_{yy})/V_{zz}$.

Quadrupolar line splittings can be observed if the q.c.c. and/or the degree of orientation are relatively small. If these conditions are not met, a minor temperature inhomogeneity over the sample causes excessive line broadening.

Suitable nuclei are ^{2}D and ^{14}N (the latter in special compounds with small field gradients, e.g. CH$_3$NC). The liquid crystal poly-γ-benzylglutamate orientates guest molecules only to a small extent; this solvent has been used to measure larger q.c.c. values.

To interpret the line splittings in terms of the q.c.c. the motional constant should be known, e.g. from the ^{1}H n.m.r. spectrum. Axial symmetry of the electric field gradient has to be assumed, as only one experimental quantity (i.e. the line splitting) is available. It should be borne in mind that this may be a poor approximation.

It has been pointed out[153] that the q.c.c. of ^{2}D bonded to a carbon atom will be reduced by electron-withdrawing substituents. This has been corroborated by experiments; the measured q.c.c. of ^{2}D in CD$_3$I, CD$_3$Br and CD$_3$CN show this tendency[154]. Similar results were obtained for nitrobenzene-2,4,6-d_3, 1,3-dinitrobenzene-2,4,6-d_3 and 1,3,5-trinitrobenzene-d_3[155] (see Table 6.2).

Table 6.2 Quadrupole coupling constants of ^{2}D and ^{14}N along the bond axes

Substance	Formula	^{2}D/kHz liq. cryst.	^{14}N/MHz liq. cryst.	^{14}N/MHz other methods
Dichloromethane-d_2	CD$_2$Cl$_2$	160*		
Methyl iodide-d_3	CD$_3$I	180±5†		
Methyl bromide-d_3	CD$_3$Br	171±4†		
Acetonitrile-d_3	CD$_3$CN	165±5†	3.60‖	3.738** (solid state)
Benzene-d_6	C$_6$D$_6$	194±4†		
2,4,6-d_3-Nitrobenzene	C$_6$H$_2$D$_3$NO$_2$	192±4‡	1.76‖	
2,4,6-d_3-1,3-Dinitrobenzene	C$_6$HD$_3$(NO$_2$)$_2$	187±4‡		
2,4,6-d_3-1,3,5-Trinitrobenzene	C$_6$D$_3$(NO$_2$)$_3$	181±4‡		
Phenylsilane-d_3	C$_6$H$_5$SiD$_3$	91±2§		
Phenylphosphine-d_2	C$_6$H$_5$PD$_2$	115±2§		
Benzenethiol-d	C$_6$H$_5$SD	146±2§		
Nitromethane	CH$_3$NO$_2$		1.45‖	
Methylisocyanide	CH$_3$NC		0.272±0.002¶	0.483±0.017†† (gas phase)

*reference 160, † reference 154, ‡ reference 155, § reference 156, ‖reference 158, ¶reference 159, **reference 162, ††reference 163.

The q.c.c. of ^{2}D in phenylsilane-d_3, phenylphosphine-d_2, and benzenethiol-d_1 have also been measured[156]. They were correlated with bond force constants and electronegativities. For covalent deuterides of elements of the first three rows of the periodic system a linear relationship between the force constants and the q.c.c. was obtained[156, 157]. The q.c.c. values show a smooth correlation with electronegativities[156].

Up to now a few ^{14}N q.c.c. have been measured in the liquid crystalline phase (see Table 6.2). For acetonitrile reasonable agreement with the solid-state value has been obtained[158]. A notable exception has been published by Yannoni[159]; the q.c.c. of methyl isocyanide measured in the liquid crystalline phase turned out to be about one half of the solid-state value. No explanation has been advanced.

Note added in proof

The account given by Lucas of the molecular vibrations in methylfluoride was followed by an analogous study of cyclopropane[82]. For this molecule, a good agreement could be established between the structural data obtained from n.m.r. in a liquid crystal and those obtained by electron diffraction, by taking the vibrational motion into account. In addition, the study includes an interesting discussion of the interdependence of the vibrational motion and the orientation of a molecule.

References

1. Brown, G. H., Doane, J. W. and Neff, V. D. (1970). *Crit. Rev. Solid State Sci.,* **1**, 303
2. Saupe, A. and Englert, G. (1963). *Phys. Rev. Lett.,* **11**, 462
3. Diehl, P. and Khetrapal, C. L. (1970). *N.M.R. Basic Principles and Progress,* Vol. 1, (Diehl, P., Fluck, E. and Kosfeld, R., editors). (Berlin: Springer Verlag)
4. Luckhurst, G. R. (1968). *Quart. Rev. Chem. Soc.,* **22**, 179
5. Meiboom, S. and Snyder, L. C. (1968). *Science,* **162**, 1337
6. Meiboom, S. and Snyder, L. C. (1971). *Accts. Chem. Res.,* **4**, 81
7. Bernheim, R. A. and Lavery, B. J. (1967). *J. Colloid Interface Sci.,* **26**, 291
8. Buckingham, A. D. and McLauchlan, K. A. (1967). *Progress in Nuclear Magnetic Resonance Spectroscopy,* **2**, 63. (Oxford: Pergamon Press)
9. Low, W. (1960). *Paramagnetic Resonance in Solids.* (New York: Academic Press)
10. Falle, H. R. and Luckhurst, G. R. (1970). *J. Mag. Res.,* **3**, 161
11. Rose, M. E. (1957). *Elementary Theory of Angular Momentum.* (New York: John Wiley and Sons, Inc.)
12. Glarum, S. H. and Marshall, J. H. (1966). *J. Chem. Phys.,* **44**, 2884
13. Luckhurst, G. R. (1971). *Chem. Phys. Lett.,* **9**, 289
14. Saupe, A. (1964). *Z. Naturforsch.,* **19a**, 161
15. Buckingham, A. D. and Pople, J. A. (1963). *Trans. Faraday Soc.,* **59**, 2421
16. Snyder, L. C. (1965). *J. Chem. Phys.,* **43**, 4041
17. Jahn, H. A. (1949). *Acta Crystallogr.,* **2**, 30
18. Lawson, K. D. and Flautt, T. J. (1967). *J. Amer. Chem. Soc.,* **89**, 5489
19. Lawson, K. D. and Flautt, T. J. (1968). *J. Phys. Chem.,* **72**, 2066
20. Black, P. J., Lawson, K. D. and Flautt, T. J. (1969). *J. Chem. Phys.,* **50**, 542
21. Black, P. J., Lawson, K. D. and Flautt, T. J. (1969). *Mol. Cryst. Liquid Cryst.,* **7**, 201
22. Anderson, J. M. and Lee, A. C. F. (1971). *J. Mag. Res.,* **4**, 160
23. Hayamizu, K. and Yamamoto, O. (1971). *J. Mag. Res.,* **5**, 94
24. Lindon, J. and Dailey, B. P. (1971). *Molec. Phys.,* **20**, 937
25. Long, R. C. and Goldstein, J. H. (1971). *J. Chem. Phys.,* **54**, 1563
26. de Vries, J. J. and Berendsen, H. J. C. (1969). *Nature (London),* **221**, 1139
27. Yannoni, C. S. (1969). *J. Amer. Chem. Soc.,* **91**, 4611
28. Francis, P. D. and Luckhurst, G. R. (1969). *Chem. Phys. Lett.,* **3**, 213
29. Seelig, J. (1970). *J. Amer. Chem. Soc.,* **92**, 3881
30. Luckhurst, G. R. and Sundholm, F. (1971). *Molec. Phys.,* **21**, 349
31. Diehl, P. and Khetrapal, C. L. (1968). *Molec. Phys.,* **14**, 283
32. Schumann, C. In the press
33. Diehl, P. and Kellerhals, H. P. (1969). *J. Mag. Res.,* **1**, 196
34. Fung, B. M. and Gerace, M. J. (1970). *J. Chem. Phys.,* **53**, 1171
35. Woodman, C. M. (1967). *Molec. Phys.,* **13**, 365
36. Diehl, P., Khetrapal, C. L. and Kellerhals, H. P. (1968). *Molec. Phys.,* **15**, 333
37. Castellano, S. and Bothner-By, A. A. (1964). *J. Chem. Phys.,* **41**, 3863
38. Diehl, P., Kellerhals, H. P. and Niederberger, W. (1971). *J. Mag. Res.,* **4**, 352
39. Musher, J. I. (1967). *J. Chem. Phys.,* **46**, 1537
40. Gutowsky, H. S., McCall, D. W. and Slichter, C. P. (1953). *J. Chem. Phys.,* **21**, 279

41. Saupe, A. and Nehring, J. (1967). *J. Chem. Phys.*, **47**, 5459
42. Musher, J. I. (1967). *J. Chem. Phys.*, **47**, 5460
43. Grimley, T. B. (1963). *Molec. Phys.*, **6**, 329
44. Woodman, C. M. (1966). *Molec. Phys.*, **11**, 109
45. Longuet-Higgins, H. C. (1963). *Molec. Phys.*, **6**, 445
46. Altmann, S. L. (1967). *Proc. Roy. Soc. A*, **298**, 184
47. Watson, J. K. G. (1971). *Molec. Phys.*, **21**, 577
48. Altmann, S. L. (1971). *Molec. Phys.*, **21**, 587
49. Serre, J. (1968). *Int. J. Quantum Chem.*, **28**, 107
50. Woodman, C. M. (1970). *Molec. Phys.*, **19**, 753
51. Mortimer, F. S. (1969). *J. Mag. Res.*, **1**, 1
52. Buckingham, A. D., Burnell, E. E. and de Lange, C. A. (1969). *Molec. Phys.*, **16**, 299
53. Whiffen, D. H. (1971). *Chem. Britain*, **7**, 57
54. Gerritsen, J. (1971). *Thesis*, Free University, Amsterdam
55. Bulthuis, J., Gerritsen, J., Hilbers, C. W. and MacLean, C. (1968). *Rec. Trav. Chim.*, **87**, 417
56. Diehl, P., Khetrapal, C. L. and Lienhard, U. (1969). *Org. Mag. Res.*, **1**, 93
57. Barili, P. L. and Veracini, C. A. (1971). *Chem. Phys. Lett.*, **8**, 229
58. Dereppe, J.-M., Degelaen, J. and van Meersche, M. (1970). *J. Chim. Phys.*, **67**, 1875
59. Fung, B. M. and Gerace, M. J. (1970). *J. Chem. Phys.*, **53**, 1171
60. Khetrapal, C. L., Patankar, A. V. and Diehl, P. (1970). *Org. Mag. Res.*, **2**, 405
61. Khetrapal, C. L., Kunwar, A. C., Kanekar, C. R. and Diehl, P. (1971). *Mol. Cryst. Liq. Cryst.*, **12**, 179
62. Khetrapal, C. L., Kunwar, A. C. and Kanekar, C. R. (1971). *Chem. Phys. Lett.*, **9**, 437
63. Long, R. C. and Goldstein, J. H. (1971). *J. Chem. Phys.*, **54**, 1563
64. Pawlizeck, J. B. and Günther, H. (1971). *J. Amer. Chem. Soc.*, **93**, 2050
65. Wüthrich, K., Meiboom, S. and Snyder, L. C. (1970). *J. Chem. Phys.*, **52**, 230
66. Diehl, P., Henrichs, P. M. and Niederberger, W. (1971). *Org. Mag. Res.*, **3**, 243
67. Meiboom, S. and Snyder, L. C. (1970). *J. Chem. Phys.*, **52**, 3857
68. Gazzard, I. J. and Sheppard, N. (1971). *Molec. Phys.*, **21**, 169
69. Huckerby, T. N. (1971). *Tetrahedron Lett.*, **38**, 3497
70. Bucci, P., Franchini, F., Serra, A. M. and Veracini, C. A. (1971). *Chem. Phys. Lett.*, **8**, 421
71. Swinton, P. F. and Gatti, G. (1970). *Spectrosc. Lett.*, **3**, 259
72. Yannoni, C. S. (1970). *J. Amer. Chem. Soc.*, **92**, 5237
73. Yannoni, C. S. (1969). *J. Chem. Phys.*, **51**, 1682
74. Diehl, P., Khetrapal, C. L., Niederberger, W. and Partington, P. (1970). *J. Mag. Res.*, **2**, 181
75. Diehl, P., Kellerhals, H. P. and Niederberger, W. (1970). *J. Mag. Res.*, **3**, 230
76. Diehl, P., Henrichs, P. M. and Niederberger, W. (1971). *Molec. Phys.*, **20**, 139
77. Diehl, P., Henrichs, P. M., Niederberger, W. and Vogt, J. (1971). *Molec. Phys.*, **21**, 377
78. Diehl, P. and Henrichs, P. M. (1971). *J. Mag. Res.*, **5**, 134
79. Bovée, W., Hilbers, C. W. and MacLean, C. (1969). *Molec. Phys.*, **17**, 75
80. Cyvin, S. J. (1968). *Molecular Vibrations and Mean Square Amplitudes.* (Amsterdam: Elsevier)
81. Bulthuis, J. and MacLean, C. (1971). *J. Mag. Res.*, **4**, 148
82. Lucas, N. J. D. (1971). *Molec. Phys.*, **22**, 147, 233
83. Spiesecke, H. (1970). *Liquid Crystals and Ordered Fluids*, 123. (New York: Plenum Press)
84. Englert, G., Saupe, A. and Weber, J-P. (1968). *Z. Naturforsch.*, **23a**, 152
85. Sackmann, E. (1969). *J. Chem. Phys.*, **51**, 2984
86. Kato, Y. and Saika, A. (1967). *J. Chem. Phys.*, **46**, 1975
87. Barfield, M. (1970). *Chem. Phys. Lett.*, **4**, 518
88. Blizzard, A. C. and Santry, D. P. (1970). *Chem. Commun.*, 87 and 1085
89. Blizzard, A. C. and Santry, D. P. (1971). *J. Chem. Phys.*, **55**, 950
90. Buckingham, A. D. and Love, I. (1970). *J. Mag. Res.*, **2**, 338
91. Love, I. (1970). *Molec. Phys.*, **19**, 733
92. Nakatsuji, H., Kato, H., Morishima, I. and Yonezawa, T. (1970). *Chem. Phys. Lett.*, **4**, 607
93. Nakatsuji, H., Hirao, K., Kato, H. and Yonezawa, T. (1971). *Chem. Phys. Lett.*, **6**, 541
94. Snyder, L. C. and Anderson, E. W. (1965). *J. Chem. Phys.*, **42**, 3336
95. Saupe, A. (1965). *Z. Naturforsch.*, **20a**, 572

96. Gerritsen, J. and MacLean, C. (1971). *J. Mag. Res.*, **5**, 44
97. Gerritsen, J. and MacLean, C. (1971). *Mol. Cryst. Liq. Cryst.*, **12**, 97
98. Gerritsen, J. and MacLean, C. (1971). *Spectrochim. Acta*, **27a**, 1495
99. Yim, C. T. and Gilson, D. F. R. (1969). *Can. J. Chem.*, **47**, 1057
100. Gerritsen, J., Koopmans, G., Rollema, H. S. and MacLean, C. (1972). *J. Mag. Res.*, **7**
101. Buckingham, A. D. and Dunn, M. B. (1970). *Molec. Phys.*, **19**, 721
102. Spiesecke, H. and Saupe, A. (1970). *Mol. Cryst. Liq. Cryst.*, **6**, 287
103. Krugh, T. R. and Bernheim, R. A. (1969). *J. Amer. Chem. Soc.*, **91**, 2385
104. Krugh, T. R. and Bernheim, R. A. (1970). *J. Chem. Phys.*, **52**, 4942
105. Carrington, A. and Luckhurst, G. R. (1964). *Molec. Phys.*, **8**, 401
106. Falle, H. R. and Luckhurst, G. R. (1966). *Molec. Phys.*, **11**, 299
107. Glarum, S. H. and Marshall, J. H. (1966). *J. Chem. Phys.*, **44**, 2884
108. Möbius, K., Haustein, H. and Plato, M. (1968). *Z. Naturforsch*, **23a**, 1626
109. Coppinger's radical has been measured by Luckhurst, G. R. (1966). *Molec. Phys.*, **11**, 205
110. McConnell, H. M. and Strathdee, J. (1959). *Molec. Phys.*, **2**, 129
111. Falle, H. R., Luckhurst, G. R., Horsfield, A. and Ballester, M. (1969). *J. Chem. Phys.*, **50**, 258
112. Falle, H. R. and Luckhurst, G. R., Lemaire, H., Marechal, Y., Rassat, A. and Rey, P. (1966). *Molec. Phys.*, **11**, 49
113. Lemaire, H. (1967). *J. Chim. Phys.*, **64**, 599
114. Forrester, A. R., Thomson, R. H. and Luckhurst, G. R. (1968). *J. Chem. Soc. B*, 1311
115. Agranat, I., Rabinovitz, M., Falle, H. R., Luckhurst, G. R. and Ockwell, J. N. (1970). *J. Chem. Soc. B*, 294
116. Krebs, P., Sackmann, E. and Schwarz, J. (1971). *Chem. Phys. Lett.*, **8**, 417
117. Lemaire, H., Rassat, A., Rey, P. and Luckhurst, G. R. (1968). *Molec. Phys.*, **14**, 441
118. Hill, T. L. (1956). *Statistical Mechanics*. (New York: McGraw Hill)
119. Maier, W. and Saupe, A. (1959, 1960). *Z. Naturforsch*, **14a**, 882; **15a**, 287
120. Chen, D. H. and Luckhurst, G. R. (1969). *Trans. Faraday Soc.*, **65**, 656
121. Chen, D. H., James, P. G. and Luckhurst, G. R. (1969). *Mol. Cryst. Liq. Cryst.*, **8**, 71
122. Fryburg, G. C. and Gelerinter, E. (1970). *J. Chem. Phys.*, **52**, 3378
123. Schwerdtfeger, C. F. and Diehl, P. (1969). *Molec. Phys.*, **17**, 417
124. Diehl, P. and Schwerdtfeger, C. F. (1969). *Molec. Phys.*, **17**, 423
125. James, P. G. and Luckhurst, G. R. (1970). *Molec. Phys.*, **19**, 489
126. Rowell, J. C., Phillips, W. D., Melby, L. R. and Panar, M. (1965). *J. Chem. Phys.*, **43**, 3442
127. Chen, D. H. and Luckhurst, G. R. (1969). *Molec. Phys.*, **16**, 91
128. Carr, F. E., Hoar, E. A. and McDonald, W. T. (1968). *J. Chem. Phys.*, **48**, 2822
129. Maier, W. and Meier, G. (1961). *Z. Naturforsch*, **16a**, 470
130. Schara, M. and Šentjurc, M. (1970). *Solid State Commun.*, **8**, 593
131. de Vries, A. (1970). *Mol. Cryst. Liq. Cryst.*, **10**, 31
132. Glarum, S. H. and Marshall, J. H. (1967). *J. Chem. Phys.*, **46**, 55
133. Brooks, S. A., Luckhurst, G. R. and Pedulli, G. F. (1971). *Chem. Phys. Lett.*, **11**, 159
134. de Gennes, P. G. (1969). *Mol. Cryst. Liq. Cryst.*, **7**, 325
135. Hubbell, W. L. and McConnell, H. M. (1969). *Proc. Nat. Acad. Sci. U.S.*, **63**, 16
136. Hubbell, W. L. and McConnell, H. M. (1969). *Proc. Nat. Acad. Sci. U.S.*, **64**, 20
137. Hubbell, W. L. and McConnell, H. M. (1971). *J. Amer. Chem. Soc.*, **93**, 314
138. Saupe, A. (1966). *Mol. Cryst.*, **1**, 527
139. Nehring, J. and Saupe, A. (1969). *Mol. Cryst. Liq. Cryst.*, **8**, 403
140. Robertson, J. C., Yim, C. T. and Gilson, D. F. R. (1971). *Can. J. Chem.*, **49**, 2345
141. Anderson, J. M. (1971). *J. Mag. Res.*, **4**, 231
142. Buckingham, A. D., Burnell, E. E. and de Lange, C. A. (1968). *J. Amer. Chem. Soc.*, **90**, 2972
143. Buckingham, A. D., Burnell, E. E. and de Lange, C. A. (1971). *J. Chem. Phys.*, **54**, 3242
144. Diehl, P., Khetrapal, C. L., Kellerhals, H. P., Lienhard, U. and Niederberger, W. (1969). *J. Mag. Res.*, **1**, 527
145. Yannoni, C. S., Dailey, B. P. and Ceasar, G. P. (1971). *J. Chem. Phys.*, **54**, 4020
146. For example, Mehring, M., Griffin, R. G. and Waugh, J. S. (1970). *J. Amer. Chem. Soc.*, **92**, 7222
147. Yannoni, C. S. (1971). *I.B.M. J. Res. Dev.*, **15**, 59
148. Yim, C. T. and Gilson, D. F. R. (1969). *J. Amer. Chem. Soc.*, **91**, 4360
149. Nehring, J. and Saupe, A. (1970). *J. Chem. Phys.*, **52**, 1307

150. Millett, F. and Dailey, B. P. (1971). *J. Chem. Phys.*, **54,** 5434
151a. Yannoni, C. S. and Powers, J. V. (1969). *I.B.M. Research Report,* R.C. 2608
151b. Yannoni, C. S. (1970). *J. Chem. Phys.*, **52,** 2005
152. Ceasar, G. P. and Dailey, B. P. (1969). *J. Chem. Phys.*, **50,** 4200, and references quoted
     therein
153. Olympia, P. L., Wei, I. Y. and Fung, B. M. (1969). *J. Chem. Phys.*, **51,** 1610
154. Caspary, W. J., Millett, F., Reichbach, M. and Dailey, B. P. (1969). *J. Chem. Phys.*, **51,**
     623
155. Wei, I. Y. and Fung, B. M. (1970). *J. Chem. Phys.*, **52,** 4917
156. Fung, B. M. and Wei, I. Y. (1970). *J. Amer. Chem. Soc.*, **92,** 1497
157. Merchant, S. Z. and Fung, B. M. (1969). *J. Chem. Phys.*, **50,** 2265
158. Gerace, M. J. and Fung, B. M. (1970). *J. Chem. Phys.*, **53,** 2984
159. Yannoni, C. S. (1970). *J. Chem. Phys.*, **52,** 2005
160. Klein, M. P., Gill, D. and Kotowycz, G. (1968). *Chem. Phys. Lett.*, **2,** 677
161. Nakatsuji, H., Morishima, I., Kato, H. and Yonezawa, T. (1971). *Bull. Chem. Soc. Jap.*,
     44, 2010
162. Negita, H., Casabella, P. A. and Bray, P. J. (1960). *J. Chem. Phys.*, **32,** 314
163. Kemp, M. K., Pochan, J. M. and Flygare, W. H. (1967). *J. Phys. Chem.*, **71,** 765
164. Lemaire, H., Rassat, A., Rey, P. and Luckhurst, G. R. (1968). *Molec. Phys.*, **14,** 441
165. Hayamizu, K. and Yamamoto, O. (1970). *J. Mag. Res.*, **2,** 377
166. Hayamizu, K. and Yamamoto, O. (1971). *J. Chem. Phys.*, **54,** 3243
167. Lindon, J. and Dailey, B. P. (1970). *Molec. Phys.*, **19,** 285
168. Morishima, I., Mizunno, A. and Yonezawa, T. (1970). *Chem. Phys. Lett.*, **7,** 633
169. James, P. G. and Luckhurst, G. R. (1971). *Molec. Phys.*, **20,** 761

7
Electron Resonance of Gaseous Free Radicals

JOHN M. BROWN
University of Southampton

7.1 INTRODUCTION

Small open-shell molecules constitute one of the most important groups of molecules in chemistry. The fact that relatively few of them have been studied with the precision afforded by microwave and radio-frequency spectroscopy does not contradict this statement; it is rather an indication of the experimental difficulties involved. For the theoretician, the magnetic hyperfine interactions which are peculiar to open-shell molecules provide an extremely sensitive and specific probe of the electronic wave function. This information is, of course, additional to that provided by the electric quadrupole coupling constant and the molecular electric dipole moment, parameters which can be measured for both open- and closed-shell molecules. However, the magnetic interaction constants yield rather more detailed information. This is in part because there are at least three constants which may be experimentally determined and partly because the magnetic hyperfine structure depends only on the disposition of those electrons which carry angular momentum rather than the entire charge distribution about the nucleus.

Open-shell molecules are also extremely important intermediates in gas-phase chemical reactions. In a few favourable cases they can be detected directly, but for the most part their participation has to be inferred from indirect observations. There is thus a need for as detailed a spectroscopic classification of these molecules as possible, partly to provide values for the parameters on which reaction schemes can be based and partly to yield new methods for detecting and monitoring the reaction intermediates. Gas-phase electron resonance has certainly justified itself on the second count; it has been the basis of the study of a wide variety of simple chemical reactions in discharge-flow systems over the last 5 years.

In addition to providing useful data for tangential branches of chemistry, the spectroscopy of open-shell molecules is itself an intrinsically interesting and challenging subject for study. These molecules are usually characterised by very short lifetimes, and it is rarely easy to generate them in sufficient concentration for detection in the microwave or radio-frequency spectral regions. Furthermore, the wide variety of intramolecular interactions which they display has turned up several interesting problems in both applied quantum mechanics and spectral analysis.

Most of this review is relevant to all branches of the spectroscopy of open-shell molecules in the microwave and radio-frequency regions. It is regrettable that subjects in science are too often classified by techniques rather than problems and the present author apologises for adopting the same easy terms of reference. This review will therefore be concerned with the detection of electric dipole transitions in open-shell molecules using a conventional e.s.r. spectrometer, operating with a fixed microwave frequency and a variable magnetic field. Experiments which are identical in principle but fall outside this frequency range will not be covered, although their relationship to gas-phase electron resonance should be appreciated. For example, in the radiofrequency region both electric[1] and magnetic resonance[2] experiments have been performed on open-shell molecules in a beam, whereas in the far infrared, the new technique of laser electron resonance has scored a triumphant success with the detection of the CH radical[3]. In addition, the material presented in Section 7.4 will show that the distinction which is often drawn between electron resonance and zero-field microwave spectroscopy is extremely artificial; electron resonance spectroscopy is in fact just a specialised form of microwave spectroscopy and fundamentally different from conventional electron spin resonance. Finally, this review will be concerned with the spectroscopy rather than the applications of electron resonance. In particular, no mention will be made of the growing mass of kinetic data[4] obtained with the technique; this represents sufficient material for a review of its own.

This article is being written in the shadow of a very authoritative review of the same subject by Carrington et al.[5]. Rather than undertake the futile exercise of duplicating their material, the present author has decided to build on their work and to concentrate on those aspects which have been investigated subsequent to their review. To this extent, the present review is far from comprehensive, but it is designed to be read in conjunction with that of Carrington et al.[5].

In Section 7.2, several points relating to theoretical developments in the subject are discussed. Section 7.3 deals with recent experimental results on diatomic open-shell molecules and includes tables of all the parameters obtained by electron resonance at the time of writing. Sections 7.4 and 7.5 deal with the most exciting development of the last 3 years, the detection of triatomic open-shell molecules by electron resonance. Examples of both linear and bent molecules have now been studied. The former are subject to sizeable vibronic interactions (the Renner–Teller effect); the theory and experimental results for this class of molecules are discussed in Section 7.4. Bent triatomic open-shell molecules present their own peculiar difficulties from the electron resonance standpoint, and these are discussed in Section 7.5. Finally, in Section 7.6, some suggestions for future development are made.

7.2 THEORETICAL ASPECTS

A proper understanding of the theoretical principles necessary for the description of molecular systems in the presence of external fields adds greatly to the satisfaction derived from the study of electron resonance spectra. More importantly, however, familiarity with the theory is often an essential precursor to the successful and incisive experiment. Most aspects of the theory have been well covered in the review by Carrington et al.[5], and, since it is not the intention of the present article to duplicate their material, the newcomer to the field is directed there in the first place. This section merely covers developments in the theory that have been published subsequent to their review and attempts to clear up some rather ill-defined but minor points relating to its application.

The general lines along which the theory is developed is represented diagrammatically in Figure 7.1. We start out with a particular model for our system of molecules and relate this model by a series of mathematical steps to signals traced out on a pen recorder. This serves two purposes. First, it is required for the analysis of the spectrum and, second, it provides the means by which the spectral lines can be related back to what one hopes are more fundamental molecular parameters. The model commonly adopted is that of a collection of isolated (i.e. non-interacting) molecules, each of which consists of an assembly of electrons and nuclei.

The first stage in the theoretical development involves the derivation of an 'exact' Hamiltonian operator for the molecule. Unfortunately, the use of the word 'exact' has always to be qualified since the mathematical and physical theories on which the development is based have their own deficiencies. In practice, one is content with a Hamiltonian operator which

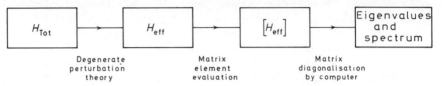

Figure 7.1 The step-wise development of the theory employed in the analysis and interpretation of molecular spectra

omits only those terms which are not experimentally detectable. This requirement has been met for spectroscopists working in the radiofrequency and microwave regions in two recent papers by Howard and Moss[6,7]; their approach and results are discussed briefly in Section 7.2.1.

The second stage involves the reduction of the complete molecular Hamiltonian to an effective Hamiltonian operator. If we construct a matrix representation of the complete Hamiltonian, it will in general be infinite, encompassing a complete range of electronic and vibrational states. For the most part, we shall be concerned with the spectroscopy of molecules in their ground vibronic states; the rotational states of the ground vibronic state are approximately represented by a block of the infinite matrix. Unfortunately, however, for most convenient choices of basis set, there are rather large matrix elements linking the states in this block with states in excited vibrational and/or electronic states. The effects of these matrix elements have traditionally been included by use of the second-order perturbation theory[8]. This leads to an effective Hamiltonian operator which, though it has lost some of its precision, is very much easier to handle since it now operates only within the rotational sub-space of the ground vibronic state. Once again, one's criterion of acceptable accuracy is dictated by experimental results and some recent work probing this point is presented in Section 7.2.2.

The third link in the chain joining our theoretical model with the experimental measurements involves the evaluation of a matrix representation of the effective Hamiltonian. This is a task which is now performed almost invariably by the experimental spectroscopist rather than the theoretician. Again, this topic has been well covered in the review by Carrington *et al.*[5], using the methods of irreducible tensors. All spectroscopists should be exhorted to acquire a good working knowledge of irreducible tensor theory and some familiarity with the full rotation group. Many are familiar with the well-tried procedures described by Condon and Shortley[9], but few appear to have taken the small extra step required to solve the same problems using irreducible tensor methods. Admittedly, the latter do not offer any real advantages if one is only concerned with the coupling of two angular momentum operators, but as soon as one has to handle three or more angular momenta, one quickly appreciates how much more powerful the irreducible tensor methods are. Section 7.2.3 deals with a few miscellaneous points concerning the evaluation of the matrix elements of the effective molecular Hamiltonian which either have not been covered or are unclear elsewhere.

The final step is the diagonalisation of the Hamiltonian matrix and the determination of the eigenvalues and eigenfunctions (if required). Since the eigenvalues of a molecule in an external field can very rarely be obtained in closed form, there was a time when this step was both difficult and time-consuming. Electronic computers, however, have rendered this task so trivial that it is now almost the exclusive preserve of graduate students!

The logical sequence from model to experiment has been presented in what might be called the forward direction. In practice, of course, one has to reverse this process and relate an experimental spectrum back to molecular parameters. Although the really crucial step back from the effective to the full Hamiltonian has only been attempted in a very few cases, it seems that the overall reverse process is not so bedevilled with indeterminancies and redundancies as the corresponding process in other branches of spectroscopy.

7.2.1 The molecular Hamiltonian

Papers which derive the Hamiltonian operator for a molecule have been appearing in the literature since the early 1930s. However, it is true to say that these have, in the main, been derivations of partial Hamiltonians, developed to explain particular types of experimental result obtained in particular frequency regions. Outstanding examples are the papers of Van Vleck[10] (on the electronic Hamiltonian), Wilson and Howard[11] (on the vibration–rotation Hamiltonian), Frosch and Foley[12] (on the nuclear hyperfine Hamiltonian) and Curl[13] (on the spin–rotation and Zeeman Hamiltonians). If one is interested in interpreting results of high precision, one requires a Hamiltonian which will more than match the experimental results in accuracy. In such a situation, one can justifiably question whether the procedure adopted in the earlier work leads to the correct form for the terms of interest in the Hamiltonian. For example, might they not be modified by cross-terms in the full Hamiltonian which could be neglected if only part of the Hamiltonian is developed?

It must be stressed that these doubting thoughts have so far arisen only in the minds of theoreticians; there does not seem to be any important experimental reason for calling the molecular Hamiltonian to question. However, the doubts have been sufficient for several theoreticians to go back over the ground and to attempt to derive as nearly complete a form for the molecular Hamiltonian as possible. In other words, the aim has been to carry through all parts of the Hamiltonian to the end, at which stage one can readily see which terms are important and which are not.

Even this laudable aim has so far been frustrated. The reason is that the people who are currently interested in molecules are chemists but that there are few chemists who have the knowledge of quantum electrodynamic theory required for an acceptable solution to the problem. A solution which is rigorous within the confines of relativistic quantum mechanics has been given by Howard and Moss for both non-linear[6] and linear[7] molecules. The most important terms that fail to appear in such a treatment are the radiative correction to the electron spin g-factor and the Lamb shift; these are therefore included in a purely phenomenological way. In addition, since nuclei are rather hard to describe in relativistic quantum mechanics, the assumption has to be made that the nuclei behave as Dirac particles, but with magnetic moments given by experiment, and the higher-order nuclear multipole interactions can be included phenomenologically. Not surprisingly, the final form for the Hamiltonian is quite lengthy and complicated; the interested reader is referred to Howard and Moss' original papers for the details. In this section, we shall merely present the logical skeleton of their work and point out the novel terms in their 'complete' Hamiltonian.

Under the influence of the electrostatic fields that exist within a molecule, the electrons move with velocities which are of the order of $\frac{1}{10}$ of the velocity

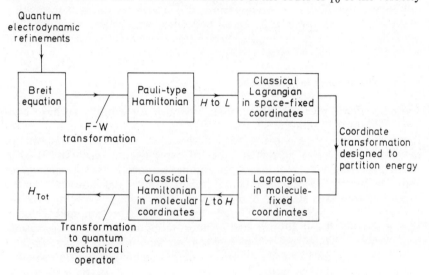

Figure 7.2 The stages in the development of a 'complete' molecular Hamiltonian, following Howard and Moss[6].

of light. It is therefore essential to use a relativistic theory to describe the mechanics of a molecule and the Dirac equation forms the logical starting point for the description of a single electron[14]. Regrettably, relativistic quantum mechanics does not provide us with a rigorous analogue for a many-electron (and nuclei) system, but the Breit equation[15] describes two-particle interactions correctly to order $\alpha^4 mc^2$.* Fortunately, all interactions involving three or more particles can be neglected to this accuracy so that the Breit equation forms the starting point for Howard and Moss' work (see Figure 7.2). The yardstick by which the Breit equation is judged is, of

---

*The terms in the relativistic Hamiltonian can be conveniently ordered in magnitude by expressing their size as $\alpha^n mc^2$ where m is the electron (rest) mass, c is the velocity of light and α is the fine structure constant, a dimensionless parameter equal to $\mu_B e^2 c/2h = 7.297 \times 10^{-3}$. The leading term is thus of order $\alpha^0 mc^2$.

course, provided by quantum electrodynamics so once again we see the need for chemists to master this subject if they are not to accept too much on faith.

The Dirac equation, although formulated to describe the electron, has the rather unexpected property of describing two types of particles, the electron and the positron whose states are separated by $\sim 2\,mc^2$ where m is the electron/positron mass and c is the velocity of light; the Breit equation is similar. Since we are only going to be interested in the relatively closely-spaced electronic states of our molecule, it is desirable to reduce the Breit operator so that it describes only these states. This step can be performed in two, essentially equivalent, ways; either by the method of small components[16] or by using a Foldy–Wouthysen transformation[17, 18]. This reduction of the Breit equation to a non-relativistic form of the Pauli type is directly analogous to the reduction of the complete Hamiltonian to an effective Hamiltonian operating in the ground vibronic state of the molecule (discussed in Section 7.2.2). The transformation introduces several characteristic terms and for our purposes perhaps the most interesting are those involving the electron 'spin'. This concept of electron spin is required to explain the intrinsic angular momentum properties of the electron. The physical origin of this angular momentum lies in a very-high-frequency oscillatory motion which is executed by the electron but cannot be described explicitly by an equation of the non-relativistic form; the transformation is thus rather like an averaging process.

The reduction of the Breit equation in this way yields a molecular Hamiltonian with particle positions and momenta expressed in an arbitrary, laboratory-fixed coordinate system (Figure 7.2). This Hamiltonian is next simplified and made more useful by means of suitably chosen coordinate transformations. Experiment suggests that the total molecular energy can be approximately partitioned into translational, rotational, vibrational and electronic energies. The coordinate transformations are chosen to effect this same partitioning in the molecular Hamiltonian, in so far as it is possible. The three transformations are:

(a) the coordinate origin is changed from a space-fixed one to the molecular centre of mass, keeping the axes parallel to those of the original axis system. This separates off the translational motion approximately.

(b) Keeping the same origin, the axes are next changed to a set rotating with the molecule to separate off rotational motion approximately.

(c) The origin is next changed to one at the nuclear centre of mass in the molecule so as to separate vibration from electronic motion approximately.

It must be stressed that this partitioning of the Hamiltonian is only approximate; the coordinate transformations are chosen to keep the cross terms between the various types of energy as small as possible. For reasons of mathematical convenience, the transformations are usually performed on the classical molecular Lagrangian rather than the Hamiltonian (the two functions are quite easily interconverted).

The coordinate transformation (a) above introduces a number of terms into the resultant operator which are labelled 'mass-polarisation' corrections, for example the mass polarisation correction to the spin–orbit coupling term. These terms are of the same form as the leading terms of each type in the Hamiltonian, but are qualified by a factor of order m/M where M is the total molecular mass. Most of these mass polarisation terms do not appear to have been detected experimentally.

The final stage in Howard and Moss' development is the introduction of explicit forms for the electric and magnetic potentials and a decoding of the various pieces of mathematical shorthand used in the development. This results in a Hamiltonian expressed in terms of symbols with which the experimental spectroscopist is more or less familiar and the final results are quite accessible.

The conclusion of their work is that the piecemeal development of the molecular Hamiltonian does not appear to be too much in error. Apart from some of the mass polarisation terms mentioned above, the only new terms to emerge are explicit expressions for the orbit–vibration interaction, the spin–vibration interaction and the vibrational Zeeman term.

The second paper by Howard and Moss extends their molecular Hamiltonian to cover the special case of linear molecules. Molecules of this type pose a particular difficulty in that they are characterised by two rotational coordinates rather than three, the third angle becoming a vibrational coordinate in the linear configuration. This 'absence' of the third Euler angle leads to problems later on in the theoretical treatment when one comes to compute matrix elements, since much of the theory, and in particular irreducible tensor methods, require there to be three rotational coordinates. Hougen[19], and later Watson[20], have provided a solution to this problem by introducing the third Euler angle as a redundant coordinate which leads to an isomorphic Hamiltonian. This isomorphic Hamiltonian can then be handled in the normal way, except that only certain of its eigenvalues and eigenfunctions are acceptable, the other solutions being a consequence of the redundancy introduced into the Hamiltonian. Howard and Moss[7] have shown how the external field potential terms are affected by this redundancy. Once again, the final results are listed in an accessible form at the end of their paper. This Hamiltonian agrees in most details with that given for a diatomic molecule by Carrington et al.[5]

7.2.2 The effective Hamiltonian

The general need for, and role of, the effective Hamiltonian in gas-phase electron resonance spectroscopy has been mentioned in the introduction to this section. The aim is to reduce the total Hamiltonian to one which operates within the vibronic state of interest only, but which has the same eigenvalues and eigenfunctions as the total Hamiltonian to within experimental error. Born and Oppenheimer[21] and Van Vleck[8, 10] have shown that this can be achieved using perturbation theory. The total Hamiltonian, H_{tot}, is divided into a zero order Hamiltonian, H_0, and a perturbation Hamiltonian, H':

$$H_{tot} = H_0 + \lambda H' \tag{7.1}$$

where λ is a dimensionless parameter in the range $0 \leqslant \lambda \leqslant 1$. The detailed way in which H_{tot} is partitioned is a matter of convenience. Some authors have taken H_0 as the electronic Hamiltonian (in the Born–Oppenheimer sense), but for diatomic molecules there is some advantage in taking H_0 as the sum of the electronic and vibrational Hamiltonians, so that both electronic and vibrational degeneracies are lifted in zero order. In this case, the zero-order states are labelled by different values of the electronic (and possibly vibrational) quantum numbers. The interactions involving rotational motion and electron and nuclear spin moments are usually included in the perturbation Hamiltonian, H'. Thus, each zero-order eigenvalue is degenerate in the rotational and spin quantum numbers and we must use degenerate perturbation theory to derive higher-order contributions to the eigenvalues. In addition, with our choice of H_0, the zero-order basis set is a simple product of an electronic (or vibronic) function and a function of rotational and spin coordinates. It is this property which enables the perturbation theory expressions to be factorised and the electronic (or vibronic) operators to be treated separately from the rotational and spin operators (which is, of course, the essential feature of an effective Hamiltonian).

Van Vleck[8] was the first author to derive an effective Hamiltonian of any generality in this way. He used the so-called Van Vleck transformation which is, in effect, degenerate perturbation theory carried through to second order. Miller[22], using the formulation of degenerate perturbation theory due to Bloch[23], has extended this treatment in principle to any desired order (in practice, most problems become too laborious by about 4th order). In addition, he has considered the special case of open-shell molecules. Several other procedures have also been suggested, but they are tailored for specific problems[24, 25]. Although they do not have the generality of the degenerate perturbation theory approach, they may well be more appropriate for the problems for which they were devised.

The procedure for deriving an effective Hamiltonian, H_{eff}, is thus well defined and second-order expressions for most cases of experimental interest have been published in the literature[8, 26-29]. It is regrettable that there still appears to be some confusion about the correct form of H_{eff} in many of these cases[30] and workers in the field are strongly recommended not to use previously published expressions without first checking them for themselves. This requirement of accuracy is particularly important in the field of gas-phase electron resonance, where the problem is almost invariably under-determined. In this situation, the achievement of a good fit of the experimental spectrum does not necessarily mean that the form of the effective Hamiltonian is correct.

A more interesting recent development in this area is the realisation that, in certain cases, some third-order terms in the effective Hamiltonian are experimentally significant. The prime quantity measured in an electron resonance experiment is usually the molecular magnetic moment in the state under study (the only exceptions to this statement involve transitions between levels with large zero-field separations). Thus the terms in H_{eff} most likely to show deviation from the second-order expressions are the terms in the Zeeman Hamiltonian linear in the external magnetic field, since these are the terms which are most searchingly tested by the experiment.

Two types of molecule have been studied which show anomalies in their magnetic moments (i.e. g-factors) when interpreted in terms of a second-order effective Hamiltonian. First, there is a group of diatomic molecules containing at least one heavy nucleus, with atomic number greater than say 16 (sulphur), and secondly there is the class of linear triatomic molecules in $^2\pi$ states displaying Renner–Teller effects. We shall deal with the first case in this section and postpone discussion of the triatomic molecules to Section 7.4, where we deal with their electron resonance spectra in some detail. We merely note here that there are special reasons for the breakdown of the second-order effective Hamiltonian for this type of molecule.

The necessity for inclusion of 3rd and higher-order terms in H_{eff} indicates that there are terms in the perturbation Hamiltonian H' which are comparable in magnitude with the separation between the electronic (or vibronic) eigenvalues defined by H_0. The largest term in H' is the electron spin–orbit interaction and since it increases rapidly with atomic number, the mixing of zero-order eigenfunctions is likely to become more pronounced as heavier nuclei are involved. Hence, the explanation of the magnetic moment anomalies of diatomic molecules with a heavy nucleus in terms of sizeable third-order contributions to the effective Hamiltonian from spin–orbit interactions, although not yet proven, is eminently plausible. In effect, the large spin–orbit interaction mixes the ground electronic state with excited electronic states, which may have markedly different magnetic moments.* This explanation has been suggested independently by two authors, Howard[31] and Miller[32].

The zero-order wave-functions (derived from H_0) may be characterised by two good quantum numbers, Λ and Σ. These are associated with the components of the orbital (L) and spin (S) angular momenta along the internuclear axis (z) respectively. It is not difficult to show[33] that the matrix elements of the spin–orbit Hamiltonian, H_{so}, in this basis set are subject to the following selection rules:

$$\Delta S = 0, \pm 1 \tag{7.2a}$$

$$\Delta \Lambda = -\Delta \Sigma = 0, \pm 1 \tag{7.2b}$$

From equation (7.2a), we see that H_{so} mixes states of different multiplicities, and from equation (7.2b) we see that, although Λ and Σ cease to be good quantum numbers if the mixing is significant, their sum (which is designated by the quantum number Ω) is conserved. This corresponds to Hund's coupling case (c)[34].

Howard[31] and Miller[32] both give details of the various third-order contributions to the molecular magnetic moment (i.e. the terms linear in the external magnetic field). Even when, one makes use of the constraints afforded by the selection rules on the various matrix elements, there is a large number of contributing third-order pathways and any hope of performing an *a priori* calculation of the third-order g-factor corrections is soon dashed. (The order of magnitude of these corrections can be assessed from these expressions, however, being about a 0.1% correction to the free electron orbital and spin g-factors for a spin–orbit interaction of 10^3 cm^{-1}.)

---

*It is perhaps worth mentioning that this phenomenon has long been appreciated, both theoretically and experimentally, in liquid phase and solid state e.s.r. Under the conditions of these experiments, the electron orbital angular momentum is quenched and so the higher-order effects are not masked by the much larger first-order spin–orbit interaction and orbital Zeeman effect, as they tend to be in gas-phase electron resonance.

The important point for the experimentalist is how these considerations affect his analysis. Let us take the case of a molecule in a $^2\pi$ state as an example. If we accept the second-order effective Hamiltonian of Carrington et al.[27] for the moment, we see that the magnetic moment of the molecule in a particular rotational level depends on four parameters, the electron spin and orbital g-factors (g_S and g_L), the molecular rotational g-factor (g_r) and the separation between the two spin components of the $^2\pi$ state (A). This statement assumes, of course, that any nuclear hyperfine effects present are small. The pertinent question is how many of these parameters are independently determinable from measurements of the electron resonance spectrum. The answer to this question can be obtained by determining the Zeeman energy of the molecule in the rotational level J to second order in a case (a) basis set[5, 27], namely:

$$W_z^{(1)} + W_z^{(2)} = \frac{\mu_B H_0 M_J}{J(J+1)} \{g_L \Lambda + g_S \Sigma)\Omega - g_r J(J+1) \mp \frac{g_S B_0}{(A-2B_0)}[J(J+1)-\tfrac{3}{4}]\}$$

$$+ \text{ second order Zeeman interaction} \qquad (7.3)$$

In equation (7.3), μ_B is the Bohr magneton, H_0 is the applied magnetic field, J is the quantum number associated with the overall angular momentum of the rotational state in question, Ω and M_J are its projections on the molecule- and space-fixed z-axes respectively, Λ and Σ are the projections of the electron orbital and spin angular momenta on the internuclear axis and B_0 is the rotational constant in ground vibronic state. The upper and lower sign choices in front of the second-order term refer to the $|\Omega| = \tfrac{3}{2}$ and $\tfrac{1}{2}$ components respectively; the conventional second-order Zeeman interaction, arising from mixing with the adjacent rotational levels, is not given explicitly. We see from equation (7.3) that, if measurements are confined to one spin component (that with $|\Omega| = \tfrac{3}{2}$, usually) of the $^2\pi$ state, there are essentially two determinable combinations of parameters and that these two combinations can be separated by making measurements on at least two rotational levels. If reasonable a priori estimates of g_L and g_S can be made, then the two determinable combinations yield values for g_r and A (in principle, at least); the value for B_0 is obtained from the magnitude of the second order Zeeman splitting.

We have seen that, if third-order terms in H_{eff} are significant, it is not possible to obtain reliable a priori values for g_L and g_S. In this case, the most that one can hope to determine from measurements on the $|\Omega| = \tfrac{3}{2}$ component are values for the two determinable combinations of the four molecular parameters; in other words, however many rotational levels are studied, it is not possible to determine values for the individual parameters. However, an examination of equation (7.3) reveals the interesting fact that magnetic moment measurements on two rotational levels in each spin component would allow one to determine values for g_S, g_L, g_r and A separately. The only example of the determination of the magnetic moment of a molecule in the $|\Omega| = \tfrac{1}{2}$ component of a $^2\pi$ state by electron resonance is the rather special case of OH [35], which is subject to considerable rotational distortion (i.e. it departs considerably from the limit of case (a) coupling). However, it should be generally possible to measure the magnetic moment of a molecule in the $(\Omega) = \frac{1}{2}$ component by electron resonance methods if the fixed microwave frequency is chosen to be close to a zero field Λ-doubling frequency. Alternatively, one could measure the frequency shifts of the Zeeman components of lines on the application of quite modest magnetic fields under very high resolution (molecular beam) conditions.

Similar considerations for a molecule in a $^1\Delta$ state reveal that the magnetic moment in any rotational state depends on two parameters, g_L and g_r [28]. If g_L can be reliably estimated, measurements on one rotational level will provide a value for g_r; however, if g_L has to be regarded as a parameter to be determined by experiment, then measurements on at least two rotational levels are necessary to obtain separate values for g_L and g_r.

We see therefore that there is a need for fairly extensive measurements on higher rotational levels of several diatomic open-shell molecules. The likely candidates with current experimental precision are given in Table 7.1; improvement in resolution will add lighter molecules to this list.

Table 7.1 Diatomic open-shell molecules whose electron resonance spectra are likely to exemplify breakdown of the second-order effective Hamiltonian

In $^2\pi_{\frac{3}{2}}$ states	ClO
	BrO
	IO
	SeF
	SeH
	TeH
In a $^1\Delta$ state	SeO

7.2.3 Calculation of matrix elements

The rotational invariance of many physical systems is an extremely basic and useful concept and leads directly to the ideas of angular momentum conservation. When these ideas are specialised to deal with quantum mechanical problems, their most natural technical outlet is provided by the methods of irreducible tensor operators. The newcomer to the subject is recommended to first read one or more of the basic texts (Rose[36], Edmonds[37] or Brink and Satchler[38]) in order to acquire the necessary background. The special modifications of the techniques which render them applicable to molecular (as opposed to fundamental particle or atomic) problems have been discussed by Freed[39] and Carrington et al.[5].

For the actual practice of evaluating matrix elements, both Edmonds and Brink and Satchler have commendable features. The former contains all the necessary basic equations and very useful tables of the Wigner three- and six-j symbols. (The present author prefers to use the Wigner n-j symbols rather than the Clebsch–Gordan and Racah coefficients owing to their easier symmetry properties on interchange of momenta, but this is largely a matter of taste.) However, Edmond's definition of the rotation operator, although appealing in some respects, is found to be rather difficult to use.

The definition of Brink and Satchler (and Rose) is much easier and is there-fore recommended. Brink and Satchler's book also contains a very fine set of appendices which summarise and interrelate all the important functions.

Like many other aspects of spectroscopy, irreducible tensor methods are made unnecessarily harder and more forbidding by a general lack of accepted convention and nomenclature. The rotation operator mentioned above is a case in point. Fortunately much of the variety is of a very trivial nature, and is easily mastered. However, the author would like to make one further recommendation on a point which relates specifically to molecular problems, namely that the components of an irreducible tensor measured in a space-fixed (non-rotating) coordinate system are labelled generically by p, whereas the components in the molecule-fixed coordinate system are labelled by q. This association can easily be remembered with the help of the rather dread-full mnemonic, *space* and *molecule*. Thus, for example, the rotation operator matrix elements are written $\mathscr{D}_{pq}^{(k)}(\alpha, \beta, \gamma)$ where (α, β, γ) are the Euler angles relating the molecule- to the space-fixed axes; since α is the magnitude of the rotation about the space-fixed Z-axis while γ is the magnitude of the rotation about the molecule-fixed z-axis, the first of the two subscripts on the symbol always refers to a space-fixed component and the second to a molecule-fixed component. Similarly, the relationship between the components of a kth-rank tensor operator $T^k(A)$ in space- and molecule-fixed coordinate systems is given by

$$T_p^k(A) = \sum_q \mathscr{D}_{pq}^{(k)*}(\alpha, \beta, \gamma) T_q^k(A) \tag{7.4}$$

It is as well to emphasise that irreducible tensor methods can be used to calcu-late the matrix elements of the Hamiltonians of both linear and non-linear molecules with equal ease. For the latter class of molecules, it is most con-venient to employ a set of symmetric rotor eigenfunctions for the rotational part of the basis set, since they are directly proportional to the irreducible representations of the rotation group:

$$|N, K, M\rangle = \left[\frac{(2N+1)}{8\pi^2}\right]^{\frac{1}{2}} \mathscr{D}_{MK}^{(N)*}(\alpha, \beta, \gamma) \tag{7.5}$$

where N is the quantum number associated with molecular rotation and K and M are its molecule- and space-fixed projections respectively.

Some difficulty has also been experienced in marrying irreducible tensor methods with the technique of reversed angular momentum introduced by Van Vleck[8]. The latter is an ingenious solution to the problem of calculating matrix elements of molecule-fixed operators which draws heavily on analo-gous angular momentum coupling schemes in atomic systems. Unfortu-nately, the point of these analogies is often lost on a generation of spectro-scopists who are much more familiar with coupling schemes in molecules than in atoms, and it is more sensible to tackle the molecular problems afresh and in their own right. Because the components of the angular momentum operator commute with the anomalous sign of i when referred to the molecule-fixed axis system:

$$[J_x, J_y] = -iJ_z, \text{ etc.} \tag{7.6a}$$

whereas irreducible tensor methods are based on the normal commutation relations between space-fixed components:

$$[J_X, J_Y] = iJ_Z \text{ etc.} \tag{7.6b}$$

incredible mathematical contortions are required to use irreducible tensor methods on the molecule-fixed components of J. In addition, J is the operator which relates to an infinitesimal rotation of the molecule-fixed axis system in laboratory space and so it is more correctly regarded as a space-fixed operator.

It is therefore recommended that all terms in the Hamiltonian involving the rotational angular momentum operator J (or N) are expanded in the first instance as space-fixed irreducible tensor components and then those parts that are physically appropriate (e.g. electronic coordinates or orbital angular momentum) are referred to the molecule-fixed axis system by means of the rotation matrix, equation (7.4). To see how this is done in practice, let us evaluate the matrix elements of the term $BJ.L$ from the rotational Hamiltonian of a linear molecule in a case (a) basis set, $|\eta \Lambda S \Sigma J \Omega M\rangle$. η labels the particular vibronic state, Λ is the projection of L on the internuclear axis, S and Σ are the quantum numbers for the total spin angular momentum and its projection on the molecule axis, and J, Ω and M are the quantum numbers for the total angular momentum and its projections on the molecule- and space-fixed z axes respectively.

Now $BJ.L \equiv T^1(J) \cdot T^1(BL)$

$$= \sum_p (-1)^p T_p^1(J) T_{-p}^1(BL)$$

$$= \sum_{p,q} (-1)^p T_p^1(J) \mathscr{D}_{-pq}^{(1)*}(\omega) T_q^1(BL) \tag{7.7}$$

where (ω) stands for (α, β, γ). We shall also need the results

$$\langle J \Omega M | \mathscr{D}_{pq}^{(k)*}(\omega) | J', \Omega', M' \rangle = (-1)^{J-\Omega}(-1)^{J-M}[(2J+1)(2J'+1)]^{\frac{1}{2}}$$
$$\times \begin{pmatrix} J & k & J' \\ -\Omega & q & \Omega' \end{pmatrix} \begin{pmatrix} J & k & J' \\ -M & p & M' \end{pmatrix} \tag{7.8}$$

and

$$\sum_{M'',p} \begin{pmatrix} J & 1 & J \\ -M & p & M'' \end{pmatrix} \begin{pmatrix} J' & 1 & J \\ -M' & p & M'' \end{pmatrix} = \delta_{JJ'} \delta_{MM'} (2J'+1)^{-1} \tag{7.9}$$

(see Brink and Satchler[38], Appendix I). Thus we have

$$\langle \eta \Lambda S \Sigma J \Omega M | BJ. L | \eta' \Lambda' S' \Sigma' J' \Omega' M' \rangle$$

$$= \delta_{SS'} \delta_{\Sigma \Sigma'} \sum_{p,q} \langle \eta \Lambda | B T_q^1(L) | \eta' \Lambda' \rangle \langle J \Omega M | (-1)^p T_p^1(J) \mathscr{D}_{-pq}^{(1)*}(\omega) | J' \Omega' M' \rangle$$

$$= \delta_{SS'} \delta_{\Sigma \Sigma'} \sum_{p,q} \langle \eta \Lambda | B T_q^1(L) | \eta' \Lambda' \rangle \sum_{J'' \Omega'' M''} \langle J \Omega M | (-1)^p T_p^1(J) | J'' \Omega'' M'' \rangle \times$$
$$\langle J'' \Omega'' M'' | \mathscr{D}_{-pq}^{(1)*}(\omega) | J' \Omega' M'' \rangle$$

$$= \delta_{SS'} \delta_{\Sigma \Sigma'} \sum_{p,q} \langle \eta \Lambda | B T_q^1(L) | \eta' \Lambda' \rangle \sum_{\Omega'' M''} (-1)^p (-1)^{J-M} \begin{pmatrix} J & 1 & J \\ -M & p & M'' \end{pmatrix}$$
$$\times \langle J \Omega \| T^1(J) \| J \Omega \rangle \delta_{JJ''} (-1)^{J-\Omega''}(-1)^{J-M''}[(2J'+1)(2J+1)]^{\frac{1}{2}}$$
$$\times \begin{pmatrix} J & 1 & J' \\ -\Omega'' & q & \Omega \end{pmatrix} \begin{pmatrix} J & 1 & J' \\ -M'' & -p & M' \end{pmatrix} \text{ from (7.8)}$$

$$= \delta_{SS'}\delta_{\Sigma\Sigma'}\sum_q \langle \eta\Lambda | BT_q^1(L) | \eta'\Lambda'\rangle (-1)^p (-1)^{J-M}(-1)^{J-\Omega}(-1)^{J-M''}$$

$$\times \begin{pmatrix} J & 1 & J \\ -\Omega & q & \Omega' \end{pmatrix} \langle J \| T^1(J) \| J\rangle \delta_{JJ'}\delta_{MM'}$$

from (7.9)

$$= \delta_{SS'}\delta_{\Sigma\Sigma'}\,\delta_{JJ'}\delta_{MM'}\sum_q (-1)^{J-\Omega}\langle \eta\Lambda | BT_q^1(L) | \eta'\Lambda'\rangle \begin{pmatrix} J & 1 & J \\ -\Omega & q & \Omega' \end{pmatrix} \langle J \| T^1(J) \| J\rangle$$

(7.10)

The last line follows since $M'' + p - M = 0$. The choice of phases implicit in this calculation is consistent with that laid down by Condon and Shortley[9] and, if the procedure suggested here is adopted, a comparison of matrix elements derived by different workers should be a worthwhile exercise since they should agree in all respects, at least to within a phase factor.

7.3   RESULTS FOR DIATOMIC OPEN-SHELL MOLECULES

The past 5 years have seen an impressive consolidation in the study of diatomic molecules by electron resonance. A number of new species have been detected and several of the older ones have been subjected to more careful examination. The extent of these results can be appreciated by reference to Tables 7.2, 7.3 and 7.4; Table 7.2 is concerned with molecules in $^2\pi$ states, Table 7.3 with molecules in $^3\Sigma$ states and Table 7.4 with molecules in $^1\Delta$ states. It is hoped that the reader is familiar with the nomenclature used; if he is in any doubt, the parameters are almost invariably defined in the references cited. The values of parameters in brackets, although used to fit the electron resonance spectra, were obtained from other work.

Some care should be exercised in using the results quoted in these tables. We have already seen in Section 7.2.2 how some of the parameters for the heavier molecules can be in error (in particular the A value for $^2\pi$ molecules and the g_r-value for $^1\Delta$ molecules) because of some doubtful assumptions in the analysis. For $^2\pi$ molecules this can result in errors in the B_0 values quoted since the effective B value in the $|\Omega| = \frac{3}{2}$ component is given by

$$B_{\text{eff}} = B_0(1 + A/B_0 + \dots) \tag{7.11}$$

and it is B_{eff} which determines the magnitude of the 'second order' Zeeman splitting. Furthermore, these tables by no means represent the state of our knowledge of these open-shell molecules since they have been studied by several other techniques, such as molecular beam electric resonance, zero field microwave spectroscopy and high-resolution electronic spectroscopy. Comparison of the results of zero-field microwave spectroscopy with those of electron resonance is particularly interesting, since the two experiments are so similar. In some cases, the parameters determined duplicate each other (for example, the molecular electric dipole moment[46]), whereas in others (for example, the magnetic hyperfine interactions), they complement each other. However, it remains true to say that an electron resonance experiment is usually characterised by a much smaller search problem but

Table 7.2 Experimental parameters for diatomic molecules in ²π states

Molecule	J	B_0/cm⁻¹	A/cm⁻¹	ν_Λ/MHz	Magnetic hyperfine/MHz	e^2q_0Q/MHz	μ_e/D	References
¹⁶OH	$\frac{3}{2}$	(18.515)	(−139.032)	1666.34±0.10	$|A_1|$ = 27.01±0.05, $|A_2|$ = 0.51±0.05	—	(1.66)	
	$\frac{5}{2}$			6033.5±1.0	$|A_1|$ = 5.39±0.05, $|A_2|$ = 0.68±0.05	—	—	40
	$\frac{7}{2}$			13437.8±2.0	$|A_1|$ = 1.00±0.05, $|A_2|$ = 0.87±0.05	—	—	
	$\frac{3}{2}(\Omega=\frac{1}{2})$			(7797.6)	$|A_1|$ = 20.55±0.06, $|A_2|$ = 14.56±0.05	—	—	35
	$\frac{5}{2}(\Omega=\frac{1}{2})$			(8166.1)	$|A_1|$ = 15.14±0.03, $|A_2|$ = 9.03±0.03	—	—	
¹⁶OD	$\frac{3}{2}$	(10.016)		310.12±0.08	$|A_1|$ = 4.84±0.03, $|A_2|$ = 0.03±0.03	—	—	40
¹⁷OH	$\frac{3}{2}$	(18.451)		1650.0±0.8	A_1 = −117.5±0.1, (¹⁷O), A_2 = −3.48±0.1	−4.2±1.0	—	41
¹⁷OD	$\frac{3}{2}$	(9.937)		305.5±0.8	A_1 = −114.1±0.2, A_2 = −1.08±0.1	−4.5±1.2	—	41
¹⁶OH†	$\frac{3}{2}(v=1)$	(17.807)	(−139.199)	1537.8±0.3	A_1 = 24.22±0.1, A_2 = 0.44±0.2 (d = 53.3)	—	—	42, 44
	$\frac{5}{2}(v=1)$			5593.3±1.5	A_1 = 3.92±0.5, A_2 = 0.73±0.03	—	—	43
	$\frac{3}{2}(v=2)$	(17.108)	(−139.440)	1413.5±0.3	A_1 = 21.06±0.1, A_2 = 0.45±0.04	—	—	
	$\frac{5}{2}(v=2)$			—	A_1 = 2.9±0.5, A_2 = 0.65±0.05	—	—	
	$\frac{3}{2}(v=3)$	(16.414)	(−139.688)	1293.2±0.3	A_1 = 18.00±0.15, A_2 = 0.40±0.04	—	—	42
	$\frac{5}{2}(v=3)$	(15.728)	(−140.014)	4751.3±2.0		—	—	
	$\frac{3}{2}(v=4)$			1176.4±0.3	A_1 = 14.22±0.15, A_2 = 0.31±0.05	—	—	
	$\frac{5}{2}(v=4)$			4346.3±2.0	—	—	—	
³²SH	$\frac{3}{2}$	(9.4611)	(−376.96)	111.42±0.04	$|A_1|$ = 5.61±0.04, $|A_2|$ = 0.08±0.04	—	0.62±0.01	45, 46

Species	I				Parameters			Ref.
^{33}SH	3/2	7.77±0.01		440.9±0.2	$A_1 = 71.1\pm1.0$ (^{33}S)	13.5±2.0	—	[47, 48]
SeH	3/2	7.80±0.02	(−1685)	14.48±0.03	$A_1 = 312.7\pm0.2$ (^{77}Se)	—	—	[49]
				57.4±0.3			0.49±0.05	[50, 46]
SeD	3/2	3.940±0.005		1.90±0.01	$A_1 = 311.8\pm0.2$ (^{77}Se)	—	0.483±0.003	[50]
TeH	3/2	5.56±0.15	−2250±200	6.6±0.2	$A_1 = 1.8\pm0.2$	—	—	[50, 46]
^{35}ClO	3/2	0.622±0.001	−282±9		$h = 111\pm2$	−88±6	1.26±0.04	[51]
^{37}ClO	3/2	0.611±0.001	−282±9		$h = 93\pm2$	−69±6	1.61±0.04	[52, 46]
^{79}BrO	3/2	0.4282±0.0005	−815±120		$h = 504.5\pm1.0$	649.8±1.9	2.45±0.04	[52]
^{81}BrO	3/2	0.4264±0.0005	−815±120		$h = 543.9\pm1.0$	542.7±1.9	0.87±0.05	[53, 46]
^{127}IO	3/2	0.3389±0.0007	−446±70		$h = 582.1\pm2.3$	−1907.0±13	1.52±0.05	[53]
SF	3/2	0.5527±0.0005	−387±25		$h = 428.4\pm1.5$	—	(0.1587)	[53, 46]
SeF	3/2	0.3625±0.0013	−560±70		$h = 325.6\pm3.9$	−1.8±0.3		[27, 46]
^{14}NO	3/2	(1.6957)	(123.16)	0.906±0.01 ; 3.601±0.025	$A_1 = 29.836\pm0.025$, $A_2 = 0.019\pm0.005$; $A_1 = 12.440\pm0.04$, $A_2 = 0.031\pm0.010$	—		[27, 46]
^{15}NO	3/2 ; 5/2			0.814±0.01 ; 3.224±0.025	$A_1 = −41.886\pm0.020$, $A_2 = −0.024\pm0.005$; $A_1 = −17.480\pm0.035$, $A_2 = −0.040\pm0.009$	—		[54]
NS	3/2 ; 5/2	0.7722±0.001	(223.03)		$h = 57.0\pm0.2$	−2.86±0.31	1.86±0.03	[55, 56, 46]
CF	3/2	(1.40827)	(77.11)		$h = 662.9\pm3$, $b = 190\pm50$	—	0.65±0.05	[57]

Table 7.3 Experimental parameters for diatomic molecules in $^3\Sigma$ states

Molecule	J-values	B/cm^{-1}	λ/cm^{-1}	γ/cm^{-1}	g_S	g_r	g_L^e	Magnetic hyperfine /MHz	References
O$_2$	N = 1–13	(1.43777)	(1.9848)	(−0.0084)	2.001997 ±0.00002	$(1.26\pm0.12)\times10^{-4}$	$(2.813\pm0.03)\times10^{-3}$		58*
^{32}SO	1	(0.71791)	(5.2766)	(−0.0052)	(2.0023)	$-(2.0\pm2.0)\times10^{-4}$	$(4.0\pm0.3)\times10^{-3}$		59*
					2.00197 ±0.0001	$(1.9\pm2.0)\times10^{-4}$	$(3.7\pm0.5)\times10^{-3}$		60
	0, 1, 2				2.00199 ±0.0001	$(1.1\pm0.7)\times10^{-4}$	$(3.64\pm0.2)\times10^{-3}$		61
^{33}SO	1	(0.71067)						$b = (+)58\pm3$ $c = (-)101\pm3$	62
SeO	1	(0.466)	(86.4)	−0.03				$|h| = 360$ (^{77}Se)	28

*The signs of the g-factors in references 58 and 59 have been changed so that they are consistent with the usage of subsequent authors

Table 7.4 Experimental parameters for diatomic molecules in $^1\Delta$ states

Molecule	J	B/cm^{-1}	g_L	g_r	a/MHz	μ_e/D	References
O$_2$	2,3	1.41808±0.00020	0.999866±0.000010	$(-1.234\pm.025)\times10^{-4}$	–	–	63
^{16}O^{17}O	2,3		0.999860	(-1.70×10^{-4})	−424	–	64
^{32}SO	2	0.709±0.001	(1.00)	–	–	1.31±0.04	65,46
^{33}SO	2				145.1±3.0		49
SeO	2	0.461±0.002		$(11\pm3)\times10^{-4}$		2.01±0.06	28,46
NF	2	(1.2225)			109 (^{14}N) 753 (^{19}F)	0.37±0.06	66

that the zero-field experiment provides the more precise results since it is much easier to measure frequencies accurately than magnetic fields.

It is not possible to discuss all the results given in the tables within the space allocated for this review. However, it seems to the author that there are three aspects which are of especial interest and these form the subject matter of the rest of this section.

7.3.1 Comparison of experimental parameters with those predicted by *ab initio* calculations

We intimated earlier that one of the important aspects of electron resonance studies of open-shell molecules is that they provide values for parameters which can be used to test *ab initio* wave functions. Some comparisons that have been made between experimentally determined parameters and their values computed from Hartree–Fock self consistent field functions are given in Table 7.5. The numbers given in this table prompt a few general remarks.

Table 7.5   Comparison between observed and calculated parameters

Molecule	Parameter	Experiment	Theory	References
^{16}OH	a	86.0 ± 0.6 MHz		35,67
	b	-119.0 ± 0.4 MHz	see	35
	c	133.2 ± 1.0 MHz	text	35
	d	56.5 ± 0.4 MHz		35
^{17}OH	A_1	-117.5 ± 0.1 MHz	see	41,67
	A_2	-3.48 ± 0.1 MHz	text	41
	$e^2 q_0 Q$	-4.2 ± 1.0 MHz	-2.6 MHz	41,67
^{32}SH	μ_e	0.62 ± 0.01 D	0.86 D	46,68
^{33}SH	A_1	71.1 ± 1.0 MHz	see text	49
	$e^2 q_0 Q$	13.5 ± 2.0 MHz	12.25 MHz	49,69
ClO	μ_e	1.26 ± 0.04 D	0.81 D	46,70
SF	μ_e	0.87 ± 0.05 D	1.395 D	46,71
SeF	μ_e	1.52 ± 0.05 D	2.21 D	46,71
NS	μ_e	1.86 ± 0.03 D	1.732 D	46
CF	μ_e	0.65 ± 0.05 D	0.48 D·	46,72
$NF(^1\Delta)$	μ_e	0.37 ± 0.06 D	0.17 D	93,94

First, it can be seen that, for the molecular dipole moment and the nuclear quadrupole coupling constant, the agreement between experiment and theory is really very good. However, it is well known that these parameters depend primarily on the electronic distribution of the dominant electronic configuration in the $^2\pi$ state, and it is only this aspect of the wave function which is being tested. The second point is that even though very good wave functions are now available for the molecules listed, only in a few cases has the interesting comparison of magnetic hyperfine parameters derived from experiment and theory been made[41, 49]. The third point is that there are several other molecules which are now quite well documented experimentally, but for which good wave functions have not yet been published; the obvious examples in this class are $^3\Sigma$ and $^1\Delta$ ^{33}SO [49, 62], and $^1\Delta^{16}O^{17}O$ [64]. In other words, there is a lot of scope for theoretical work in this area.

All the cases where magnetic hyperfine parameters have been investigated[35, 41, 54] suggest that the approximation that the electronic wave function can be written as a simple configuration with all paired electrons in identical orbitals is a very poor one for the purposes of interpreting these parameters. For example, the picture of a molecule in a $^2\pi$ state in this approximation is of one electron carrying both orbital and spin angular momenta. This picture is unreliable and it should not be used to effect a complete separation of hyperfine parameters when measurements in both spin components of the $^2\pi$ state are not available[35, 40]. The necessity for inclusion of configuration interaction in the wave function can often be appreciated from consideration of the Fermi contact interaction; if the dominant $^2\pi$ configuration has the unpaired electron in a p-type orbital, this interaction would vanish in the single configuration approximation. With regard to theoretical calculations of the Fermi contact interaction, the results for molecules are not expected to be very good, since the estimates even for atoms are often rather poor at the moment.

Finally, it is intriguing to note that the *ab initio* calculations suggest the electric dipole moment of CF in its ground state has the *positive* end on the fluorine atom. It would be interesting to have an experimental confirmation of this unexpected result[72].

7.3.2 The detection of vibrationally excited OH by electron resonance

Electron resonance spectra are in general quite insensitive to the vibrational state of the molecule being studied. The reason for this is that the principal terms in the Zeeman Hamiltonian are wholly electronic in origin and virtually independent of the internuclear distance(s). From some points of view, this can be an advantage. For example, parameters in H_{eff} such as the spin–orbit coupling constant, A, which contribute to the molecular magnetic moment in second order, can often be determined quite reliably; by contrast, the second-order effect of the same terms on the zero field microwave spectrum is often masked by vibration–rotation interaction terms since the rotational Hamiltonian shows a quite pronounced vibrational dependence.

Thus, electron resonance spectra which are characterised by quite large zero-field splittings will usually show a greater dependence on vibrational state. A case in point is ^{32}SO in its $X^3\Sigma^-$ state whose electron resonance spectrum has been detected in the $v = 1$ state with an intensity which corresponds to thermal equilibrium[59]. The linear triatomic molecules NCO and NCS have also been detected in thermal equilibrium in vibrationally excited states; in this case, there is a large vibronic interaction which makes the electron resonance spectrum very sensitive to one of the three vibrational quantum numbers. This is discussed more fully in Section 7.4. Very recently, there have been three independent studies of the electron resonance spectrum of vibrationally excited OH. These studies differ from the two other examples cited in that the molecules are generated by chemical reactions which produce significant over-population of higher vibrational levels corresponding to Boltzmann distribution temperatures of several thousand degrees absolute. In this way, OH has been detected in levels $v = 0$–4 (but no higher) from the reaction of H atoms with F_2O [43] and in levels $v = 0-9$ from the reaction of H atoms with ozone[42, 44]. In the case of OH, the reason why the vibrational states can be resolved is that the spectra are characterised by a large zero-field splitting. This splitting is caused by Λ-doubling and shows quite a large dependence on vibrational state for this molecule (see Table 7.2).

Churg and Levy[43] concentrated their attention on the determination of the zero field frequencies of the $J = \frac{3}{2}$ transition in the $v = 1$ state. These frequencies are important to radio astronomers seeking an explanation for the anomalous emission characteristics of OH molecules in the interstellar medium (in $v = 0$ states). It should be pointed out that there is some uncertainty in the values for the magnetic hyperfine constant d in both $v = 0$ and 1 states[42, 43] which causes a corresponding uncertainty in the zero field frequencies.

Clough et al.[42] have presented a more complete spectroscopic analysis for the levels $v = 1$–4, and their results are listed in Table 7.2. The general conclusion is that the Λ-doubling intervals and the hyperfine interaction parameters decrease steadily as v increases. These authors have also presented a consistent interpretation of their experimental parameters in terms of vibronic matrix elements that occur in the effective Hamiltonian. Their results are essentially duplicated by Lee, Tam, Larouche and Woonton[44].

Finally, it is interesting to note that the conventional electron resonance spectrometer, as it stands, is not a very useful tool for the study of vibrational relaxation processes. Although it has the great advantage of providing information on the relative (or absolute) populations of all detectable states, the size of the microwave cavity is such that relaxation by wall collisions is the dominant process at typical operating pressures.

7.3.3 The electron resonance spectrum of NF in its $^1\Delta$ state[66]

It is slightly invidious to select one spectroscopic study out of a number of very competent pieces of work published during the last few years (see Tables 7.2, 7.3 and 7.4). However, the spectrum of $^1\Delta$ NF is of peculiar interest, since it is the first recorded example of an open-shell molecule involving two nuclear spins for which a complete determination of the hyperfine coupling parameters is available (that of ^{17}OH was the first example involving two nuclear spins[41]). A reproduction of the spectrum arising from molecules in the lowest rotational level, $J = 2$, is given in Figure 7.3. The spectrum is characterised by a large ^{19}F hyperfine interaction and a small ^{14}N interaction; since the molecule is being studied in a singlet state, the magnetic hyperfine structure arises solely from the interaction between the nuclear spins and the electronic orbital moments. The ^{19}F hyperfine interaction is large enough for off-diagonal matrix elements in a decoupled basis set to be important and this accounts for the different spacings between the groups of transitions with $M_I = +\frac{1}{2}$ and $-\frac{1}{2}$, which can be seen in Figure 7.1.

A detailed analysis and interpretation of the spectrum has yet to be published. However, preliminary results[66] show that one can have considerable confidence in an LCAO description of the electronic wave function for this type of molecule; the two hyperfine parameters provide independent estimates of the two normalised coefficients in the wave function which are almost exactly consistent. This is important because the LCAO description

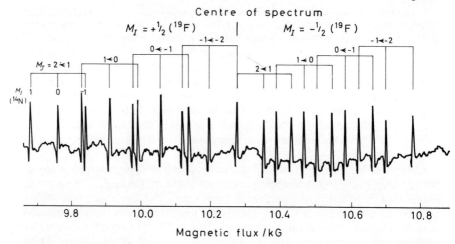

Figure 7.3 The gas-phase electron resonance spectrum of NF in the $J = 2$ rotational level of its $^1\Delta$ state. The spectrum has been recorded using Stark modulation at an operating frequency of 9.5 GHz

is often very useful in the interpretation of less favourable spectra where the hyperfine parameters are not completely separated (for example, CF in its $^2\pi$ state[57]).

7.4 ELECTRON RESONANCE OF LINEAR TRIATOMIC OPEN-SHELL MOLECULES

The most important development in the field of electron resonance over the last 3 years is undoubtedly the extension of its range to embrace triatomic molecules. In principle, there is no reason why still larger molecules should not be studied by this technique. This section is concerned with linear triatomic molecules (specifically in $^2\pi$ electronic states) and the following section deals with non-linear molecules.

The electron resonance spectra of diatomic molecules are characteristic of the coupling between three angular momenta, L and S, the electronic orbital and spin angular momenta respectively and R, the angular momentum of the two nuclei in laboratory space. Additional structure (and information) arises from nuclear spin angular momenta I, if present. Diatomic molecules have only one vibrational degree of freedom, a totally symmetric stretching mode. Although interactions between vibrational and electronic motions do occur, they are rarely the dominant feature in any branch of spectroscopy, and almost insignificant in electron resonance.

The linear triatomic molecule, however, has three normal coordinates, two non-degenerate stretching modes (Q_1 and Q_3) and one doubly degenerate bending mode (Q_2). The former vibrational motions manifest themselves spectroscopically in much the same way as the single mode of the diatomic molecule and hence are rather uninteresting from our point of view. However, there is a vibrational angular momentum G associated with the degenerate bending mode and the coupling between this angular momentum and L can have pronounced effects on the electron resonance spectrum. The physical origin of this coupling can be appreciated from the following picture: suppose we first clamp the three nuclei in a linear configuration. For a molecule in an electronic π state, there is an associated twofold electronic degeneracy for this configuration. However, if we now permit the nuclei to depart from the linear axis to execute the bending motion, this electronic degeneracy is lifted and the orbital angular momentum is partially or wholly quenched. Thus we see that the orbital contribution to the magnetic moment of the molecule depends strongly on the form and magnitude of this bending motion and consequently the electron resonance spectra of such molecules in different vibrational (more strictly, vibronic) states are widely separated and easily resolved. This phenomenon of vibronic coupling between G and L is called the Renner–Teller effect[73].

At the time of writing, two linear triatomic molecules in $^2\pi$ states have been detected by electron resonance, NCO[74] and NCS[75, 78]; spectra have been obtained for ground and excited vibronic states of both molecules. Even though the states which have been studied are only affected by the Renner coupling in second order, deviations from the predictions of the existing Renner–Teller theory[76, 77] were sufficient to prompt a re-examination and extension of this theory[74, 31].

7.4.1   The effective Hamiltonian for a linear triatomic molecule in a $^2\pi$ state

We have mentioned the derivation of H_{eff} for a diatomic molecule in an earlier Section (7.2.2). The general aim and approach is exactly the same for a triatomic molecule, but the additional complication of the Renner–Teller effect causes some differences in the details of the derivation.

The first difference arises in the choice of basis set, that is the eigenfunctions of H_0 in equation (7.1). For diatomic molecules one can, with advantage, use a Born–Oppenheimer type basis set, that is one in which the electronic basis functions are functions of the internuclear distance, R. Such a basis set provides a very good description of the electronic motion, even in the presence of the vibrational motion of the nuclei. It would be desirable if one could adopt a similar basis set for linear triatomic molecules, but regretably its use leads to singularities in some of the integrals involved. In the face of such difficulties it is easiest to fall back on the 'crude adiabatic' basis set introduced by Longuet-Higgins[76]. This is the set of eigenfunctions of a zero-order Hamiltonian obtained by fixing the nuclei at their (linear) equilibrium configuration; the functions are simple products of electronic and nuclear wave functions and the electronic factor is independent of both nuclear coordinates and momenta. The advantage of this basis set is that the matrix elements of the total Hamiltonian are simple to derive (we have chosen a nuclear reference configuration which preserves the twofold electronic degeneracy) but its disadvantage is that finite motion of the nuclei off-axis causes some of the off-diagonal matrix elements to be rather large.

Second, because there is not too much interest in the effects of vibration on the electron resonance spectra of diatomic molecules, it has not been necessary to be very careful about distinguishing between an effective Hamiltonian in which the effects of matrix elements off-diagonal in electronic state are included to a certain order of perturbation theory and one in which the effects of matrix elements off-diagonal in vibrational quantum number are included. For triatomic molecules there is a coupling between L and G which profoundly affects the electron resonance spectrum, and we cannot afford to be so imprecise. In this situation, the most appealing approach (which is adopted implicitly by Carrington *et al.*[74]) is to break the derivation of an effective rotational Hamiltonian down into two stages. The first step involves the elimination of matrix elements off-diagonal in electronic quantum number to the desired precision; this is greatly facilitated by use of the crude adiabatic basis set since many of the nuclear operators, in particular the kinetic-energy operator, have only diagonal matrix elements. At the end of this process, we are left with an effective Hamiltonian which operates only within the vibration, rotation and spin sub-space of the ground electronic state. The second stage consists in the derivation of a convenient matrix representation of this effective Hamiltonian and the elimination of the

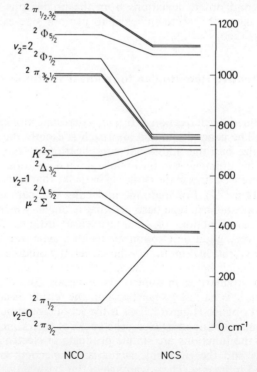

Figure 7.4 The lowest vibronic states of NCO and NCS

effects of matrix elements off-diagonal in vibrational quantum number to the desired precision. This leads to a second effective Hamiltonian which now operates only within the sub-space of one particular vibrational state. Note that at this stage we still have not lifted the electronic and vibrational degeneracies of the state; in other words, one vibrational state is in general spanned by several vibronic, and an infinite number of rotational, states.

The third distinction between diatomic and triatomic molecules arises when one comes to consider what is an acceptable precision for H_{eff}, i.e. to what order in perturbation theory must the derivation be taken? For triatomic molecules, there are additional terms in the perturbation Hamiltonian; also the choice of basis set leads to a less rapid convergence of the perturbation treatment than for diatomic molecules. Almost certainly the last word has not been said yet on this subject, but Carrington et al.[74] have found experimental evidence for significant third-order contributions to the magnetic moment. This evidence derives from measurements of the magnetic moment in different vibronic states and suggests that there is a vibronically dependent contribution to the magnetic moment; Carrington et al.[74] have shown that such a term can arise from vibronic mixing of excited Σ and Δ states into the π state and can be included in the effective Hamiltonian by taking an effective orbital g-factor, g_L':

$$g_L' = g_L + (v_2 + 1)\Delta g_L \qquad (7.12)$$

where v_2 is the vibrational quantum number associated with the degenerate bending mode, Q_2. The interested reader is referred to their paper for more details.

7.4.2 The vibronic energy levels and experimental results for NCO and NCS in their $^2\pi$ states

The lowest vibronic energy levels of NCO and NCS are illustrated in Figure 7.4. The levels are labelled by their values of v_2, which is still a reasonably good quantum number, and also their appropriate vibronic symmetry species. With regard to the latter, the main symbol is Σ, π, Δ, etc. as $|K| = 0, 1, 2, \ldots$ where $K = \Lambda + 1$ (l is the quantum number associated with the projection of G on the internuclear axis); the subscript in each state symbol denotes the value of $|P|$, where P is the quantum number associated with the component of the total angular momentum along the axis. The pattern of levels drawn in each case represents the combined effects of the Renner and spin–orbit coupling terms in the second of the two effective Hamiltonians discussed in the last section.

The NCO and NCS radicals have been produced in the gas phase by the abstraction of a hydrogen atom from their parent acids[75]. Electron resonance spectra of NCO have been obtained in the lowest rotational levels of the $^2\pi_{\frac{3}{2}}$ ($v_2 = 0$), $^2\Delta_{\frac{5}{2}}$ ($v_2 = 1$) and $^2\Phi_{\frac{7}{2}}$ ($v_2 = 2$) vibronic states. The relative intensities of the three spectra are consistent with a room temperature Boltzmann distribution between the states, and signal averaging techniques were required to detect the $^2\Phi_{\frac{7}{2}}$ spectrum[74]. For NCS, spectra from the lower two vibronic states, $^2\pi_{\frac{3}{2}}$ and $^2\Delta_{\frac{5}{2}}$, only have been obtained at the time of writing[78].

It should be noted that all the states so far observed are characterised by $v_2 + 1 = |K|$ and hence are subject to the effects of Renner–Teller coupling in second and higher orders only[77]. Nevertheless, the coupling is sufficiently large to produce easily detectable contributions to the magnetic moments of both molecules and values for the vibrationally dependent component of the orbital g-factor, Δg_L are reliably determined (see Table 7.6). The spectrum of the $J = \frac{3}{2}$, $^2\pi_{\frac{3}{2}}$ ground vibronic state is shown in Figure 7.5. It shows the characteristic 'second-order' Zeeman splitting and ^{14}N magnetic hyperfine and electric quadrupole structure. Values of the parameters used

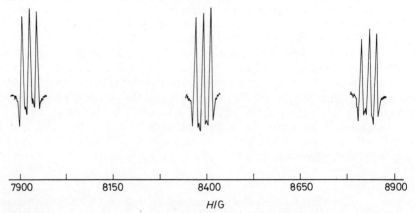

7900   8150   8400   8650   8900

*H/*G

Figure 7.5 The gas-phase electron resonance spectrum of NCO in the $J = \frac{3}{2}$ level of its ground vibronic state. The spectrum was recorded using Stark modulation at an operating frequency of 9.27 GHz.
(From Carrington *et al.*[74], reproduced by permission of Taylor & Francis Ltd.)

to fit the spectra are listed in Table 7.6 for all the detected vibronic states of both NCO and NCS; where available, parameters which affect the spectrum in second order only are taken from the analyses of the electronic spectra of these molecules. Since there is considerable uncertainty in the Renner parameter, ε, for the ground state of NCS, this parameter has been determined from the electron resonance spectrum.

Careful examination of Table 7.6 reveals that the value for Δg_L in NCO shows an apparent dependence on vibronic state. This is not expected from theoretical considerations and may point to a small error in the value for ε obtained from the electronic spectrum. If ε is changed from -0.159 to -0.150, the three spectra can all be fitted by a single value of Δg_L, -6.3×10^{-3}. A very recent re-analysis of the electronic spectrum[79] supports the need for this change.

Finally, it is interesting to ask whether any other of the vibronic states shown in Figure 7.4 are susceptible to study by electron resonance methods. The other spin components of the vibronic states studied ($^2\pi_{\frac{1}{2}}$, $v_2 = 0$; $^2\Delta_{\frac{3}{2}}$, $v_2 = 1$; $^2\Phi_{\frac{5}{2}}$, $v_2 = 2$) are characterised by very small magnetic moments ($\sim 10^{-3} \mu_B$) and hence are not detectable by conventional electron resonance. The $\mu^2\Sigma^{(+)}$, $v_2 = 1$ state would seem to be a very good candidate from population considerations. For NCO, the parameters in H_{eff} are of such a magnitude that the angular momentum coupling scheme is much closer to the

Table 7.6 Experimental parameters for linear triatomic molecules in $^2\pi$ states

Molecule	Vibronic state	J	A_v/cm^{-1}	B_v/cm^{-1}	h/MHz	e^2q_0Q/MHz	Δg_L	μ_e/D	References
NCO	$v_2 = 0,\ ^2\pi_{\frac{1}{2}}$	$\frac{3}{2}\ \frac{5}{2}$	(-96.20)	(0.38940)	54.1 ± 0.4	-2.2 ± 0.4	$-(5.8\pm0.6)\times10^{-3}$	0.742 ± 0.05	74
	$v_2 = 1,\ ^2\Delta_{\frac{5}{2}}$	$\frac{5}{2}\ \frac{7}{2}$	(-95.90)	(0.39046)	54.1	-2.2	-5.3×10^{-3}		
	$v_2 = 2,\ ^2\Phi_{\frac{7}{2}}$	$\frac{7}{2}$	(-95.80)	(0.3915)	54.1	-2.2	-4.8×10^{-3}		
NCS	$v_2 = 0,\ ^2\pi_{\frac{3}{2}}$	$\frac{3}{2}$	(-319.9)	(0.2036)	25.4 ± 0.6	3.3 ± 0.8	$-(5.5\pm2.0)\times10^{-3}$	2.5 ± 0.3	78
	$v_2 = 1,\ ^2\Delta_{\frac{3}{2}}$	$\frac{5}{2}$							

The values in parentheses were taken from the analyses of the electronic spectra of NCO and NCS. Two other parameters were used to fit the electron resonance spectra, again taken from the electronic spectra; the vibrational frequency, ω_2 (538.94 cm^{-1} for NCO and 387 cm^{-1} for NCS) and the Renner parameter, ε (-0.159 for NCO). The Renner parameter for NCS has been determined from the electron resonance spectrum as $\varepsilon = -0.16\pm0.03$

Hund's case (b) than the case (a) limit and the rotational energy levels display very large parity doublings. Thus, the Zeeman effects of the levels are highly non-linear and there is a very large zero-field splitting. The electron resonance spectrum of the lowest rotational level would be detectable with a spectrometer operating at a frequency near that corresponding to the zero-field splitting (20 GHz). For NCS in this vibronic state, the coupling scheme is much nearer the case (a) limit and the zero-field parity splitting is much smaller; provided a sufficient concentration of molecules can be generated, the electron resonance spectrum should be detectable with a conventional X-band spectrometer. The behaviour of the rotational levels in the $\kappa^2\Sigma^{(-)}$ state is very similar. A study of such spectra would be very informative since these states are all subject to a first-order Renner–Teller mixing, unlike those that have been detected so far.

7.5 MICROWAVE SPECTROSCOPY OF BENT TRIATOMIC OPEN-SHELL MOLECULES

This section is concerned with a very recent development in the field of what would conventionally be called electron resonance. As before, the experiments involve the detection of electric dipole transitions which are tuned into coincidence with a fixed microwave frequency by an external magnetic field. However, it will become apparent that the distinction between this experiment and zero-field microwave absorption spectroscopy is very artificial and we drop the pretence at this stage by calling both experiments microwave spectroscopy. Partly because the experiments are very recent, and partly because the spectroscopic problems posed are more complicated than the others discussed in this review, the analysis is not yet at a very advanced stage. This section is therefore chiefly concerned with the principles; these are important, since they point the way to a potentially very rich field of study.

7.5.1 Magnetic tunability of transitions in a triatomic molecule

In a non-linear triatomic open-shell molecule, the orbital angular momentum is quenched, and only the spin angular momentum contributes significantly to the molecular magnetic moment. Nevertheless, this is sufficient to produce quite a strong dependence of the molecular energy levels on an external magnetic field. However, the spin–rotation interaction which couples the spin angular momentum to the nuclear framework is never very large, being typically between 100 MHz and 1 GHz. Van Vleck[8] has shown that this spin–rotation interaction arises principally from a second-order spin–orbit interaction in the effective Hamiltonian. As the applied magnetic field is increased, there comes a point when the Zeeman interaction between the spin magnetic moment and the field is comparable with the spin–rotation interaction. From this point, the spin becomes progressively decoupled from the molecule and orientates itself with respect to the applied magnetic field instead.

It is characteristic of this type of molecule that in the spin decoupled region only magnetic dipole transitions are magnetically tuneable. Transitions of this type can readily be detected in NO_2 but it has not yet proved possible to extract any spectroscopic information[80]. Other examples are NF_2 [81] and the symmetric-top molecule, SO_3F [82]. On the other hand, the electric dipole transitions can only be tuned while the electron spin remains coupled to the nuclear framework; when it is decoupled, these transitions still have intensity but their frequencies are insensitive to the external field (i.e. the two levels involved in each transition move parallel to each other as the field is changed). The physical reason for this behaviour is easy to appreciate. The magnetic tunability derives from the electron spin magnetic moment, whilst the electric dipole intensity derives from the rotation in space of the electrons and nuclei that constitute the molecule. Once the electron spin becomes decoupled from the molecular framework, the electric dipole transition frequencies are no longer affected by the external field.

Thus, we see that electric dipole transitions in a bent triatomic molecule can only be tuned magnetically over a small frequency range (typically a few hundred MHz). If we are to detect such transitions with a conventional electron resonance spectrometer, they must be associated with large zero-field splittings, of the same order as the spectrometer microwave frequency. This means that one has to choose one's molecule and operating frequency rather carefully; even so, there are potentially many molecules which are suitable, particularly since one does not have to stop at three atoms. It should now be possible to appreciate the close connection between the sort of experiment described above and pure microwave spectroscopy. We are detecting a conventional zero-field microwave transition with a conventional electron resonance spectrometer by tuning out the small mismatch between the transition and spectrometer frequencies with an external magnetic field.

It is of course arguable that such transitions would be more sensibly detected in a zero-field microwave spectrometer. The fixed frequency and variable field experiment is certainly a rather unwieldy way of searching for new transitions and since in general the M-subcomponents of the transitions are resolved, it does not afford quite the same sensitivity, at least in principle. In practice, however, the resonant cavity fixed-frequency spectrometer appears to have slightly greater sensitivity for the detection of short-lived species. In addition, the transition frequencies in an external field have contributions which arise from second-order mixing of other spin–rotation and hyperfine states and hence contain information about the parameters defining the separation between these states. In all probability therefore, future development will involve both zero-field and variable-field experiments.

7.5.2 The microwave spectrum of the HCO molecule

A preliminary account of an experiment of the type outlined in the previous section, involving the detection of 2_{11}–2_{12} K-doubling transition in HCO, has recently been published[83]. Each component of the K-doublet is split into two by the spin–rotation interaction; there is also a further doublet splitting from the proton hyperfine interaction, leading to a pattern of eight levels (see Figure 7.6), each characterised by values of N (total angular momentum apart from electron spin), J (total angular momentum) and F (the resultant of the coupling of the proton spin and J). Each level is split into $2F+1$ components in a magnetic field and transitions between the two components of the K-doublet with selection rules $\Delta F = \Delta J = \Delta N = 0$, $\Delta M_F = \pm 1$ have been detected. These transitions are the strongest electric dipole transitions within this set of eight levels and they have been marked in Figure 7.6. A pair of typical experimental recordings is given in Figure 7.7.

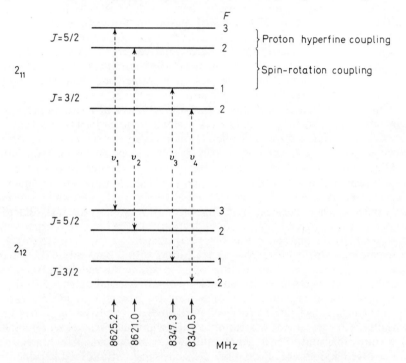

Figure 7.6 The set of eight energy levels involved in the microwave spectrum of HCO. The observed transitions and their extrapolated zero-field frequencies are given in the diagram.
(From Bowater *et al.*[83], reproduced by permission of the American Chemical Society)

For this particular molecule under the experimental conditions indicated, the second-order Zeeman splitting between transitions originating from different M_F states is too small to be resolved so that the spectrum consists of a closely spaced doublet for each of the two J-values. Because the magnetic field system can actually be made to sweep through zero field, reversing polarity in the process, the recorder trace shows each line occurring twice.

7.6   CONCLUDING REMARKS

The general message of this review is that the field of electron resonance of gas-phase molecules is in a fertile state and that, provided one has sufficient flexibility in the operating frequency of the spectrometer, there are very few polar open-shell molecules which cannot be studied by the technique — provided, of course, that one can generate a high enough concentration of molecules in the spectrometer cavity. This last condition represents in many ways the greatest barrier to progress. Simple atom–molecule reactions in

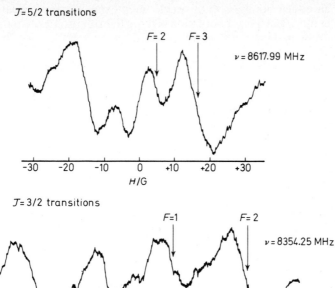

Figure 7.7 Two typical experimental recordings of the microwave spectrum of HCO. Note that the field scan passes through zero, changing polarity in the process

discharge-flow systems have proved to be much the most successful method of producing the molecule of interest. These reactions take place at a pressure of c. 1 Torr in the spectrometer cavity at room temperature. Unfortunately, relevant data from the study of reaction kinetics by other methods has long since run out and the selection of the best reaction is now very much a matter of intuition. Reactions designed around the formation of an H–F bond appear to have a higher chance of success. There are two other established methods of producing short-lived molecules for the purpose of spectroscopic study, namely elevated temperature sources and photolytic production. The former possibility has been investigated quite extensively in our laboratory, but with signal failure! The problem with this, and probably photolysis as well, is that in attempting to produce the molecules in the same spatial region as one is detecting them (i.e. in the spectrometer cavity), one is forced to compromise on design criteria for the cavity.

In discussing experimental methods, it must be mentioned that the Stark modulated cavity[84] has more than justified itself over the last few years. Not only does it give one discrimination against species which do not possess an electric dipole moment (notably, O_2 in molecular oxygen discharges) but it also provides a reliable method of measuring the molecular dipole moment. As an added bonus, a spectrometer employing Stark modulation seems to suffer much less from modulation pick-up than one using Zeeman modulation, resulting in better sensitivity. Ideally, of course, one has both types of modulation available.

A review of this field would be incomplete without at least a brief correlation of the results presented in earlier sections with those obtained by conventional zero-field microwave spectroscopy. The list of published results contains OH[85, 92], ClO[86], NS[87] and NCO[88] (which have all been detected in both spin components of the $^2\pi$ state), BrO[89] (detected in the $^2\pi_{\frac{3}{2}}$ state only) and SO (in both $^3\Sigma$ [90] and $^1\Delta$ [91] electronic states). The spectrometers used in these experiments employed long non-resonant absorption cells. There is some ground for believing that resonant cavity spectrometers[92] are better suited for the detection of short-lived species owing to their more favourable filling factor. Indeed, it is probably this factor which accounts for the sensitivity of the electron resonance technique as well. Carrington et al.[5] have shown that for saturable samples occupying a volume in space with dimensions of the order of a few centimetres, the use of a resonant cavity leads to an improvement in sensitivity of $Q^{\frac{1}{2}}$ (c. 10^2) over a non-resonant absorption cell operating at the same frequency. As far as the experimental results are concerned, the zero-field microwave studies are generally much more extensive than their electron resonance counterparts, since they are not restricted to states with sizeable magnetic moments. However, the two experiments tend to observe a given problem with a different perspective and it often happens that much greater definition can be achieved by combining the parameters from the two experiments; ClO [86] is a good case in point.

Acknowledgements

I gratefully acknowledge the help and constructive criticism of my colleagues in the preparation of this article, in particular Drs. A. Carrington, R. E. Moss and B. J. Howard. I would also like to thank Dr. Carrington for the preparation of Figures 7.5–7.7 and A. H. Curran, R. C. MacDonald, A. J. Stone and B. A. Thrush for providing me with a spectrum of NF and a value for its dipole moment before publication.

Note added in proof

Two recent papers on the electron resonance spectra of diatomic molecules in the second rotational levels ($J = 5/2$) of their $^2\pi$ states suggest slight modification of some of the parameters listed in Table 7.2. The first, by Uehara, Tanimoto and Morino[95], is a study of ^{35}ClO. Their analysis leads to values for A of $-319 \pm 5\,\mathrm{cm}^{-1}$ and for $e^2 q_0 Q$ of $-87.4 \pm 1.0\,\mathrm{MHz}$. In addition, they were able to separate the magnetic hyperfine parameters completely:

$$a = 136.35 \pm 0.4\,\mathrm{MHz}, \quad b = 19 \pm 12\,\mathrm{MHz}, \quad c = -66 \pm 12\,\mathrm{MHz}$$

In the second paper, Brown, Byfleet, Howard and Russell[96] report measurements of the electron resonance spectra of SeF, BrO and IO in their second rotational levels. These results support the suggestion that, for molecules with large spin–orbit coupling, the effective Hamiltonian should be derived to higher orders than second in perturbation theory, as discussed in Section 7.2.2 of this review. A re-analysis of the $J = 3/2$ and $5/2$ levels gives the following revised parameters:

Molecule	A/cm^{-1}	Δg_Ω	g_r
SeF	-1790	$(12.96 \pm 1.5) \times 10^{-4}$	$(1.5 \pm 1.0) \times 10^{-4}$
BrO	-980	$(2.55 \pm 1.5) \times 10^{-4}$	$(0.4 \pm 1.0) \times 10^{-4}$
IO	-2330	$(16.34 \pm 1.5) \times 10^{-4}$	$(-0.4 \pm 1.0) \times 10^{-4}$

The parameter Δg_Ω is a measure of the third order contribution to H_{eff}; it is defined in the paper of Brown et al.[96]

References

1. Gammon, R. H., Stern, R. C., Lesk, M. E., Wicke, B. G. and Klemperer, W. (1970). *J. Chem. Phys.*, **54**, 2136
2. Freund, R. S., Miller, T. A., De Santis, D. and Lurio, A. (1970). *J. Chem. Phys.*, **53**, 2290
3. Evenson, K. M., Radford, H. E. and Moran, M. M. (1971). *Appl. Phys. Lett.*, **18**, 426
4. Westenberg, A. A. and de Haas, N. (1969). *J. Chem. Phys.*, **50**, 707; and earlier references
5. Carrington, A., Levy, D. H. and Miller, T. A. (1970). *Advan. Chem. Phys.*, **18**, 149
6. Howard, B. J. and Moss, R. E. (1970). *Molec. Phys.*, **19**, 433
7. Howard, B. J. and Moss, R. E. (1971). *Molec. Phys.*, **20**, 147
8. Van Vleck, J. H. (1951). *Rev. Mod. Phys.*, **23**, 213
9. Condon, E. U. and Shortley, G. H. (1935). *The Theory of Atomic Spectra* (Cambridge: Cambridge University Press)
10. Van Vleck, J. H. (1936). *J. Chem. Phys.*, **4**, 327
11. Wilson, E. B. and Howard, J. B. (1936). *J. Chem. Phys.*, **4**, 260
12. Frosch, R. A. and Foley, H. M. (1952). *Phys. Rev.*, **88**, 1347
13. Curl, R. F. (1965). *Molec. Phys.*, **9**, 585
14. Dirac, P. A. M. (1958). *The Principles of Quantum Mechanics*, 4th Edn. (Oxford: Oxford University Press)
15. Breit, G. (1929). *Phys. Rev.*, **34**, 553
16. Moss, R. E. (1971). *Amer. J. Phys.*, **39**, 1169
17. Chraplyvy, Z. V. (1953). *Phys. Rev.*, **91**, 388; **92**, 1310
18. Barker, W. A. and Glover, F. N. (1955). *Phys. Rev.*, **99**, 317
19. Hougen, J. T. (1962). *J. Chem. Phys.*, **36**, 519
20. Watson, J. K. G. (1970). *Molec. Phys.*, **19**, 465
21. Born, M. and Oppenheimer, J. R. (1927). *Ann. Phys.*, **84**, 457
22. Miller, T. A. (1969). *Molec. Phys.*, **16**, 105
23. Bloch, C. (1958). *Nuclear Phys.*, **6**, 329
24. Dousmanis, G. C., Sanders, T. M. and Townes, C. H. (1955). *Phys. Rev.*, **100**, 1735
25. Amano, T., Saito, S., Morino, Y., Johnson, D. R. and Powell, F. X. (1969). *J. Molec. Spectrosc.*, **30**, 275
26. Tinkham, M. and Strandberg, M. W. P. (1955). *Phys. Rev.*, **97**, 937, 951
27. Carrington, A., Currie, G. N., Miller, T. A. and Levy, D. H. (1969). *J. Chem. Phys.*, **50**, 2726

28. Carrington, A., Currie, G. N., Levy, D. H. and Miller, T. A. (1969). *Molec. Phys.,* **17,** 535
29. Carrington, A., Fabris, A. R., Howard, B. J. and Lucas, N. J. D. (1971). *Molec. Phys.,* **20,** 961
30. Uehara, H. (1971). *Molec. Phys.,* **21,** 407
31. Howard, B. J. (1970). *Ph.D. Thesis* (University of Southampton)
32. Miller, T. A. (1971). *J. Chem. Phys.,* **54,** 3156
33. Kayama, K. and Baird, J. C. (1967). *J. Chem. Phys.,* **46,** 2604
34. Herzberg, G. (1950). *Spectra of Diatomic Molecules.* (Princeton, N. J.: Van Nostrand)
35. Radford, H. E. (1962). *Phys. Rev.,* **126,** 1035
36. Rose, M. E. (1957). *Elementary Theory of Angular Momentum.* (New York: John Wiley and Sons, Inc.)
37. Edmonds, A. R. (1957). *Angular Momentum in Quantum Mechanics.* (Princeton, N. J.: Princeton University Press)
38. Brink, D. M. and Satchler, G. R. (1968). *Angular Momentum.* (Oxford: Oxford University Press)
39. Freed, K. F. (1966). *J. Chem. Phys.,* **45,** 4214
40. Radford, H. E. (1961). *Phys. Rev.,* **122,** 114
41. Carrington, A. and Lucas, N. J. D. (1970). *Proc. Roy. Soc.,* **A314,** 567
42. Clough, P. N., Curran, A. H. and Thrush, B. A. (1971). *Proc. Roy. Soc.,* **A323,** 541
43. Churg, A. and Levy, D. H. (1970). *Astrophys. J.* **162,** L161
44. Lee, K. P., Tam, W. G., Larouche, R. and Woonton, G. A. (1971). *Can. J. Phys.,* **49,** 2207
45. Radford, H. E. and Linzer, M. (1963). *Phys. Rev. Lett.,* **10,** 443
46. Byfleet, C. R., Carrington, A. and Russell, D. K. (1971). *Molec. Phys.,* **20,** 271
47. Thistlethwaite, P. J. (1971). Private communication
48. Uehara, H. and Morino, Y. (1970). *J. Molec. Spectrosc.,* **36,** 158
49. Miller, T. A. (1971). *J. Chem. Phys.,* **54,** 1658
50. Carrington, A., Currie, G. N. and Lucas, N. J. D. (1970). *Proc. Roy. Soc.,* **A315,** 355
51. Radford, H. E. (1964). *J. Chem. Phys.,* **40,** 2732
52. Carrington, A., Dyer, P. N. and Levy, D. H. (1967). *J. Chem. Phys.,* **47,** 1756
53. Carrington, A., Dyer, P. N. and Levy, D. H. (1970). *J. Chem. Phys.,* **52,** 309
54. Brown, R. L. and Radford, H. E. (1966). *Phys. Rev.,* **147,** 6
55. Carrington, A., Howard, B. J., Levy, D. H. and Robertson, J. C. (1968). *Molec. Phys.,* **15,** 187
56. Uehara, H. and Morino, Y. (1969). *Molec. Phys.,* **17,** 239
57. Carrington, A. and Howard, B. J. (1970). *Molec. Phys.,* **18,** 225
58. Tinkham, M. and Strandberg, M. W. P. (1955). *Phys. Rev.,* **97,** 951
59. Carrington, A., Levy, D. H. and Miller, T. A. (1967). *Proc. Roy. Soc.,* **A298,** 340
60. Daniels, J. M. and Dorain, P. B. (1966). *J. Chem. Phys.,* **45,** 26
61. Uehara, H. (1969). *Bull. Chem. Soc. Jap.,* **42,** 886
62. Carrington, A., Levy, D. H. and Miller, T. A. (1967). *Molec. Phys.,* **13,** 401
63. Miller, T. A. (1971). *J. Chem. Phys.,* **54,** 330
64. Arrington, C. A., Falick, A. M. and Myers, R. J. (1971). *J. Chem. Phys.,* **55,** 909
65. Carrington, A., Levy, D. H. and Miller, T. A. (1966). *Proc. Roy. Soc.,* **A293,** 108
66. Curran, A. H., MacDonald, R. C., Stone, A. J. and Thrush, B. A. (1971). *Chem. Phys. Lett.,* **8,** 451
67. Cade, P. E. and Huo, W. M. (1967). *J. Chem. Phys.,* **47,** 614
68. Cade, P. E. and Huo, W. M. (1966). *J. Chem. Phys.,* **45,** 1063
69. Cade, P. E. and Huo, W. M. (1967). *J. Chem. Phys.,* **47,** 649
70. O'Hare, P. A. G. and Wahl, A. C. (1971). *J. Chem. Phys.,* **54,** 3770
71. O'Hare, P. A. G. and Wahl, A. C. (1970). *J. Chem. Phys.,* **53,** 2834
72. O'Hare, P. A. G. and Wahl, A. C. (1971). *J. Chem. Phys.,* **55,** 666
73. Renner, R. (1934). *Z. Physik,* **92,** 172
74. Carrington, A., Fabris, A. R., Howard, B. J. and Lucas, N. J. D. (1971). *Molec. Phys.,* **20,** 961
75. Carrington, A., Fabris, A. R., Howard, B. J. and Lucas, N. J. D. (1970). *Proceedings of the International Symposium on Electronic and Nuclear Magnetic Resonance,* 289 (Melbourne, Australia: Plenum Press)
76. Longuet-Higgins, H. C. (1961). *Advan. Spectrosc.,* **2,** 429
77. Hougen, J. T. (1962). *J. Chem. Phys.,* **36,** 519
78. Fabris, A. R. (1970). *Ph.D. Thesis* (University of Southampton)

79. Bolman, P. S. H. (1971). Private communication
80. Schaafsma, T. J. (1967). *Chem. Phys. Lett.,* **1,** 16
81. Piette, L. H., Johnson, F. A., Booman, K. A. and Colburn, C. B. (1961). *J. Chem. Phys.,* **35,** 1481
82. Stewart, R. A., Fujiwara, S. and Aubke, F. (1968). *J. Chem. Phys.,* **48,** 5524
83. Bowater, I. C., Brown, J. M. and Carrington, A. (1971). *J. Chem. Phys.,* **54,** 4957
84. Carrington, A., Levy, D. H. and Miller, T. A. (1967). *Rev. Sci. Instr.,* **38,** 1183
85. Dousmanis, G. C., Sanders, T. M. and Townes, C. H. (1955). *Phys. Rev.,* **100,** 1735; Radford, H. E. (1964). *Phys. Rev. Lett.,* **17,** 534
86. Amano, T., Saito, S., Hirota, E., Morino, Y., Johnson, D. R. and Powell, F. X. (1969). *J. Molec. Spectrosc.,* **30,** 275
87. Amano, T., Saito, S., Hirota, E. and Morino, Y. (1969). *J. Molec. Spectrosc.,* **32,** 97
88. Saito, S. and Amano, T. (1970). *J. Molec. Spectrosc.,* **34,** 383
89. Powell, F. X. and Johnson, D. R. (1969). *J. Chem. Phys.,* **50,** 4596
90. Powell, F. X. and Lide, D. R. (1964). *J. Chem. Phys.,* **41,** 1413; Winnewisser, M., Sastry, K. V. L. N., Cook, R. L. and Gordy, W. (1964). *J. Chem. Phys.,* **41,** 1687; Amano, T., Hirota, E. and Morino, Y. (1967). *J. Phys. Soc. Jap.,* **22,** 399
91. Saito, S. (1970). *J. Chem. Phys.,* **53,** 2544
92. Radford, H. E. (1968). *Rev. Sci. Instr.,* **39,** 1687
93. Curran, A. H., MacDonald, R. C., Stone, A. J. and Thrush, B. A. (1971). Private Communication
94. O'Hare, P. A. G. and Wahl, A. C. (1971). *J. Chem. Phys.,* **54,** 4563
95. Uehara, H., Tanimoto, M. and Morino, Y. (1971). *Molec. Phys.,* **22,** 799
96. Brown, J. M., Byfleet, C. R., Howard, B. J. and Russell, D. K. (1972). *Molec. Phys.* In press

8

Optical Detection of Magnetic Resonance in Molecular Triplet States

A. L. KWIRAM
University of Washington, Seattle

Abbreviations used in the text

e.s.r.	electron spin resonance
n.m.r.	nuclear magnetic resonance
n.q.r.	nuclear quadrupole resonance
ENDOR	electron–nuclear double resonance
ODMR	optical detection of magnetic resonance
ELDOR	electron double resonance
h.f.i.	hyperfine interaction
n.e.q.i.	nuclear electric quadrupole interaction
r.f.	radiofrequency
ISC	intersystem crossing
ZF	zerofield
SLR	spin–lattice relaxation

8.1 INTRODUCTION

8.1.1 Brief historical survey

In the 1940s magnetic resonance was detected in both the gas phase and the condensed phases[1, 2]. Most of the subsequent studies were devoted to the elucidation of ground-state magnetic properties. Very early, however, Bitter[3] proposed that the effect of radio frequencies (r.f.) on the excited state could probably be observed by monitoring the corresponding changes in the optical emission. That suggestion was quickly confirmed by Brossel and Bitter[4] using the $^3P_1$ state of the mercury atom (in the gas phase) as suggested by Brossel and Kastler[5]. Since that initial experiment a great deal of work has been done on the excited state of species in the gas phase using optical–r.f. double resonance techniques[6].

Within 6 years of the Brossel and Bitter experiment in the gas phase the analogous experiment was executed in the solid state. Geschwind, Collins and Schawlow[7] succeeded in optically detecting magnetic resonance in the ^2E excited state of Cr^{3+} in a single crystal of ruby at 1.6 K. This experiment marks the beginning of optical detection of magnetic resonance, ODMR, in the solid state.

Actually, a year earlier, Hutchison and Mangum had succeeded in measuring magnetic resonance parameters in the photo-excited triplet state of naphthalene in the solid state using *conventional* e.s.r. methods[8]. That definitive and important experiment has stimulated a great many magnetic resonance studies of photo-excited triplet states. In 1963, McConnell and Kwiram[9] tried to optically detect the e.s.r. transitions in the lowest excited triplet state of naphthalene but were unsuccessful. By 1967 such an experiment was carried out successfully and independently by Sharnoff[10] and by Kwiram[11].

This review concerns itself with ODMR experiments in the solid state. The primary focus is on the excited triplet state of organic molecules as distinct from the excited states of transition ions. The latter field is currently under review[12], and is therefore summarised only briefly here.

A number of other closely related areas are also excluded from this review. The high-field Zeeman spectroscopy experiments of Hochstrasser and co-workers[13], the pioneering efforts by Hutchison and co-workers on zero-field e.s.r. of photo-excited triplets[14], the interesting e.s.r. experiments on population inversion and on triplet excitons carried out by Wolf and co-workers[15, 16], and the many conventional optical spectroscopy experiments all have an important bearing on ODMR studies, but are beyond the scope of this review.

In focusing on the ODMR of photo-excited triplets, I shall try to provide an appreciation of the versatility of the method and the impact that ODMR studies have already had on our understanding of triplet-state properties and to emphasise thereby some of the new directions that may be promising.

A brief survey of some of the work on transition ions has been given by Geschwind[17, 18]. A short review[19] of triplet-state ODMR has appeared in Japanese and a selected survey of ODMR results has been given by El Sayed[20]. The appropriate background for the material covered in the following pages can be obtained from standard texts in magnetic resonance and optical spectroscopy. I would recommend in particular the very useful book *The Triplet State* by McGlynn, Azumi and Kinoshita[21].

8.1.2 Elementary model of the ODMR experiment

The essential features of a general ODMR experiment can be represented by the following simple model* (see Figure 8.1).

(a) An excited state consisting of magnetic states 1 and 2, at least one of which emits (optical) photons at a rate of k_i^r.

(b) The excited state is populated by pumping the transition between the ground and excited state at a rate of Λ.

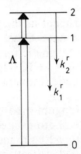

Figure 8.1 An elementary model for the ODMR experiment. Usually the pumping Λ occurs through a complex of intermediate states

For convenience we assume a Boltzmann distribution between levels 1 and 2. The observed emission intensity I is proportional to the number of photons emitted per unit time, n, and

$$n = \sum_{i=1}^{2} n_i = \sum k_i^r N_i = k_1^r N_1 + k_2^r N_2 \tag{8.1}$$

where N_i is the population of level i. If r.f. power is used to induce transitions between levels 1 and 2 the change in intensity δI is given by

$$\delta I \propto k_1^r \delta N_1 + k_2^r \delta N_2 = (k_1^r - k_2^r)\delta N$$

where we have set $\delta N = \delta N_1 = -\delta N_2$. Since δN will in general be non-zero we can immediately state several criteria sufficient to ensure that a change in n will occur. If at least one $k_i^r \neq 0$, then $\delta n \neq 0$ if either

$$k_1^r \neq k_2^r \qquad \text{(Criterion I)}$$
$$\text{or } M_{01} \neq M_{02} \quad \text{(Criterion II)}$$
$$\text{or } k_1^n \neq k_2^n \qquad \text{(Criterion III)}$$

M_{0i} designates the polarisation of the emission from level i to level 0, and k_i^n represents the non-radiative depopulating rates. If any one of the above criteria or inequalities obtains, then an ODMR effect can in principle be observed. In particular, if Criterion I obtains a change in the total emission intensity is observed. If only Criterion II holds, then a change in the intensity of one component can be observed (using a polariser). If only III obtains,

---

*For most of the discussion we ignore the details of the excitation process (see Ref. 21). We simply assume that the means for populating the magnetic levels exist.

then the steady-state emission levels may be affected. One might expect therefore, that such phenomena should be quite general.

The essential idea is that if the properties of the emitting states are different from each other in at least one respect, then one can affect energetic optical photons by inducing transitions with much less energetic r.f. photons. The fundamental question, of course, is one of sensitivity. Since the magnitude of the change in δn is proportional to δN, it is advantageous to maximise the initial difference in the populations, $N_1^o - N_2^o$, by cooling the sample to the liquid helium temperature range. All the experiments to be discussed have been carried out in this temperature regime.

The experimental arrangement is conceptually very much like that needed for conventional spectroscopic studies of the triplet state. The new requirement is that r.f. power be communicated to the sample. Standard techniques for enhancing sensitivity are typically employed[10, 11].

8.1.3 Transition ions

As already indicated, the first experiments using ODMR to study the excited states of ions in the solid state were performed by Geschwind and co-workers[7, 22-24], using single crystals of ruby. Ruby is crystalline Al_2O_3 containing small amounts ($<1\%$) of Cr^{3+}. The ground state of the Cr^{3+} is $^4A_2$. The lowest excited state is $\bar{E}(^2E)$ and lies just 29 cm^{-1} below the $2\bar{A}(^2E)$ state. By pumping the $^4A_2 \rightarrow {}^4T_{1,2}$ transitions one can populate the radiative ($\tau \approx 1$ ms) but metastable $\bar{E}$ state. In order to remove the Kramers degeneracy of the latter, a magnetic field is applied. The effect of r.f. transitions between the resulting Zeeman levels has been detected in three different ways.

The first approach depends on the fact that the light emitted by the upper and lower levels is left- and right-circularly polarised respectively when viewed along H_o. By viewing only one component (with an appropriate polariser), Geschwind et al. detected a change in its intensity whenever transitions between the two unequally populated Zeeman levels were traversed. Despite the fact that the optical line is 10 cm^{-1} wide, the ODMR line is only about 10^{-3} cm^{-1} wide. This implies that the factors which broaden the optical line have a negligible effect on the magnetic resonance parameters. The optical lines are in essence inhomogeneously broadened, and ODMR provides a similar kind of increase in resolution to that observed in ENDOR experiments where the e.s.r. line is inhomogeneously broadened.

The second method depends on the difference in the Boltzmann populations of the ground-state Zeeman levels. Of the total number of photons emitted from the excited Zeeman levels, more of those which terminate on the *upper* Zeeman level of the *ground* state will escape from the sample because more photons can be *re-absorbed* by ions in the more highly populated *lower* level before reaching the surface of the crystal. Since r.f. pumping changes the relative number of photons in the two radiative paths, a change in the level of total light intensity is detected.

The third alternative depends on the ability to resolve the two radiative components. Obviously, if the emission from a *single* level is monitored, any changes in the population of that level are directly translated into a change in the number of photons emitted ($n_i = k_i^r N_i$).

Each of the above methods has been used to study the transition ions. It is worth noting that the second method is not generally applicable to triplet-state studies since the ground state is a singlet. The first method is, however, generally applicable (see Criterion II in Section 8.1.2) since the emission from the triplet levels is linearly polarised and (usually) unique. On the other hand, it is a less essential factor for triplet states because, as we shall see later, Criterion I usually obtains ($k_i^r \neq k_j^r$) in contrast to the case for the $\bar{E}$ state of Cr^{3+} where $k_{\frac{1}{2}}^r = k_{-\frac{1}{2}}^r$. The third method is somewhat less applicable in the case of triplet states unless of course one uses high fields ($>20\,kG$) to split the emission lines.

Geschwind and co-workers have investigated the isoelectronic series V^{2+}, Cr^{3+} (using method 3 above)[22, 24] and Mn^{4+} (using method 1 above because of the wide optical lines)[23, 24]. The g values, hyperfine parameters and spin–lattice relaxation times in the $\bar{E}$ excited state of these ions were determined.

The relaxation time studies are particularly interesting. The presence of the $2\bar{A}$ level only $29\,cm^{-1}$ away (in $Cr^{3+}:Al_2O_3$) gives rise to an Orbach relaxation process in addition to the usual direct process at low temperatures. Furthermore, the cascade of $29\,cm^{-1}$ phonons cannot be transferred to the bath sufficiently rapidly and leads to a (bottle-necked) spike in the phonon spectrum of the lattice. Such an excess of phonons has since been observed directly in $Ni^{2+}:MgO$, in an elegant Brillouin scattering experiment[25]. Geschwind has noted a particularly important advantage of the ODMR technique over conventional saturation recovery methods for studying spin–lattice relaxation (SLR) processes: whereas spin diffusion is a serious problem in the latter case, it does not affect the optical measurements since the ODMR experiment monitors the *total* level population.

Another effect peculiar to non-singlet ground states is that of spin memory— the preservation of spin orientation throughout the pumping cycle. This is seen as a preferential population of the $S_z = -\frac{1}{2}$ level in the excited state due to the greater population of the $S_z = -\frac{3}{2}$ level in the ground state. This behaviour has been exploited in order to observe the optical detection of magnetic resonance in the *ground* state[26].

Finally, in this brief summary we mention the observations of Chase on coherence effects[27]. If we consider not only the diagonal matrix elements (populations) but also the off-diagonal matrix elements of the density matrix we find a coherent superposition of the Zeeman states which is equivalent to a precessing transverse magnetic moment. This coherence has the effect of modulating the intensity of the emitted light at the precession frequency and provides a means for observing electron-spin echos in the excited state.

8.1.4 The triplet state

The most visible manifestation of the triplet state is the long-lived emission that persists long after the excitation is terminated. Even though such observations date back to the late nineteenth century[25], the correct interpretation was not forthcoming until the work of Terenin[26] and of Lewis and

Kasha[27] in the mid 1940s. Even then, their conclusion that the long-lived emission originates in a triplet state was not universally accepted. Conclusive evidence for their interpretation could obviously have been provided by the then developing e.s.r. technique. Unfortunately, the simplest e.s.r. experiments were hampered by the sensitivity available with the e.s.r. spectrometers of that day. On the one hand, in pure single crystals the lifetime was too short due to triplet–triplet exciton annihilation. On the other hand, in a sample of randomly oriented molecules in a glass the signal was too weak due to the unusually large anisotropy of the dipole–dipole interaction.

The first successful e.s.r. experiment on a photo-excited triplet state was finally achieved by Hutchison and Mangum[8] in 1958. They overcame both the above problems by using oriented naphthalene molecules in a host crystal of durene. That definitive experiment has stimulated an impressive array of experiments on the magnetic resonance of triplet states[14–16, 30, 31].

The optical spectroscopists were not idle during this period and the record of their progress has been skilfully recorded. Nevertheless, in both camps there were a number of vexing problems. For the optical spectroscopists, the unresolved details of the emission from the triplet state remained. With a few notable exceptions[32, 33], most of the discussions prior to the early sixties did not contend with the unique radiative properties of the individual Zeeman levels. And measurements of the individual non-radiative rates seemed remote indeed.

Despite significant progress, the e.s.r. experiments of the early sixties were still severely limited by sensitivity considerations. Molecules with triplet-state lifetimes shorter than about 0.1 s were virtually inaccessible to e.s.r. methods. Consequently, the intimate secrets of entire classes of important molecules such as the carbonyls and the $(n\pi^*)$ azines remained hidden. Progress on the intriguing triplet exciton problem foundered for similar reasons. It was within this context that an effort was made[9] to transfer the unique advantages of the ENDOR technique to the optical domain. The sensitivity advantage seemed clearcut – if for every microwave photon absorbed one optical photon is emitted, then the enhancement in sensitivity could easily be as high as 10^3. Further, the e.s.r. experiment depends on the *population* in the triplet state which in turn is inversely proportional to the lifetime of the state. Optical detection, on the other hand, does not require extended residence in the state but only in principle a momentary visit – long enough to be affected by the applied r.f. power – and that could be short indeed.

The initial experiments[9] were performed at 2 K using a single crystal of durene containing 0.1 % naphthalene. In order to use phase-sensitive detection to enhance the signal-to-noise ratio it was essential to know the SLR times for the system so that the optimum modulation frequency could be used. Based on very limited information available at that time T_1 at 4 K was estimated to be about 1 ms. Consequently, a 300 Hz modulation frequency was used. No effect on the optical emission was detected.

In 1966, when the first systematic SLR studies on molecular crystals were carried out at low temperatures[34], the reason for the failure was clearly established. The T_1 measurements yield values as high as 100 s at 1 K for

organic free radicals[35]. Even though the relaxation mechanisms in triplet
states were expected to provide considerably faster relaxation, it was clear
that T_1 could be as long as a second[36]. The experimental conditions were
revised accordingly.

The first successful ODMR experiment in an organic triplet state was
performed by Sharnoff in 1967[10]. He reported the $\Delta m = \pm 2$ transitions for
naphthalene in a single crystal of biphenyl. Kwiram[11] reported the ODMR
for phenanthrene and these two independent experiments mark the starting
point for this Review.

8.2 MAGNETIC RESONANCE PARAMETERS

8.2.1 Theoretical summary

We begin by summarising briefly the essential magnetic resonance theory
for triplet states[21, 37].

8.2.1.1 The spin–spin Hamiltonian

The Hamiltonian representing the dipole–dipole interaction of two unpaired electrons is

$$H_{ss} = (g^2\beta^2/r^5)\left[r^2 S_1 \cdot S_2 - 3(S_1 \cdot r)(S_2 \cdot r)\right] \tag{8.2}$$

Expressed in terms of the total spin $S = S_1 + S_2$, equation (8.2) becomes

$$H_{ss} = (g^2\beta^2/2r^5)\left[(r^2-3x^2)S_x^2 + (r^2-3y^2)S_y^2 + (r^2-3z^2)S_z^2\right]$$

where we have chosen our axis system to coincide with the principal axes of
the dipole–dipole interaction thus eliminating the cross terms implicit in
equation (8.2).

The two-electron wave functions $^3\psi_\lambda^\sigma$ can be written as product functions
of orbital and spin variables r and σ respectively; $^3\psi_\lambda^\sigma = {}^3\phi \cdot \tau_\lambda$. The corresponding matrix elements of H_{ss} are

$$(H_{ss})_{\lambda\lambda'} = \langle\phi|H_{ss}(r)|\phi\rangle\langle\tau_\lambda|H_{ss}(\sigma)|\tau_{\lambda'}\rangle$$

which leads to an effective spin Hamiltonian

$$H_{ss} = S \cdot D \cdot S \tag{8.3}$$

where

$$D_{qq} = \langle 3q^2-r^2\rangle(g^2\beta^2/2r^5) \equiv Q$$

and $q = x, y, z; Q = X, Y, Z$. Note that $Tr D \equiv 0$. Using the above definitions
and relations we can also write H_{ss} in the form

$$H_{ss} = -XS_x^2 - YS_y^2 - ZS_z^2 \tag{8.4a}$$

or

$$H_{ss} = D(S_z^2 - \tfrac{1}{3}S^2) + E(S_x^2 - S_y^2) \tag{8.4b}$$

where

$$X = (D/3) - E$$
$$Y = (D/3) + E \qquad (8.5)$$
$$Z = -2D/3$$

This is the main contribution to the zero-field splitting. Spin–orbit interactions provide an additional splitting which though small ($<0.1\%$ for $\pi\pi^*$ states and $<10\%$ for $n\pi^*$ states) is not entirely negligible. Since it is virtually impossible to separate the two contributions in the normal experiment we will defer the discussion on spin–orbit splittings to the Section on excitons.

A convention that is frequently followed is to choose $D \geqslant 3E$. For many planar aromatic $\pi\pi^*$ states this usually means that the z axis is perpendicular to the molecular plane. Thus, this magnetic resonance convention conflicts with the molecular spectroscopy convention[38] which specifies that the x axis be chosen perpendicular to the plane (for C_{2v} and D_{2h}). We will adhere to the latter convention*.

If only the spin-independent Hamiltonian is considered, then the four spin functions are degenerate. Hence, we choose those linear combinations which transform according to the irreducible representations of the point group (in C_{2v} or higher). The (antisymmetric) singlet spin function is

$$\sigma_o = (\alpha\beta - \beta\alpha)/\sqrt{2}$$

The triplet spin functions are generated by the equations[39]

$$S_{1q}\sigma_o(1, 2) = \tfrac{1}{2}\tau_q(1, 2) \qquad (8.6)$$

The spin operators S_{1q} transform as rotations R_q. Since σ_o is totally symmetric, τ_q must transform like R_q. Thus the orthogonal spin functions τ_q for $S = 1$ are

$$\tau_x = (\beta\beta - \alpha\alpha)/\sqrt{2} \equiv [|\bar{1}\rangle - |1\rangle]/\sqrt{2}$$
$$\tau_y = i(\beta\beta + \alpha\alpha)/\sqrt{2} \equiv i[|\bar{1}\rangle + |1\rangle]/\sqrt{2} \qquad (8.7)$$
$$\tau_z = (\alpha\beta + \beta\alpha)/\sqrt{2} \equiv |0\rangle$$

The spin operators acting on these functions give

$$S_\lambda\tau_\mu = i\varepsilon_{\lambda\mu\nu}\tau_\nu \qquad (8.8)$$

where

$$\varepsilon_{\lambda\mu\nu} = \begin{cases} 0 \text{ if any two of } \mu, \lambda, \nu \text{ are identical} \\ 1 \text{ if } \lambda, \mu, \nu \text{ are cyclic} \\ -1 \text{ if } \lambda, \mu, \nu \text{ are non-cyclic} \end{cases}$$

$$S_\lambda^2\tau_\mu = (1 - \delta_{\mu\lambda})\tau_\mu \qquad (8.9)$$

---

*The z axis is chosen as the in-plane axis of highest symmetry. In the D_{2h} symmetry group where some ambiguity exists, z is taken to be the long axis. In pyrazine and pyrimidine, the z axis lies along and normal to (respectively) the direction of the N—N direction.

where $\delta_{\mu\lambda}$ is the Kronecker delta. Hence $S^2\tau_\lambda = 2\tau_\lambda$ as it must for a triplet state $(S = 1)$, and

$$H_{ss}\tau_q = Q\tau_q \qquad (8.10)$$

This implies that even in ZF, H_{ss} can give rise to a splitting of the Zeeman levels provided the symmetry is lower than spherical. This is the so-called zero-field splitting (ZFS).

Transitions moments *within* the triplet state are calculated using the (electron) magnetic dipole moment operator $\boldsymbol{\mu}_e \cdot \boldsymbol{h}_1(t)$ where h_{1q} represents an oscillatory field h_1 polarised in the q direction. Hence by equation (8.8) the $\tau_\mu \to \tau_\nu$ transition is polarised perpendicular to the $\mu\nu$ plane. In other words an oscillatory field polarised in the x direction can induce only the $\tau_y \leftrightarrow \tau_z$ transition. In essence, the *plane* of spin polarisation is rotated about the $\boldsymbol{h}_1$ axis (which must have a component in the plane to be effective).

In the presence of a strong static magnetic field, H_o, the Zeeman levels are mixed and change in energy. If $g\beta H_o \gg D$, one has the well-known high-field Zeeman states $|\bar{1}\rangle, |0\rangle, |1\rangle$. If the field is applied along one of the principal axis directions of $\boldsymbol{D}$, the energy of that level remains unchanged while the other two levels are mixed and diverge under the influence of the field. If $H_o \| z$, then τ_z remains pure and τ_x and τ_y mix. If we designate the resulting mixed states by $\tau_\pm = (\tau_x \pm i\tau_y)/\sqrt{2}$ in high field, then the transitions $\tau_z \to \tau_\pm$ will be allowed only if h_{1y} or h_{1x} are non-zero — i.e. only if h_1 is perpendicular to H_o for these $\Delta m = \pm 1$ transitions. Similarly the so-called $\Delta m = \pm 2$ transitions $\tau_+ \leftrightarrow \tau_-$ require h_{1z} — i.e. h_1 parallel to H_o. If H_o is in an arbitrary direction all three transitions are excited.

8.2.1.2 The hyperfine Hamiltonian

If we assume that the h.f.i. is diagonal in the xyz coordinate system we can write

$$H_{hf} = \boldsymbol{S} \cdot \boldsymbol{A} \cdot \boldsymbol{I} = A_{xx}S_x I_x + A_{yy}S_y I_y + A_{zz}S_z I_z \qquad (8.11)$$

Since $\langle S_q \rangle \equiv 0$ in zero field, the contribution of H_{hf} is zero to first order. The second-order contribution has the form

$$-\Delta E_{Qq}^{(2)} = \sum_s \sum_v (E_v - E_Q)^{-1} |\langle \tau_v t_s | H_{hf} | \tau_Q t_q \rangle|^2 \qquad (8.12)$$

where τ_Q and t_q represent the electron spin and nuclear spin functions respectively. As a result of this second-order mixing, the zero-order wave functions acquire first-order correction terms. These corrections are relatively unimportant for calculating allowed e.s.r. transition moments but are very important in calculating forbidden e.s.r. transition moments, as will become apparent.

The transition moment for nuclear transitions *within* a Zeeman level is proportional to $\boldsymbol{h}_2 \cdot (\gamma_e \boldsymbol{S} - \gamma_n \boldsymbol{I})$ where h_{2q} represents an oscillatory magnetic field polarised in the q direction[2]. In general, one can write the transition probability for a 'nuclear' transition between levels i and j, for example, as

$$P_{ij} \propto 4\chi^2 \gamma_e^2 + 4\chi\gamma_e\gamma_n + \gamma_n^2 \qquad (8.13)$$

where γ_e, γ_n are the gyromagnetic ratios for electron and nucleus respectively ($\gamma_n/\gamma_e = 10^{-3}$). Consequently even a small mixing coefficient χ will suffice to make the electron spin transition moment operator completely dominate the nuclear transition moment operator. This effect is important in the calculation of ENDOR transition probabilities.

The effect of the H_{hf} in the presence of a strong magnetic field has been discussed elsewhere[37]. At intermediate fields the hyperfine *splitting* is reduced by the factor $\cos 2\theta$ where $\cotan 2\theta = 2g\beta H_o/(X-Y)$, for example. If splittings are recorded at fixed frequency by sweeping the field, then the *measured* splittings are equal to the level separation because $dE/dH_o = g\beta \cos 2\theta$.

8.2.1.3   The nuclear electric quadrupole Hamiltonian

If the nuclear spin $I \geqslant 1$, then the n.e.q. term in the Hamiltonian will be non-zero. If we set

$$Q' \equiv \frac{e^2qQ}{4I(2I-1)} \tag{8.14}$$

we can define

$$Q_{xx} = Q'(1+\eta)$$
$$Q_{yy} = Q'(1-\eta) \tag{8.15}$$
$$Q_{zz} = -2Q'$$

The conventional n.e.q. Hamiltonian

$$H'_Q = Q'[(3I_z^2-I^2)+\eta(I_x^2-I_y^2)] \tag{8.16a}$$

can, therefore, be written as

$$H_Q = -Q_{xx}I_x^2-Q_{yy}I_y^2-Q_{zz}I_z^2 \tag{8.16b}$$

which has exactly the same form as H_{ss} of equation (8.4a). Consequently, the entire formalism of Section 8.2.1 can be carried over directly.

8.2.2   Experimental results

8.2.2.1   High-field ODMR

(a) *Single crystal studies* — The earliest ODMR experiments on the triplet state were carried out in a magnetic field of about 3 kG using conventional e.s.r. spectrometers[10, 11, 40]. The essential ideas can best be illustrated by an example. Figure 8.2 gives the energy levels for phenanthrene in the presence of a field aligned in turn along each of the principal axes of D. The labelling of the levels is consistent with an orbital symmetry of B_2.

Consider $H_o \| z$. We assume that $W_{ij} > k_i$ such that a Boltzmann distribution is maintained at all times. Given that the τ_z level is the principal emitter, we expect that inducing the low field $\Delta m = \pm 1$ transition (thus transferring some net number of molecules *from* τ_z to τ_y) will cause a *decrease* in the emission intensity — the high-field transition will increase. This agrees with the observed spectra, as do the results for $H_o \| x$ and $H_o \| y$. If A_1 had been

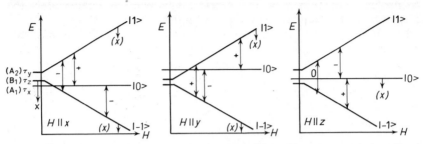

Figure 8.2 Triplet state energies versus H for phenanthrene. The state symmetries reflect a choice of B_2 for the orbital symmetry

used instead of B_2 there would be no correlation between predicted and observed results.

The $\Delta m = \pm 2$ transition is not expected to give rise to an intensity change for $H_o \| z$ unless τ_x has a non-negligible radiative rate constant. (Since τ_y is not coupled to the vibrationless ground state it carries no intensity — see Section 8.3.2.1.) For $H_o \| x$ or y the $\Delta m = 2$ transitions are observed with the indicated sign. Only the *sign* of the intensity change has been indicated above. The relative magnitude can be calculated using equation (8.30) in Section 8.3.3. The nine pieces of information that result are more than adequate and lead to the unambiguous conclusion that the orbital symmetry of the triplet state is indeed B_2. This confirms the original assignment made by Azumi and McGlynn[41]. The orbital symmetry for naphthalene[11] and quinoxaline[11, 40] is also B_2*†.

The $\pi\pi^*$ systems discussed above had already been investigated by conventional e.s.r. methods. However, apart from the efforts of Bowers and co-workers[43], no $n\pi^*$ systems had been observed by e.s.r. largely because of the short (~ 1 ms) lifetimes of $n\pi^*$ systems. ODMR seemed well suited to such problems[11], and the first investigation of this kind was conducted by Sharnoff[44]. He studied pyrazine in *p*-dichlorobenzene using a *K*-band (24 GHz) spectrometer‡. The observation of strong signals for pyrazine ($\tau \approx 1$ ms) illustrates the distinct sensitivity advantage of ODMR over e.s.r. for such studies. Sharnoff obtained values of 0.309 cm^{-1} for D and 0.006 cm^{-1} for E at 1.3 K. This large D value is a consequence of the one-centre interactions that occur at the nitrogen positions.

In a subsequent experiment[45], Sharnoff found that a pure durene crystal, after irradiation with u.v. light for a period of time, exhibits a green phosphor-

---

*These details, though available, were not included in the article on phenanthrene[11] because of contradictory data in the literature. (For a discussion of this problem see Ref. 42.) Although the problem has not yet been entirely resolved, it is clear that the original model for the interpretation of ODMR results[11] is generally valid.

†For the axis system defined in the footnote on p. 279.

‡The energy quantum in an *X*-band spectrometer is typically *less* than the ZFS. To convert cm^{-1} to Hz simply multiply the value in cm^{-1} by the speed of light.

escence. He concluded that this is due to the $n\pi^*$ triplet state of aldehyde molecules generated by the irradiation process. The measured ZF splitting parameters are $D = 0.335_3$ and $E = -0.043_2$ cm^{-1} at 1.8 K.

(b) *Randomly oriented molecules* — Forman and Kwiram[46] measured the first ODMR spectrum of a sample of randomly oriented molecules. This experiment is analogous to the conventional e.s.r. studies of triplet-state molecules in glasses first carried out by van der Waals and de Groot[31]. Forman and Kwiram investigated primarily the $\Delta m = \pm 2$ transitions of naphthalene, quinoxaline and phenanthrene in EPA at 4.2 K. From Figure 8.2 it is seen that the $\Delta m = \pm 2$ transitions for phenanthrene occur in the order $H_0 \parallel z$, $H_0 \parallel x$, $H_0 \parallel z$ as H_0 increases. The anticipated intensity change for $H_0 \parallel z$ is very small. However, for $H_0 \parallel x$ we expect a decrease and for $H_0 \parallel y$ an increase. Hence the symmetry of the state can be determined from the $\Delta m = \pm 2$ transition alone in this case. This qualitative argument has been placed on a much more rigorous footing[46, 47] by performing computer simulations of the ODMR spectra utilising information on the radiative rate constants, polarisation, e.s.r. transition probability, SLR rates, the value of H_0, and so on. Not only D and E, but the absolute signs are obtained as well from the $\Delta m = \pm 2$ transitions. More recently, Sharnoff has reported the $\Delta m = \pm 1$ transitions for cyclopentanone and pyrimidine[48]. In particular, Chan and Sharnoff[49] were able to show on the basis of high-field studies of pyrimidine in a glass that the earlier interpretation of Kuan, Tinti and El Sayed[50] regarding the principal emitting level of pyrimidine was inadequate. Chan and Sharnoff found that τ_z is actually the principal emitter. This has also been confirmed subsequently by Burland and Schmidt[51].

Whether one investigates a single crystal or a sample of randomly oriented molecules, one requires an additional piece of information in order to correlate the ZFS parameters with the molecular axes. If the crystal structure of the host is known and *if* the relative orientation of the guest can be inferred with reasonable confidence, then the axes can be assigned in the usual manner[37]. If the orientation of the guest cannot be established *a priori*, then one can use polarised exciting light (photoselection)*. If the polarisation of the $S_0 - S_1$ transition is known, then the orientation of the guest can be deduced. This method was used recently by Clements and Sharnoff[52] to show that the $N - N$ direction in pyrazine corresponds to the largest principal element of D. Cheng and Kwiram had inferred this earlier on the basis of the h.f. splittings measured at low field[53]. The use of h.f. data removes the need to know the polarisation of the singlet–singlet transition. However, if randomly oriented molecules are used, then photoselection may be the best alternative.

8.2.2.2   *Zero-field ODMR*

(a) *Randomly oriented molecules* — It was clear from the time of the earliest ODMR experiments that the utility of the ODMR method could in many instances be greatly enhanced by elimination of the magnetic field. There are a number of reasons for this.

---

*For a description of this method consult Ref. 21 under 'magneto-photoselection'.

(a) Whereas for conventional e.s.r. experiments sensitivity considerations dictate that magnetic fields of several kilogauss be used (to increase both the Boltzmann factor and the size of the detected microwave quantum), such fields have a negligible effect on the energy of the visible photons.

(b) The absence of an external magnetic field leads to substantially longer SLR times. This fact combined with the nature of the populating and de-populating mechanisms for the triplet state leads to significant spin align-ment especially at lower temperatures. Consequently, applying a magnetic field may in some cases actually reduce the sensitivity of the ODMR experiment.

(c) Since there is no unique direction in the sample when $H = 0$, one can dispense with the troublesome (mixed) single-crystal requirement. Moreover, for randomly oriented molecules in ZF the total spectral breadth will be on the order of, or $<$, 10 MHz (since H_Q and H_{hf} are not expected to contribute to the line-width in first order). This is in contrast to the situation in high field where the total spectral range may be >1000 MHz due to the ZFS. Therefore, the ODMR sensitivity does not suffer when randomly oriented molecules are studied in ZF, and the lines are expected to be quite narrow.

(d) Finally, in ZF one deals with the pure triplet spin states which trans-form according to the irreducible representations of the symmetry group of the molecule. Consequently, both optical and kinetic results can be more directly interpreted.

Schmidt and van der Waals[54] reported the first ODMR experiments in zero field. They observed the effect of microwave transitions between the ZF levels of quinoxaline. The sensitivity was excellent even at low r.f. power levels. At somewhat higher power levels, additional lines began to appear. The individual components were as narrow as 1 MHz. On the other hand, the first observations made about the same time of an $n\pi^*$ triplet state (pyrazine) using ZF ODMR revealed asymmetric lines about 10 MHz wide[53]. The origin of the splitting, or the asymmetry, of the lines lies in the quadrupole and second-order h.f. interactions. This matter was first treated semi-quanti-tatively by Schmidt and van der Waals for quinoxaline[55], and by Cheng and Kwiram for pyrazine[53]. Similar observations had been made by Harris et al.[56] and discussed in connection with ZF ENDOR measurements.

We illustrate the nature of the argument by considering pyrazine in some detail. The ZF parameters are* $X = 3.60_5$, $Y = 3.17_5$, $Z = -6.78_0$ GHz. For simplicity we consider an 'idealised' pyrazine molecule in which one of the nitrogen atoms is spinless (i.e. ^{16}N with $I = 0$) and thus will be ignored (as will the protons) in the discussion that follows. The zero-order wave functions for electron and nuclear spin are the nine product functions $|Xx\rangle, |Xy\rangle, |Xz\rangle, |Yx\rangle$, etc. where X represents the electron spin function τ_x and $|x\rangle$ by analogy represents the nuclear spin function t_x for the $(I = 1)^{14}N$ nucleus. The hyperfine interaction mixes these in second order. We have,

---

*These values are for pyrazine in a single crystal of p-dichlorobenzene. They differ somewhat from the values reported by Sharnoff[44] as well as those obtained by Harris (private communica-tion). Cheng and Kwiram[57] find that a variety of values can be obtained depending on how the host crystal is aged and cooled. The values listed above are for a melt-grown crystal aged for more than a month at room temperature and cooled slowly to 4.2 K.

using equation (8.12), that

$$-\Delta E_{Xx}^{(2)} = \sum_s \sum_v (E_v - E_X)^{-1} |\langle \tau_v t_s | H_{hf} | \tau_x t_x \rangle|^2 \tag{8.17a}$$

Since $(X - Z) \ll (X - Y)$ or $(Z - Y)$ the only large term in the sum will be for $v = Z$. But by (8) only S_y can mix τ_z and τ_x, hence the only term in H_{hf} that contributes is $A_{yy}S_y I_y$.

$$+\Delta E_{Xx}^{(2)} = E_{ZX}^{-1} |\langle Zz | A_{yy}S_y I_y | Xx \rangle|^2 = A_{yy}^2/(X - Z) \tag{8.17b}$$

To evaluate $\Delta E_{Xx}^{(2)}$ we need a value for A_{yy}. However, since hyperfine coupling constants for $n\pi^*$ triplet states were unknown, Cheng and Kwiram[53] studied the ODMR transitions in pyrazine at *low* field (< 500 G). The hyperfine interactions determined in this way led to a value for $A_{yy} \approx 30$ MHz. Thus $\Delta E_{Xx}^{(2)} = (30)^2/430 = 1.8$ MHz. If the second-order shifts for the remaining levels in τ_x and τ_z are calculated one finds

$$\alpha \equiv \Delta E_{Xx}^{(2)} = \Delta E_{Xz}^{(2)} = A_{yy}^2/(X - Z) = -\Delta E_{Zx}^{(2)} = -\Delta E_{Zz}^{(2)} \tag{8.17c}$$

and

$$\Delta E_{Xy}^{(2)} = 0 = \Delta E_{Zy}^{(2)} \tag{8.17d}$$

The resulting energy levels are shown in Figure 8.3.

The e.s.r. transition moments μ_{ij} are given by

$$\mu_{ij} \propto \langle i | \mu_e \cdot H(t) | j \rangle \propto \langle i | h_1 \cdot S | j \rangle \tag{8.18}$$

where i, j represent any of the nine states. Since (non-zero) spin matrix elements are of order unity, the allowed e.s.r. transitions (for favourable polarisation of h_1) are simply those for which no nuclear spin-flip occurs. If the molecules are randomly oriented, then each transition such as $Yx \to Xx$ (nuclear spin state unchanged) is equally probable. Hence it is clear from Figure 8.3 that two of the three allowed transitions between τ_y and τ_x will have equal energy and only two lines appear in the $(\tau_y \leftrightarrow \tau_x)$ e.s.r. spectrum. Consequently no quadrupole information is obtained. This is generally the case — unless quadrupole interactions are quite large, the e.s.r. spectrum is not affected thereby and thus yields no quadrupole parameters in first order.

The h.f. interaction, however, shifts the levels (the high (low) frequency ODMR line is broadened to the high (low) frequency side) and concomitantly modifies the wave functions. The first-order wave functions are listed in Figure 8.3. If μ_{ij} is evaluated for these corrected wave functions a number of additional 'forbidden' transitions are found. For example $\mu_{5,9}$ is non-zero because of the H_{hf} contribution. Actually $|\mu_{5,9}|^2 \propto \chi^2$ and corresponds to a transition in which roughly speaking an electron and a nucleus undergo simultaneous spin-flips. The forbidden transitions are lower in intensity by an order of magnitude or more depending on the mixing due to H_{hf}. Nevertheless, insofar as these transitions broaden (or split) the ZF ODMR lines, they provide a measure of the n.e.q.i. and the h.f.i. For example, the splitting between lines 1 and 3 provides a measure of α and thus A_{yy}. Similarly the splitting between 2 and 4 is just $2(x - z) = 6Q'(1 + \eta)$.

Thus in principle, an analysis of the ZF ODMR structure can lead to values for the n.e.q. and h.f. parameters. Unfortunately, the ubiquitous problem of correlating lines with transitions remains, and, unless the spectrum is very simple, or independent information on the h.f.i. is available, it may not be possible to interpret the spectrum unambiguously. For example, pyrazine with two equivalent ^{14}N nuclei gives rise to a ZF ODMR spectrum that is not readily interpreted. Nevertheless, using h.f. and n.e.q. parameters, obtained from low-field ODMR and ENDOR studies, the ZF spectrum has been simulated quantitatively[57].

For the quinoxaline molecule, Schmidt and van der Waals used the hyper-

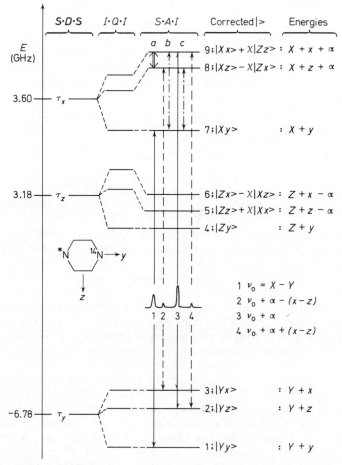

Figure 8.3 The ZF states of pyrazine including n.e.q.i. and h.f.i. (for one N nucleus). The corrected wave functions and the corresponding energies are shown on the right. Some of the allowed e.s.r. (——), forbidden e.s.r. (---) and ENDOR (— — —) transitions are also depicted

fine data obtained by Vincent and Maki[58] from a high-field e.s.r. experiment. The quadrupole parameters were estimated from n.q.r. measurements by Schempp and Bray[59] on pyridine and pyrazine. Using these data, Schmidt and van der Waals were able to obtain an approximate simulation of the ZF spectrum. Detailed values of the parameters were not extracted from the

ODMR spectrum itself for several reasons. First, the h.f. parameters are not really known accurately for quinoxaline, and, therefore, it is not possible to accurately determine other parameters which are based on these data. Second, only one h.f. element makes a significant contribution to the state mixing and thus only the product $Q'(1+\eta)$ can be extracted. Third, it is not possible to rigorously determine the principal axis directions for the various tensors from ZF studies alone. Thus the description is only qualitative or at best semi-quantitative.

A more extensive assignment of ZF ODMR lines has been made by Buckley and Harris[60]. For example, they have identified roughly a dozen transitions in both ZF multiplets of the 8-chloroquinoline molecule. These correspond variously to (a) pure electron spin transitions, (b) simultaneous electron and N, (c) simultaneous electron and Cl and (d) simultaneous electron, N and Cl transitions. Both ^{35}Cl and ^{37}Cl transitions are identified. ENDOR measurements are used to confirm (and aid in) the assignments. Nevertheless, the parametrisation is again only semi-quantitative.

For molecules containing only nuclei with $I < 1$ and having negligible h.f.i., one would anticipate very narrow ZF ODMR lines (< 0.5 MHz or 10^{-5} cm^{-1})—broadened mainly by factors such as disorder (multiple sites) or crystal field effects on the ZFS. Even somewhat wider lines still provide excellent resolution, and this fact has been used to advantage by Owens et al.[61] to study 8-chloroquinoline in durene. They found that emission occurs from two almost identical sites A and B. These sites give rise to a small splitting of the vibronic bands. For the A sites they determine $D_A = 0.09849$ cm^{-1} and $E_A = 0.01417$ cm^{-1}. For the B sites, $D_B = 0.09857$ cm^{-1} and $E_B = 0.01434$ cm^{-1}. The ODMR lines for the former sites tend to be more intense and the optical emission exhibits anomalous polarisation. They argue that this is probably due to a slight rotation of the molecule in the A site and suggest that this may explain the anomalous results observed for naphthalene and quinoxaline in durene*.

The ZF experiments can clearly provide very precise information regarding the energies of the various transitions. As we have just seen, given line-widths of a MHz or so for transitions in the 1–10 GHz range one can determine the ZF parameters to four or five significant figures. Actually the significance of these figures must be viewed with care. In Section 8.2.1.1 it was pointed out that the ZF splitting consists not only of the H_{ss} contribution but also the H_{so} contribution. In addition, the ZF parameters are sensitive to environment as well as to temperature. These effects can be as large as 0.1% or more. For these reasons, the experimental conditions should be clearly stated whenever precise data are reported.

The strong dependence of the ZF splitting on the environment may be of considerable value in studying solvent structure and the effect of temperature and pressure. For example, Yamanashi and Kwiram have detected ODMR signals from the tryptophan moieties in lysozyme. It appears that several sites can be distinguished[62].

(b) Single crystals—If single crystals are studied in ZF then one can determine the r.f. transition moments in addition to the r.f. transition fre-

---

*See footnote (*) page 282.

quency. For example, if a non-zero transition moment is observed for the $\tau_i \leftrightarrow \tau_j$ transition when the oscillatory magnetic field, h_1, is polarised in the molecular z direction (h_{1z}), then by the arguments presented in Section 8.2.1.1 we conclude that i and j correspond to x and y. Similar measurements in the other canonical directions may lead to an unambiguous assignment of the axes.

Such a procedure was used by Chen and El Sayed[63] in a somewhat more general form to assign the magnetic axes in tetrachlorobenzene. Such an analysis clearly requires a knowledge of the direction of polarisation of the field, h_1. Kalman and El Sayed have discussed the field distribution in a helix (the most commonly used r.f. structure for ZF studies)[64]. They used a single crystal of durene doped with 2,3-dichloroquinoxaline in order to justify their conclusions regarding the field distribution[65].

If a single crystal is not available one may still be able to extract the desired information by using the photoselection method. Even if a single crystal is used, it may be that only two (or one) of the three transitions are observed in the ODMR spectrum and these data may be insufficient for a unique assignment of the axes. In that event it may be possible to break the impasse by employing multiple resonance techniques. As a last resort, one can apply a magnetic field and make the assignment in the usual manner.

(c) *Transferred h.f.i.* – An interesting effect has been reported by Fayer, Harris and Yuen[66]. They observed satellites on the ODMR lines of the *guest* molecule and concluded that there were due to Cl nuclei located on the neighbouring (ground state) diamagnetic *host* molecules. This *inter-molecular* contribution to the structure of ZF ODMR lines is analogous to the *intramolecular* contributions discussed under (a) above. The inter-molecular case requires a finite h.f.i. between the unpaired electrons of the (excited) guest molecule and the Cl nuclei of the host molecule. From the positions of the satellite lines they determine the n.e.q.i. for the Cl nuclei in tetrachlorobenzene and dichlorobenzene. This effect is not equivalent to the distant ENDOR effect discussed by McCalley and Kwiram[67]. In the latter case there is no (transferred) h.f.i. whatsoever with the distant nuclei.

(d) *Orbitally degenerate triplet states* – Yamanashi, Kwiram and Gouterman[68] detected the ZF ODMR transitions in the ^3E state of Zn etioporphyrin. This is an orbitally degenerate state. Consequently the usual treatment for non-degenerate triplet states is not applicable. The magnetic resonance parameters for the ^3E state have been derived by Gouterman, Yamanashi and Kwiram[69]. Despite the fourfold symmetry of the porphyrin structure, at least three parameters, D, E, and E' appear. The effect of spin–orbit and crystal-field interactions on the ODMR spectrum and line-widths is discussed. Such an analysis may also account for the unusual character of the triplet state e.s.r. spectrum of the perinaphthene negative ion[70].

8.2.2.3   ODMR of triplet excitons

The spin Hamiltonian for triplet excitons in a typical (two-site) organic crystal is given by

$$2H_{ex} = (H_{ss})_A + (H_{ss})_B \tag{8.19a}$$

where H_{ss} is given by equation (8.4b). Equation (8.19a) can be re-expressed in the form

$$2H_{ex} = D^+ S_{z'}^2 + E^+(S_{x'}^2 - S_{y'}^2)$$

where D^+ and E^+ represent averaged ZF parameters for the A and B sites of the exciton system. The conventional e.s.r. spectrum of photo-excited triplet excitons has been observed by Schwoerer, and Wolf[16]. The spectra are characterised by narrow lines exhibiting no hyperfine (or quadrupole) structure, and by ZF parameters which are related to the crystallographic axes a,b,c rather than the molecular axes. Furthermore, the SLR times tend to be very short compared to those for the isolated molecules.

Sharnoff[71, 72] was the first to report ODMR signals due to triplet excitons (in benzophenone ($H \neq 0$)). In fact, two kinds of ODMR signals were observed – relatively broad lines characterised by long SLR times, and narrow Lorentzian lines characterised by (anisotropic) SLR times of about 10^{-7} s at 4.2 K. The former were assigned to excitations on isolated benzophenone traps, and the latter to triplet excitons. The measured fine-structure parameters for the excitons are $A = -0.0273$, $B = 0.0873$ and $C = -0.0605$ cm^{-1} compatible with $D = 0.152$ and $E = -0.021$ cm^{-1} for the isolated molecule. Emission is primarily from τ_b (75%) with a 25% contribution from τ_a. The behaviour of the exciton signals leads to the tentative conclusion that the exciton motion is coherent with anisotropic times characterising the motional narrowing processes[72].

Francis and Harris[73, 74] have carried out some very interesting ZF ODMR experiments on the excitons in 1,2,4,5-tetrachlorobenzene (TCB). The crystal structure of TCB is C_i which provides for essentially an ideal one-dimensional exciton system. For such a system, one can describe the triplet state energies of the exciton bands in the same manner as for isolated molecules except that the energies are 'modulated' by a term $\Delta_T \cos ka$ where $2\Delta_T$ is the triplet exciton band width, k is the wave vector and a is a characteristic length. If one assumes, in addition, that spin–orbit coupling affects the τ_z level (e.g.) then we obtain a correction to the energy of E^z, namely

$$E_{so}^z(k) = \frac{-\xi^2}{E_{ST}^0}\left(1 - \frac{(\Delta_S - \Delta_T)}{E_{ST}^0}\cos ka\right)$$

where ξ is the molecular spin–orbit coupling parameter and E_{ST}^0 is the energy difference between the singlet and the triplet state involved. Hence the difference in energy between two of the exciton bands is just $D \pm E - E_{so}^z(k)$. Imposing the restriction that only transitions having $\Delta k = 0$ are allowed we find that the $D \pm E$ transitions of the isolated molecule will be broadened to a width of order $E_{so}^z(\pi/a) - E_{so}^z(0) = 2\xi^2/E_{ST}^0$.

The transition probability of the interband e.s.r. transitions is proportional to $\rho(\omega)\exp[-E(\omega)/kT]$ where $\rho(\omega)$ is the density of states function for the triplet excitons and the exponential represents the Boltzmann factor. Hence Figure 8.4 represents the various contributions to the final line-shape function.

The actual line-shape of the curve in Figure 8.4(c) depends on the values of $\rho(\omega)$ and Δ_T, and these can be extracted from the data. Further, the line-width (from $k = 0$ to $k = \pi/a$) is directly proportional to ξ^2. Hence this provides a means of evaluating the spin–orbit contribution Δ_{ST}^{ξ} to the ZF splitting. For TCB, Francis and Harris obtain the following preliminary results.

$$\Delta_{ST}^{\xi} = 6 \text{ MHz (estimated)}$$

$$2\Delta_T = 1.3 \text{ cm}^{-1} = 39 \text{ GHz}$$

(The associated coherence time is $\sim 10^{-9}$ s corresponding to a minimum coherence length of 50–100 Å.) These studies provide a unique insight into the properties of triplet excitons and, in addition, furnish the only method, as far as I know, of separating the contributions of H_{ss} and H_{so} to the ZFS.

Sharnoff and Iturbe[75] and more recently Francis and Harris[76] have used high field and ZF ODMR (respectively) to show that spin alignment is

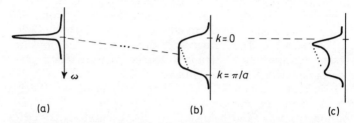

$k = 0$

$k = \pi/a$

(a)　　　　　　　　　(b)　　　　　　　　　(c)

Figure 8.4 The line-shape for one of the ZF e.s.r. transitions: (a) in the isolated molecule, and (b) as broadened by $E_{so}^z(k)$ via the exciton interactions. The contribution due to the Boltzmann factor is shown by the dotted lines and that due to the density of states function $\rho(\omega)$ is added in (c)

preserved in the exciton manifold and in the transfer process from exciton to triplet trap. In the high-field studies the trap phosphorescence in benzo-phenone was monitored while the triplet exciton e.s.r. transitions were traversed. The signs of the resulting (trap-monitored) ODMR signals were consistent with the assumption that the spin alignment is preserved in the exciton-to-trap energy transfer process. In the zero-field studies, the trap phosphorescence in TCB was monitored while the triplet exciton e.s.r. transitions were traversed. In this case (for trapping rates large compared to the inverse exciton lifetime) the change in signal intensity is shown to be independent of $\rho(\omega)$. This implies that the ZF ODMR exciton resonance will not show a minimum between $k = 0$ and $k = \pi/a$ and only the Boltzmann contribution to the signal intensity will be observed. (See Figure 8.4(b).) In other words the spin alignment within the exciton manifold is transferred without (significant) alteration to the traps. If spin alignment were not pre-served, then even though an ODMR signal might be detectable it would not retain the peculiar line-shape which characterises the exciton-monitored ODMR signal. The conservation of spin direction in triplet–triplet energy transfer between a pair of molecules had been considered earlier by El Sayed et al.[77].

The properties of the exciton-trap energy transfer processes described above may provide a method for optically detecting the triplet state magnetic resonance of non-phosphorescent molecules like pyridine. The exciton band of the host would act as the (alignment conserving) conduit for the energy transferred from 'pyridine' to a phosphorescing 'detector' (guest) molecule which has a lower triplet state energy.

8.2.2.4 Optical detection of ENDOR

Electron nuclear double resonance (ENDOR) experiments have played an important role in unravelling complex magnetic resonance problems[78]. ENDOR refers to that class of experiments in which the effect of inducing n.m.r. transitions is detected by monitoring the e.s.r. transitions. The main advantage of the technique is the substantial increase in resolution which it affords.

The application of the ENDOR technique to the study of excited triplet states was first made by Ehret, Jesse and Wolf, using conventional e.s.r. detection[30]. The success of that experiment implies that inducing nuclear transitions affects the e.s.r. transitions and, therefore, the population of the triplet state levels. Consequently, if the radiative rate constants are non-zero the optical emission will be affected as well.

An ENDOR experiment using optical detection was successfully performed on (ground-state) F-centres in KCl[80]. However, the first ZF optically detected ENDOR (ODENDOR) in triplet states was detected in 2,3-dichloroquinoxaline ($\pi\pi^*$) by Harris, Tinti, El-Sayed and Maki[56]. This was soon followed by the work of Chan, Schmidt and van der Waals[81] on quinoxaline ($\pi\pi^*$) and by the ODENDOR ($H \neq 0$) experiments of Cheng, van Zee and Kwiram[82] on the $n\pi^*$ triplet state of pyrazine.

The most extensive and interesting ZF ODENDOR experiments and analysis have been carried out by Harris and co-workers[56, 60, 83, 84] on the $\pi\pi^*$ states of aza-aromatics. The essential features can be illustrated by considering Figure 8.3 again. The ENDOR spectrum corresponds to transitions (a, b, c) between the nuclear states *within* the τ_λ manifold. For example, in ZF, where the nuclear spin states are substantially mixed due to the (second-order) h.f.i., it is essential to calculate the transition moment $(\mu_n)_{ij}$ for the n.m.r. transition correctly. In an early ODENDOR report the use of an improper transition moment operator led to an incorrect interpretation of the quinoxaline ENDOR[81] spectrum. This was subsequently corrected by Buckley, Harris and Maki[83]. The transition moment operator is[2] $h_2 \cdot (\gamma_e S - \gamma_n I)$. If this operator is used to calculate $(\mu_n)_{ij}$ for nuclear transitions between the *zero-order* states in pyrazine one finds that all the transitions are roughly equally probable and weak. However, using the *first-order* functions,

$$| (\mu_n)_{8, 9} |^2 \propto 4\chi^2\gamma_e^2 + 4\chi\gamma_e\gamma_n + \gamma_n^2 \qquad (8.20)$$

But, since $\gamma_n/\gamma_e \approx 10^{-3}$ the last two terms can be neglected (as long as $\chi > 10^{-2}$ say). Thus it is the electron rather than the nuclear dipole moment operator which confers intensity on ENDOR transitions between states affected by the h.f.i. Consequently, 'nuclear' transitions which borrow intensity from μ_e may be orders of magnitude more probable than pure nuclear transitions and may in fact be the only ENDOR lines detectable.

In the pyrazine example, A_{yy} is the only element of the h.f. tensor that causes significant mixing of the ZF states (even though $A_{yy} < A_{xx}, A_{zz}$) because of the small energy difference ΔE_{xz}. Consequently, only one ENDOR transition involving $\tau_\lambda t_x \rightarrow \tau_\lambda t_z$ is expected where $\lambda \neq y$. This will generally be the case – if A_{ii} alone induces mixing then the most probable ENDOR transition is expected to involve the nuclear states (within τ_λ, $\lambda \neq i$) in the plane perpendicular to i [60]. If more than one h.f. element is important in mixing the ZF states then more than one ENDOR transition within the τ_λ manifold is expected. The situation will, of course, be more complicated if, for example, two or more nuclei are present, or if the principal axes of the various interaction tensors are not coincident.

In intermediate or high fields ($H \neq 0$), where the mixing coefficient χ is much smaller, the above effects are less dramatic though still important.

The frequency of the ENDOR transitions depends on the second-order h.f.i. and also on the n.e.q.i. (cf. the discussion in Section 8.2.2.2). From Figure 8.3 it is seen that $\Delta E_{8,9} = 3Q'(1+\eta)$ and similarly $\Delta E_{7,8} = 3Q'(1-\eta) + A_{yy}^2/(X-Z)$. Consequently if only the $8 \rightarrow 9$ transition is observed (as happens when only A_{yy} is effective in mixing the states) one cannot obtain Q' and η independently but only their product. Even if one of the other transitions is observed Q' is not determined unless a value for A_{yy} is obtained.

These features of the ZF ODENDOR spectrum have been clearly delineated by Harris and co-workers[60, 83]. For $\pi\pi^*$ triplet states of aza-aromatics it is known that $A_{xx} \gg A_{yy}, A_{zz}$ for the ^{14}N nucleus[58]. (The proton h.f.i. is small and can be ignored for this qualitative discussion.) Therefore, using equations (8.8) and (8.12) it is seen that h.f. mixing will be important only within τ_y and τ_z and detectable ENDOR transitions are expected only within these two manifolds. (This is analogous to the pyrazine case if the lowest manifold of Figure 8.3 is re-labelled τ_x.) In quinoline, one ENDOR transition is observed for ^{14}N and this yields a value for $Q'(1+\eta)$[83]. This value is approximate, based on the assumption that the principal axes for $D, A,$ and Q are coincident and that the contributions due to A_{yy} and A_{zz} are negligible. (The proton h.f.i. has been ignored throughout.) For quinoxaline, with two equivalent ^{14}N nuclei, three relatively intense ENDOR lines are observed in the τ_z manifold, but again this provides a value for only $Q'(1+\eta)$[83].

The results for the chlorine nucleus in 2,3-dichloroquinoxaline and 8-chloroquinoline are similar[56, 60]. From the forbidden transitions in the ODMR spectrum, it is seen that only one h.f. element is dominant (A_{xx}, just as for N). Consequently, ENDOR transitions are observed in the τ_y and τ_z manifolds and only $Q'(1+\eta)$ can be determined for the Cl nucleus in ZF. ENDOR transitions involving the N nucleus alone, the Cl nucleus alone and N and Cl nuclei simultaneously are observed in 8-chloroquinoline. Both ^{35}Cl and ^{37}Cl data are obtained. The data are only approximate because of the neglect of the smaller elements of the hyperfine tensors and assumptions about the principal axis directions for the various interactions.

A similar study has been made by Buckley and Harris[84] on p-dichloro-benzene. They find $e^2qQ = -64.5$ MHz for Cl in the excited state by assuming $\eta = 0$. The ground-state value of $e^2qQ = -69.6$ MHz and these numbers represents a significant difference between the electron distribution in the ground and excited state. The difference is attributed to an altered field gradient (although it has not been rigorously established that the change in

η is negligible). Buckley and Harris conclude from their results that the molecule is probably distorted in the triplet state.

Buckley and Harris[84] have also indicated the utility of ENDOR-induced ODMR experiments (analogous to the ENDOR-induced e.s.r. experiments of Hyde[85]) for assigning the ZF ODMR transitions. Alternatively, one can also assign the ZF ODMR transitions by noting which ENDOR transitions are observed as the various ODMR lines are individually saturated[60].

Pyrazine[82] and pyrimidine[86] have been studied (in single-crystal hosts) at low field and intermediate field using ODENDOR in order to obtain complete h.f. and n.e.q. tensors in the excited state. Quantitative results can be obtained in this way although data reduction is non-trivial and requires diagonalising the 27×27 Hamiltonian matrix for each orientation of the crystal. In this way tentative values for the N n.e.q. tensor in pyrazine have been obtained:

$$Q_{xx} = 2.0 \text{ MHz}; \qquad Q_{yy} = -2.7 \text{ MHz}; \qquad Q_{zz} = 0.7 \text{ MHz}.$$

These values represent a dramatic change from the ground-state parameters – the major element (Q_{yy}) changes sign. Since the n.e.q. tensor depends on the distribution of all the electrons, whereas the h.f. tensor depends on only the unpaired electrons in the environs of the nucleus of interest, one can extract very detailed information about the electronic wave function from ENDOR data. To obtain such detailed information, however, it will be necessary in general to study (mixed) single crystals in the presence of a magnetic field. Although such experiments are difficult, the information obtained is unique. It is worth noting that the ENDOR line-widths in pyrazine can be as narrow as 30 kHz which represents an increase in resolution of six orders of magnitude over that achieved optically. The unusually high resolution available with the ENDOR technique suggests another area of investigation – the measurement of the change in magnetic resonance parameters under the influence of a (static) electric field. Preliminary calculations indicate that such effects should indeed be measurable*.

8.2.2.5   Optical detection of ELDOR

For many cases of interest emission from the triplet state occurs predominantly from one Zeeman level. That implies that in ZF only two of the three possible ZF ODMR transitions will be detected since the third transition will not be coupled to the emitting level. Consequently, it is not always possible to determine *a priori* whether the frequency of the third transition (1–3) is the sum (Fig. 8.5(a)) of the first two or the difference (Fig. 8.5(b)). In such cases it is convenient to introduce a second microwave frequency in order to pump two electron-spin transitions simultaneously (compare the ENDOR technique) as illustrated in Figure 8.5(c). For example, the frequency of one oscillator can be fixed at the known transition frequency for 2–3 while the frequency of the second oscillator is varied. When the 1–3 transition is traversed the steady-state emission level will change since levels 1 and 3 are now effectively coupled to level 2, the emitting level. Such

*Kwiram, A. L., unpublished results.

an electron double resonance (ELDOR) experiment was first reported in ZF ODMR work by Kuan, Tinti and El-Sayed[50].

High-field ELDOR experiments employing conventional e.s.r. detection had been performed previously[87], but they are considerably more difficult

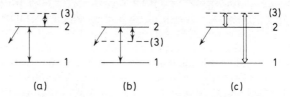

(a)   (b)   (c)

Figure 8.5 Given only two transition frequencies (and only one radiative level), the energy level disposition can be represented by either (a) or (b). The third transition can be observed by electron–electron double resonance as shown in (c)

experimentally than optically detected ELDOR experiments. In the latter case both r.f. frequencies can be fed into the same slow wave structure and, except for the addition of the second r.f. source, the experiment is virtually equivalent (in execution) to the standard ODMR experiment. Nevertheless, it provides the experimentalist with an additional avenue for exploring the properties of the system.

8.3 SPIN–ORBIT COUPLING AND THE TRIPLET STATE WAVEFUNCTIONS

In the foregoing discussion we have simply asserted that the triplet–singlet (as well as the intersystem crossing) transition probability is non-zero. We now wish to investigate in some detail the validity of this assertion. Considerable insight can be gained using simple group theoretical arguments.

8.3.1 The triplet–singlet transition moment: zero field

Within the framework of the Born–Oppenheimer approximation, the Hamiltonian for the triplet state can be written

$$H = H_o + H_{so} + H_{ss} + H_{hf} + H_Q \qquad (8.21)$$

where H_o is the zero-order spin-independent electronic Hamiltonian for fixed nuclei, and the remaining terms represent the spin-dependent terms — H_{so} is the spin–orbit term, and the last three terms have already been discussed in connection with the magnetic resonance parameters (see Section 8.2.1).

The electric-dipole transition moment, M_{ij}, between electronic states i and j (determined by H_o) is given by

$$M_{ij} \propto \langle i \mid er \mid j \rangle \qquad (8.22)$$

where er is the electric-dipole moment operator. The magnetic-dipole moment operator can generally be ignored[88]. The wave functions of interest to us are eigenfunctions of H_o and are represented by $^1\psi^o$ for the ground singlet state and $^3\psi^o$ for the lowest triplet state. The total wavefunctions including spin are designated $^1\psi_o^o$ and $^3\psi_\lambda^o$ where the subscripts o and λ represent σ_o and τ_λ respectively. Since er is a function of space variables alone we anticipate that transitions between zero-order states of different multiplicity will be zero, i.e. $M_{ij}^o = M_{o\lambda}^o \equiv 0$.

The triplet–singlet transitions $M_{o\lambda}$ are, however, not zero, because of perturbations which mix triplet states with singlet states and vice versa*. The perturbation which figures most prominently in singlet–triplet mixing is the spin–orbit interaction. However, we must consider not only H_{so}, which represents the interaction for fixed nuclei, but also the contributions that arise when the nuclei are allowed to vibrate. If the nuclei are permitted to execute small displacements then the Hamiltonian in equation (8.21) can be expanded in a Taylor series about the equilibrium configuration[90]. Expressed in terms of the normal coordinates, Q_i, the modified Hamiltonian (keeping only first-order terms) is[21, 90]

$$H(Q_i) = H + \frac{\partial H_o}{\partial Q_i} Q_i + \frac{\partial H_{so}}{\partial Q_i} Q_i \qquad (8.23)$$

where the vibrational effect on the last three terms of equation (8.21) has been assumed negligible. Consequently, there are two terms in (8.23), H_{so} and $(\partial H_{so}/\partial Q_i)Q_i \equiv H_{sv}$, which can cause intersystem mixing and they are referred to as spin–orbit and spin–vibronic perturbations respectively[90]. Each term in equation (8.23) must transform as the totally symmetric representation, Γ_o, of the symmetry group of the molecule.

Using these spin–orbit interaction terms we can write a first-order wave function for the triplet manifold as

$$^3\psi_\lambda = {}^3\psi_\lambda^o + \sum_n \frac{\langle {}^3\psi_\lambda | H' | {}^1\phi_n^o \rangle}{E_\lambda - E_n} {}^1\phi_n^o \qquad (8.24a)$$

which we abbreviate

$$^3\psi_\lambda = {}^3\psi_\lambda^o + \sum_n E_{\lambda n}^{-1} H'_{\lambda n} | n \rangle \qquad (8.24b)$$

where H' is either H_{so} or H_{sv}. Using (8.22) the transition moment between the first-order $^3\psi_\lambda$ state and $^1\phi_o^o$, the ground state, is

$$M_{o\lambda} = \sum_n E_{\lambda n}^{-1} H'_{\lambda n} \langle o | er | n \rangle \equiv \sum_n E_{\lambda n}^{-1} H'_{\lambda n} m_{on} \qquad (8.25)$$

where m_{on} is the transition moment for the singlet manifold. In other words, the triplet–singlet transition acquires some intensity due to the partial admixture of an excited singlet state. Clearly, for $M_{o\lambda}$ to be non-zero, the corresponding singlet–singlet transition moment m_{on} must be non-zero. Not infrequently, $m_{on} \equiv 0$ is zero order because of the orbital symmetry of S_n. Nevertheless a weak $S_n \to S_o$ transition may be observed as a result of first-order vibronic mixing [due to $(\partial H_o/\partial Q_i)Q_i$] within the singlet manifold.

---

*We will consider only the admixture of excited singlet states into the lowest triplet state. The effect of mixing excited triplet states into the ground singlet state has also been discussed[21, 89].

In any case, for a Zeeman level to have a radiative component, it must contain an admixture of a singlet state having some finite transition probability to the ground state, regardless of the mechanism whereby the latter is acquired.

8.3.1.1 The 0–0 band

Consider $H' = H_{so}$. Since H_{so} does not involve the vibrational operator, mixing will occur only with vibrationless states and the emission will be to the vibrationless ground state (0–0 band). Re-writing (8.25) we get

$$M_{o\lambda} = \sum E_{\lambda n}^{-1} \langle \lambda | H_{so} | n \rangle \langle n | er | o \rangle \qquad (8.26)$$

Since H_{so} and $| o \rangle$ transform as Γ_o, equation (8.26) implies that for $M_{o\lambda}$ to be non-zero $\Gamma(\lambda) = \Gamma(er)$ or

$$\Gamma(\lambda) \otimes \Gamma(er) \subset \Gamma_o \qquad (8.27a)$$

where the symbol $\subset$ reads *contains*.

Note that the polarisation corresponds to the *translation* which transforms as $\Gamma(\lambda)$. But $\Gamma(\lambda)$ transforms as the direct product representation $\Gamma_s \otimes \Gamma_\sigma = \Gamma(\lambda)$ where Γ_s and Γ_σ correspond to space and spin functions respectively. The spin functions transform as the rotations, and only one-dimensional representations occur for the C_{2v} and D_{2h} groups. Therefore, each Zeeman level must transform according to a different irreducible representation and will exhibit a unique polarisation. This very important conclusion is equivalent to the requirement embodied in Criterion II. Frequently, as we shall see shortly, only one of the Zeeman levels has a significant transition probability to the vibrationless ground state. In that case, the polarisation of that vibronic transition is pure. If more than one Zeeman level is significantly coupled to the same vibronic ground state, then a mixed polarisation will result.

Since the ODMR experiment provides information on both $\Gamma(\lambda)$ (the polarisation) and Γ_σ (the principal axes of H_{ss}) one can readily deduce Γ_s, the orbital symmetry of the lowest triplet state. Such a procedure based on ODMR data was first carried out for phenanthrene[11].

8.3.1.2 The vibronic bands

The only first-order perturbation which can give rise to the vibronic bands is $H_{sv} \equiv (\partial H_{so}/\partial Q_i)Q_i$. This perturbation will mix triplet levels with excited singlet states which possess a quantum of vibrational energy. Thus

$$H''_{\lambda m} = \sum_i \langle \lambda | \partial H_{so}/\partial Q_i | m \rangle \langle \chi_o | Q_i | \chi_{Q_i} \rangle \qquad (8.28)$$

where χ_o represents the totally symmetric vibration(s) of the triplet state. Hence λ is mixed with an excited singlet state transforming according to

$\Gamma(Q_i) \otimes \Gamma(m)$. On the other hand, for $M_{o\lambda}$ to be non-zero m_{om} must be non-zero which means that

$$\Gamma(er) \otimes \Gamma(\lambda) \otimes \Gamma(Q_i) \subset \Gamma_o \qquad (8.27b)$$

Note that this expression reduces to (8.27a) if the 0–0 band or any totally symmetric vibronic band is monitored. Alternatively, equation (8.27) can be re-written in the form $\Gamma(er) \otimes \Gamma(\lambda) \subset \Gamma_g$ where the ground state transforms as Γ_g and $\Gamma_g = \Gamma_o \otimes \Gamma(Q_i)$.

We have emphasised only first-order mechanisms above. In many cases second-order contributions may be significant or even dominant. As stated earlier, many singlet–singlet transitions are vibronically induced, or alternatively, vibronic perturbations will mix singlet states with other singlets (or triplets with triplets). Consequently second-order contributions arise from such vibronic mixing *within* the singlet or *within* the triplet manifold, the intersystem mixing being provided by H_{so} or H_{sv}.

8.3.2   Assignment of the orbital symmetry, Γ_s

Group theory will, of course, only inform us whether certain matrix elements are zero or non-zero. For those in the latter category, it provides us with no information about the magnitudes of the individual matrix elements. Those can, of course, be obtained only by an evaluation of the corresponding integrals. Fortunately, a few general rules emerge from such calculations. In particular, it can be shown that the (direct) spin–orbit matrix elements are greatest when the two coupled states differ in the electronic character of one electron[21]. In other words, $\sigma\pi^*$ states are coupled preferentially with $\pi\pi^*$ states (and vice-versa). If the σ orbital is symmetric to reflection in the molecular plane, then the total ($\sigma\pi^*$) wave function must be antisymmetric to such an operation. Hence in C_{2v} such functions will transform like A_2 or B_1. Since fluorescence from A_2 is forbidden, only $\Gamma(\lambda) = B_1$ states can lend intensity to the vibrationless $T_1 - S_0$ transition. This very important result is equivalent to the requirement embodied in Criterion I and has been confirmed for quinoxaline[40, 42, 91], dichloroquinoxaline[92] and phenazine[93] – typical aza-aromatic molecules with a lowest $\pi\pi^*$ triplet state.

For those molecules with a lowest $n\pi^*$ triplet state such as pyrazine and pyrimidine, one must, of course, consider mixing with $\pi\pi^*$ singlet states. These must be *symmetric* to reflection in the molecular plane and thus transform like A_g, B_{1u}, B_{2u} or B_{3g} in D_{2h} and A_1 or B_2 in C_{2v}. In the case of pyrimidine (C_{2v}) this implies that both the A_1 and the B_2 level may have intensity, as is indeed observed[49, 51]. For pyrazine (D_{2h}) one would expect emission from B_{1u} and B_{2u}. However, a more detailed consideration of the properties of the pyrazine wave functions shows that H_{so} interactions will dominate for B_{1u}. This again agrees with experiment[44, 51, 53].

In the case of planar aromatic hydrocarbons with $\pi\pi^*$ triplet states lowest, it is found that spin–orbit coupling is much less effective (since one-centre terms vanish), and hence second-order perturbations may compete effectively with the first-order spin–orbit perturbation and obscure the simple rules outlined above. Nevertheless, naphthalene and phenanthrene each show dominant emission from the B_1 level[11, 42] whereas benzene (D_{2h}) exhibits a dominant emission from $\Gamma(\lambda) = B_{3u}$ consistent with our rules. Similar results obtain for p-dichlorobenzene[84] and tetrachlorobenzene[95]. One can summarise these results as follows.

$$\text{For } C_{2v}: \begin{cases} \text{if } \pi\pi^* \text{ then } B_1\,(x) \\ \text{if } n\pi^* \text{ then } A_1(z),\, B_2(y) \end{cases}$$
$$\text{For } D_{2h}: \begin{cases} \text{if } \pi\pi^* \text{ then } B_{3u}\,(x) \\ \text{if } n\pi^* \text{ then } B_{1u}\,(z),\, B_{2u}\,(y) \end{cases} \tag{8.29}$$

In other words, for C_{2v}, if T_1 is $\pi\pi^*$ then the 0–0 band will originate from the B_1 level and be x polarised. Other contributions to the emission intensity are to be expected. Nevertheless, the characteristics tabulated in (8.29) should be present and should in fact dominate if one-centre terms contribute significantly to the H_{so} matrix elements. The extension of (8.29) to the other vibronic bands is straightforward using (8.27b).

To illustrate these points consider the phenanthrene molecule and the C_{2v} character table[11]. The lowest triplet state is $\pi\pi^*$ and the orbital wave-function must transform either like A_1 or B_2. Therefore $\Gamma(\lambda) = \Gamma_s \otimes \Gamma_\sigma$ will transform either like

$$A_1 \otimes \tau_x = B_2 : y\text{-polarised emission}$$
$$A_1 \otimes \tau_y = B_1 : x\text{-polarised emission}$$
$$A_1 \otimes \tau_z = A_2 : \text{—}$$

or

$$B_2 \otimes \tau_x = A_1 : z\text{-polarised emission}$$
$$B_2 \otimes \tau_y = A_2 : \text{—}$$
$$B_2 \otimes \tau_z = B_1 : x\text{-polarised emission}$$

But the rules above suggest that emission (to the 0–0 band) will be from a B_1 level. Hence τ_y will be the principal emitter if $\Gamma_s = A_1$, or τ_z if $\Gamma_s = B_2$. ODMR results clearly show[11] that τ_z is the principal emitter leading to a B_2 assignment for the orbital symmetry. In the same way it was found that quinoxaline[40, 42] (as well as naphthalene[42]) emission is primarily from τ_z with x polarisation corresponding to B_2.

A nice application of this method of assigning Γ_s has recently been made in a study of benzene by Gwaiz, El-Sayed and Tinti[94]. Benzene in its lowest excited ($\pi\pi^*$) triplet state is distorted and possesses only D_{2h} symmetry. Using the D_{2h} character table and equation (8.29), we deduce that if the orbital symmetry is B_{1u} then τ_y will be most radiative (and of x polarisation), if B_{2u} then τ_z would be most radiative. Gwaiz et al. found that indeed τ_y was most radiative and concluded that the lowest triplet state of benzene is of B_{1u} symmetry.

The number of independent measurable parameters can be increased by monitoring various non-symmetric vibronic bands. Space does not permit an extension of the discussion to these cases. Although the analysis follows that outlined above for the vibrationless band, the number of possible paths increases significantly. Nevertheless, by a systematic comparison of the experimental measurements with the multitude of possible paths, a unique solution can frequently be found. Analyses of this sort have recently been carried out by Chen and El-Sayed[63] for 1,2,4,5-tetrachlorobenzene (TCB) and by Buckley, Harris and Panos[96] for p-dichlorobenzene. The result in both cases is $\Gamma_s = B_{1u}$.

8.3.3   The transition moment: $H \neq 0$

In the presence of a magnetic field the Zeeman levels of the triplet state will become mixed. Consequently, the radiative parameters of the ZF states will also become mixed. It is convenient to express the radiative properties of the high-field Zeeman states in terms of the ZF parameters[57]. Since the intensity I_λ of the emission from any ZF level λ is proportional to the square of the transition moment $M_{o\lambda}$ we can write using equation (8.25)

$$I_\lambda = cM_{o\lambda}^* M_{o\lambda} = c\left\{ \sum_n E_{\lambda n}^{-2} |H_{\lambda n}|^2 |m_{on}|^2 + \right.$$

$$\left. \sum_{n \neq m} E_{\lambda n}^{-1} E_{\lambda m}^{-1} (H_{\lambda n}^* H_{\lambda m})(m_{on} m_{om}) \right\} \qquad (8.30a)$$

where c is simply the proportionality constant. If $H_o \| z$, then τ_z remains unchanged and τ_x and τ_y become mixed. For arbitrary values of H_o the states $\tau_\pm$ will be given by $\tau_\pm = a\tau_x \pm b\tau_y$ where b/a is a function of $(g\beta H_o)/(X-Y)$.

Therefore

$$I_\pm \propto a^2 I_x + b^2 I_y \pm (\Delta E)^{-2} \sum_{n,m} iab\{H_{ny}^* H_{xm} - H_{nx}^* H_{ym}\} m_{on} m_{om} \quad (8.30b)$$

where I_x and I_y are given by (8.30a) with $\lambda = x, y$ respectively. At sufficiently high fields $|b/a| \rightarrow 1$, and $I_\pm = \frac{1}{2}(I_x + I_y)$ if coherence effects are ignored. For arbitrary field directions one must first calculate the weighting coefficients, a_λ^2, for the basis states and obtain for the high-field state k

$$I_k \propto a_x^2 I_x + a_y^2 I_y + a_z^2 I_z$$

8.3.4   Optical-r.f. double resonance

Tinti et al.[97] introduced a variation of the ZF ODMR experiment which is particularly useful for studying the characteristics of the vibronic bands. Instead of monitoring the change in emission intensity at some fixed wavelength while sweeping the r.f., they invert the usual procedure and continuously saturate a particular ODMR transition while recording an otherwise conventional phosphorescence spectrum. The effect of the r.f. power can be sufficiently large so that the phosphorescence spectrum in the presence and absence of r.f. power is distinctly different.

To illustrate the utility of the method, consider a system with the properties depicted in Figure 8.6. Emission from τ_z which terminates on the 0–0 band (or some totally symmetric vibronic band) is said to belong to sub-spectrum I. Vibronic bands originating on τ_y which involve the a_2 mode belong to sub-spectrum II. The polarisation of the emission in I and II is unique since only one ZF level is involved in each case. Sub-spectrum III is of mixed polarisation and involves the b_1 mode.

In the absence of r.f. power, the phosphorescence spectrum may look like (b). If we assume that entry is primarily to τ_y and that the SLR rates W_{ij} are very slow, then in steady state the population of $N_y > N_x$, N_z. Therefore, in the presence of r.f. power pumping the $\tau_y \leftrightarrow \tau_z$ transition, we might expect the lines belonging to sub-spectrum I to be more intense, and those in II to be less intense than before. If the change is small it may be difficult to detect the difference between the two spectra. Therefore, El-Sayed, Owens and Tinti[98] introduced amplitude modulation of the r.f. with subsequent

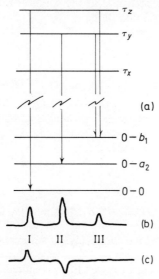

Figure 8.6 The radiative properties shown in (a) give rise to the phosphorescence spectrum (b). (c) The result if spectrum (b) is recorded (after phase sensitive detection) while pulsed r.f. power is continuously applied to the appropriate ZF transition

phase-sensitive detection of the phosphorescence. When the vibronic spectrum is recorded in this manner the lines are either positive or negative as shown in (c), depending on whether the corresponding ODMR signal corresponds to an increase or a decrease (or no change). In the event that a vibronic band originates on a level not pumped by the r.f., it will not appear in the (amplitude-modulated) spectrum. Similarly, for vibronic bands of mixed polarisation, such as III, the signal will depend on the ratio of the rate constants. If, however, a polariser is inserted in front of the detector the two components may be resolved giving a negative (or positive) going signal[11]*.

This method provides a particularly convenient means of assigning the

---

*Alternatively, one can monitor the decay curve after extinguishing the exciting light. If the rate constants for the two levels are different (and $W_{ij} = 0$) then the polarisation will change with time if the vibronic band originates from more than one ZF level[92].

various vibronic bands in the phosphorescence spectrum. Tinti and El-Sayed[92] have studied the optical r.f. double resonance spectrum of 2, 3-dichloro-quinoxaline in some detail. By combining these results with decay times measured as a function of temperature and magnetic field, H, they were able to assign the symmetry of the vibronic bands and determine the ZF levels from which they originate. The spectra recorded for TCB[63] provide a good example of this approach which will doubtless find numerous applications. It should be possible, by an extension of this method, to distinguish between various *rates* of decay by pulsing the exciting light and recording the resulting (phase-sensitive detected) spectra for different repetition rates. Alternatively, Harris and Hoover have suggested the use of phase-shift methods, in connection with amplitude modulation of the r.f., in order to determine rate constants[99].

8.3.5 Singlet–triplet intersystem crossing

Measurements of the rate constants for entry into the Zeeman levels (to be discussed in Section 8.4) can provide significant insight into the intersystem crossing (ISC) processes. There are, of course, a number of ways in which ISC can take place. For example, the initial excitation may lead to the excited singlet state of the *host* or to the host singlet exciton band, with a variety of subsequent non-radiative processes ultimately leading to the lowest triplet state. Considerable caution should, therefore, be exercised in interpreting the ISC rate constants. With this in mind we shall discuss an idealised ISC process – excitation to the guest singlet state S_1 followed by ISC to some vibrational level of T_i and internal conversion to T_1. As in the case of T–S transitions, the perturbations which provide for intersystem mixing are H_{so} and H_{sv}. Since spin–orbit matrix elements will be largest when one-centred terms are present, we will first turn our attention to some aza-aromatics.

The lowest excited singlet state of pyrazine is $n\pi^*$ and of symmetry $^1B_{3u}$. Therefore, direct spin–orbit mediated ISC can occur only to a $(\pi\pi^*)$ state having total symmetry of B_{3u}. The $\pi\pi^*$ triplet states presumed to lie below $^1B_{3u}$ are thought to be of orbital symmetry B_{1u}, B_{2u} and B_{3u}. Only $B_{2u} \otimes \tau_z$ and $B_{1u} \otimes \tau_y$ yield $\Gamma(\lambda) = B_{3u}$. Therefore, ISC could occur to either of these two states. Experimentally[11, 51], however, it is found that entry is primarily to the τ_y level of T_1. Since the quantisation direction of the spin is conserved in the internal conversion process, it can be concluded that ISC takes place from $^1B_{3u}$ to $^3B_{1u} \otimes \tau_y$. Burland and Schmidt have carried out a similar analysis for pyrimidine[51] as well, but the results are not clearly understood at this time. Schmidt *et al.* have presented results for quinoxaline[91] and phenazine[93] (S_1 is $\pi\pi^*$) which are again in agreement with expectation. In general, for aza-aromatics direct spin–orbit coupling to the Zeeman level transforming like $\Gamma(S_1)$ will provide the dominant ISC mechanism. Other effects, including vibronically induced perturbations, are expected to contribute to a lesser extent in aza-aromatics.

In the case of planar aromatic hydrocarbons where one- and two-centre spin–orbit interactions vanish, one might expect other perturbations to be more important. Recently, El-Sayed and Chen have studied TCB in this connection[95]. From their ODMR results they find the direct spin–orbit coupling cannot be the dominant factor in ISC. Vibronically induced perturbations must be invoked, and it remains to be seen whether this is a general feature of the aromatic hydrocarbons.

8.4 THE TRIPLET STATE RATE CONSTANTS

It has been taken for granted throughout the foregoing discussion that the various rate constants were at least qualitatively known. Using this knowledge we were able to confirm the rather sweeping generalisations provided by group theory regarding the intersystem crossing processes and the perturbation schemes which break the multiplicity forbiddenness. In fact, the very success of the ODMR experiment depends upon the relative magnitudes of the rate constants associated with the individual Zeeman levels of the triplet state, as implied by the three criteria of Section 8.1.2. We now wish to consider in more detail how these rate constants affect the ODMR experiment, and in particular to show how they can be quantitatively determined using the ODMR technique.

8.4.1 The general rate equations

The rate constants associated with the triplet state are shown in Figure 8.7. The corresponding rate equations are

$$\dot{N}_0 = -(\lambda_1+\lambda_2+\lambda_3)N_0+k_1N_1+k_2N_2+k_3N_3$$
$$\dot{N}_1 = \lambda_1N_0-(k_1+P_{12}+P_{13}+W_{12}+W_{13})N_1+(P_{12}+W_{21})N_2+ (W_{31}+P_{13})N_3$$
$$\dot{N}_2 = \lambda_2N_0+(P_{12}+W_{12})N_1-(k_2+P_{12}+P_{23}+W_{21}+W_{23})N_2+ (W_{32}+P_{23})N_3$$
$$\dot{N}_3 = \lambda_3N_0+(P_{13}+W_{13})N_1+(P_{23}+W_{23})N_2- (k_3+P_{13}+P_{23}+W_{31}+W_{32})N_3 \tag{8.31}$$

(or $\dot{N}_i = \sum_j \alpha_{ij}N_j$ and α is referred to as the rate matrix)

where $k_i = k_i^r + k_i^n$, the sum of radiative and non-radiative rates for level i. It should be noted that k_i^n includes not only the direct relaxation to the ground state but any other non-radiative depopulating processes as well. The latter may in many cases be very small. In the same vein we note that the λ_i ($\Lambda \equiv \sum_i \lambda_i$) represent composite rates, because generally we are not pumping the $S_o \rightarrow \tau_i$ transition directly but rather through a complex of (unknown) intermediate states. Considerable care must, therefore, be exercised when discussing these parameters, for reasons already outlined.

The general solutions of the differential equations in equation (8.31) are rather unwieldly, and it is desirable, therefore, to choose experimental conditions which will simplify the problem*. We will assume that (a) $W_{ij} = 0$, (b) $\lambda_3 = k_3 = 0$, and (c) $P_{ij} = P_{12} \equiv P$. The first condition, which states that all the SLR rates can be neglected, is the most critical. Fortunately, this

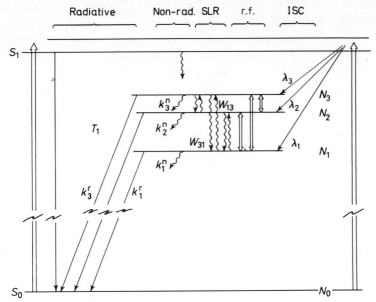

Figure 8.7 The rate constants associated with the triplet state

condition can generally be met at sufficiently low temperatures ($T \leqslant 1.5$ K). With these constraints equation (8.31) becomes

$$\dot{N}_0 = -\Lambda N_0 + k_1 N_1 + k_2 N_2$$
$$\dot{N}_1 = \lambda_1 N_0 - (k_1 + P)N_1 + PN_2 \qquad (8.32)$$
$$\dot{N}_2 = \lambda_2 N_0 + PN_1 - (k_2 + P)N_2$$

8.4.2   The steady-state equations

In the steady-state approximation ($\dot{N}_i = 0$) we can solve for N_i.

$$N_0/N = [k_1 k_2 + kP]/D$$
$$N_1/N = (k_2 \lambda_1 + \Lambda P)/D$$
$$N_2/N = (k_1 \lambda_2 + \Lambda P)/D \qquad (8.33)$$

where $D = k_1 k_2 + k_2 \lambda_1 + k_1 \lambda_2 + (k + 2\Lambda)P$ and $N = \Sigma N_i$ is the total population and $k = k_1 + k_2$.

---

*If that is not feasible, then one can use a method similar to the one presented by Gordon[100], diagonalise the rate matrix numerically and thus simulate the dynamic behaviour of the system. Such an approach has been used in a study of the pyrazine rate constants (van Zee, J. and Kwiram, A., unpublished results).

We designate by N_i^o and $\mathcal{N}_i^o$ the populations of level i for the cases $P = 0$ and $P \gg k_i$, λ_i respectively. The former simply gives the steady state populations in the absence of r.f. power (and relaxation effects) and the latter in the presence of sufficient r.f. power to dominate all the other triplet state rate constants. For the latter case we let

$$\lim_{P \to \infty} D \equiv d = (k + 2\Lambda)P$$

and

$$\lim_{P \to 0} D \equiv D^o = k_1 k_2 + k_2 \lambda_1 + k_1 \lambda_2$$

The change in the population of the levels under the influence of r.f. power is given by $\Delta N_i \equiv \mathcal{N}_i^o - N_i^o$

$$\Delta N_0/N = (k_1 - k_2)(k_1 \lambda_2 - k_2 \lambda_1)P/dD^o$$
$$\Delta N_1/N = (\Lambda + k_2)(k_1 \lambda_2 - k_2 \lambda_1)P/dD^o$$
$$\Delta N_2/N = -(\Lambda + k_1)(k_1 \lambda_2 - k_2 \lambda_1)P/dD^o \tag{8.34}$$

The intensity of the light emitted by a single level is simply $I_i = k_i^r N_i$ and the fractional change $\Delta I_i/I_i^o$ is

$$\Delta I_1/I_1^o = (k_2 + \Lambda)(k_1 \lambda_2 - k_2 \lambda_1)/(k + 2\Lambda)k_2 \lambda_1 \tag{8.35a}$$
$$\Delta I_2/I_2^o = -(k_1 + \Lambda)(k_1 \lambda_2 - k_2 \lambda_1)/(k + 2\Lambda)k_1 \lambda_2 \tag{8.35b}$$

The fractional change in the total intensity is

$$\Delta I/I^o = (k_1^r \Delta N_1 + k_2^r \Delta N_2)/(k_1^r N_1^o + k_2^r N_2^o)$$
$$= \frac{(k_1 \lambda_2 - k_2 \lambda_1)[k_1^r(k_2 + \Lambda) - k_2^r(k_1 + \Lambda)]}{(k_1^r k_2 \lambda_1 + k_2^r k_1 \lambda_2)(k + 2\Lambda)} \tag{8.36}$$

Whether we monitor a single component or the total intensity, the change in intensity under steady-state conditions will be zero if $\lambda_2/\lambda_1 = k_2/k_1$. But this is just the condition for the initial populations to be equal ($N_1^o = N_2^o$) and one would not expect any effect. Apart from this particular condition, a change will always occur and *with opposite sign* if the individual components are monitored. If the total intensity is monitored, $\Delta I/I^o$ may be zero if

$$k_1^r(k_2 + \Lambda) = k_2^r(k_1 + \Lambda)$$

or

$$k_2^r/k_1^r = (1 + k_2/\Lambda)/(1 + k_1/\Lambda)$$

Thus if $\Lambda_i \gg k_i$, k_j the condition $k_1^r = k_2^r$ implies $\Delta I = 0$. At the other extreme, if $\lambda_i \ll k_i$, k_j then the condition for zero intensity change is $k_2^r/k_1^r = k_2/k_1$

In order to illustrate the kind of steady-state intensity changes that might be expected in general we re-write equation (8.36) in terms of ratios of rate constants. Let $\alpha \equiv \lambda_2/\lambda_1$,

$$1+\lambda_2/\lambda_1 \equiv 1+\alpha \equiv \alpha'$$

$$1+k_1^r/k_1^n \equiv 1+m \equiv m'$$

$$1+k_2^r/k_2^n \equiv 1+n \equiv n'$$

$$1+k_2^r/k_1^r \equiv 1+\rho \tag{8.37}$$

Substituting these relations into (8.36) gives

$$\Delta I/I^\circ = \frac{(m'\alpha - n'\rho)[(n-m)k_2^r+(1-\rho)\alpha'\lambda_1]}{\rho(m'\alpha+n')[(m'+n'\rho)k_1^r+2\alpha'\lambda_1]} \tag{8.38}$$

8.4.2.1   Special cases

Consider the following idealised cases.

(a) $\lambda_1 \gg k_i (\rho \neq 1)$

$$\Delta I/I^\circ = (m'\alpha - n'\rho)(1-\rho)/2\rho(m'\alpha+n')$$

and

$$d(\Delta I/I^\circ)/d\rho = (\rho^2 - \alpha m'/n')/2\rho^2(m'\alpha+n')$$

The intensity goes to zero for $\rho = \alpha m'/n'$. If $m = n$ then $\Delta I/I^\circ = 0$ for $\rho = \alpha$ which is equivalent to the condition $N_1^\circ = N_2^\circ$. Given typical values for n, m, α then $\Delta I/I^\circ$ is always large (and positive) for $\rho \ll 1$, as one would expect. However, if $m \neq n$ then ΔI can change sign and the maximum *negative* effect occurs for $\rho = (\alpha m'/n')^{\frac{1}{2}}$. Note that the presence of the non-radiative paths can lead to intensity changes the exact opposite of those one would have predicted on the basis of the simple model we have used hitherto.

If instead of the total intensity $\Delta I/I^\circ$ we consider a single component then

$$\Delta I_1/I_1^\circ = (m'\alpha - n'\rho)/2n'\rho$$

The slope does not change sign.

(b) $\lambda_1 \ll k_i; (|n-m| > 1)$

$$\Delta I/I^\circ = (n-m)(m'\alpha - n'\rho)/(m'\alpha+n')(m'+n'\rho)$$

and

$$d(\Delta I/I^\circ)/d\rho = \alpha' m' n'(m-n)/(m'\alpha+n')(m'+n'\rho)^2$$

which says that the slope does not change sign (since for $n = m$ this approximate equation is not valid). Similarly, if only one component is monitored

$$\Delta I_1/I_1^\circ = (m'\alpha - n\rho)/(m'+n'\rho)$$

and again the slope does not change sign. The non-radiative rates have a significant effect on the intensity changes to be expected in a steady-state experiment, and in principle one should be able to determine k_i^n from such experiments.

An enlightening comparison can be made between ODMR results and those of conventional spectroscopy. For the latter a measurement of the ratio of the intensity for the two individual components gives

$$\frac{I_2^o}{I_1^o} = \frac{k_2^r}{k_1^r} \tag{8.39}$$

However, the ODMR experiment measures

$$\frac{\Delta I_2}{\Delta I_1} = -\frac{k_2^r(k_1 + \Lambda)}{k_1^r(k_2 + \Lambda)} = -\frac{(k_1 + \Lambda)}{(k_2 + \Lambda)} \frac{I_2^o}{I_1^o} \tag{8.40}$$

and the two measurements will agree only if $k_1 = k_2$. This result again emphasises the important role of the non-radiative rate constants*.

The special cases exhibited above (with $W_{ij} = 0$ and $k_3 = \lambda_3 = 0$) provide a rough idea of how the steady-state intensity is expected to change under the influence of r.f. power. The total population in the excited state is also affected. It can readily be shown from (8.33) that $\mathcal{N}_1^o + \mathcal{N}_2^o = N_1^o + N_2^o$ only if one λ_i is much larger than all the other rate constants. If that is not the case, then the equality is not fulfilled. Since in fact the ground-state population is also affected one might expect that the fluorescence intensity will be altered. This is the basis for a recent suggestion by Harris and Hoover that ODMR might be detected by monitoring the fluorescence[101]†.

Steady-state equations for a variety of multiple resonance experiments have been presented in a lengthy treatment by El-Sayed[102]. He assumes at the outset that the sum of the steady-state populations in the triplet state is independent of whether r.f. is present or not. In view of the comments above, some caution is appropriate when applying those equations.

8.4.2.2   Extraction of rate constants

If at some time $t = 0$ the exciting light is turned off, the time-dependent intensity in the *absence* of r.f. power is given by

$$I(t) = c\Sigma k_i^r N_i^o \exp(-k_i t)$$

Under favourable circumstances it is possible to decompose the resulting decay curve into its (three) components and thus obtain each of the k_i terms[92]. In addition Tinti and El-Sayed[92] have decomposed the analogous decay curves in the *presence* of r.f. power to obtain the relative radiative decay constants k_i^r. These, in conjunction with the values for k_i, can be used to determine the relative steady-state populations and ultimately the relative values of the λ_i[102]. Despite the fact that all the rate constants can be extracted in this fashion[102], the method is subject to significant error, and may in many instances not be feasible. Hence, even though steady-state methods may be useful in qualitative or even semi-quantitative sense for determining rate constants, transient methods provide in general a more fruitful approach.

---

*The above arguments were presented in an abbreviated form at the International Conference on Luminescence, Twente, The Netherlands, 1968.

†It may also be possible to detect ODMR by monitoring the *delayed* fluorescence, since exciton annihilation is a spin-dependent process.

The transient method has been exploited most extensively by van der Waals and co-workers[51, 91, 93, 103] and we shall consider this approach in some detail.

8.4.3 Transient ODMR in zero field

The analysis of transient phenomena can again be simplified considerably by choosing $W_{ij} = 0$. An additional simplification can be achieved in either of two ways. If the excitation source is turned off at some time, $t = 0$, then *after* that time $\lambda_i = 0$ for all i since none of the levels are being populated from the ground state. Alternatively, if the experiment is carried out on a time scale which is short compared to all the rate constants, then appropriate rates can be ignored. Thus even though the light stays on the effect of λ_i, for example, can be ignored. Such a procedure is possible in adiabatic inversion experiments.

8.4.3.1 $\lambda_i = 0$ (after the excitation has ceased)

We assume that the light was on for a time sufficient to achieve steady-state conditions. The time at which the light is extinguished is designated as $t = 0$. The total intensity at that instant of time is therefore

$$I(0) = c\Sigma k_i^r N_i(0) = c\Sigma k_i^r N_i^o \qquad (8.41)$$

since $N_i(0)$ is just equal to the steady-state populations N_i^o for $P_{ij} = 0$.

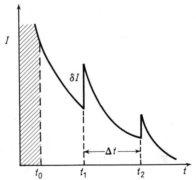

Figure 8.8 The transient change in intensity, δI, upon application of r.f. power at $t = t_1$ (and at $t = t_2$)

Since for $t > 0$, $\lambda_i = 0$ (light off) the populations begin to decay exponentially and we have

$$I(t) = c\Sigma k_i^r N_i(t) \qquad (8.42)$$

But since $\lambda_i = 0$, we can solve (8.31) explicitly and for $P_{ij} = 0$

$$N_i(t) = N_i^o \exp(-k_i t) \qquad (8.43)$$

Therefore

$$I(t) = c\Sigma k_i^r N_i^o \exp(-k_i t) \qquad (8.44)$$

If at some time $t = t_1$ we apply (strong) r.f. power to one of the ZF transitions we would expect to see a change in the intensity of the phosphorescence. This is illustrated schematically in Figure 8.8 for a case in which $k_3 \gg k_2, k_1$.

Because of this disparity in decay rates, the population N_3 will decrease (for $t>0$) relative to N_1 and N_2. Therefore inducing the $3j$ ZF transition will increase N_3, resulting in an increase in the phosphorescence intensity*. From measurements of this sort van der Waals and co-workers have developed a systematic method for determining the rate constants. The procedure is reduced to three special cases depending on whether the decay is predominantly from one, two, or all three of the Zeeman levels. Following the analysis of van der Waals and co-workers[91, 103] we present the method for extracting the rate constants for the case in which one of the rate constants k_i is much greater than the other two. Choose $k_3 \gg k_1, k_2$.

8.4.3.2 Extraction of rate constants

(a) *The total decay rates, k_1 and k_2* — If at time t_1 r.f. power is applied to one of the ZF transitions, say $3j$, the populations are altered and the change in the total intensity $\delta I_{3j}(t_1)$ of the phosphorescence is given by

$$\delta I_{3j}(t_1) = c(k_3^r \Delta N_3 + k_j^r \Delta N_j) \qquad (j = 1, 2) \tag{8.45}$$

If the r.f. transition is induced in a sufficiently short time interval we can ignore the effects of the various relaxation constants. (This assumption can easily be tested.) If a fraction f of the population is transferred, then the population N_i^+ immediately after the application of r.f. power will be related to that immediately before, N_i^-, by

$$N_3^+ = (1-f)N_3^- + fN_j^- \tag{8.46a}$$

$$N_j^+ = fN_3^- + (1-f)N_j^- \tag{8.46b}$$

Hence

$$-\Delta N_3 = +\Delta N_j = f(N_3^- - N_j^-) \tag{8.46c}$$

Thus equation (8.45) becomes

$$\delta I_{3j}(t_1) = -cf(N_3^- - N_j^-)(k_3^r - k_j^r) \tag{8.47}$$

But, for this case $[k_3 \gg k_1, k_2]$, the population N_3^- of level 3 will have decayed to almost zero for $t \geqslant 5k_3^{-1} \equiv t_o$. Therefore if we measure $\delta I_{3j}(t_1)$ at times $t_1 > t_o$ we can set $N_3^- \approx 0$. Therefore equation (8.47) reduces to

$$\delta I_{3j}(t_1) = cfN_j^-(k_3^r - k_j^r) = cfN_j^0(k_3^r - k_j^r)\exp(-k_jt) \tag{8.48}$$

where the last equality follows from (8.43). If $\ln \delta I_{3j}(t_1)$ v. t is plotted, the slope of the resulting straight line yields the total decay rate, k_j. In this way k_1 and k_2 are obtained.

The assumptions that $W_{ij} = 0$ and $N_3^- = 0$ can be tested by plotting k_i v. T and $\ln \delta I_{3j}$ v. t respectively as illustrated in Figures 8.9(a) and (b). In the regime $T < T_o$, $W_{ij} \simeq 0$. In the regime $t > t_o$, $N_3^- \simeq 0$.

---

*Schmidt et al.[103] have called this effect 'microwave induced delayed phosphorescence'. We shall refer to it as transient ODMR by analogy to transient n.m.r. experiments[2].

(b) *The relative radiative rates*, k_i^r — From equation (8.42), given that $N_3^- = 0$, we have

$$I(t') = c(k_j^r N_j + k_i^r N_i) \qquad (i = 1, j = 2) \qquad (8.49)$$

If, on the other hand, r.f. power is *continuously* applied to transition $3j$ starting at $t = 0$, then for $t > t_o$ both N_3^- and N_j^- will be essentially zero and

$$I_{c3j}(t') = ck_i^r N_i \qquad (i \neq j) \qquad (8.50)$$

where subscript c designates the continuous application of r.f. power starting at $t = 0$. Hence

$$I(t') - I_{c3j}(t') \equiv \Delta I_{c3j}(t') = ck_j^r N_j \qquad (8.51)$$

Combining equations (8.48) and (8.51) we get

$$\delta I_{3j}(t')/\Delta I_{c3j}(t') = f(k_3/k_j - 1) \qquad (8.52)$$

and we obtain directly the ratios of the radiative rate constants, given f. If the $3j$ transition is *saturated*, $f = 1/2$. In general f can be evaluated independently (see Section 8.4.3.2f).

(c) *The determination of* k_3 — From equation (8.50) we note that the total intensity $I_{c3j}(t')$ under continuous r.f. pumping is $I_{c3j}(t') = ck_i^r N_i$. On the

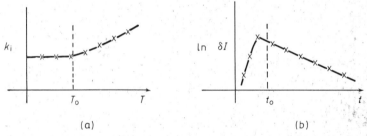

k_i T_o T $\ln \,\, \delta I$ t_o t

(a)   (b)

Figure 8.9 A method of verifying the assumptions, $W_{ij} = 0$ and $N_3^- = 0$. (a) In the regime $T < T_o$, W_{ij} can be neglected; (b) in the regime $t > t_o$, N_3^- can be neglected

other hand, if r.f. power is applied only *momentarily* at $t = t_1$ then the total intensity is [cf. equations (8.46) and (8.43) with $N_j - N_3 \approx N_j$]

$$I_{3j}(t') = c[k_i^r N_i + fN_j^-(t')k_3^r + (1 - f)k_j^r N_j^-(t')]$$
$$= c[k_i^r N_i^- \exp(-k_i t') + fk_3^r N_j^- \exp(-k_3 t') + (1-f)k_j^r N_j^- \exp(-k_j t')] \qquad (8.53)$$

where $t' = t - t_1$. The first term is just the normal contribution from level i, and is equivalent to $I_{c3j}(t')$. The third term is the contribution due to level j *after* applying r.f. power at t_1. Hence the first and third terms constitute the *time-dependent baseline* for the decay of level 3. Given that $\Delta I_{c3j}(t') = ck_j^r N_j$, all the quantities are known for the first and third terms. By subtracting the baseline from $I_{3j}(t')$ we obtain an expression for the intensity due to level 3 alone, $I_3(t')$

$$I_3(t') = cfN_j^- k_3^r \exp(-k_3 t') \qquad (8.54)$$

The slope of the line from a plot of $\ln I_3(t')$ v. t' yields k_3. Clearly if $k_1^r \approx k_2^r \approx 0$ the baseline would be time-independent and k_3 is obtained directly from the phosphorescence decay curve.

(d) *The steady-state populations* — At $t = 0$, $I(0) \approx c k_3^r N_3^o$*. Combining this with (8.48) gives

$$\delta I_{3j}(t')/I(0) = f(N_j^o/N_3^o) \exp(-k_j t) \tag{8.55}$$

(since $k_j \ll k_3$) which determines the ratios N_i^o/N_j^o.

(e) *The relative pumping rates* λ_i/λ_j — From equation (8.32) we obtain

$$\lambda_i N_o^o = k_i N_i^o \qquad (i = 1, 2, 3)$$

Hence

$$N_i^o = \frac{\lambda_i}{k_i} N_o^o \tag{8.56}$$

Since N_i^o/N_j^o are given by (8.55) and the k_i values are known, equation (8.56) provides us with λ_i/λ_j

$$\lambda_i/\lambda_j = \frac{N_i^o}{N_j^o} \frac{k_j}{k_i} \tag{8.57}$$

(f) *The determination of* f — In several of the equations above the parameter f appeared. If the r.f. power saturates the two levels involved $f = \frac{1}{2}$. However, it is not safe to assume that $f = \frac{1}{2}$. Schmidt *et al.* have described a method for measuring f[91].

If r.f. power is applied briefly to the $3j$ transition at t_1, the populations are given by (8.46) and the time development for level j is

$$N_j(t') = (1-f)N_j^-(t_1) \exp(-k_j t') \tag{8.58}$$

If at a later time t_2 r.f. power is again briefly applied, the intensity change will be ($\Delta t \equiv t_2 - t_1$)

$$\begin{aligned} \delta I_{3j}(t_2) &= cf N_j^-(t_2)(k_3^r - k_j^r) \\ &= cf(1-f)N_j^-(t_1)(k_3^r - k_j^r) \exp(-k_j \Delta t) \end{aligned} \tag{8.59}$$

Hence $\delta I_{3j}(t_2)/\delta I_{3j}(t_1) \equiv r = (1-f) \exp(-k_j \Delta t)$

and

$$f = 1 - r \exp(-k_j \Delta t) \tag{8.60}$$

Since all the parameters on the right have been measured, the value of f is established.

8.4.4　Transient ODMR and the rotating frame

8.4.4.1　Adiabatic inversion

Throughout the discussion thus far we have handled the magnetic resonance phenomenon strictly in terms of the effect r.f. fields have on the population

---

*This assumption may not be valid, for if $k_3^r \gg k_i^r$, and if $N_i^o \gg N_3^o$ as well, then the terms in $I(0)$ are comparable. In that case pulsed (optical) excitation methods can be used[91].

of the Zeeman levels. In order to discuss the dynamic behaviour of the spin system including coherence effects it is necessary to introduce the concept of the rotating frame (or the interaction representation) which has been so helpful in understanding magnetic resonance experiments[2].

Consider a collection of free spins precessing at the Larmor frequency ω_L about the direction of the external magnetic field $\boldsymbol{H} = H_o\boldsymbol{k}$ in the laboratory reference frame $\mathscr{S}$. If we view the spin system from a reference frame $\mathscr{S}'$ which is also rotating about $\boldsymbol{k}$ at ω_L then the spins will appear stationary. The net magnetisation vector $\boldsymbol{M} = M_o\boldsymbol{k}$ lies along $\boldsymbol{k}$. If a small oscillatory field $\boldsymbol{h}_1$ also at ω_L is applied perpendicular to $\boldsymbol{k}$ it will also appear to be stationary when viewed from $\mathscr{S}'$. However, the presence of h_1 in $\mathscr{S}'$ will cause the magnetisation vector $\boldsymbol{M}$ to precess in $\mathscr{S}'$ about h_1. The angle θ through which it will have precessed in time t is simply given by $\theta = \gamma h_1 t$. If we choose the proper values of h_1 and t we can make $\theta = \pi$, for example. This means that $\boldsymbol{M}$ will end up in the $-\boldsymbol{k}$ direction in $\mathscr{S}'$. Back in the laboratory frame, this corresponds to a population inversion and this procedure is referred to as adiabatic inversion with a 180 degree pulse. (The adiabatic constraints have been discussed elsewhere.) Instead of applying h_1 for a given time t one can sweep through the resonance at a rate consistent with the adiabatic conditions but fast relative to certain relaxation rates. This latter variation is referred to as the rapid passage method.

Harris[104] has given a clear discussion of the conditions for adiabatic rapid passage in excited triplet states in ZF. Using dichloroquinoxaline, Harris and Hoover[101] have achieved an inversion factor of 0.9 (cf. f of equation (8.46)) with one watt of r.f. power. Further they have outlined a procedure (not unlike that of Section 8.4.3.2) for determining various rate constants from adiabatic inversion experiments. Although the method has not yet been extensively studied, they argue that it may in some cases be preferable to saturation experiments since the time scale of the experiment (i.e. the duration of the r.f. pulse) can be very short ($\sim \mu s$). Consequently various rate constants can be ignored in the solution of the rate equations.

8.4.4.2 Coherence effects

As is well known from nuclear magnetic resonance, the interaction of a coherent r.f. field with a spin system leads to a variety of interesting coherence effects. When such coherence exists in an excited radiative state it gives rise to amplitude modulation of the optical emission. Such modulation at microwave frequencies has been observed by Chase[27] in the radiative emission from the excited quartet of Eu^{2+} in CaF_2. Harris[104] has also discussed coherence effects in excited triplet states in ZF and has observed[105] modulation of the phosphorescence intensity. More recently Schmidt, van Dorp and van der Waals[106] have detected coherence effects indirectly. Harris[104] employs the language of the interaction representation whereas Schmidt et al. use a vector model, first introduced by Feynman et al.[107], to describe the effect. The essential features can easily be described in terms of the rotating frame transformations.

Consider the perdeutero-quinoline molecule. Since the SLR rates are very small at the temperature Schmidt *et al.* used, we can treat the system as a two-level problem. The two levels of interest are τ_z and τ_y where we assume for simplicity that $k_z^r \gg k_y^r = 0$. The exciting light is maintained for a time sufficient to establish steady-state conditions and is then turned off. The population of τ_z quickly approaches zero ($N_z \to 0$), since τ_z is the rapidly decaying level. Hence after a short interval most of the excited molecules are in τ_y and therefore the emission intensity is virtually zero (cf. Section 8.4.3.2). If r.f. power is now applied to the $\tau_z \leftrightarrow \tau_y$ transition for a time t, the emission intensity is expected to increase. However, since the final state of the magnetisation vector depends on the duration of the r.f. pulse, through the relation $\theta = \gamma h_1 t$, the actual population in τ_z depends on t. Consequently the intensity of the phosphorescence also depends on t and is, in fact, proportional to $\cos^2 h_1 t$. (The conditions necessary to observe modulation at *microwave* frequencies (or optical phase precession) have also been discussed by Harris[104]) Schmidt *et al.* made point-by-point measurements of the change in intensity $\delta I_{zy}(t)$ v. t and thus obtained indirect evidence of the existence of coherence effects in the triplet manifold of quinoline-d_7.

It may, in fact, turn out that excited triplet states will provide an ideal system in which to study electron spin dynamics. These systems are unique in that the ZF splitting provides a means of studying such phenomena in the absence of an external field.

8.4.5   Spin–lattice relaxation rates

Until recently, relatively little was known about spin–lattice relaxation in molecular crystals. By and large, our understanding of SLR in triplet states is still very meagre, even though a few measurements have been made recently[11, 16, 108–111] on the lowest excited triplet state of aromatic molecules. The related free-radical SLR studies are fraught with pitfalls[35], and these pitfalls are all the more subtle in excited-state studies. A great deal of caution is necessary, therefore, in order to ensure that extraneous contributions due to cross-relaxation (to other triplets or free radicals), radiative and non-radiative decays, Zeeman effects, exciton effects, localised rotation or tunnelling and so on, do not invalidate the SLR measurements or the conclusions. The use of electron spin-echo methods, when possible, is recommended.

In the related free-radical systems ($S = \frac{1}{2}$) modulation of the spin–orbit interaction rather the h.f.i. appears to be the dominant relaxation mechanism[35] even though the g-tensor is almost isotropic. The g-tensor anisotropy in triplet-state systems is comparable to that found in free radicals. Since the SLR rates in triplet states are faster (at $T \leqslant 4.2$ K) than in typical free-radical systems, one might infer that phonon modulation of the spin–spin interaction will provide the dominant relaxation mechanism for triplet-state systems. This point remains to be established.

A systematic and careful investigation of SLR rates in excited triplet states as a function of magnetic field and temperature should prove to be an interesting area of research, despite the many difficulties associated with such studies. Some of the unique features of excited-state SLR studies have already been noted by Geschwind[18].

8.5 LEVEL CROSSING

It is not necessary to use r.f. power in order to alter the populations of the triplet-state levels. Any perturbation, such as temperature change or a (d.c.) magnetic field change, which influences the populations can cause a change in the emission intensity. Level-crossing experiments, which have been used effectively in gas-phase studies[112], represent a particularly interesting example of such perturbations. A very general argument suggests that similar experiments should also be possible in the solid state.

Consider a molecule in an excited state, e, where e is higher in energy than the lowest excited triplet state, T_1. We designate two of the levels in T_1 by i and j. Let the probability amplitude for decay via i and j to the ground state, g, be represented symbolically by ϕ_{gie} and ϕ_{gje}. If these are two alternate and distinct paths, then the probability P for decay is simply the sum of the squares of the probability amplitudes:

$$P = |\phi_{gie}|^2 + |\phi_{gje}|^2$$

However, if the two alternate paths are equivalent (i.e. when i and j 'cross') then the probability for decay is the square of the sum of the probability amplitudes:

$$P = |\phi_{gie} + \phi_{gje}|^2$$

The additional (interference) terms in the latter expression will affect the intensity of the emission from i, j.

Cheng and Kwiram searched for such level anticrossing signals in several triplet state systems, by scanning the magnetic field (in the absence of r.f. power). An intense signal was observed[53] from pyrazine in a p-dichlorobenzene host at $H \approx 100\,\text{G}$. The line-width was a few gauss. Unfortunately, the value of the field at which the signal is observed is only about one-half of what one would expect, based on the ZF parameters of pyrazine in its lowest triplet state. The origin of this signal is not understood.

More recently, similar phenomena have been observed by Veeman and van der Waals[113] in benzophenone crystals and identified as level anticrossing signals. Signals due to cross-relaxation coincidences are also reported. This provides unambiguous evidence for level-crossing effects in the solid state.

Level-crossing methods may provide a means of studying the ZFS parameters at higher temperatures. It is also conceivable that the magnetic resonance parameters of higher excited triplet states involved in the ISC process could be studied using such methods.

Acknowledgement

Because of the rapid developments in this field, every effort has been made to include the most recent published and unpublished results obtained in various laboratories. I wish to express my appreciation to those authors who were kind enough to send preprints of their recent work. The literature was surveyed through the first half of 1971.

References

1. Ramsey, N. F. (1956). *Molecular Beams*, (Oxford: Oxford University Press)
2. Abragam, A. (1961). *Principles of Magnetic Resonance*, (Oxford: Oxford University Press)
3. Bitter, F. (1949). *Phys. Rev.*, **76**, 833
4. Brossel, J. and Bitter, F. (1952). *Phys. Rev.*, **86**, 308
5. Brossel, J. and Kastler, A. (1949). *Compt. Rend.*, **229**, 1213
6. See, for example, Kastler, A. (1967). *Physics Today*, **20**, (9), 34
7. Geschwind, S., Collins, R. J. and Schawlow, A. L. (1959). *Phys. Rev. Lett.*, **3**, 545
8. Hutchison, C. A. and Mangum, B. W. (1961). *J. Chem. Phys.*, **34**, 908
9. McConnell, H. M. and Kwiram, A. L. Unpublished results
10. Sharnoff, M. (1967). *J. Chem. Phys.*, **46**, 3263
11. Kwiram, A. L. (1967). *Chem. Phys. Lett.*, **1**, 272
12. Geschwind, S. (1972). *Paramagnetic Resonance*, 353 (New York: Plenum Press)
13. Hochstrasser, R. M. and Castro, G. (1965). *Solid State Commun.*, **3**, 425
14. Brandon, R. W., Gerkin, R. E. and Hutchison, C. A., Jr. (1964). *J. Chem. Phys.*, **41**, 3717
15. Schwoerer, M. and Sixl, H. (1968). *Chem. Phys. Lett.*, **2**, 14
16. Haarer, D. and Wolf, H. C. (1970). *Mol. Cryst. and Liquid Cryst.*, **10**, 359
17. Geschwind, S. (1967). *Hyperfine Interactions*, 225 (Amsterdam: North Holland)
18. Geschwind, S. (1969). *Proc. XV Colloq. AMPERE*, 61 (Amsterdam: North Holland)
19. Morigaki, K. and Hayashi, H. (1970). *Kotai Butsuri*, 5, 528
20. El-Sayed, M. A. (1971). *Accounts Chem. Res.*, **4**, 23
21. McGlynn, S. P., Azumi, T. and Kinoshita, M. (1969). *Molecular Spectroscopy of the Triplet State*, (New Jersey: Prentice-Hall)
22. Geschwind, S., Devlin, G. E., Cohen, R. L. and Chinn, S. R. (1965). *Phys. Rev.*, **137**, A1087
23. Imbusch, G. F. and Geschwind, S. (1965). *Phys. Lett.*, **18**, 109
24. Imbusch, G. F., Chinn, S. R. and Geschwind, S. (1967). *Phys. Rev.*, **161**, 295
25. Brya, W. J., Geschwind, S. and Devlin, G. E. (1968). *Phys. Rev. Lett.*, **21**, 1800
26. Imbusch, G. F. and Geschwind, S. (1966). *Phys. Rev. Lett.*, **17**, 238
27. Chase, L. L. (1968). *Phys. Rev. Lett.*, **21**, 888
28. Terenin, A. (1944). *Zhur. Fiz. Khim.*, **18**, 1
29. Lewis, G. N. and Kasha, M. (1945). *J. Amer. Chem. Soc.*, **67**, 994
30. Ehret, P., Jesse, G. and Wolf, H. C. (1968). *Z. Naturforsch.*, **23A**, 195
31. van der Waals, J. H. and deGroot, M. S. (1959). *Mol. Phys.*, **2**, 333
32. McClure, D. S. (1949). *J. Chem. Phys.*, **17**, 665
33. Weissman, S. I. (1950). *J. Chem. Phys.*, **18**, 232
34. Kwiram, A. L., Dalton, L. R. and Cowen, J. A. (1968). *Bull. Amer. Phys. Soc.*, **13**, 458
35. Dalton, L. R. (1971). *Thesis*, Harvard University
36. Lessinger, L. and Kwiram, A. L. Unpublished results
37. Carrington, A. and McLachlan, A. D. (1967). *Introduction to Magnetic Resonance*, 115 (New York: Harper and Row)
38. Mulliken, R. S. (1955). *J. Chem. Phys.*, **23**, 1997
39. van der Waals, J. S. and deGroot, M. S. (1967). *The Triplet State*, 101 (Cambridge: University Press)
40. Schmidt, J., Hesselmann, I. A. M., deGroot, M. S. and van der Waals, J. H. (1967). *Chem. Phys. Lett.*, **1**, 434
41. Azumi, T. and McGlynn, S. P. (1962). *J. Chem. Phys.*, **37**, 2413
42. Forman, A. and Kwiram, A. L. (1968). *J. Chem. Phys.*, **49**, 4714
43. Bowers, K. W. and Mazza, C. (1968). *Bull. Amer. Phys. Soc.*, **13**, 170
44. Sharnoff, M. (1968). *Chem. Phys. Lett.*, **2**, 498 and (1969). ibid, **4**, 162
45. Sharnoff, M. (1969). *Mol. Cryst. and Liquid Cryst.*, **5**, 297
46. Forman, A. and Kwiram, A. L. (1969). *Molecular Luminescence*, 321 (New York: Benjamin)
47. Dalton, L. R. and Kwiram, A. L. Unpublished results
48. Sharnoff, M. (1969). *Mol. Cryst. and Liquid Cryst.*, **9**, 265
49. Chan, Y. and Sharnoff, M. (1970). *J. Lumin.*, **3**, 155
50. Kuan, T. S., Tinti, D. S. and El-Sayed, M. A. (1970). *Chem. Phys. Lett.*, **4**, 507
51. Burland, D. M. and Schmidt, J. (1971). *Mol. Phys.* In press
52. Clements, R. F. and Sharnoff, M. (1970). *Chem. Phys. Lett.*, **7**, 4

53. Cheng, L.-T. and Kwiram, A. L. (1969). *Chem. Phys. Lett.,* **4,** 457
54. Schmidt, J. and van der Waals, J. H. (1968). *Chem. Phys. Lett.,* **2,** 640
55. Schmidt, J. and van der Waals, J. H. (1968). *Chem. Phys. Lett.,* **3,** 546
56. Harris, C. B., Tinti, D. S., El-Sayed, M. A. and Maki, A. H. (1969). *Chem. Phys. Lett.,* **4,** 409
57. Cheng, L.-T. and Kwiram, A. L. Unpublished results
58. Vincent, J. S. and Maki, A. H. (1963). *J. Chem. Phys.,* **38,** 3088
59. Schempp, E. and Bray, P. J. (1967). *J. Chem. Phys.,* **46,** 1186
60. Buckley, M. J. and Harris, C. B. (1970). *Chem. Phys. Lett.,* **5,** 205
61. Owens, D., El-Sayed, M. A. and Ziegler, S. (1970). *J. Chem. Phys.,* **52,** 4315
62. Yamanashi, B. S. and Kwiram, A. L. Unpublished results. Kwiram, A. L. (1970). *Bull. Amer. Phys. Soc.,* **15,** 373
63. Chen, C. R. and El-Sayed, M. A. (1971). *Chem. Phys. Lett.,* **10,** 307
64. El-Sayed, M. A. and Kalman, O. F. (1970). *J. Chem. Phys.,* **52,** 4903
65. Kalman, O. F. and El-Sayed, M. A. (1971). *J. Chem. Phys.,* **54,** 4414
66. Fayer, M. D., Harris, C. B. and Yuen, D. A. (1970). *J. Chem. Phys.,* **53,** 4719
67. McCalley, R. C. and Kwiram, A. L. (1970). *Phys. Rev. Lett.,* **24,** 1279
68. Kwiram, A. L., Yamanashi, B. S. and Gouterman, M. (1971). *Bull. Amer. Phys. Soc.,* **16,** 832
69. Gouterman, M., Yamanashi, B. S. and Kwiram, A. L. (1972). *J. Chem. Phys.,* **56,** 4073
70. deGroot, M. S., Hesselmann, I. A. M. and van der Waals, J. H. (1966). *Mol. Phys.,* **10,** 241
71. Sharnoff, M. (1969). *J. Chem. Phys.,* **51,** 451
72. Sharnoff, M. (1969). *Symp. Faraday. Soc.,* **3,** 137
73. Francis, A. H. and Harris, C. B. (1971). *Chem. Phys. Lett.,* **9,** 181
74. Francis, A. H. and Harris, C. B. (1971). *Chem. Phys., Lett.,* **9,** 188
75. Sharnoff, M. and Iturbe, E. B. (1971). *Phys. Rev. Lett.,* **27,** 576
76. Francis, A. H. and Harris, C. B. (1971). *J. Chem. Phys.,* **55,** 3595
77. El-Sayed, M. A., Tinti, D. S. and Yee, E. M. (1969). *J. Chem. Phys.,* **51,** 5721
78. See Kwiram, A. L. (1971). *Ann. Rev. Phys. Chem.,* **22,** 133 and references therein
79. Ehret, P., Jesse, G. and Wolf, H. C. (1968). *Z. Naturforsch,* **23A,** 195
80. Parry, P. D., Carver, T. R., Sari, S. O. and Schnatterly, S. E. (1969). *Phys. Rev. Lett.,* **22,** 326
81. Chan, I. Y., Schmidt, J. and van der Waals, J. H. (1969). *Chem. Phys. Lett.,* **4,** 269
82. Cheng, L.-T., van Zee, J. A. and Kwiram, A. L. (1970). *Bull. Amer. Phys. Soc.,* **15,** 268, 373
83. Buckley, M. J., Harris, C. B. and Maki, A. H. (1970). *Chem. Phys. Lett.,* **4,** 591
84. Buckley, M. J. and Harris, C. B. (1972). *J. Chem. Phys.,* **56,** 137
85. Hyde, J. S. (1965). *J. Chem. Phys.,* **43,** 1806
86. Lin, T.-S. and Kwiram, A. L. Unpublished results
87. See Hyde, J. S., Chien, J. C. W. and Freed, J. H. (1968). *J. Chem. Phys.,* **48,** 4211
88. Sharnoff, M. (1967). *The Triplet State,* 165 (Cambridge: University Press)
89. Hameka, H. F. (1965). *Advanced Quantum Chemistry,* 188 (New York: Addison-Wesley)
90. Albrecht, A. C. (1963). *J. Chem. Phys.,* **38,** 354
91. Schmidt, J., Antheunis, D. A. and van der Waals, J. H. (1971). *Mol. Phys.* In press
92. Tinti, D. S. and El-Sayed, M. A. (1971). *J. Chem. Phys.,* **54,** 2529
93. Antheunis, D. A., Schmidt, J. and van der Waals, J. H. (1970). *Chem. Phys. Lett.,* **6,** 255
94. Gwaiz, A. A., El-Sayed, M. A. and Tinti, D. S. (1971). *Chem. Phys. Lett.,* **9,** 454
95. El-Sayed, M. A. and Chen, C. R. (1971). *Chem. Phys. Lett.,* **10,** 313
96. Buckley, M. J., Harris, C. B. and Panos, R. M. Unpublished results
97. Tinti, D. S., El-Sayed, M. A., Maki, A. H. and Harris, C. B. (1969). *Chem. Phys. Lett.,* **3,** 343
98. El-Sayed, M. A., Owens, D. and Tinti, D. S. (1970). *Chem. Phys. Lett.,* **6,** 395
99. Harris, C. B. and Hoover, R. J. Unpublished results
100. Gordon, R. G. (1967). *J. Chem. Phys.,* **46,** 4399
101. Harris, C. B. and Hoover, R. J. (1972). *J. Chem. Phys.,* **56,** 2199
102. El-Sayed, M. A. (1970). *J. Chem. Phys.,* **52,** 6438
103. Schmidt, J., Veeman, W. S. and van der Waals, J. H. (1969). *Chem. Phys. Lett.,* **4,** 341
104. Harris, C. B. (1970). *J. Chem. Phys.,* **54,** 972
105. Harris, C. B. Unpublished results

106. Schmidt, J., van Dorp, W. G. and van der Waals, J. H. (1971). *Chem. Phys. Lett.,* **8,** 345
107. Feynman, R. P., Vernon, F. L. and Hellwarth, R. W. (1957). *J. Appl. Phys.,* **28,** 49
108. Fischer, P. H. H. and Denison, A. B. (1969). *Mol. Phys.,* **17,** 297
109. Maruani, J. (1970). *Chem. Phys. Lett.,* **7,** 29
110. Hall, L. H. and El-Sayed, M. A. (1971). *J. Chem. Phys.,* **54,** 4958
111. Hochstrasser, R. M. and Prasad, P. N. (1971). *Chem. Phys. Lett.,* **9,** 113
112. Eck, T. G., Foldy, L. L. and Wieder, H. (1963). *Phys. Rev. Lett.,* **10,** 239
113. Veeman, W. S. and van der Waals, J. H. (1970). *Chem. Phys. Lett.,* **7,** 65

9
E.S.R. Study of
Irradiated Organic Crystals

MACHIO IWASAKI
Government Industrial Research Institute, Nagoya, Japan

9.1 INTRODUCTION

Electron spin resonance (e.s.r.) studies on irradiated organic single crystals have been an important source of knowledge on the hyperfine and g tensors of organic radicals, from which a number of fundamental concepts of the hyperfine interaction and g anisotropy have been derived together with information on the geometrical and electronic structure of organic radicals. Some of the results for such fundamental problems have been obtained in earlier work and have already been reviewed by a number of workers[1-3].

The purpose of this article is to review papers published during the past 5 years (1966–August 1971) except for some earlier work from which the recent studies stem. However, it is not feasible to cover all the papers which appeared during that period, so some selected topics will be discussed mainly in connection with the radiation damage process in organic acids studied through the structural information obtained from the oriented radicals trapped in single crystals. The recent studies in this field have been primarily concerned with the elucidation of the radiation-damage process in organic acids and related compounds, so most of the papers quoted in this review will be in that field. The most serious omission may be the recent progress in single-crystal work on the nucleic acid bases. However, the results obtained with these materials may have more biological significance and may be better reviewed by the appropriate worker in the field. Most of the fluorine-containing compounds are also omitted because these have been already summarised in the author's recent review[4].

9.2 BASIC PRINCIPLES

The basic principle of single-crystal e.s.r. has been already described in a number of reviews[1,2] and books[3,5] and only an outline will be given. The spin Hamiltonian of the systems under consideration may be expressed by

$$\mathscr{H} = \beta HS \cdot \boldsymbol{g} \cdot \boldsymbol{h} + \sum_i \boldsymbol{S} \cdot \boldsymbol{A}_i \cdot \boldsymbol{I}_i - \sum_i g_{ni}\beta_{ni}H\boldsymbol{h} \cdot \boldsymbol{I}_i \tag{9.1}$$

where the first term represents the electronic Zeeman energy, the second the hyperfine interaction, the third the nuclear Zeeman energy, h the unit vector along the external magnetic field, and the other symbols refer to the usual conventions. The first-order solution with the high-field approximation can be easily obtained[6-10]. First one assumes that the electron spin is quantised along the vector $g \cdot h$, the unit vector of which is given by

$$u = g \cdot h / g_{hh} \tag{9.2}$$

$$g_{hh} = (h \cdot g^2 \cdot h)^{\frac{1}{2}} \tag{9.3}$$

Assuming that the components of S perpendicular to u can be neglected, the following spin Hamiltonian is obtained:

$$\mathscr{H} = g_{hh}\beta H S_u - \sum_i K_i(S_u) \cdot h \cdot I_i \tag{9.4}$$

where $S_u = S \cdot u$ and

$$K_i(S_u) = g_{ni}\beta_{ni}H E - g_{hh}^{-1} S_u g \cdot A_i \tag{9.5}$$

where E is the unit tensor. Since $K_i(S_u) \cdot h$ is the effective field felt by the ith nucleus, the nuclear spin I_i is quantised along the vector $K_i(S_u) \cdot h$, the unit vector of which is given by

$$k_i(S_u) = K_i(S_u) \cdot h / K_i(S_u) \tag{9.6}$$

$$K_i(S_u) = (h \cdot K_i^2(S_u) \cdot h)^{\frac{1}{2}} \tag{9.7}$$

Then the spin Hamiltonian is expressed by

$$\mathscr{H} = g_{hh}\beta H S_u - \sum_i K_i(S_u) I_{ki} \tag{9.8}$$

where $I_{ki} = k_i(S_u) \cdot I_i$. Finally, the first-order energy is

$$E(M_s, M_I) = g_{hh}\beta H M_s - \sum_i K_i(M_s) M_I \tag{9.9}$$

For a single coupling nucleus with $I = \frac{1}{2}$, the two allowed ($\Delta M_s = 1, \Delta M_I = 0$) and the two forbidden ($\Delta M_s = 1, \Delta M_I = 1$) transitions are

$$h\nu_a = g_{hh}\beta H \pm (K_+ - K_-)/2 \tag{9.10}$$

$$h\nu_f = g_{hh}\beta H \pm (K_+ + K_-)/2 \tag{9.11}$$

where K_+ and K_- denote $K(+\frac{1}{2})$ and $K(-\frac{1}{2})$, respectively. The separations of the two allowed and the two forbidden lines are then

$$h\Delta\nu_a = K_+ - K_- \tag{9.12}$$

$$h\Delta\nu_f = K_+ + K_- \tag{9.13}$$

The transition probabilities for the allowed and forbidden transitions are proportional to $\cos^2(\theta/2)$ and $\sin^2(\theta/2)$, respectively, where θ is the angle between the two vectors corresponding to the directions of the nuclear quantisations, $k(+\frac{1}{2})$ and $k(-\frac{1}{2})$, before and after the transition, respectively. Then

$$\cos\theta = h \cdot K_+ \cdot K_- \cdot h / K_+ K_- \tag{9.14}$$

Thus, the transition probabilities are expressed by

$$I_a = |\Delta v_a^2 - (2v_n)^2|/|\Delta v_a^2 - \Delta v_f^2| \qquad (9.15)$$

$$I_f = |\Delta v_f^2 - (2v_n)^2|/|\Delta v_a^2 - \Delta v_f^2| \qquad (9.16)$$

where $v_n = g_n \beta_n H/h$

The $\boldsymbol{g}$ tensor can be determined from the angular variation of the centre of hyperfine lines as is seen in equations (9.10) and (9.11), and the hyperfine tensor from the angular variation of the hyperfine separations by the use of least-square fitting to equations (9.12) and (9.13)[9]. In organic free radicals, the $\boldsymbol{g}$ anisotropy is not so large that the deviation of the electron spin quantisation from the external magnetic field may be ignored in the conventional first-order approximation. In this case, $\boldsymbol{g}/g_{hh}$ in $\boldsymbol{K}(S_u)$ may be equated by the unit tensor. The observation of the forbidden lines makes it possible to determine the relative sign of the principal elements of the hyperfine tensor[7,9]. In the case with large hyperfine coupling values, second-order effects should be taken into consideration. The second-order solution of the spin Hamiltonian has been obtained by Maruani et al.[11] for the non-axially symmetric hyperfine and $\boldsymbol{g}$ tensors which are not coaxial with each other.

The origin of the g anisotropy is the mixing of the excited orbital energy levels with the ground level via the spin–orbit couplings. The g shift from the free spin values is given by[5,12]

$$\Delta g_{hh} = 2\sum_n \sum_{k,j} \frac{\langle \psi_0 | \zeta_k L_{hk} \delta_k | \psi_n \rangle \langle \psi_n | L_{hj} \delta_j | \psi_0 \rangle}{E_0 - E_n} \qquad (9.17)$$

where the sum runs over all pairs of atoms k, j; ζ_k is the spin–orbit coupling constant of the kth atom, L_h the component along the field direction of the orbital angular momentum operator, and δ_k means that when the L_{hk} operator inside the bracket acts on some atomic orbital it gives zero, unless the orbital belongs to atom k. The wave functions ψ_0 and ψ_n may be regarded as the molecular orbitals occupied by the unpaired electron in the ground state and in the nth excited state, respectively, and $E_0 - E_n$ is the difference in energy between these states. A marked g shift is seen in radicals in which the unpaired electron is largely localised on a heavy atom with a large spin–orbit coupling constant. Likewise another factor may be the existence of orbital levels closely separated from the unpaired electron orbital, providing that the matrix element in equation (9.17) is non-vanishing.

The trace of the hyperfine coupling tensor corresponds to the isotropic coupling constant which is given by

$$a = (8\pi/3)g_n\beta_n g\beta |\psi(0)|^2 \qquad (9.18)$$

The anisotropic component, $\boldsymbol{B} = \boldsymbol{A} - a\boldsymbol{E}$, arises from the dipole–dipole interaction between the unpaired electron and nuclear magnetic moment. The element of the anisotropic hyperfine tensor along the external field direction is given by

$$B_{hh} = g_n\beta_n g\beta \langle \psi_0 | (3\cos^2\theta - 1)/r^3 | \psi_0 \rangle \qquad (9.19)$$

The isotropic component arises from the spin density at the nucleus concerned, so that the spin density in the orbitals having a node at the nucleus does not contribute to the isotropic component. The major contribution arises from the spin in the s orbital belonging to the nucleus concerned. The anisotropic component arises from the spin densities in the orbitals having non-spherical symmetry with respect to the nucleus concerned. Consequently, the major contribution comes from the spin in the p or d orbitals.

Thus the hyperfine and g tensors provide us with detailed information about the unpaired electron wave function. In addition, the principal axes of the hyperfine and g tensors reflect the geometrical structure of the radicals as well as the symmetry of the unpaired electron orbital, so that direct information on the geometry of radicals and the orientation of the trapped radicals in the crystalline lattice can be obtained.

9.3   STRUCTURE OF SOME σ-ELECTRON RADICALS

While π-electron radicals in irradiated organic crystals have been extensively studied and their structures were well characterised in earlier work[2] only a few σ-electron radicals have been studied as oriented radicals in single crystals during recent years. Since the nature of the hyperfine coupling and g anisotropy is very different from that in the π-electron radicals, σ-electron radicals may be of considerable interest.

9.3.1   Carbon-based σ-radicals

9.3.1.1   Formyl radical

One of the most interesting σ-radicals is the formyl radical $H\dot{C}{=}O$ with an unusually large proton hyperfine coupling as compared with that found in π-radicals. Holmberg[13] has determined the proton and ^{13}C hyperfine tensors together with the g tensor using an irradiated single crystal of formic acid. The results are tabulated in Table 1 together with those obtained for other σ-radicals. The large proton hyperfine coupling indicates that the spin density of 0.249 is delocalised on the H_{1s} orbital as a result of the contribution of the no-bond structure $\cdot H\ddot{C}{=}O$. The negative anisotropic component in the direction normal to the radical plane ensures a positive value for the isotropic coupling. The g tensor exhibits a negative g shift similar to that for CO_2^- in sodium formate[14], indicating the existence of a low-lying π antibonding orbital, and an appreciable amount of spin density on the oxygen atom having a large spin–orbit coupling constant. The ^{13}C hyperfine coupling tensor is also similar to that found for CO_2^-. The large isotropic component 130 G indicates that this radical has a bent structure. The spin densities in the C_{2s}, C_{2p_y}, and C_{2p_x} orbitals are estimated to be 0.115, 0.432, and 0.086, respectively.

9.3.1.2   Substituted formyl radicals

In an irradiated single crystal of malonic acid, McCalley and Kwiram[15] have found an extremely interesting radical $RCH_2\dot{C}{=}O$, where the hydrogen atom in the formyl radical is substituted by the RCH_2 group ($R = COOH$).

Table 9.1 Hyperfine and g tensors for some carbon-based σ-radicals

Radical	Host crystal	g Tensor	¹³C Hyperfine tensor* isotropic	anisotropic			¹H Hyperfine tensor* isotropic	anisotropic			Reference
HC—O	formic acid	2.0037 2.0023 1.9948	+130.5	+26.0	−8.7	−17.3	+126.3	+8.7	−2.8	−6.0	13
RCH₂Ċ = O	malonic acid	2.0040 2.0019 1.9964	α +149.6	+29.0	−12.5	−16.4	−2.7	+4.5	−1.6	−3.0	15
			β +52.0	+9.4	−4.5	−4.9					
RCH=ĊH	maleic acid	2.0032 2.0026 2.0018	+90.4	+43.4	−15.1	−28.2	α +13.5	+13.5	−3.9	−9.6	16
							β +57.9	+1.6	−0.4	−1.1	19
ĊO₂⁻	sodium formate	2.0032 2.0014 1.9975	+167.2	+27.9	−11.4	−16.4					14

*The coupling values are given in gauss.

The unusually large hyperfine coupling observed for the methylene ^{13}C atom indicates that a considerable amount of the unpaired spin is delocalised on the β ^{13}C atom even though the hydrogen atom in the formyl radical is replaced by a carbon atom. The isotropic component leads to a spin density of c. 0.2 in the single bond between the carbon atoms. The ^{13}C hyperfine coupling tensor of the radical carbon atom and the g tensor are quite similar to those for the formyl radical as shown in Table 9.1. The hyperfine couplings with two methylene protons are also determined. However, the nature of the β proton couplings in the $R\dot{C}=O$ σ-radicals is not well elucidated.

9.3.1.3   Vinyl-type radicals

One of the most important organic σ-radicals may be a vinyl-type radical. Recently, Iwasaki and Eda[16] have found the substituted vinyl radical HOOC—CH=CH in a single crystal of maleic acid irradiated at 77 K. The g tensor exhibits typical anisotropy of σ-electron radicals, that is, the g_{max} axis normal to the radical plane, the g_{min} value with a negative shift, and the g_{int} value close to the free-spin value. The negative shift of the g_{min} value indicates excitation to the low-lying π antibonding orbital. However, the magnitude of the shift is smaller than that in the $R\dot{C}=O$ or CO_2^- radicals, where part of the unpaired spin density on the oxygen atom contributes to the large negative shift.

   The experimental evidence for the positive sign of the isotropic component of the α proton coupling was obtained on the basis of the anisotropic component perpendicular to the radical plane being of negative value; this is easily understood from simple consideration of the dipole–dipole interaction. The magnitudes and the directions of the anisotropic components are consistent with the theoretical values calculated by the method of McConnell and Strathdee[17], with the correction pointed out by Barfield[18]. The ^{13}C hyperfine tensor was also determined[19]. The relatively small isotropic component indicates that the s nature of the unpaired electron orbital is smaller than that of the sp^2 hybrid orbital. The rough estimate of the C=C—H bond angle obtained from the sp hybridisation ratio is 140–150 degrees. This is consistent with the calculated bond angle and reveals the positive isotropic coupling of α proton[20]. The positive value is in marked contrast to the negative value found for many π-radicals.

9.3.2   Oxygen- and nitrogen-based σ-radicals

9.3.2.1   Carboxyl radicals

The e.s.r. evidence for the existence of the carboxyl radical $RCOO\cdot$, which is often assumed in organic solution chemistry, has been obtained by Iwasaki and his group in irradiated crystals of a number of unsaturated dicarboxylic acids such as maleic acid[21, 22], potassium hydrogen maleate[23], sodium hydrogen maleate[24], potassium hydrogen fumarate[25], and acetylene dicarboxylic acid[26]. For example, maleic acid gives radical (1).

$$
\begin{array}{ccc}
\text{H} & & \text{H} \\
\diagdown & & \diagup \\
\text{C} & = & \text{C} \\
\diagup & & \diagdown \\
\text{O}{=}\text{C} & & \text{C}{-}\text{O}\bullet \\
| & & \\
\text{OH}\cdots\text{O} & &
\end{array}
$$

(1)

by loss of the side OH proton. The observed g tensors are highly aniso-tropic with a large positive shift ($g_{max} \approx 2.02$), indicating that the unpaired electron is mainly localised in the carboxyl oxygen atoms. The principal direction of the element which is close to the free spin value is found to be in the COO plane. They have concluded that the unpaired electron occupies the antibonding orbital, $\psi = c_1 p_{01} - c_2 p_{02}$, between the two in-plane p orbital of the carboxyl oxygen atoms, forming σ-electron radicals. Similar carboxyl radicals were also found in succinic acid[27] and potassium hydrogen malonate[28] irradiated at 4.2 K.

It is interesting to note that potassium hydrogen maleate, which is known to have a symmetrical *intramolecular* hydrogen bond, gives a symmetrical carboxyl radical (2) in which the unpaired electron is shared by the two ring oxygen atoms[23].

$$
\begin{array}{ccc}
\text{H} & & \text{H} \\
\diagdown & & \diagup \\
\text{C} & = & \text{C} \\
\diagup & & \diagdown \\
\text{O}{=}\text{C} & & \text{C}{=}\text{O} \\
\diagdown & & \diagup \\
\text{O} & \bullet & \text{O}
\end{array}
$$

(2)

A similar radical was found in potassium hydrogen malonate[28], which is known to have a symmetrical *intermolecular* hydrogen bond. In this case the unpaired electron is shared by the oxygen atoms belonging to the two neighbouring molecules (see Structure (3)).

$$
\begin{array}{ccc}
& \text{O}^{\frac{1}{2}-} \quad {}^{\frac{1}{2}-}\text{O} & \\
& \diagup \qquad \diagdown & \\
-\text{CH}_2{-}\text{C} \quad \bullet \quad \text{C}{-}\text{CH}_2{-} \\
& \diagdown\diagdown \qquad \diagup\diagup & \\
& \text{O} \qquad \text{O} &
\end{array}
$$

(3)

Both the radicals (2) and (3) indicate similar g anisotropies which are con-siderably smaller than that of the localised carboxyl radicals like (1).

9.3.2.2 Iminoxy radicals

Kurita and his group have studied a number of iminoxy radicals in single crystals. For example, glyoxime forms σ-radicals by loss of the OH hydrogen atom[29]. The nitrogen hyperfine coupling tensor (39.9, 22.0, 20.4 G) indicates that the unpaired electron occupies the antibonding orbital between the two in-plane lone-pair orbitals of the nitrogen and oxygen atoms. The spin density on the nitrogen atom is 0.41 and the rest of the spin is mainly on the oxygen atom, giving rise to the fairly large g anisotropy (2.0025, 2.0064, 2.0118). The bond angle C=N—O was estimated to be 142 degrees from the spin density in the nitrogen 2s and 2p orbitals. The anisotropic component of the hydrogen hyperfine tensor (24.8, 21.4, 19.9 G) indicates that the radical has the *syn* configuration. The isotropic component is also consistent with this configuration in comparison with the β-proton coupling in vinyl radicals. The other examples are found in *syn-* and *anti-p-*chlorobenzaldoxime[30], and potassium 1-oximinopropionate[31].

9.3.2.3   Others

In irradiated malonamide Cyr and Lin[32] have found a nitrogen-centred radical in which the proton hyperfine coupling is unusually large (81.3 G). They have assigned this to be the σ-radical RCON̊H formed by loss of the NH hydrogen atom. Similar radicals having a large proton hyperfine coupling are also found in cyanoacetylurea (86 G) [33] and dicyanodiamide (75 G)[34]. However, Neta and Fessenden[35] have suggested that the single-crystal data are more in accord with a radical structure of the type RCH=N̊, because of the relatively small isotropic component of the nitrogen coupling tensor and of the small anisotropic component of the hydrogen coupling tensor. Very recently Symons[36] has suggested the possibility that the radical is $\overset{+}{N}H_3$.

9.4   RADIATION DAMAGE PROCESSES

In earlier work, single-crystal studies were carried out mainly for stable radicals surviving at room temperature[2]. However, recent attention has been focused on the unstable species which are detectable only at low temperatures. Usually irradiation at 77 or 4 K gives entirely different species from those produced at room temperature. Furthermore, it is gradually becoming clear that the stable radicals are formed as secondary products from the initial unstable species, through the radical reactions in the irradiated solids. Consequently, to detect the primary species produced by ionising radiation is the most important problem in understanding the radiation damage process.

Recent studies on irradiated single crystals of organic acids and related compounds have shown that the main primary process induced by ionising radiation is the ejection of an electron from a molecule to leave a positive hole trapped in a preferred site (molecular cation), followed by electron capture and the formation of a molecular anion. Both the cationic and anionic species should be detectable by e.s.r. as paramagnetic centres. It was found that such ionic primaries are often stabilised at low temperatures. It was further found that a number of neutral radicals are created by a sequence of events from the ionic primaries until finally they are stabilised at room temperature, or they disappear to form diamagnetic species by recombination or disproportionation. A number of attempts have been made to elucidate the reaction paths from the positive and negative primaries to the secondary radicals. The fates of ionic primaries are now becoming fairly clear. It seems that the sequence of the events from the positive primary is initiated by proton donation to the neighbouring molecule and that from the negative primary by proton acceptance, as will be demonstrated in the following sections.

9.4.1   Electron-trapping centres (molecular anions)

9.4.1.1   Sulphur-containing compounds

The first example of an ionic primary was found independently by two groups of workers[37, 38] in the case of L-cystine dihydrochloride. In their first observation of a radical having a *g* tensor (2.0024, 2.0174, 2.0178), Box and Freund[37] intuitively assigned the spectrum to a cation and Akasaka et al.[38] assigned the same spectrum to an anion on the basis of the theoretical aspect of its *g* anisotropy. Shortly after the latter assignment, Box et al.[39] detected an extra absorption having a *g* tensor (2.005, 2.028, 2.033) in the crystal irradiated at 4.2 K, and found that the extra absorption exhibits *g* anisotropy consistent with assignment to the cation. The large *g* anisotropy indicates that the unpaired electron is localised on the sulphur atom because of its large spin–orbit coupling constant. Furthermore, the symmetry properties of the paramagnetic centre in the crystalline lattice clearly show that there is no scission of the S—S bond, and that the unpaired electron should equally occupy the two sulphur atomic orbitals of the S—S bond. If the species is to be an anion, the unpaired electron is expected to occupy the lowest vacant orbital, that is, the S—S antibonding σ-orbital. This results in the axially symmetric *g* tensor with respect to the S—S bond, along which the free spin value for the component of the *g* tensor should be observed. If the species is a cation, the unpaired electron is expected to occupy the highest filled orbital of the undamaged molecule, that is, the antibonding orbital located between the two sulphur lone-pair orbitals. This results in a non-axially symmetric *g* tensor in which the free-spin value component is to be along the normal to the S—S bond. Furthermore, it is expected that the cation having an unpaired electron in sulphur lone-pair orbitals should have the larger *g* anisotropy because of the excitation of the electron from the other adjacent lone-pair orbital. The observed *g* tensors for the two primary species exhibit just these expected properties. The species having the axially symmetric *g* tensor with the lesser anisotropy was assigned to structure (4) and the species having the non-axially symmetric *g* tensor with the larger anisotropy was assigned to structure (5). The 3×3 line hyperfine structure of the anion, which arises from the two methylene groups, is also consistent with this interpretation.

A number of other examples have been obtained in irradiated crystals of sulphur-containing compounds. Disulphides such as dithiodiglycolic acid[40], benzoyldisulphide[27] and cystine dihydrobromide[27] were found to give a similar anion structure (4). However, it should be noted that sulphides such as thiodiglycolic acid[41, 42] give a carboxyl anion (6) similar to that found for carboxylic acids (see Section 9.4.1.2). The lower energy resulting from de-localisation of the electron over both sulphur atoms in the disulphide mole-cule probably accounts for the difference in behaviour. Thiourea and its derivatives[27, 43] give an anion (7) in which the electron is trapped in the C—C π-antibonding orbital resulting in the small g anisotropy. These results indicate that an ejected electron is captured in a low-lying vacant orbital of the surrounding molecule.

9.4.1.2   Carboxylic acids

In a single crystal of succinic acid irradiated at 77 K, Box et al.[44] found a four-line spectrum which is attributable to the molecular anion (8).

$$R-\overset{-}{S}-S-R \qquad R-\overset{+}{S}-S-R \qquad R-\overset{\cdot}{\underset{OH}{C}}\overset{O^-}{\diagup}$$

$$(4) \qquad\qquad\qquad (5) \qquad\qquad\qquad (6)$$

$$\overset{\backslash}{\underset{/}{\overset{\cdot}{C}}}-S^- \qquad HOOC-CH_2-CH_2-\overset{\cdot}{\underset{OH}{C}}\overset{O^-}{\diagup}$$

$$(7) \qquad\qquad\qquad\qquad (8)$$

The hyperfine tensors of the two methylene protons are consistent with this interpretation[42]. The ^{13}C hyperfine tensor and the g tensor indicate that the unpaired electron is mainly localised in the $2p_\pi$ orbital on the carboxyl carbon atom. However, the fairly large isotropic component (112 G) of the ^{13}C hyperfine tensor suggests that the radical carbon has a pyramidal structure in contrast to the planar structure found in many alkyl radicals ($a_c = 33$ G for HOOCĊHCOOH in malonic acid)[2]. Although the possibility that the spectrum arose from the molecular cation was not entirely excluded in this single-crystal work, the assignment to the anion was confirmed by Bennett and Gale[45] who used sodium deposits on a series of monocarboxylic acids. A number of other examples of similar molecular anions have been found in saturated carboxylic acids and their salts, such as propylmalonic acid[46], dimethylmalonic acid[27], tartaric acid[47], disodium succinate[48, 49], strontium acetate[50], and potassium hydrogen malonate[28].

In contrast to saturated carboxylic acids, unsaturated acids give the molecular anions in which the unpaired electron occupies the delocalised π orbital. The molecular anion found by Iwasaki et al.[21] in irradiated single crystals of maleic acid exhibits typical α proton couplings due to the two vinylene protons; this indicates that the unpaired electron is delocalised throughout the anion which preserves the framework of the parent molecule. Similar delocalised anion radicals were also found in potassium hydrogen fumarate[51, 52], dipotassium fumarate dihydrate[52], and fumaric acid-doped succinic acid[53]. In the last example it is interesting to note that an ejected electron is selectively captured by the guest molecule of fumaric acid rather than the host molecule of succinic acid. This may mean that the electron can migrate in the crystal and is captured by the low-lying π antibonding orbital of the unsaturated acid which is supposed to be a deeper trap for the electron.

9.4.1.3 Amino acids

The electron-trapping centres found in a number of amino acids are essentially the same as those in saturated carboxylic acids. The most thoroughly studied anion is the one obtained in a single crystal of alanine irradiated at 77 K [54-58]. Alanine is known to exist in the zwitterion form in the crystal; therefore the molecular anion should have the form $CH_3CH(\overset{+}{N}H_3)\dot{C}OO^{2-}$. However, the observed spectrum shows that the anion is protonated which results in a form of structure (9):

$$H_3C-\underset{\overset{|}{+}NH_3}{CH}-\overset{\displaystyle O^-}{\underset{\displaystyle OH}{\dot{C}}}$$

(9)

The principal directions of the OH proton coupling indicate that the proton lies in the COO plane. It was also found that the added proton originates from the $\overset{+}{N}H_3$ group in the neighbouring molecule[56]. The ^{13}C hyperfine coupling tensor with the isotropic component of 89.7 G indicates that the radical carbon atom has the pyramidal structure[58] as is the case with the anion in succinic acid. Similar anions were also found in glycine[59, 60], glycine hydrochloride[61], β-alanine[57, 62], α-aminobutyric acid[57], α-aminoisobutyric acid[57,63,64], aspartic acid[57], cysteic acid[57], valine[65], lysine hydrochloride[64], and serine[105].

9.4.2 Positive hole centres (molecular cation)

9.4.2.1 Sulphur-containing compounds

The existence of a positive hole centre in an irradiated single crystal of L-cystine dihydrochloride is well established and is described in Section 9.4.1.1. Other examples have been found for a number of sulphur-containing compounds. Disulphides such as dithiodiglycolic acid[40] and benzyldisulphide[27] give a similar cation (see structure (5)) to that of L-cystine dihydrochloride where the positive hole is trapped in the antibonding orbital between the two sulphur lone-pair orbitals. It is, however, to be noted that cystine dihydrobromide gives a bromine atom as a positive hole centre resulting from the lesser electron affinity of bromine compared with chlorine[27]. Sulphides such as thiodiglycolic acid[41, 42] and methionine[66, 67] give a cation $R-\overset{+}{S}-R$ where the positive hole is trapped in the sulphur non-bonding orbital. In thiourea derivatives the hole is delocalised throughout the π orbital system of the cation where the unpaired spin density on the sulphur atom is fairly high. These hole-trapping centres are characterised by their highly anisotropic g tensor.

Recently, Kominami et al.[68] have found a photo-induced reversible hole transfer between two types of hole traps in an irradiated single crystal of N-acetyl-DL-methionine. It was suggested from the e.s.r. spectral change that the deeper trap is a sulphur non-bonding orbital while the shallower is a carboxyl π orbital. The deeper trap exhibits a typical g anisotropy (2.023, 2.013, 2.004) for sulphide cations such as thiodiglycolic acid (2.022, 2.011, 2.004)[41] and DL-methionine (2.022, 2.013, 2.002)[67]. The shallower trap exhibits a nearly isotropic g factor (2.003 $\pm$ 0.001) with the hyperfine coupling tensor (15, 11, 6 G) due to the polar hydrogen atom. The trapped hole in the deeper site is transferred to the shallower trap by photo-excitation with 350–600 nm light. The process is reversed when the crystal is warmed to c. 140 K for a few minutes, or kept at 77 K for a few days. The results are explained by intramolecular hole transfer.

Akasaka et al.[69] studied the photo-induced hole–electron recombination in irradiated L-cystine dihydrochloride and reached the conclusion that the photo-excitation of the cation leads to recombination with the anion.

9.4.2.2 Carboxylic acids

In a crystal of succinic acid irradiated at 77 K the broad single-line spectrum was found together with the well-resolved four-line spectrum due to the molecular anion. Although the broad spectrum was tentatively assigned to the cationic species by Box et al., the structure is not well characterised[44]. In a further study of this crystal irradiated at 4.2 K an extra absorption having large g anisotropy was detected and was first attributed to the cationic species (10) [70].

$$HOOC-CH_2-CH_2-C\overset{\overset{\displaystyle \dot{O}^+}{\|}}{\underset{\displaystyle OH}{\diagdown}}$$

(10)

$$HOOC-CH_2-CH_2-C\overset{\overset{\displaystyle O}{\|}}{\underset{\displaystyle O\cdot}{\diagdown}}$$

(11)

However, on the basis of the recent analysis of its g tensor it has been reassigned to the species shown by structure (11) formed from the molecular cation by transferring the acidic proton to the neighbouring molecule. Iwasaki et al.[21] have also found a similar carboxyl radical in maleic acid and its related compounds, as mentioned in Section 9.3.3.1.

Now at this stage we may raise the following question: is there a possibility of a positive hole itself being trapped as a similar type of carboxyl radical if the carboxylic acid salts are irradiated? Very recently, Iwasaki et al.[25] have found that there are two sorts of carboxyl radical in a single crystal of potassium hydrogen fumarate irradiated at 77 K. The COO⁻ group in this molecule twists 35 degrees from the plane of the rest of the molecule containing the COOH group. This situation together with the large g anisotropy of the carboxyl radical made it possible to determine which COO group is occupied by the unpaired electron. The comparison of the g tensor with the x-ray crystallographic data indicates that one of the two carboxyl radicals is formed by the ejection of an electron from the COO^- group (12) and the other is formed by a loss of acidic hydrogen (13).

$$\cdot OOC—CH{=}CH—COOH \qquad OOC—CH{=}CH—COO\cdot$$
$$\qquad (12) \qquad\qquad\qquad\qquad\qquad (13)$$

This clearly indicates that the positive hole can be trapped in an oxygen non-bonding orbital as is the case with the sulphur-containing compounds. The results may also suggest that the trapping of the positive hole in the COOH group is followed by spontaneous proton transfer to the neighbouring molecule through intermolecular hydrogen bonding.

9.4.2.3 Amino acids

Although the electron-trapping centre in irradiated amino acids is well characterised, until recently the existence of the positive hole centre had not been clarified. Very recently, Friday and Miyagawa[71] have analysed the complex spectrum overlapping with the anion absorption in irradiated alanine by using a series of deuterated alanines such as $CH_3CHNH_3^+CO_2^-$, $CH_3CHND_3^+CO_2^-$, $CH_3CDNH_3^+CO_2^-$ and $CH_3CDND_3^+CO_2^-$. The complex spectrum of normal alanine was very much simplified in the latter crystal, and the hyperfine structure was attributed to the CH_3 group. The extra hyperfine structures found for the other crystals are systematically interpreted as arising from the C—H proton and the nitrogen atom couplings. They proposed that, on the basis of the anomalous hyperfine coupling of the CH_3 protons, the spectra can be assigned to the positive hole centre trapped in the $(H_3)C—C$ bond. This result is in marked contrast to the case of sulphur-containing compounds or carboxylic acids, in which the positive hole is trapped in the non-bonding orbital, which is supposed to be the highest filled orbital of the parent molecule.

On the other hand, a similar complex absorption in α-aminoisobutylic acid has been assigned to (14) by Fujimoto et al.[64]. However, Cadena et al.[72]

$$(CH_3)_2C(NH_2)CO_2^- \qquad (CH_3)_2CCO_2^-$$
$$\qquad (14) \qquad\qquad\qquad\qquad (15)$$

have re-assigned this to (15) because $(CH_3)_2C(\overset{+}{ND_3})CO_2^-$ gives rise to the identical spectrum. Minegishi et al.[73] have argued against this interpretation on the basis of their determination of its g tensor and have proposed that the spectrum is of the cationic species in which the positive hole is trapped at the carboxylic oxygen atom. The origin of the complex hyperfine structure was attributed to the neighbouring proton couplings in the adjacent molecule. In connection with the positive hole centre in alanine the complete analysis of the complex spectra may be an important factor in elucidating the nature of the hole trapping in irradiated amino acids.

9.4.3 Triplet–radical pairs formed from ionic primaries

Because of the inherent nature of the creation of the positive hole- and electron-trapping centres, the two centres should be produced in pairs. The large polarisation due to their formal charge may stabilise the counter ion in the vicinity forming a triplet–radical pair.

In an irradiated single crystal of maleic anhydride Iwasaki and Eda[74] have found an interesting triplet e.s.r. absorption which is interpretable by the radical pair between the anionic and cationic primaries. The fine structure constants are determined to be $|D| = 134$ G, and $|E| = 10$ G, respectively. The small hyperfine coupling having anisotropy characteristic of the α proton indicates that the species is the delocalised π radical which preserves the framework of the parent molecule. The intensity of the signal of the radical pair aligned along the c axis with a separation of one unit (5.39 Å) is enhanced by illumination with visible light at the expense of some absorptions arising from the distant pairs. The prolonged illumination leads to the disappearance of all the signals indicating the recombination of the positive hole and the electron.

A similar radical pair formed from the ionic primaries was also found in potassium hydrogen fumarate[51]. The hyperfine and $\boldsymbol{g}$ tensors of this pair are consistent with the delocalised π radicals having the geometry of the undamaged molecule. The fine structure constants are determined to be $|D| = 103$ G and $|E| = 4$ G, respectively. The fine structure tensor indicates that the paired radicals are aligned along the c axis with a separation of one unit (6.46 Å).

9.4.4 Secondary radicals from negative primaries

9.4.4.1 Amino acids

The fate of the anionic primary in irradiated amino acids has been studied most extensively. The most common radical, often stabilised at room temperature or at intermediate temperatures, is the one formed by rupture of the C—N bond[6,54,57,63–65,75,105]. From controlled warming experiments it was suggested by Box et al.[63] and by Sinclair and Hanna[54] that this radical originates from the anionic species. Alanine irradiated at 77 K gives a protonated anion as described in Section 9.4.1.3. Upon warming a loss of the NH_3 group gives rise to quantitative conversion to the $CH_3\dot{C}HCOOH$ radical[54]. It was, however, found that this radical has a different orientation from that of the stable $CH_3\dot{C}HCOOH$ radical[6], and that further warming to room temperature causes reorientation to the stable form. A similar change of the radical orientation in the crystal was also observed in irradiated α-aminoisobutylic acid[63,76] and lysine hydrochloride[64]. The mechanism involved in the formation of the stable radical from the anion is a sort of dissociative electron capture, although the protonation to the anion precedes the dissociative process.

$$RCH(\overset{+}{N}H_3)CO_2^- + e^- \rightarrow RCH(\overset{+}{N}H_3)\dot{C}O_2^{2-}$$
$$\xrightarrow{+H^+} RCH(NH_3)\dot{C}{\overset{O^-}{\underset{OH}{\big\langle}}} \rightarrow R\dot{C}HCOOH + NH_3$$

The most conclusive evidence for this process was obtained from an irradiated crystal of glycine hydrochloride in which the positive hole is trapped with the formation of Cl_2^- [61]. The stability of Cl_2^- at elevated temperatures makes it possible to observe the fate of the anion without the complication introduced by the degradation path developing from the positive primary. The quantitative formation of $\dot{C}H_2COOH$ from the anion was unequivocally observed in this system.

Besides this deamination radical, in some rare cases another type of stable radical is formed by C—H bond rupture, that is, $(\overset{+}{N}H_3)\dot{C}HCO_2^-$ in glycine[59], $(CH_3)_2\dot{C}CH(\overset{+}{N}H_3)CO_2^-$ in valine[65, 77] and $HOOC\dot{C}HCH_2CH$ $(\overset{+}{N}H_3)COOH$ in glutamic acid hydrochloride[78]. The origin of this type of radical in valine was studied in some detail by Horan et al.[79] using polycrystalline samples. The kinetics of the radical conversion process indicates that the anion quantitatively converts into the deamination radical by a first-order reaction, and that upon further warming the radical in question is formed from the deamination radical by first-order kinetics. However, it was found that there are two separate processes in the formation of this radical, since the build-up of this radical exceeds 1.5 times the original amount of the anion. This shows that the radical in question originates mainly from the anion, and partly from the cationic primary.

9.4.4.2 Saturated carboxylic acids

A mysterious single-line spectrum has often been observed in irradiated carboxylic acids when the crystal is warmed after irradiation. Schwartz et al.[80] have found that the intensity of this single-line spectrum in succinic acid is greatly enhanced after disappearance of the molecular anion. They have assigned this to CO_2^- on the basis of the similarity of the g tensor (2.0046, 2.0010, 1.9973) with the literature value (2.0032, 2.0014, 1.9975) for CO_2^- in sodium formate[14]. The the following process was suggested:

$$[HOOCCH_2CH_2COOH]^- \longrightarrow HOOCCH_2CH_3 + CO_2^-$$

On the other hand, McCalley and Kwiram[15] have studied this single-line species in malonic acid, and have obtained conclusive evidence that this species is the substituted formyl radical, $HOOCCH_2C{=}O$, based on their observation of the methylene ^{13}C hyperfine coupling as described in Section 9.3.2.2. They have further pointed out that the principal directions of the g tensor for the single-line spectrum in succinic acid are consistent with the substituted formyl radical[81]. It was also found that the disappearance of the single-line species correlates with the appearance of the $R\dot{C}HCOOH$ radical, which is the most common end-product in the radiation-damage process in carboxylic acids. Thus the reaction path from the negative primary is considered as dissociative electron capture followed by a hydrogen-abstraction reaction from the surrounding molecules.

$$R—CH_2—\dot{C}{\underset{OH}{\overset{O^-}{<}}} \longrightarrow R—CH_2\dot{C}{=}O + OH^- \longrightarrow R—\dot{C}HCOOH$$
$$(16) \qquad\qquad\qquad\qquad (17) \qquad\qquad\qquad\qquad (18)$$

The direct path from (16) to (18) was also suggested by McCalley and Kwiram[15, 81] on the basis of the deuterium substitution effects on the formation of (17) and (18). The rate-determining steps for the formation of (17) and (18) are assumed to be the C—O bond scission and the hydrogen transfer from the neighbouring molecule, respectively.

$$ H\overset{\cdot}{C}\begin{smallmatrix} \diagup OH \\ \diagdown OH \end{smallmatrix} $$
(19)

$$ RCH_2\overset{+}{C}\begin{smallmatrix} \diagup OH \\ \diagdown OH \end{smallmatrix} $$
(20)

It should be mentioned here that Holmberg[13] suggests that in irradiated formic acid the precursor of the formyl radical is (19), and that the formyl-type radicals are not formed in the sodium deposits on formic acid and other carboxylic acids studied by Bennett and Gale[45]. These results might mean that the formation of the formyl-type radicals requires protonation to the anion followed by loss of H_2O. This is analogous to the deamination process in amino acids (Section 9.4.4.1).

$$ R-CH_2-\overset{\cdot}{C}\begin{smallmatrix} \diagup O^- \\ \diagdown OH \end{smallmatrix} \xrightarrow{+H^+} R-CH_2-\overset{\cdot}{C}\begin{smallmatrix} \diagup OH \\ \diagdown OH \end{smallmatrix} \rightarrow R-CH_2\overset{\cdot}{C}=O+H_2O $$

The protonation might be due to the migration of H^+ from the diamagnetic cation (20) formed in a reaction path involving a cationic primary (see Section 9.4.5.1). An alternative explanation is given by Ayscough et al. who have reported the formation of a substituted formyl radical in irradiated polycrystalline acetic acid[82, 83]. They postulate a recombination of the electron from the molecular anion with a diamagnetic cation, liberating a hydrogen atom which forms $CH_3C(OH)_2$ and then decomposes to $CH_3\overset{\cdot}{C}=O+H_2O$.

Recently, Fujimoto and Janecka[84] have studied ^{13}C enriched sodium acetates, $^{13}CH_3CO_2^-$ and $CH_3{}^{13}CO_2^-$, and found that CO_2^- is formed at elevated temperature. They have also assigned a single-line species found in disodium succinate·$6H_2O$ to CO_2^- [48, 49]. However, the mechanism of the formation of CO_2^- is not clear. Further study may be needed to establish the relation between the formyl-type radical and the formation of the CO_2^- radical.

9.4.4.3   Unsaturated carboxylic acids

In the case of unsaturated carboxylic acids, stable radicals at room temperature are usually formed by hydrogen addition to the unsaturated bond. For example, maleic[85] and fumaric[86] acids give $HOOCCH_2-\overset{\cdot}{C}HCOOH$ and acrylic acid[87] gives $CH_3\overset{\cdot}{C}HCOOH$. Hydrogen-addition radicals have been also studied by Griffith and his group[88, 89] who used urea inclusion compounds of a number of unsaturated compounds. Recently, Iwasaki et al.[53] have found that the hydrogen-addition radical originates from the anionic primary. The spectrum of the molecular anion of succinic acid is replaced by that of the fumaric acid anion when the small amount of fumaric acid is added as a dopant. Upon warming the crystal to room temperature the spectrum of the fumaric acid anion is changed into the deuterium-addition radical if the acidic-proton deuterated crystal is used. This suggests that the origin of the added hydrogen is the acidic proton and the hydrogen-addition radical is formed from the anion. It has been postulated[53] that the origin of the added proton is the protonated diamagnetic cation where the proton affinity of the oxygen atom is supposed to be reduced by the positive formal charge, which results in the liberation of a proton. The proton from the diamagnetic cation migrates through the intermolecular hydrogen bond until it recombines with the molecular anion, forming the hydrogen-addition radical. Thus the following mechanism was proposed:

$$R\text{---}COOH \rightsquigarrow [R\text{---}COOH]^+ + e^-$$

$$[R\text{---}COOH]^+ + R\text{---}COOH \rightarrow R\text{---}COO\cdot + R\text{---}\overset{+}{C}\!\!<^{OH}_{OH}$$

$$e^- + R\text{---}COOH \rightarrow R\text{---}\dot{C}\!\!<^{O^-}_{OH}$$

$$R\text{---}\dot{C}\!\!<^{O^-}_{OH} + H^+ \text{ from } R\text{---}\overset{+}{C}\!\!<^{OH}_{OH} \rightarrow \dot{R}H\text{---}COOH + R\text{---}COOH$$

where R is the unsaturated group and $\dot{R}H\text{---}COOH$ denotes the hydrogen-addition radical. It is interesting to note that the anionic primary is stabilised by protonation to form the stable radical, in contrast to the case of saturated carboxylic acids and amino acids where the dissociative process follows the protonation.

9.4.4.4 Sulphur-containing compounds

Stable radicals often formed in irradiated sulphur-containing compounds are neutral sulphur radicals R—S· in which the unpaired electron is mainly localised in a non-bonding 3p orbital of the sulphur atom[90-92]. In irradiated Lcystine dihydrochloride, Akasaka et al.[38], found that on warming to $-78\,°C$ the S—S bond ruptured in the anionic primary giving an unstable R—S· radical in which the C—S bond has the same orientation as that in the undamaged molecule. Further warming to room temperature causes the C—S bond to rotate giving a stable R—S· radical. The process is considered to be a form of dissociative electron capture as is the case in other compounds.

$$R\text{---}S\overline{\text{---}}S\text{---}R \rightarrow R\text{---}\overline{S} + \cdot S\text{---}R$$
(21)

However, this does not mean that all the R-S· radicals are produced from the negative primary. It was found that a variety of the sulphur radicals having different conformations are formed upon warming cystine dihydrochloride[27] and penicillamine hydrochloride[93] after they have been irradiated at 4.2 K, and from cysteine irradiated at 77 K [94, 95]. Some of these are found to be formed from the positive primary as will be described in Section 9.4.5.4.

In contrast to the case of disulphides, thiodiglycolic acid gives a different stable radical, namely, HOOC—CH_2—S—$\dot{C}H$—COOH at room temperature[96]. The formation of this radical is more like the case of dicarboxylic acid. As mentioned in Section 9.4.1.1 thiodiglycolic acid[41, 42] gives the carboxyl anion in contrast to the disulphides. The difference in the formation of the stable radical is perhaps related to the different structure of the anion.

9.4.5 Secondary radicals from positive primaries

9.4.5.1 Carboxylic acids

The fate of the positive primary is one of the most serious problems in the radiation damage process and little is known about the sequence of events originating from the positive primary. Successful results were only obtained by Iwasaki et al.[21, 22] from maleic acid irradiated at 77 K. They have found that the carboxyl radical and a small amount of a vinyl-type radical are produced together with the molecular anion. The amount of the former two radicals is nearly equal to that of the anion, suggesting that the former two radicals originate from the positive primary. They have further found that the carboxyl radical converts into a vinyl-type radical when the crystal is warmed slightly above 77 K or is exposed to radiation from a tungsten light. During this change the signal due to the anion remains unchanged. The hyperfine anisotropy indicates that the radical carbon atom in the vinyl-type radical is the one which has the COO· group in the carboxyl radical. These results strongly indicate that the fate of the positive primary is a transfer of the acidic proton to the neighbouring molecule followed by a loss of CO_2, forming a decarboxylation radical:

Thus it may be suggested that the radical R·, often produced in saturated carboxylic acids by loss of COOH, also originates from the positive primary.

$$[R-COOH]^+ \rightarrow R-COO· \rightarrow R·+CO_2$$

The proton transferred from the positive primary is believed to give rise to the formation of the protonated diamagnetic cation

$$R-\overset{+}{C}\begin{smallmatrix} \nearrow OH \\ \searrow OH \end{smallmatrix}$$

(22)

It is interesting to note that the positive primary liberates the proton while the negative primary accepts the proton to form protonated anions or hydrogen-addition radicals. The marked contrast between these cases may be related to the effect of the formal charge on the proton affinity of the carboxyl oxygen atom[22, 53, 56].

In potassium hydrogen maleate, Toriyama et al.[97] have found that the carboxyl radical fragments into CO_2^- through an intermediate radical which is supposed to be a vinyl-type radical

$$\cdot OOC\!-\!CH\!=\!CH\!-\!COO^- \rightarrow \dot{C}H\!=\!CH\!-\!COO^- + CO_2 \rightarrow HC\!\equiv\!CH + CO_2^-$$

Usually the radical yield at room temperature from the unsaturated carboxylic acids is considerably smaller than that from the saturated carboxylic acids[86]. This may be due to this fragmentation into small radicals which can diffuse and recombine at elevated temperature forming a non-paramagnetic species. The formation of carboxyl radicals was also found in sodium hydrogen maleate[24], potassium hydrogen fumarate[25], acetylene dicarboxylic acid[26], succinic acid[27], and potassium hydrogen malonate[28]. Since the carboxyl radical is very unstable, the decarboxylation radicals are often observed immediately after irradiation[27].

In the case of saturated carboxylic acids, the decarboxylation radical is fairly stable and often stabilised at room temperature. For example, CH_2COOH in malonic acid is detectable at room temperature[98] and $CH_3\dot{C}HCOOH$ in succinic acid is probably formed from $\dot{C}H_2CH_2COOH$ by intramolecular hydrogen transfer[99, 100]. The decarboxylation radicals were also found in sodium acetate[101, 102] and some hydroxylated carboxylic acids such as tartronic acid[103] and tartaric acid[104].

9.4.5.2   Amino acids

Recently, Sinclair has studied the spectral change upon warming irradiated glycine and its deuterium compounds[60]. It was suggested that the spectrum overlapping with the anion absorption in irradiated $(\overset{+}{N}D_3)CH_2CO_2^-$ is attributable to the decarboxylation radical, $(\overset{+}{N}D_3)\dot{C}H_2$. If the mechanism established for carboxylic acids is operative, this radical may be formed from the positive primary by the loss of CO_2. It was also found that the spectrum which is attributable to $\cdot\overset{+}{N}H_2CH_2CO_2^-$ appears upon warming without loss of the anion absorption. The following process was proposed:

$$\overset{+}{N}H_3\dot{C}H_2 + (\overset{+}{N}H_3)CH_2CO_2^- \rightarrow NH_3CH_3 + \cdot\overset{+}{N}H_2CH_2CO_2^-$$

There is no evidence of such a reaction sequence originating from the cationic primary in other α-amino acids. Very recently, however, Box and Budzinski[62] have obtained more conclusive evidence of the existence of the $-CH_2-\dot{C}H_2$ radical in β-alanine and ε-aminocaproic acid irradiated at 4.2 K. They have attributed this radical more specifically to $\overset{+}{N}H_3CH_2\dot{C}H_2$ formed by loss of CO_2 from the positive primary.

9.4.5.3   Sulphur-containing compounds

The fate of the positive primary in cystine dihydrochloride was studied by Box et al.[27]. Upon warming the crystal to 100 K the rupture of the S—S bond in the cation takes place to give the R—S· radical, which is different from that originating from the anion. It is to be noted that the conversion from the cation to R—S· takes place quantitatively without causing any loss of the anionic species. This means that the process involved is the self-decomposition of the cationic species. It was also found that the neutral sulphur radical which is ascribed to the R—S—S· radical is formed even at 4.2 K together with the ionic primaries. From stoichiometrical considerations it was suggested that this neutral radical is formed from the dissociation of some other cationic species. This may also indicate the self-dissociative nature of the cationic primary.

The fate of the positive primary in sulphhydryls was also studied by Budzinski and Box[93] using L-penicillamine hydrochloride. They have found the R—S· radical originating from the positive primary and the conformational change similar to that found in R—S· originating from the anionic primary of cystine (Section 9.4.4.4).

9.4.5.4   OH *Radicals from water of crystallisation*

In some unsaturated carboxylic acid salts with hydrated water such as sodium hydrogen maleate trihydrate[106] and dipotassium fumarate dihydrate[107], the OH radical has been found to be formed from the hydrated water molecule. It was suggested that the OH radical is formed from the positive hole trapped in the hydrated water molecule followed by proton transfer to the neighbouring carboxylic group similar to that in the formation of the carboxyl radical in unhydrated acids. It should be noted that the OH radical in dipotassium fumarate dihydrate is formed by loss of the hydrogen atom which participates in the shortest hydrogen bonding with the neighbouring fumarate ion[107].

9.5   STEREOSPECIFICITY AND CRYSTALLINE FIELD EFFECTS IN RADICAL FORMATION

A number of experimental results indicate that the formation of radicals and their reaction in the crystalline solids exhibit remarkable stereospecificity and seem to be strongly affected by the crystalline field, namely, the molecular packing, relative orientation of the surrounding molecules, intermolecular hydrogen bonding and other weak interactions in these crystalline solids. These effects may be of key importance for the proper understanding of the chemical reactions occurring in the solid state. Miyagawa *et al.*[56] have suggested that the problem may be related to the selectivity in the enzymatic reactions. Although at present there is not sufficient knowledge for solving the problem, it may be of some interest to describe some experimental results related to this topic.

9.5.1   Pairwise trapping of radicals

One of the most interesting findings in the recent studies of irradiated organic crystals is the pairwise trapping of radicals. The two radicals produced in a pair are usually separated by a distance of 5–10 Å, and give rise to a triplet state as a whole because of the interaction between the two unpaired electrons. Therefore, the spectrum of the radical pairs has characteristics of triplet state e.s.r.[108–110], that is, for the $\Delta M_s = 1$ transition it shows the zero-field splitting due to the magnetic interaction of the two unpaired spins, and it also gives the weak signal at $g \approx 4$ arising from the forbidden transition $\Delta M_s = 2$ which is weakly allowed by the mixing of the states $M_s = 0$, and ± 1. The hyperfine separations of the radical pairs are approximately half of those observed for the isolated radicals, because of the rapid exchange of the two electron spins. Two sorts of hyperfine anomalies have been found. One arises from the second-order effects giving rise to the asymmetric hyperfine pattern, from which the absolute sign of J in the JS_1S_2 term in the spin Hamiltonian can be determined, if the absolute sign of the fine structure constant D is known[110]. The other is associated with the nuclear-forbidden transition[111], giving rise to the anomaly in the hyperfine structure for one of the two electronic transitions. It has been pointed out that the absolute sign of D can be determined from this anomaly if the absolute sign of the hyperfine coupling constant is known.

The radical pairs found so far may be classified into two groups. One group, which may be called the two-molecule type, consists of a pair of radicals formed between the two adjacent molecules as is seen in a pair of iminoxy radicals in irradiated oximes[108, 109, 112–114]. Other examples have also been found in a variety of compounds such as oxalic acid[115], mono-fluoroacetamide[116], hydroquinone clathrate[117], hydroxyurea[118], resorcinol[119] and potassium hydrogen malonate[28]. Although most of the two-molecule type pairs are a similar pair formed by dehydrogenation from the two adjacent molecules, monofluoroacetamide gives a dissimilar pair [$\dot{C}H_2$ $CONH_2 \cdots \dot{C}HFCONH_2$] formed by dehydrogen fluoride[116]. The ionic radical pairs described in Section 9.4.3 also belong to this category. The other group, which may be called the single-molecule type, consists of a pair of radicals formed by the decomposition of a single molecule, as in the case of tetraphenylhydrazine[120]. Although the direct observation has not been made until recently, the existence of the single-molecule type pair can be reasonably understood if direct bond scission is assumed. However, the mechanism of the formation of the two-molecule type pair has not been completely elucidated. The most characteristic feature of the formation of this type of pair is the stereospecificity, that is, the paired radicals are produced in a specific distance and orientation in the crystalline lattice. For example, in dimethylglyoxime the yield of the radical pair aligned along the a axis with a spacing of one unit is much higher than that of other pairs having different distances and orientations[108, 121–123]. A number of other examples also indicate the preferential formation of some specific radical pairs in the lattice.

The mechanism of pairwise trapping was discussed by Iwasaki et al.[124] in their paper on irradiated n-hydrocarbons. There are three possible explanations; the first is the hot hydrogen atom reaction with the neighbouring molecule, the second, the ion–molecule reaction followed by the charge neutralisation reaction forming the two radicals closely in pair, and the third, the charge neutralisation reaction between the anionic and cationic primaries induced by the charge migration in the lattice. Ionic processes like the latter two were thought to be more probable. To understand the stereospecific formation of the paired radicals one has to assume the stereospecific preference in the ion–molecule reaction or the charge migration in the crystal. Levedev et al.[121–123] have found that a variety of radical pairs with different orientations and spacings from those of the major pair are formed as minor products in dimethylglyoxime. They have studied the interconversion between these radical pairs.

9.5.2 Stereospecific formation of isolated radicals

In single-crystal e.s.r. studies, two chemically equivalent species can be distinguished if their orientations in the crystal are different from each other. For example, the two distinguishable $\dot{C}H_2COOH$ radicals with different orientations are formed from malonic acid by loss of the COOH group, depending on which COOH group is eliminated. In spite of the chemical equivalence of the two COOH groups, Horsfield et al.[98] have found that one of the two radicals predominates in the irradiated crystal. According to the x-ray crystallographic data, one of the two COOH groups twists 15 degrees from the plane of the three carbon atoms and the other twists approximately 90 degrees. The e.s.r. result indicates that the predominant radical is formed by the loss of the latter COOH group. The result was further confirmed by Hyde et al.[125] using the electron–electron double resonance technique. Such a selective formation of radicals suggests that the crystalline field affects strongly the radical formation. As was described in Section 9.4.5.1 the $\dot{C}H_2COOH$ radical is formed from a positive primary via a carboxyl radical. The x-ray data show that the COOH group with the 90 degree twist has a shorter hydrogen bonding than the other. The proton transfer forming the carboxyl radical is considered to be preferable in the COOH group with the shorter hydrogen bonding. This may be the cause of the selective formation of the $\dot{C}H_2COOH$ radical.

As is described in Section 9.4.1.3, the molecular anion of L-alanine was found to be the protonated form. Despite the chemical equivalence of the two oxygen atoms in the fully ionised CO_2^{2-} group, the protonation takes place selectively at only one of the two oxygen atoms. This stereospecific proton transfer was interpreted as quantum tunnelling of the proton through the hydrogen bond[56]. In a double-minimum potential curve of the proton in a hydrogen bond, $N—H \cdots O$, the minimum at the oxygen side is higher in energy than the one at the nitrogen side. As a result of an increased attraction between O and H caused by the formal negative charge on O in the anion, the energy of the minimum at the oxygen side would be reduced, resulting in the tunnelling of the proton from the nitrogen side to the oxygen side. In the crystal of alanine one of the two carboxyl oxygen atoms has a shorter hydrogen bond than the other oxygen atom. The principal direction of the hyperfine coupling tensor of the added hydrogen atom indicates that the protonated oxygen atom is indeed the one having the shorter hydrogen bonding, and is the one in which the tunnelling frequency is very much higher than the other. The selective formation of OH radicals from water of crystallisation in irradiated dipotassium fumarate dihydrate is also considered to be due to hydrogen bonding, as mentioned in Section 9.4.5.4.

Miyagawa and Itoh[126] have pointed out that stereospecific addition takes place when the hydrogen-addition radical is formed in fumaric acid-doped succinic acid crystal. The radical produced, namely, $HOOC—CH_\beta H_{\beta'}$ $CH—COOH$ has two non-equivalent β proton couplings, among which the proton giving the larger coupling was found to be the added hydrogen. Iwasaki et al.[53] suggested that this stereospecificity is interpretable as a proton transfer from a neighbouring molecule. The crystal structure of the host molecule and the orientation of the fumaric acid anion, which is the precursor of the hydrogen-addition radical, indicate that there is an acidic proton located right above the radical carbon of the anion. Therefore, the acidic proton may transfer from the direction of the density axis of the un-paired electron. This mode of addition is expected to produce radicals having the larger coupling for the added proton.

Another type of stereospecific addition was found in irradiated sodium hydrogen maleate trihydrate, in which the two kinds of hydrogen-addition radical can be formed, depending on which vinylene carbon atom adds a hydrogen. The hydrogen maleate ion is known to have a symmetrical structure with respect to the plane perpendicular to the $C{=}C$ bond, be-cause of the symmetrical intramolecular hydrogen bonding. In spite of the equivalence of the two vinylene protons it was found that the hydrogen addition takes place at only one of the two vinylene carbon atoms[24]. The reason for this is not clear since the crystal structure is not known. It is, how-ever, to be noted that potassium hydrogen maleate gives a symmetrical delocalised π-radical formed by the loss of the bridge hydrogen atom in contrast to the above-mentioned sodium salt[127, 128]. It is surprising that sodium and potassium salts give entirely different radicals. Such a metal-substitution effect has also been found in irradiated crystals of acetic acid salts. Strontium acetate hemihydrate[50] produces $CH_3CO_2^{2-}$ rather than the methyl radical as is the case with other acetates[101, 129]. Succinic acid gives two stable radicals, $HOOCCH_2\dot{C}HCOOH$ and $CH_3\dot{C}HCOOH$, while di-sodium succinate gives the latter predominantly[130]. Bales et al. have ascribed this to the difference between the molecular packing in the two crystals[130].

9.5.3 Proton–deuteron exchange reactions

Miyagawa and Itoh[131] have found an interesting proton–deuteron exchange reaction for a stable radical $CH_3\dot{C}HR$ in an irradiated crystal of alanine grown from a heavy water solution. It was found that the C—H bonds in the radical are successively replaced by the deuterons of ND_3 groups of the surrounding molecules. They have further found two other exchange reactions for the $CH_3\dot{C}DR$ radical in the crystal of irradiated α-deutero-L-alanine[132]. One is the exchange between a deuteron in a radical and the CH_3 proton in the adjacent molecule, and the other is the exchange between the methyl proton in the radical and an α-deuteron in the adjacent molecule. In the same crystal the exchange with the NH proton also takes place but the rate is slower than that of the former two exchanges. In an aqueous solution the NH proton is much more easily replaced by a deuteron while the CH proton is not so readily replaced by a deuteron. Therefore, if the exchange reaction takes place in solution the rate of the former two types of exchange would be much lower than that of the latter exchange. Consequently the observations of Miyagawa and Itoh[131, 132] indicate that the crystal field dominates the exchange rates and that the dependence of the exchange rate on the nature of the chemical bond is much less important. They have also found the similar proton–deuteron exchange for $DOOCCH_2CHCOOD$ trapped in deuterated aspartic acid[133]. A very similar radical trapped in succinic acid does not show this type of exchange. This may also indicate that the crystal field affects the chemical reaction in the irradiated solids.

9.6   HINDERED MOTION IN TRAPPED RADICALS

Since the temperature change of the e.s.r. spectra due to the hindered internal motion of the $CH_3\dot{C}HR$ radical in an irradiated single crystal of L-alanine was successfully interpreted by the modified Bloch treatment[134], a number of examples of the temperature change due to molecular motions have been found by many workers. For example, methyl groups undergo restricted internal motions about a C—C bond in such radicals as $CH_3\dot{C}(COOH)CH_2$ COOH and (23) in irradiated itaconic acid[135] and itaconic anhydride[136], respectively.

$$CH_3\overset{\displaystyle .}{C}(\overset{\displaystyle \overset{\text{O}}{\frown}}{C=O})CH_2(C=O)$$

(23)

The effect of the restricted internal motion of a methylene group on the β-proton coupling has also been studied for $^-OOC\dot{C}H(CH_2)_3COO^-$ in hexamethylenediammonium adipate[137], $HOOC\dot{C}HCH_2COOH$ in urea adduct of fumaric acid[138], $CH_3(CH_2)_2\dot{C}(COOH)_2$ in n-propylmalonic acid[139], and $HOOCCH_2\dot{C}(OH)COOH$ in ammonium malate monohydrate[140]. Recently, $^-OOCCF_2\dot{C}FCOO^-$ in sodium perfluorosuccinate has been found to exhibit a remarkable motional effect which averages both the α and β fluorine coupling tensors at room temperature[141].

Reorientation of the terminal methylene group has been also found for $\dot{C}H_2COO^-$ in glycine[142], triglycinesulphate[143], zinc acetate[144–146] and sodium acetate[84]. Among these $\dot{C}H_2COO^-$ in zinc acetate has been studied most extensively by two groups. Hayes et al.[146] have pointed out that in the exchange of the α protons the variation of the principal axes of the coupling tensor due to the restricted motion should be taken into consideration to describe accurately the motional effect on the spectral line position and broadening. They have analysed the spectral change on the basis of the density matrix description. Fujimoto and Janecka have determined the ^{13}C hyperfine tensors of $\dot{C}H_2COO^-$ in sodium acetate at both 77 and 300 K [84]. Since the principal direction of A_C (max) is fixed in space during the reorientation, proton exchange rather than internal rotation was suggested as an alternative mechanism for the reorientation. Toriyama and Iwasaki have studied the motional average of the g tensor in $CF_2(OO\cdot)CONH_2$ [147].

In these examples the motion is quenched at 77 K and the spectral change can be interpreted by a site-exchange phenomenon, or by a restricted oscillation of large amplitude. However, in some cases the motion is not quenched even at 4.2 K and a significant quantum mechanical effect on the spectral change has been discovered. The spectral change due to a rapid tunnelling of a methyl group was first demonstrated by Freed[148] who predicted theoretically a seven-line hyperfine structure with the intensity ratio of $1:1:1:2:1:1:1$ instead of a four-line one with the intensity ratio of $1:3:3:1$. This was subsequently observed at 4.2 K for $R\dot{C}CH_3R'$ radicals in acetylalanine[149], and chloroacetylalanine[150] and for the radical in 4-methyl-2, 6-di-t-butylphenol[151, 152]. Further studies have been carried out at low temperature for the unstable $CH_3\dot{C}HR$ radical in freshly-irradiated alanine[153] and $CH_3\dot{C}(COOH)_2$ in methylmalonic acid[154].

If one assumes internal motion in the $\dot{C}$—CH_3 fragment with C_{3v} symmetry, the torsional levels may split into the two levels with A symmetry and doubly degenerate E symmetry, and the separation of the two levels depends on the tunnelling frequency. The nuclear spin states which can be associated with the A and E torsional states must have A and E symmetry, respectively, because the spin–rotor product functions must be of A symmetry[148]. The usually observed four-line spectrum $(1:3:3:1)$ of the CH_3 group consists of the equally spaced four lines $(1:1:1:1)$ associated with the A nuclear spin state and the two lines $(2:2)$ associated with the E nuclear spin state. It was, however, shown that the degeneracy of the E torsional state is removed by the hyperfine interaction for the ground torsional level, resulting in the splitting of the hyperfine lines with the E nuclear spin state into three, or more generally four, lines[148, 153]. Consequently the seven- or eight-line spectrum may be observed for the ground torsional level in marked contrast to the four-line spectrum for the excited torsional levels. In fact, the population of the ground torsional level is predominant at 4.2 K, so the seven- or eight-line spectrum is observed.

At temperatures intermediate between 4.2 and 77 K the population in the excited torsional level contributes to the hyperfine structure, resulting in spectral changes into the usual four-line pattern (statistical model). It was, however, pointed out that the quantitative spectral change observed at the intermediate temperatures cannot be explained simply by the population change of the torsional levels predicted from the isolated-rotor model. The observed spectral change occurs in a fairly narrow temperature range and is interpreted by the onset of thermally activated random motion, that is, the Brownian process with no spin correlation, resulting in the usual four-line hyperfine pattern[148, 153]. More recently it was shown that the statistical model, when corrected for the Brownian process, appears to explain all the published experimental results satisfactorily[155]. However, the problem is not settled completely, especially as to the estimates of the potential barrier, which gives the experimentally observed energy difference between the ground and first excited torsional levels as well as the tunnelling splitting in the ground level. Although all the proposed models for this problem are based on the threefold potential, one may need to use a more realistic potential, which is represented by a Fourier series, $\Sigma_{n=1}(V_{on}/2)\cos 3n\theta$, to describe the problem more accurately[155].

9.7 ENDOR STUDIES

The electron nuclear double resonance (ENDOR) technique[156] is advantageous not only because of its high resolution but also because of the simplicity of its spectra as compared with those of e.s.r. Applications of ENDOR to the study of irradiated organic crystals are now becoming of great importance, since the inherent wide line-width of the e.s.r. spectra of irradiated solids makes it difficult to resolve the small hyperfine couplings and the overlapping of the hyperfine lines which originate from a number of different radicals.

One of the most beautiful applications may be seen in the study of rotating methyl groups in the trapped radicals. As described in Section 9.6, the rotating methyl group an additional complication to the e.s.r. spectra measured at low temperatures because of the quantum mechanical effect. It was, however, shown by Clough and Poldy[151] that the ENDOR spectrum exhibits only two lines corresponding to the A and E symmetry species and that this provides unescapable evidence for the rapid reorientation of the CH_3 group even at 4.2 K. Similar applications were used by Wells and Box for the rotating methyl groups of $(CH_3)_2CCOOH$ radical in α-aminoisobutyric acid[76] and of $R\dot{C}CH_3R'$ radical in chloroacetyl alanine[150]. In the former case one of the two methyl groups was found to be locked at 4.2 K when the crystal is annealed. The accurate hyperfine coupling tensors of the methyl protons of $(CH_3)_2CCH(NH_3)COO^-$ radical in valine[65] were determined together with the nitrogen coupling tensor, leading to the conclusive identification of this radical. Read and Whiffen also studied the rotating methyl group of $CH_3CHCOOH$ radical in succinic acid[99]. Dalton and Kwiram[157] studied the ring-puckered motion in cyclo-$(CH_2)_3CCOOH$.

The hyperfine coupling tensors of the β protons for the anion in succinic and thioglycolic acids are determined with high precision by Wells[42]. The results confirm that the molecular anions preserve the same geometry as the undamaged molecules. The complicated e.s.r. hyperfine structure of thiodiglycolic acid cation is simplified in the ENDOR spectrum to give the conclusive evidence for the molecular cation[41]. Irradiated glycine[59], glutamic acid hydrochloride[78], histidine hydrochloride[158], acetylglycine[159], β-alanine[62] and anthracene[160] have also been studied by ENDOR. Sodium hyperfine coupling with CO_2^- ion in sodium formate is determined together with the quadrupole splitting[161]. Very recently a detailed analysis has been performed by Kwiram[162] for irradiated glutaric acid taking the second-order shift into consideration. Complication of the hyperfine lines due to the nuclear forbidden transition in the triplet–radical pair in potassium hydrogen malonate was elucidated by Box et al.[28].

Besides these applications, Cook and Whiffen[163] have shown that the absolute sign of the β-proton coupling constant can be determined by the ENDOR frequency shift arising from the small perturbation due to the nuclear–nuclear dipole interaction between the two methyl protons. Cook[164] has studied the triplet–radical pair in dimethylglyoxime and has determined the absolute signs of the hydrogen and nitrogen coupling constants using the situation that only a high- or low-frequency line is observed in the triplet state ENDOR.

References

1. Whiffen, D. H. (1961). *Free Radicals in Biological Systems,* 227 (M. S. Blois Jr., H. W. Brown, R. M. Lemmon, R. O. Lindblom and M. Weissbluth, editors). (New York: Academic Press)
2. Morton, J. R. (1964). *Chem. Rev.,* **64,** 453
3. Ayscough, P. B. (1967). *Electron Spin Resonance in Chemistry,* 58. (London: Methuen)
4. Iwasaki, M. (1971). *Fluorine Chem. Rev.,* **5,** 1
5. Carrington, A. and McLachlan, A. D. (1967). *Introduction to Magnetic Resonance,* 99. (New York: Harper)
6. Miyagawa, I. and Gordy, W. (1960). *J. Chem. Phys.,* **32,** 255
7. McConnell, H. M., Heller, C., Cole, T. and Fessenden, R. W. (1960). *J. Amer. Chem. Soc.,* **82,** 766
8. Atherton, N. M. and Whiffen, D. H. (1960). *Molec. Phys.,* **3,** 1
9. Iwasaki, M., Noda, S. and Toriyama, K. (1970). *Molec. Phys.,* **18,** 201
10. Poole, C. P. and Farach, H. A. (1971). *J. Mag. Resonance,* **4,** 312
11. Maruani, J., McDowell, C. A., Nakajima, H. and Raghunathan, P. (1968). *Molec. Phys.,* **14,** 349
12. Stone, A. J. (1963). *Proc. Roy. Soc. (London),* **A271,** 424
13. Holmberg, R. W. (1969). *J. Chem. Phys.,* **51,** 3255
14. Ovenall, D. W. and Whiffen, D. H. (1961). *Molec. Phys.* **4,** 135
15. McCalley, R. C. and Kwiram, A. L. (1970). *J. Amer. Chem. Soc.,* **92,** 1441
16. Iwasaki, M. and Eda, B. (1970). *J. Chem. Phys.,* **52,** 3837
17. McConnell, H. M. and Strathdee, J. (1959). *Molec. Phys.,* **2,** 129
18. Barfield, M. (1970). *J. Chem. Phys.,* **53,** 3836
19. Eda, B. and Iwasaki, M. (1971). *10th Symposium on Electron Spin Resonance,* 132, (Osaka: Chem Soc. Japan)
20. Cochran, E. L., Adrian, F. J. and Bowers, V. A. (1964). *J. Chem. Phys.,* **40,** 213
21. Iwasaki, M., Eda, B. and Toriyama, K. (1970). *J. Amer. Chem. Soc.,* **92,** 3211
22. Eda, B. and Iwasaki, M. (1971). *J. Chem. Phys.,* **55,** 3442
23. Toriyama, K. and Iwasaki, M. (1971). *J. Chem. Phys.,* **55,** 2181
24. Toriyama, K., Muto, H. and Iwasaki, M. (1971). *J. Chem. Phys.,* **55,** 1885
25. Iwasaki, M., Minakata, K. and Toriyama, K. (1971). *J. Amer. Chem. Soc.,* **93,** 3533
26. Muto, H., Toriyama, K. and Iwasaki, M. (1970). *13th Symposium on Radiation Chemistry,* 92. (Tokyo: Chem. Soc. Japan)
27. Box, H. C., Freund, H. G., Lilga, K. T. and Budzinski, E. E. (1970). *J. Phys. Chem.,* **74,** 40
28. Box, H. C., Budzinski, E. E. and Potter, W. (1971). *J. Chem. Phys.,* **55,** 315
29. Kurita, Y., Kashiwagi, M. and Saisho, H. (1965). *Nippon Kagaku Zasshi,* **86,** 578
30. Kashiwagi, M. and Kurita, Y. (1966). *J. Phys. Soc. Japan,* **21,** 558
31. Hayashi, H., Itoh, K. and Nagakura, S. (1967). *Bull. Chem. Soc. Japan,* **40,** 284
32. Cyr, N. and Lin, W. C. (1969). *J. Chem. Phys.,* **50,** 3701
33. Lau, P. W. and Lin, W. C. (1969). *J. Chem. Phys.,* **51,** 5139
34. Lau, P. W. and Lin, W. C. (1971). *J. Chem. Phys.,* **54,** 823
35. Neta, P. and Fessenden, R. W. (1970). *J. Phys. Chem.,* **74,** 3362
36. Symons, M. C. R. (1971). *J. Chem. Phys.,* **55,** 1493
37. Box, H. C. and Freund, H. G. (1964). *J. Chem. Phys.,* **40,** 817
38. Akasaka, K., Ohnishi, S., Suita, T. and Nitta, I. (1964). *J. Chem. Phys.,* **40,** 3110
39. Box, H. C. and Freund, H. G. (1964). *J. Chem. Phys.,* **41,** 2571
40. Box, H. C., Freund, H. G. and Frank, G. W. (1968). *J. Chem. Phys.,* **48,** 3825
41. Box, H. C., Freund, H. G. and Budzinski, E. E. (1968). *J. Chem. Phys.,* **49,** 3974
42. Wells, J. W. (1970). *J. Chem. Phys.,* **52,** 4062
43. Box, H. C., Budzinski, E. E. and Gorman, T. (1968). *J. Chem. Phys.,* **48,** 1748
44. Box, H. C., Freund, H. G. and Lilga, K. T. (1965). *J. Chem. Phys.,* **42,** 1471
45. Bennett, J. E. and Gale, L. H. (1968). *Trans. Faraday Soc.,* **64,** 1174
46. Tamura, N., Collins, M. A. and Whiffen, D. H. (1966). *Trans. Faraday Soc.,* **62,** 2434
47. Moulton, G. C. and Cernansky, M. P. (1969). *J. Chem. Phys.,* **51,** 2283
48. Vyas, H. M., Janecka, J. and Fujimoto, M. (1970). *Can. J. Chem.,* **48,** 2804
49. Fujimoto, M. and Seddon, W. A. (1970). *Can. J. Chem.,* **48,** 2809
50. Tolles, W. M., Sanders, R. A. and Gisch, R. G. (1971). *J. Chem. Phys.,* **54,** 1532

51. Iwasaki, M., Minakata, K. and Toriyama, K. (1971). *J. Chem. Phys.*, **55**, 1472
52. Iwasaki, M., Eda, B., Toriyama, K., Muto, H. and Nunome, K. (1971). *10th Symposium on Electron Spin Resonance*, 126. (Osaka: Chem. Soc. Japan)
53. Iwasaki, M., Muto, H. and Toriyama, K. (1971). *J. Chem. Phys.*, **55**, 1894
54. Sinclair, J. W. and Hanna, M. W. (1967). *J. Phys. Chem.*, **71**, 84
55. Minegishi, A., Shinozaki, Y. and Meshitsuka, G. (1967). *Bull. Chem. Soc. Japan*, **40**, 1549
56. Miyagawa, I., Tamura, N. and Cook, J. W. Jr. (1969). *J. Chem. Phys.*, **51**, 3520
57. Ayscough, P. B. and Roy, A. K. (1968). *Trans. Faraday Soc.*, **64**, 582
58. Sinclair, J. and Hanna, M. W. (1969). *J. Chem. Phys.*, **50**, 2125
59. Collins, M. A. and Whiffen, D. H. (1966). *Molec. Phys.*, **10**, 317
60. Sinclair, J. (1971). *J. Chem. Phys.*, **55**, 245
61. Box, H. C., Budzinski, E. E. and Freund, H. G. (1969). *J. Chem. Phys.*, **50**, 2880
62. Box, H. C. and Budzinski, E. E. (1971). *J. Chem. Phys.*, **55**, 2446
63. Box, H. C. and Freund, H. G. (1966). *J. Chem. Phys.*, **44**, 2345
64. Fujimoto, M., Seddon, W. A. and Smith, D. R. (1968). *J. Chem. Phys.*, **48**, 3345
65. Box, H. C., Freund, H. G. and Budzinski, E. E. (1967). *J. Chem. Phys.*, **46**, 4470
66. Cadena, D. G. and Rowlands, J. R. (1968). *J. Chem. Soc. B*, 488
67. Kominami, S., Akasaka, K. and Hatano, H. (1970). *9th Symposium on Electron Spin Resonance*, 86. (Tokyo: Chem. Soc. Japan); *J. Phys. Chem.*, In the press
68. Kominami, S., Akasaka, K. and Hatano, H. (1971). *Chem. Phys., Lett.*, **9**, 510
69. Akasaka, K., Kominami, S. and Hatamo, H. (1971). *J. Phys. Chem.*, **75**, 3746
70. Snipes, W., Editor. (1966). *Electron Spin Resonance and the Effects of Radiation on Biological Systems*, 123. (Washington, D.C.: National Academy of Sciences)
71. Friday, E. A. and Miyagawa, I. (1971). *J. Chem. Phys.*, **55**, 3589
72. Cadena, D. G., Linder, R. E. and Rowlands, J. R. (1969). *Can. J. Chem.*, **47**, 3249
73. Minegishi, A., Shinozaki, Y. and Meshitsuka, G. (1972). *J. Chem. Phys.*, **56**, 2481
74. Iwasaki, M. and Eda, B. (1968). *Chem. Phys. Lett.*, **2**, 210
75. Box, H. C., Freund, H. G. and Budzinski, E. E. (1966). *J. Amer. Chem. Soc.*, **88**, 658
76. Wells, J. W. and Box, H. C. (1967). *J. Chem. Phys.*, **46**, 2935
77. Shields, H., Harrick, P. and DeLaigle, D. (1967). *J. Chem. Phys.*, **46**, 3649
78. Whelan, D. J. (1968). *J. Chem. Phys.*, **49**, 4734
79. Horan, P. K., Henriksen, T. and Snipes, W. (1970). *J. Chem. Phys.*, **52**, 4324
80. Schwartz, R. N., Hanna, M. W. and Bales, B. L. (1969). *J. Chem. Phys.*, **51**, 4336
81. McCalley, R. C. and Kwiram, A. L. (1970). *J. Chem. Phys.*, **53**, 2541
82. Ayscough, P. B., Mach, K., Oversby, J. P. and Roy, A. K. (1971). *Trans. Faraday Soc.*, **67**, 360
83. Ayscough, P. B. and Oversby, J. P. (1971). *Trans. Faraday Soc.*, **67**, 1365
84. Fujimoto, M. and Janecka, J. (1971). *J. Chem. Phys.*, **55**, 5
85. Cook, J. B., Elliott, J. P. and Wyard, S. J. (1967). *Molec. Phys.*, **12**, 185
86. Cook, R. J., Rowlands, J. R. and Whiffen, D. H. (1963). *J. Chem. Soc.*, 3520
87. Shioji, Y., Ohnishi, S. and Nitta, I. (1963). *J. Polymer Sci.*, **A1**, 3373
88. Wedum, E. E. and Griffith, O. H. (1967). *Trans. Faraday Soc.*, **63**, 819
89. Griffith, O. H. and Wedum, E. E. (1967). *J. Amer. Chem. Soc.*, **89**, 787
90. Kurita, Y. and Gordy, W. (1961). *J. Chem. Phys.*, **34**, 282
91. Kurita, Y. (1967). *Bull. Chem. Soc. Japan*, **40**, 94
92. Hahn, Y. H. and Rexroad, R. N. (1963). *J. Chem. Phys.*, **38**, 1599
93. Budzinski, E. E. and Box, H. C. (1971). *J. Phys. Chem.*, **75**, 2564
94. Wheaton, R. F. and Ormerod, M. G. (1969). *Trans. Faraday Soc.*, **65**, 1638
95. Akasaka, K. (1965). *J. Chem. Phys.*, **43**, 1182
96. Kurita, Y. and Gordy, W. (1961). *J. Chem. Phys.*, **34**, 1285
97. Toriyama, K., Iwasaki, M., Noda, S. and Eda, B. (1971). *J. Amer. Chem. Soc.*, **93**, 6415
98. Horsfield, A., Morton, J. R. and Whiffen, D. H. (1961). *Molec. Phys.*, **4**, 327
99. Read, S. F. J. and Whiffen, D. H. (1967). *Molec. Phys.*, **12**, 159
100. Klinck, R. E. (1968). *J. Chem. Phys.*, **49**, 4722
101. Rogers, M. T. and Kispert, L. D. (1967). *J. Chem. Phys.*, **46**, 221
102. Janecka, J., Vyas, H. M. and Fujimoto, M. (1971). *J. Chem. Phys.*, **54**, 3229
103. Moulton, G. C. and Crenshow, H. T. (1965). *Radiation Res.*, **25**, 139
104. Moulton, G. C. and Cernansky, B. (1970). *J. Chem. Phys.*, **53**, 3022
105. Castleman, B. W. and Moulton, G. C. (1971). *J. Chem. Phys.*, **55**, 2598
106. Toriyama, K. and Iwasaki, M. (1971). *J. Chem. Phys.*, **55**, 1890

107. Minakata, K., Toriyama, K. and Iwasaki, M. (1971). *14th Symposium on Radiation Chemistry,* 170. (Sapporo: Chem. Soc. Japan)
108. Kurita, Y. (1964). *J. Chem. Phys.,* **41,** 3926
109. Kurita, Y. (1964). *Nippon Kagaku Zasshi,* **85,** 833
110. Itoh, K., Hayashi, H. and Nagakura, S. (1969). *Molec. Phys.,* **17,** 561
111. Iwasaki, M. and Minakata, K. (1971). *J. Chem. Soc.,* **54,** 3225
112. Kurita, Y. and Kashiwagi, M. (1966). *J. Chem. Phys.,* **44,** 1727
113. Kashiwagi, M. and Kurita, Y. (1966). *J. Phys. Soc. Japan,* **21,** 558
114. Hayashi, H., Itoh, K. and Nagakura, S. (1967). *Bull. Chem. Soc. Japan,* **40,** 284
115. Moulton, G. C., Cernansky, S. M. P. and Straw, D. C. (1967). *J. Chem. Phys.,* **46,** 4292
116. Iwasaki, M. and Toriyama, K. (1967). *J. Chem. Phys.,* **46,** 4693
117. Ohigashi, H. and Kurita, Y. (1969). *J. Mag. Resonance,* **1,** 464
118. Reiss, K. and Shield, H. (1969). *J. Chem. Phys.,* **50,** 4368
119. Cambell, D. and Symons, M. C. R. (1969). *J. Chem. Soc. A,* 428
120. Wiersma, D. A., Lichtenbelt, J. H. and Kommandeur, J. (1969). *J. Chem. Phys.,* **50,** 2794
121. Yakimechenko, O. E., Doroshina, G. P. and Lebedev, Ya. S. (1969). *Khim. Vys. Energ.,* **3,** 242
122. Lebedev, Ya. S. (1969). *Radiation Effects,* **1,** 213
123. Yakimechenko, O. E. and Levedev, Ya. S. (1971). *Int. J. Radiat. Phys. Chem.,* **3,** 17
124. Iwasaki, M., Ichikawa, T. and Ohmori, T. (1969). *J. Chem. Phys.,* **50,** 1991
125. Hyde, J. S., Kispert, L. D. and Sneed, R. C. (1969). *J. Chem. Phys.,* **48,** 3824
126. Miyagawa, I. and Itoh, K. (1966). *Nature (London),* **209,** 504
127. Heller, H. C. and Colle, T. (1962). *J. Amer. Chem. Soc.,* **84,** 4448
128. Iwasaki, M. and Itoh, K. (1964). *Bull. Chem. Soc. Japan,* **37,** 44
129. Rogers, M. T. and Kispert, L. D. (1968). *Amer. Chem. Soc. Advan. Chem. Ser.,* **82,** 327
130. Bales, B. L., Schwartz, R. N. and Hanna, M. W. (1969). *J. Chem. Phys.,* **51,** 1974, 5178
131. Itoh, K. and Miyagawa, I. (1964). *J. Chem. Phys.,* **40,** 3328
132. Itoh, K., Miyagawa, I. and Chen, C. S. (1970). *J. Chem. Phys.,* **52,** 1822
133. Miyagawa, I. and Itoh, K. (1965). *J. Chem. Phys.,* **43,** 2915
134. Miyagawa, I. and Itoh, K. (1962). *J. Chem. Phys.,* **36,** 2157
135. Fujimoto, M. (1963). *J. Chem. Phys.,* **39,** 846
136. Eda, B. and Iwasaki, M. (1966). *5th Symposium on Electron Spin Resonance,* 4, (Sendai: Chem. Soc. Japan)
137. Kashiwagi, M. and Kurita, Y. (1963). *J. Chem. Phys.,* **39,** 3165
138. Corvaja, C. (1966). *J. Chem. Phys.,* **44,** 1958
139. Tamura, N., Collins, M. A. and Whiffen, D. H. (1966). *Trans. Faraday Soc.,* **62,** 1037
140. Corvaja, C. (1967). *Trans. Faraday Soc.,* **63,** 26
141. Kispert, L. D. and Rogers, M. T. (1971). *J. Chem. Phys.,* **54,** 3326
142. Morton, J. R. (1964). *J. Amer. Chem. Soc.,* **86,** 2325
143. Kato, T. and Abe, R. (1970). *J. Phys. Soc. Japan,* **29,** 389
144. Ohigashi, H. and Kurita, Y. (1968). *Bull. Chem. Soc. Japan,* **41,** 275
145. Tolles, W. M., Crawford, L. P. and Valenti, J. L. (1968). *J. Chem. Phys.,* **49,** 4745
146. Hayes, R. G., Steible, D. J., Tolles, W. M., Hunt, J. W. (1970). *J. Chem. Phys.,* **53,** 4466
147. Toriyama, K. and Iwasaki, M. (1969). *J. Phys. Chem.,* **73,** 2663
148. Freed, J. H. (1965). *J. Chem. Phys.,* **43,** 1710
149. Gamble, W. L., Miyagawa, I. and Hartman, R. L. (1968). *Phys. Rev. Lett.,* **20,** 415
150. Wells, J. W. and Box, H. C. (1968). *J. Chem. Phys.,* **48,** 2542
151. Clough, S. and Poldy, F. (1967). *Phys. Lett.,* **24A,** 545
152. Clough, S. and Poldy, F. (1969). *J. Chem. Phys.,* **51,** 2076
153. Davidson, R. B. and Miyagawa, I. (1970). *J. Chem. Phys.,* **52,** 1727
154. Clough, S., Starr, M. and McMillan, N. D. (1970). *Phys. Rev. Lett.,* **25,** 839
155. Davidson, R. B. and Miyagawa, I. (1972). *Phys. Rev.* In the press
156. Feher, G. (1956). *Phys. Rev.,* **103,** 834
157. Dalton, L. R. and Kwiram, A. L. (1970). *Bull. Amer. Phys. Soc.,* **15,** 93
158. Box, H. C., Freund, H. G. and Lilga, K. T. (1967). *J. Chem. Phys.,* **46,** 2130
159. Piazza, R. E. and Pattern, R. A. (1969). *Molec. Phys.,* **17,** 213
160. Bohme, U. R. and Jesse, G. W. (1969). *Chem. Phys. Lett.,* **3,** 329
161. Cook, R. J. and Whiffen, D. H. (1967). *J. Phys. Chem.,* **71,** 93
162. Kwiram, A. L. (1971). *J. Chem. Phys.,* **55,** 2484
163. Cook, R. J. and Whiffen, D. H. (1965). *J. Chem. Phys.,* **43,** 2908
164. Cook, R. J. (1968). *Proc. of the 15th Colloque Ampère,* 269. (Amsterdam: Averbuch)

10
Biological Applications of Electron Spin Resonance Spectroscopy

J. R. BOLTON
University of Western Ontario, London

and

J. T. WARDEN
University of Minnesota, Minneapolis, Minnesota

10.1 INTRODUCTION

Electron spin resonance spectroscopy has found a wide application in the study of biological systems as evidenced by the several books and reviews which have been written on the subject[1-6]. It has been a difficult subject to review, not only because of the breadth of the field but also because of the significant fraction of papers of questionable value. We have scanned nearly 1000 references in preparing this review, and it has not been easy to select those that should be included. We have tried to include most of the pertinent review articles, significant papers where no review covers the field, and papers of outstanding merit. At the end of each section a guide to further reading has been provided for those interested in a particular subject. In general, we have surveyed the literature from January 1966 to June 1971.

The selection of topics was certainly biased by our own interests; nevertheless, we hope that we have not inadvertently omitted any important topics.

We trust that this review will serve two purposes. For the biologist it should provide a stimulus to learn more about the e.s.r. technique. For this purpose some of the general treatments of e.s.r.[7, 8] as well as specific techniques[9-11] are recommended. For the physical scientist this review will furnish a key to the literature should he wish to explore a particular biological area.

10.2 TISSUES AND CELLS

Although there are many problems in the use of whole cells or tissues in e.s.r. studies of the role of free radicals in cellular processes, such studies have been quite common, sometimes useful, and have encompassed every field of biological endeavour[12-17]. The quality of published articles dealing with e.s.r. cell and tissue measurements has generally been poor, due to lack of sophistication of the experimenters. However, it should be noted that excellent studies do exist in which the authors have realised the shortcomings of experimental techniques as applied to tissue experiments.

One of the chief problems in the use of tissues in e.s.r. spectrometers is the non-resonant absorption of microwave energy by cellular or other water. Techniques developed to avoid this difficulty have been quasi-successful; however, some of these involve alterations to the sample that are definitely non-physiological. Thus, it is quite difficult in many cases to draw reliable conclusions concerning the role of an observed e.s.r. signal in a biological process.

Tissue e.s.r. methodology most commonly invokes three techniques for sample manipulation: (i) lyophilisation, (ii) the rapid freeze technique, and (iii) the 'surviving-tissue' technique. The practice of lyophilisation has been frequently criticised on the grounds that results obtained are not reproducible and that the sample does not reflect the physiological state of the functioning tissue[16]. Vanin, Chetverikov, and Blyumenfel'd have compared lyophilised and frozen samples and have concluded that the lyophilisation process changes the e.s.r. spectrum of tissues and cannot be regarded as reliable for physiological measurements[18]. Heckly in summarising lyophilisation applications in e.s.r. warns that free radical content in lyophilised preparations does not reflect the free radical concentration in the tissue before drying[16]. Furthermore, free radicals observed in dried samples do indicate the capacity of the preparation to react with oxygen. Hence, the concentration of free radicals in lyophilised samples is critically dependent on oxygen concentration. Likewise, the moisture content of dry preparations influences the yield of radicals.

Because of the inherent disadvantages of lyophilisation, most experimentalists have employed rapid freezing or 'surviving-tissue' techniques. Although rapid freezing is advantageous in the study of transition metal constituents in the cell and provides an improvement in the signal-to-noise ratio; it does, however, require that the tissue be in a non-physiological state.

The remaining preparative method denoted as the 'surviving-tissue' technique by Commoner and Ternberg[19], has been more appropriately designated the 'dying-tissue' technique by Mallard and Kent[13]. The tissue is normally studied at room temperature, while held in a thin sample holder to prevent non-resonant absorption by water. Since the sample is probably under anaerobic conditions and often lacks a protective or nutrient medium, the tissue can be easily considered near death. Likewise, free radical concentration in tissues appears to be dependent on preparation procedures and the time elapsed between isolation of the sample and recording of the spectrum[20, 21]. Thus, although the 'surviving-tissue' method provides the closest approximation to a physiologically significant experiment, care should still be exercised in the interpretation of e.s.r. results. It should also be recognised that the signal-to-noise ratio is often poor with 'surviving-tissue' samples, and as a consequence, free radicals present in the sample may be missed unless computer averaging is available.

The following example is presented to illustrate how pitfalls in tissue e.s.r. studies can trap even the conscientious researcher. Rats fed on a diet containing hepatic carcinogens were found to develop an unusual e.s.r. signal in liver tissue with a g-factor of 2.035. This signal appeared within 5–45 days after the carcinogenic diet commenced and disappeared within 10–40 days after its initial conception, although results were often variable. After 4–12 months tumours appeared lacking both the abnormal $g \approx 2.035$ signal and the common $g \approx 2.005$ resonance usually present in liver tissue[22]. Recently, it has been shown that liver incubated in nitrite gives the $g \approx 2.035$ signal and this resonance is tentatively identified as a free radical complex of iron, nitric oxide and a thiol protein[23]. Nitrite enhances the formation of the $g \approx 2.035$ resonance and appears to inhibit the activity of the carcinogen[24]. The positive results assigned to the carcinogenic diet in the early experiments have now been attributed to presence of nitrate in the St. Louis tap water on which the rats were maintained. Fluctuations of the nitrate concentration in the feeding water undoubtedly produced the variability observed in earlier experiments.

A recent novel application of e.s.r. in tissue research involves the quantitative monitoring of the resorption rate of bone grafts[25]. The stable free radical label was obtained by γ-irradiation of bone graft material. By determining the number of spins present in the graft material as a function of time after implantation, the resorption rate could easily be measured. Another bio-medical use of spin resonance is the investigation of the reaction of the gaseous pollutant NO_2 with lung and blood components[26].

Further innovations utilising e.s.r. to explore the impact of pollution on man and his biological environment can be anticipated in the future.

References 13, 16 and 17 are recommended for further reading.

10.3 RADIATION BIOLOGY

The introduction of electron spin resonance techniques to radiation biology has prompted many of the recent advances in our understanding of radiation-induced cellular damage. The literature is quite extensive with a number of journals being devoted solely to radiobiology; however, comprehensive review articles are available[27-32]. Although space considerations do not allow an extensive analysis of the current state of e.s.r. investigations in radiation biology, shortcomings of radiobiological e.s.r. and some important areas of application will be summarised.

Since the genesis of e.s.r. in radiation research, free radicals have been implicated in the process of radiation-induced damage. Resonances attributed to cysteine-like sulphur and a glycylglycine-type doublet were observed and energy-transfer mechanisms were invoked[33]. Owing to the inherent reactivity of free radicals, experimental considerations have required lyophilisation of the sample or freezing to liquid nitrogen temperatures to insure that the primary radicals are stabilised. Hence, the fundamental problem has been the assignment of biological relevance to physical observations under extremely non-physiological states.

Experimental difficulties readily arise in the practice of radiobiological e.s.r. Copeland indicates that quantitative studies are extremely difficult and all findings and conclusions should be subjected to rigorous examination[31]. Saturation of radical concentration should be avoided by a judicious choice of dosage from the linear portion of the yield curve. Likewise, microwave power saturation effects become prominent at low temperatures; hence, power saturation curves should always be determined. Variation of radical yield as a function of particle size and manner of preparation has been reported[34]. Likewise, caution should be exercised in relating experiments performed on a mixture of components to results obtained from the individual components alone. However, even when precautions have been taken, Wyard indicates that interlaboratory comparisons of equivalent samples have produced disturbingly large variations in the calculated radical yield[29]. These findings indicate that quantitative radiobiological e.s.r. should be practised only with great caution and reserve.

Amino acids, nucleic acids, proteins, and tissues or cellular organisms are among the systems that have been investigated by radiobiological e.s.r. Amino acids and nucleic acids have been often selected because as essential components of cellular systems, they can be easily used in model system investigations. Single-crystal irradiation of amino acids and nucleic acid components have provided precise information on the structure of radicals produced during exposure to ionising radiation, and correlations have been drawn to radical transformations observed during annealing of irradiated protein samples. Thus, single-crystal experiments have allowed the assignment of the cysteine-like sulphur resonance in proteins and the thymine octet in DNA [28].

The mechanism for energy migration and radioprotection in proteins has always been of great interest among practitioners of radiobiological e.s.r. [27]. Evidence for both intra- and inter-molecular transfer of energy in model protein systems has been obtained; however, the precise mechanism dominant in cellular tissue is only conjecture. Milvy and Farcasiu have recently monitored the variation of energy transfer in the DNA–cysteamine system as a function of cysteamine concentration, pH and pyrimidine content[35]. Energy transfer was found to be a linear function of thymine content in partially depyrimidinated DNA. Likewise, reducing the pH led to diminished transfer to cysteamine, but increased the radical yield of thymine. The maximum migration distance was calculated to be 6 Å in this system. Transfer in a Sephadex–trypsin matrix also indicates that the energy acceptor must be in intimately close proximity to the donor[36]. In a related experiment, Pullman, Blumenthal, and Handler-Bernich[37] have investigated the interaction of sulphhydrylamine radioprotectors with thymidylic acid and have concluded that significant spin transfer can only take place when the radioprotector is bound by the enolate form of the nucleotide base. Hence, energy transfer in irradiated cellular systems may be dependent on physical contact between interacting constituents.

Although long-lived, radiation-induced radicals have been found in various plant and animal tissue, transient free radicals have been recently observed in x-irradiated animal tissues[38]. A functional relationship is claimed to exist between the sensitivity of a tissue to radical damage and the type of free radical induced by radiation. Likewise, various types of radiation-induced free radicals have been observed in dried spores from *Osmunda regalis*[39]. These radical types can be differentiated on the basis of their reactivities to biologically significant gases (NO, NO_2, O_2) which are able to modify radiation damage.

References 28 and 31 are recommended for further reading.

10.4 PROTEINS CONTAINING TRANSITION METAL IONS

10.4.1 Haem proteins

Haem proteins are characterised by one or more prosthetic groups consisting of a porphyrin tetrapyrrole with iron (either Fe^{2+} or Fe^{3+}) at the centre. The optical spectrum of these proteins consists of a strong 'Soret' absorption at 400–450 nm plus weaker components in the 520–560 nm region. The haem proteins were among the first biological systems to be investigated by e.s.r. The elegant work of Ingram *et al.*[40] on methaemoglobin and metmyoglobin single crystals still stands as one of the most important biological e.s.r. investigations. E.S.R. signals are seen only for proteins in the ferric state. Both high-spin ($S = \frac{5}{2}$) and low-spin ($S = \frac{1}{2}$) states are found, with some proteins exhibiting an equilibrium mixture of the two states.

Several reviews are available in this field[41-45]. We will restrict our coverage to single-crystal and polycrystalline studies. Both low-spin and high-spin cases will be considered.

10.4.1.1 Single-crystal studies

Since the pioneering work of Ingram *et al.* there have been surprisingly few single-crystal studies on haem proteins[46]. This is probably due to the difficulty in growing single crystals of proteins and the involved e.s.r. measurements and calculations that are required. However, the structural detail obtainable from these studies is well worth the effort. Most studies have been carried out at cryogenic temperatures, but recently successful studies of metmyoglobin fluoride and methaemoglobin fluoride were reported at room temperature[46]. This technique avoids the potentially damaging action of freezing the crystals.

An interesting model system was reported by Scholes[47] where haemin and haematin were doped in single crystals of perylene. Not only was a thorough study of the g anisotropy undertaken but also nitrogen superhyperfine structure was observed from the porphyrin.

10.4.1.2 High-spin polycrystalline studies

The high-spin ($S = \frac{5}{2}$) form of haem proteins is found only when the sixth coordination position to the iron is occupied by a 'weak' ligand such as H_2O. A 'strong' ligand such as OH^- or F^- forces the iron into the low-spin ($S = \frac{1}{2}$) state. Most haemoglobins and myoglobins are found in the high-spin state whereas most cytochromes are found in the low-spin state. Under conditions of a large zero-field splitting ($D > 2$ cm^{-1}) high-spin Fe^{3+} exhibits an axial or nearly axial e.s.r. absorption characterised by $g_{\parallel} \approx 2$ and $g_{\perp} \approx 6$. In a polycrystalline sample the major derivative feature occurs at $g \approx 6$ with a small component at $g \approx 2$. These features, especially at $g \approx 6$, are highly sensitive to the detailed magnetic environment around the iron. Since conformational changes can alter this environment, the e.s.r. spectrum becomes a sensitive probe of these changes[43]. This technique has been used to probe the conformational changes which occur when α and β chains of haemoglobin bind together[48-50].

Quantitative studies have been attempted on high-spin haem proteins but theoretical[51] and practical difficulties[43] make such measurements of questionable value.

10.4.1.3 Low-spin polycrystalline studies

The low-spin e.s.r. spectra of haem proteins are characterised by three derivative features near $g \approx 2$. The symmetry is usually strongly rhombic. Blumberg and Peisach[42] have found that a wide variety of low-spin haem proteins can be placed on a plot of the rhombicity v. the crystal tetragonal field. On such a diagram similar classes of proteins group together. Such a correlation diagram should be useful in characterising a new protein.

Space does not permit a discussion of the scores of e.s.r. studies in the field of haem proteins. Needless to say it is an important field in e.s.r. biological studies.

References 42 and 43 are recommended for further reading.

10.4.2 Iron–sulphur proteins

Of all the proteins on which structural studies have been made, it is fair to say that our knowledge of the active centre of iron–sulphur (or non-haem) proteins would be most greatly lacking if it were not for the extensive applications of the e.s.r. technique coupled with the results of other techniques such as electron nuclear double resonance (e.n.do.r.), Mössbauer spectroscopy, circular dichroism and optical spectroscopy. Several reviews are available covering this field[52-57]; hence, we shall confine our discussion to the structural details which have been elucidated by e.s.r.

Almost all iron–sulphur proteins contain equimolar amounts of iron and labile sulphur (sulphur removed by weak acid treatment). The number of iron atoms per molecule ranges from 2 to 8. The e.s.r. spectra of the species are characterised by a large absorption at $g \approx 1.94$ and a weaker component at $g \approx 2.01$. The most detailed work concerning the nature of the active centre has been carried out on putidaredoxin from the nitrogen-fixing bacterium *Azotobacter vinelandii*. This protein contains two iron atoms per molecule. Comparison of the e.s.r. spectra of the native (^{56}Fe) protein with the spectra of a protein enriched in ^{57}Fe revealed the presence of a 1:2:1 hyperfine splitting for the ^{57}Fe protein[58]. This could only arise from the interaction of the unpaired electron with two equivalent or nearly equivalent iron nuclei.

Studies with proteins enriched in ^{33}S have shown that not only do the acid-labile sulphur atoms contribute some unresolved ^{33}S hyperfine splitting to the e.s.r. signal but so also do the sulphur atoms of the cystein moieties in the protein[59, 60].

It is possible to replace the labile sulphur atoms with either ^{80}Se or ^{77}Se. The latter has $I = \frac{1}{2}$. The e.s.r. spectrum of the ^{77}Se protein shows a 1:2:1 hyperfine splitting arising from two equivalent or nearly equivalent selenium nuclei[61]. Since the selenium protein is biologically active, it is concluded that selenium directly replaces sulphur.

It should be noted that an e.s.r. signal is only observed in the reduced state (one-electron reduction) of the protein. The oxidised state is diamagnetic while the reduced state has $S = \frac{1}{2}$. The picture of the active centre thus emerges as two iron atoms antiferromagnetically coupled to form an $S = 0$ state in the oxidised case and an $S = \frac{1}{2}$ state in the reduced case[62, 63]. The labile sulphur atoms and probably four cystein sulphur atoms are in the coordination sphere of the iron atoms. The situation is more complex than this as Mössbauer and ENDOR studies show that the two iron entities are not equivalent — one is ferric and the other ferrous in the reduced protein[64, 65].

The iron–sulphur centre is found in many proteins, in particular ferre-doxins and flavoproteins. The exact structure of this centre is still a matter of hot debate; however, it is clear that this problem is one of the most important to be studied by e.s.r. spectroscopy.

References 56 and 57 are recommended for further reading.

10.4.3 Copper proteins

Proteins containing copper play a significant role in respiration and metabolic processes. Members of this class of proteins function as oxidases, electron carriers, copper transporters and oxygen carriers. Confusion concerning the state and function of copper in these proteins led to early e.s.r. investigations, and consequently e.s.r. has played a predominant role in the characterisation of copper proteins[66–69].

Typically, two forms of paramagnetic copper have been identified in copper-containing proteins[70]. Additionally, evidence for a form of copper not detectable by e.s.r. has been presented[70]. Oxidation–reduction titrations indicate that the two copper atoms not detectable by e.s.r. in Fungal laccase must accept electrons in pairs. The lack of an e.s.r. signal has been attributed to spin pairing of the two ions in the oxidised protein[71]. Although some copper proteins contain all three forms of copper, it must not be inferred that all do. Some proteins may contain only one form of copper, e.g. azurin and amine oxidases[66].

The forms of copper observable by e.s.r. in copper proteins are easily distinguishable. Type 1 copper is responsible for the strong visible absorption near 600 nm observed in the 'blue' copper proteins. Type 2 copper ions are essentially colourless. Type 1 copper has a unique narrow hyperfine splitting $(A_\parallel \approx 0.005 \text{ cm}^{-1})$. Type 2 copper has spectra quite similar to those obtained from Cu^{2+} complexes $(A_\parallel \approx 0.02 \text{ cm}^{-1})$. Thus, copper–peptide complexes may be considered as primitive models for Type 2 copper in proteins. No ligand hyperfine splitting has been observed for Type 1 ions; however, under denaturing conditions, hyperfine structure attributable to four nitrogens has been observed[72]. When proteins containing Type 1 copper are reduced, the e.s.r. signal characteristic of Type 1 ions and the associated blue colour disappear. Denaturisation of Type 1 copper-containing proteins also leads to the elimination of the blue colour and the associated e.s.r. resonance.

Although studies on proteins containing only a Type 1 copper atom have been invaluable for elucidation of the nature of the site of the copper, the best-characterised copper proteins are those containing both Type 1 and Type 2 copper atoms. The best-investigated protein of this class is the oxidase, Fungal laccase[71]. Laccase contains four copper atoms per molecule; two are not detectable by e.s.r. One is a Type 1 ion, the other of Type 2. The Type 2 ion can be removed from the enzyme without altering the properties of the Type 1 ion; however, removal of the Type 2 copper atom results in loss of catalytic ability of the enzyme. The Type 2 moiety is also more available for chemical alteration than the Type 1 ion since Type 2 easily binds anions.

The detailed mechanism for the catalytic action of laccase is unknown, even though e.s.r.-monitored oxidation–reduction titrations have allowed some conclusions to be made. Type 1 copper appears to be the primary electron acceptor, with subsequent electron transfer to the three remaining copper atoms[73]. When all copper atoms become reduced, a four-electron transfer to oxygen would be possible. However, an alternative role for the Type 2 ion has been recently proposed. Type 2 copper has been suggested as a binding site for an intermediate formed in the reduction of O_2 to H_2O [74]. This hypothesis is supported by the ability of peroxide to form a complex with the Type 2 copper. In addition, fluoride ion bound to Type 2 copper inhibits the reduction of oxygen, rather than the oxidation of other substrates.

In addition to investigation of the role of copper in proteins bearing native copper, copper has been substituted for other protein-bound metal ions in order to obtain useful knowledge concerning the binding site[75]. The controlled addition of copper to various macromolecules, such as haemoglobin[76] and ribonuclease[77], has also provided details of metal–protein interactions.

References 66 and 67 are recommended for further reading.

10.4.4 Other metal-containing proteins

The metalloproteins that remain to be discussed are those containing Co, Mn and Mo. There have been a few reports of e.s.r. spectra of cobalt proteins, principally cobalamin (Co^{II}) (vitamin B_{12}) [78] and deoxyadenosylcobalamin (Co^{II}) [79]. Normally, the Co is in the Co^{3+} state and does not give an e.s.r. signal. However, when irradiated with visible light a conversion to Co^{2+} occurs and an e.s.r. spectrum can be seen from the low-spin state ($S = \frac{1}{2}$) at 77 K. ^{59}Co hyperfine structure ($I = \frac{1}{2}$) is clearly resolved on the $g_{\parallel}$ peaks and sometimes even superhyperfine structure from ^{14}N of nitrogen-containing substrates can be observed.

There appear to be no reports of e.s.r. studies of proteins which normally contain Mn^{2+}. Some work has been reported on proteins in which Mn^{2+} is substituted for Fe^{2+} [80], but the major work involving Mn^{2+} in biological systems has dealt with coordination of Mn^{2+} to specific binding sites on proteins[81]. When Mn^{2+} is bound to a protein it exhibits a broad 'solid-state' type of e.s.r. spectrum, whereas in solution as $Mn(H_2O)_6^{2+}$ the e.s.r. spectrum consists of six sharp lines. One can thus use e.s.r. to determine the binding constants for Mn^{2+}. Moreover, the binding constants for other ions can be determined by competition with Mn^{2+}. The e.s.r. spectrum of bound Mn^{2+} is very complex but recently it has been analysed[82].

By far the most complex protein, and yet perhaps the most interesting, to be studied by e.s.r. is xanthine oxidase[83, 84]. This protein can be obtained in high purity and large quantities. It contains flavin, iron and molybdenum. E.S.R. signals have been seen from the flavin semiquinone, iron (of the iron–sulphur type) and Mo^V. Two Mo^V e.s.r. signals can be detected [the participation of Mo was confirmed by isotopic substitution with ^{95}Mo ($I = \frac{5}{2}$) [85]]. The first, called Mo-γ,δ, is obtained in the very early stages of reduction by xanthine, especially at high pH (~ 10). The second, called

Mo-α,β, is obtained at low pH with xanthine. Both signals exhibit ^{95}Mo hyperfine structure; in addition, the Mo-α,β signal shows a doublet e.s.r. signal arising from a nearby proton. This proton is exchangeable in D_2O [83] and when 8-deuteroxanthine is used in place of xanthine, the e.s.r. signal initially appears in its deuterated form[84].

Kinetic studies using the rapid-freezing technique have shown that the reaction sequence probably entails the donation of an electron from xanthine to Mo^{VI} to form the Mo-γ,δ signal. A hydrogen atom is then transferred to another Mo^{VI} to form the Mo-α,β signal; electron transfer then occurs in the sequence: flavosemiquinone $\rightarrow$ iron $\rightarrow$ oxygen. The role of oxygen in forming the O_2^- intermediate was demonstrated recently from the identification of the O_2^- e.s.r. spectrum[86] and ^{17}O hyperfine structure using an enriched sample of O_2 [87].

References 81 and 84 are recommended for further reading.

10.5 PHOTOSYNTHESIS

The initial discovery in 1956 of light-induced resonances in photosynthetic material[88] provided the impetus for e.s.r. inquiry which after 15 years is now beginning to provide insight into the primary photochemical step in photosynthesis[89, 90]. These light-generated signals have been generally assigned to two distinct classes. Resonances characteristic of the first class are generally narrow (7–10 G), lack hyperfine structure, have a g-factor of 2.0025-2.0026, and possess rapid formation and decay kinetics. Signals belonging to the second class are broad (19 G) with a g-factor of 2.0046, have partially resolved hyperfine structure, and decay slowly after termination of illumination. Photosynthetic bacteria possess only a resonance belonging to the first class. However, two light-induced signals are present in green plants and algae. One resonance is characteristic of the first class and is denoted as Signal I. The other resonance, known as Signal II, belongs to the second class of light-generated resonances. Although a new light-induced signal has been reported in the blue-green alga, *Anacystis nidulans*, its significance and universality in other species of this class have yet to be demonstrated[91].

Elucidation of the molecular identities and functional roles of these signals has been the primary objective of many laboratories. Research strategy has been directed towards both *in vitro* and *in vivo* systems.

A significant step towards the identification of the species giving rise to the bacterial e.s.r. signal came in 1969 with the assignment of the light-induced e.s.r. signal to a light-induced bleaching at 870 nm (P_{870}) in a reaction-centre fraction from *Rhodopseudomonas spheroides*. Careful determination of the spin concentration in an illuminated sample of reaction centres allowed Bolton et al.[92] to conclude that the ratio of light-induced spins to bleached reaction centre molecules was very nearly unity. Furthermore, formation and decay kinetics of the e.s.r. signal and the 870 nm bleaching were identical. Loach and Walsh[93] independently verified the concentration equivalence of spin and optical species and established that the quantum yield of formation was approximately one for both P_{870}^+ and the e.s.r. resonance. Comparison of the optical and e.s.r. kinetics from room temperature to 1.7 K also led to the conclusion that these changes arose from the same species[94]. The decay kinetics were found to be independent of temperature, suggesting a tunnelling mechanism for electron return to the primary donor. Comparison of *in vivo* e.s.r. signal characteristics and behaviour upon deuteration with e.s.r. signals generated by chemically oxidised bacteriochlorophyll lent support to the assignment of the free radical formed to a specialised form of oxidised bacteriochlorophyll[94].

Although Signal I in plant and algal systems has been ascribed to a light-generated bleaching at 700 nm (P_{700})[95], adequate confirmation of this assertion is yet to be obtained. However, *in vitro* electrolysis of chlorophyll *a* resulted in the formation of a π-cation radical having a *g*-factor of 2.0025 and a peak to peak linewidth of 9 G[96]. In addition, this species exhibited a bleaching at 667 nm; hence, the optical and e.s.r. characteristics of this cation are comparable to P_{700} and Signal I, respectively. Although the linewidth of 9 G is considerably greater than the observed linewidth of 7 G *in vivo*, the narrowing of linewidth is attributed to delocalisation of the unpaired electron over a chlorophyll dimer: $(Chl \cdot H_2O \cdot Chl)^+$[97].

Although the identity for Signal I has not been conclusively established, preliminary identification of the molecular origin of Signal II has been recently reported[98]. Removal of plastoquinone, an essential intermediate in the photosynthetic electron transport chain, resulted in a greatly diminished Signal II. However, re-addition of plastoquinone restored Signal II to its normal intensity. Replacement of plastoquinone by its perdeutero-analogue yielded a narrowed Signal II, suggesting that this broad, slowly decaying signal arises from a derivative of plastoquinone. On the basis of spectral characteristics of plastoquinone and tocopherol radicals at 77 K, this derivative has been tentatively identified as the plastochromanoxyl moiety[99].

Although the identity of the electron acceptor in the primary photo-act has been the subject of great controversy, only recently has e.s.r. provided information concerning the nature of this species. Using a light modulation detection system at 1.4 K, Feher observed a broad three-line signal in reaction centres of *R. spheroides* which he tentatively assigned to the primary acceptor[100]. Kinetic behaviour of this new resonance closely resembled the kinetics of the narrow signal assigned to P_{870}^+. Feher postulated that his broad signal may arise from tetrahedrally coordinated ferrous iron in an $S = 1$ state. Evidence for photo-reduction of a ferredoxin at liquid nitrogen temperature in spinach chloroplasts has also been reported[101]; however, the role of this species in the primary photo-act has not yet been established.

Signal I in algae has been successfully utilised to indicate electron flow from the oxygen-evolving photosystem to P_{700}[95]. By monitoring the time for induction of Signal I after dark periods of variable length, Weaver has been able to characterise some properties of the reducing pool extant between photosystem I and photosystem 2[102]. Likewise, analysis of the concentration of Signal I formed in subchloroplast particles after a laser flash provided an estimate of the size of the paramagnetic unit[103].

Although e.s.r. has contributed extensively to the understanding of primary processes *in vivo*, the application of the spin resonance technique to model photosynthetic systems has generally been overlooked. However, Wang and Tu[104] have extensively investigated chlorophyll–flavin complexes as a model for P_{700}, observing light-induced optical bleaching at 705 nm and formation of an e.s.r. signal with a g-factor of 2.0042 and a ΔH_{pp} value of 15 G. Obviously much information remains to be gleaned from model system investigations, and it is thus not unlikely that a large portion of our knowledge of the photosynthetic act will be obtained in the future from these synthetic photosystems.

Reference 89 is recommended for further reading.

10.6 SPIN LABELS

The overwhelming complexity of biological macromolecular systems has hindered the facile application of spectroscopic techniques to the solution of biophysical problems. However, the introduction of a paramagnetic probe, the spin label, by Stone and McConnell in 1965 [105] initiated the application of e.s.r. to a myriad of biologically significant systems that had previously been inaccessible. The growth of spin labelling applications has been phenomenal, as evidenced by a wealth of reviews[106–111]; and the inception of this technique ranks as one of the major innovations in the field of spin resonance.

The spin label is usually a stable nitroxide radical possessing a specific substituent group or structure that permits binding of the radical to a designated region in the biological system of interest. By synthetic tailoring of the substituent group attached to the nitroxide ring, the specificity of the label can be selected. Labels of this type have the unpaired electron predominantly delocalised on the nitrogen and oxygen atoms of the nitroxide ring, and consequently the nature of the immediate environment of the spin label will be reflected in the line shape and hyperfine splitting constants observed in the experimental spectrum. From these spectral parameters rotational correlation times (τ) can be obtained; hence the state of the immobilisation of the free radical can be determined. The observed spectrum is also dependent on solvent effects, thus the spin label can be used as a probe to indicate the degree of polarity of the immediate environment of the label. However, solvent effects can lead to erroneous determinations of τ [108].

Spin labels have been utilised to elucidate both structural and mechanistic considerations. Structural studies have been directed towards determination of enzyme active-site geometries[112], nucleic acids[113], nature of protein conformation changes[114] and membrane organisation[115–117]. Mechanistic investigations have considered the nature of oxygen binding by haemoglobin, membrane-transport phenomena[118], and molecular regulation of muscle contraction[119, 120]. This list is not all inclusive but is intended to illuminate the general applicability of spin labelling to problems of biological interest. Although the applications of spin labelling are far too diverse to cover adequately in a review of this nature, we have chosen to highlight some of the more promising uses of this technique.

The mechanism for oxygen binding by haemoglobin has puzzled biophysicists for a number of decades; however, spin-labelling studies have proved to be invaluable in their contribution to the solution of this mystery. The nitroxide label N-(1-oxyl-2,2,5,5-tetramethyl-3-pyrrolidinyl)iodoacetamide binds at cystein $\beta93$ in several haemoglobin derivatives. This position is next to the proximal histidine $\beta92$ which is bound to the haem iron. Use of this label has permitted the identification of two conformational states in carbonmonoxy, met, met-azide and met-fluoride haemoglobins[121]. These states appear to be in an equilibrium which is dependent on the nature of the ligand bound to iron and to the ionic composition of the solvent. One state (A) is more strongly immobilised than the other (B), and the relative concentrations of the two states are dependent on the spin state of the iron. Hence, the equilibrium between A and B must be very sensitive to small changes in conformation. Deoxyhaemoglobin lacks this two-state equilibrium and in addition possesses a different quatenary structure; thus, McConnell, Deal, and Ogata conclude that the local conformation which influences the e.s.r. spectra must be a reflection of the quatenary structure[121].

Haemoglobin sub-units have also been selectively labelled while isolated and then allowed to combine[122]. It was concluded that β sub-units undergo rapid conformational changes when combined with α sub-units; however, α chains do not undergo significant alteration when reunited with β sub-units. The use of spin labelled haemoglobin variants to study co-operativity of oxygen binding has been reported[123]. Hb Chesapeake which displays little co-operativity and has an amino acid substitution in the $\alpha_1-\beta_2$ sub-unit contact area, lacks the normally observed intermediate states in the e.s.r. spectrum during oxygenation. Hence, these intermediate states may reflect a co-operative phenomenon which is dependent on the nature of $\alpha_1-\beta_2$ contact. Nitroxides attached to the haeme porphyrin, allowing for the preparation of mixed-spin-state haemoglobins, have also been utilised for the investigation of co-operative effects[124].

Hsia and Piette have elegantly determined the depth of the hapten combining site for rabbit anti-DNP antibodies using a series of spin-labelled haptens. By monitoring the change in rotational correlation time with variation of the distance of separation of the nitroxide moiety from the hapten, the average depth of the hapten combining site was found to be at 10 Å [125]. Future application of this technique promises to yield essentially needed information concerning the molecular geometry of the active sites of enzymes.

The application of the spin-labelling technique to artificial and biological membrane investigations is proving to be of great benefit in probing the structure and functional roles of biological membranes[126]. Orientational studies on lecithin multilayers have indicated that spin-labelled stearic acid reflects the alignment of surrounding lecithin molecules[115]. Furthermore, motion by the lipid spin-labelled components of the multilayer is dependent on the temperature and state of hydration of the multilayer. Interaction of cations with anionic lipids results in a reduction of molecular order in lipid bilayer membranes; however, if the lipid constituent lacks net charge, then cation addition has no effect on multilayer structure[117]. Although membrane research by spin-labelling has been directed towards multilayer model systems, phospholipid vesicles[118] and biological membrane tissue[114, 116] are also active areas of investigation.

Although the use of spin labelling has traditionally been directed towards biophysical investigations, recently, the spin labelling technique has been utilised as a biomedical tool. The introduction of FRAT (free radical assay technique) as an immunochemical test for the identification of opiates[127] has stimulated the development of spin labelling as a metabolic monitor.

FRAT simply involves the use of spin-labelled morphine complexed to a morphine antibody. Since free morphine or morphine metabolites induce competitive displacement of the spin-labelled drug, the presence of morphine or morphine variants can be detected in urine by the sharpening of the e.s.r. lines due to the displacement of the spin-labelled moiety. FRAT requires only 20 µl of body fluid, can be assayed in a minute, and is one thousand times as sensitive as thin layer chromatography.

Utilisation of spin labels as *in vivo* probes has been attempted with some degree of success. Keith *et al.* introduced a spin-labelled lipid, 12-nitroxide methyl stearate, into the growth media of *Neurospora crassa*[128]. Although much of the label was destroyed, a significant portion was assimilated into a mitochondrial lipid fraction. The nitroxide label did not affect the growth rate of the organism, and although data are scarce, other studies have indicated that spin labels can be compatible with living species. This compatibility factor combined with improvement of binding specificity through advances in synthetic chemistry, may make possible the crucial comparison of *in vivo* and *in vitro* experiments.

References 109 and 110 are recommended for further reading.

10.7 MOLECULES OF BIOLOGICAL INTEREST

In this section we will consider a few of the hundreds of free radicals studied which relate more or less to biological systems. In most cases these radicals are formed *in vitro* by a one-electron oxidation or reduction of the biochemical in a suitable solvent. Since many biological reactions occur in distinct one-electron steps, it is important to characterise the e.s.r. parameters of possible free radical intermediates. This rather vast field has been well reviewed[129, 130]; thus our focus here will be on a few selected studies to illustrate some of the more interesting work.

10.7.1 Flavosemiquinones

It is now well established that flavosemiquinones occur as intermediates in the enzymatic reactions of flavoproteins[131]. Thus, the study of flavosemiquinones *in vitro* has attracted a great deal of interest. In fact, it can be

(1)

said that these radicals have been more thoroughly studied than practically any other kind of free radical. Naturally, such activity has generated several reviews[131-133]. The isoalloxazine moiety (1) has been studied most extensively. By means of deuterium (^{2}H) substitution for ^{1}H and ^{15}N substitution for ^{14}N [132], of ^{13}C substitution for ^{12}C [134] and ENDOR studies[133] a virtually complete spin-density distribution has been determined for the semiquinone in three pH regions.

The flavosemiquinones form stable chelates with metal ions such as Zn^{2+} and Cd^{2+}. These entities have also been thoroughly mapped by the isotopic substitution technique[133].

10.7.2 Porphyrin radicals

By virtue of its large conjugated system the porphyrin group (2) undergoes facile one-electron oxidation and reduction reactions. Also its ability to complex metal ions makes this entity one of the most important in bio-

(2)

chemical systems. Here we will consider some examples of porphyrin radical cations and anions where the unpaired electron is located primarily on the porphyrin part of the molecule. Iron porphyrins (haem) are considered in Section 10.4.1.

A wide variety of porphyrin radical cations and anions have been studied[130]. Of particular interest are the studies on the radical cations of chlorophyll a [96] and bacteriochlorophyll a [94] as these are implicated as intermediates in photosynthesis (see Section 10.5). E.S.R. signals attributed to the lowest excited triplet state of chlorophyll b have been reported[135]. The very low values of D (0.0306 cm^{-1}) and E (0.0034 cm^{-1}) provide further evidence for the extensive electron delocalisation in these molecules.

10.7.3 Biochemical models

The development of model systems which more or less mimic the action of biological systems has been a favourite pastime of investigators of biological systems[136]. Models involving e.s.r. measurements have been utilised mostly in photosynthesis studies. The work of Tollin et al.[137] on the chlorophyll-sensitised reduction of quinones has provided some details concerning the

MAGNETIC RESONANCE

mechanism of this reaction. Also, various solid-state models have been developed[104, 138] which do demonstrate light-induced electron-transfer reactions.

An interesting model system for photophosphorylation has been developed[139] in which light absorbed by haematoporphyrin is shown to drive the condensation of imidazole with inorganic orthophosphate to form 1-phosphoimidazole. Free radical intermediates were observed and characterised.

10.8 CONCLUSION

The wide variety of applications of e.s.r. in biological systems presented in this review should convince the reader of the value of this technique in biological investigations. However, it is only one of many techniques which can be used. It is clear from our reading that those papers which proved of most value in working towards the solution of a specific biological problem have used a combination of techniques.

Acknowledgements

This work was supported in part by a National Research Council Grant (J.R.B.), and by a National Research Council of Canada Bursary (J.T.W.).

References

 1. Ehrenberg, A., Malmström, B. G. and Vänngård, T. (editors) (1967). *Magnetic Resonance in Biological Systems*. (Oxford: Pergamon Press)
 2. Ingram, D. J. E. (1969). *Biological and Biochemical Applications of Electron Spin Resonance*. (London: Adam Hilger, Ltd.)
 3. Feher, G. (1970). *Electron Paramagnetic Resonance with Applications to Selected Problems in Biology*. (New York: Gordon and Breach)
 4. Swartz, H. M., Bolton, J. R. and Borg, D. C. (editors) (1972). *Biological Applications of Electron Spin Resonance*. (New York: John Wiley and Sons)
 5. Bray, R. C. (1969). *FEBS Lett.*, **5**, 1
 6. Ehrenberg, A. (1966). in *Instrumentation in Biochemistry*, Biochemical Society Symposium, No. 26. (London: Academic Press)
 7. Bolton, J. R. (1972). in reference 4, Chapter 1
 8. Wertz, J. E. and Bolton, J. R. (1972). *Electron Spin Resonance, Elementary Theory and Practical Applications*. (New York: McGraw-Hill)
 9. Bolton, J. R., Borg, D. C. and Swartz, H. M. (1972). In reference 4, Chapter 2
10. Poole, C. P. (1967). *Electron Spin Resonance, A Comprehensive Treatise on Experimental Techniques*. (New York: Interscience)
11. Alger, R. S. (1968). *Electron Paramagnetic Resonance: Techniques and Applications*. (New York: Interscience)
12. Duchesne, J. and Van de Vorst, A. (1970). *Bull. Acad. Roy. Belgique*, **56**, 433
13. Mallard, J. R. and Kent, M. (1969). *Phys. Med. Biol.*, **14**, 373
14. Wyard, S. J. (1968). *Proc. Roy. Soc. A*, **302**, 355
15. Wyard, S. J. (1969). in *Solid State Biophysics*. (S. J. Wyard, editor). Chapter 8. (New York: McGraw-Hill)
16. Heckly, R. J. (1972). in *Biological Applications of Electron Spin Resonance*. (H. M. Swartz, J. R. Bolton and D. C. Borg, editors). (New York: John Wiley and Sons)

17. Swartz, H. M. (1972). in *Biological Applications of Electron Spin Resonance*. (H. M. Swartz, J. R. Bolton and D. C. Borg, editors). (New York: John Wiley and Sons)
18. Vanin, A. F., Chetverikov, A. G. and Blyumenfel'd, L. A. (1968). *Biofizika*, **13**, 66
19. Commoner, B. and Ternberg, J. L. (1961). *Proc. Nat. Acad. Sci. U.S.*, **47**, 1374
20. Dettmer, C. M., Driscoll, D. H., Wallace, J. D. and Neaves, A. (1967). *Nature (London)*, **214**, 492
21. Commoner, B., Woolum, J. C. and Larsson, E. (1969). *Science*, **1965**, 703
22. Vithayathil, A. J., Ternberg, J. L. and Commoner, B. (1965). *Nature (London)*, **207**, 1246
23. Woolum, J. C. and Commoner, B. (1970). *Biochim. Biophys. Acta*, **201**, 131
24. Commoner, B., Woolum, J. C., Senturia, B. H. and Ternberg, J. L. (1970). *Cancer Res.*, **30**, 2091
25. Ostrowski, K., Dziedzic-Goclawska, A., Stachowicz, W., Michalik, J., Tarsoly, E. and Kommender, A. (1971). *Calc. Tiss. Res.*, **7**, 58
26. Rowlands, J. R. and Gause, E. M. (1971). *Arch. Int. Med.*, **128**, 94
27. Henriksen, T. (1968). *Scand. J. Clin. Lab. Invest.*, **22**, 7
28. Henriksen, T. (1969). in *Solid State Biophysics*. (S. J. Wyard, editor). Chapter 6. (New York: McGraw-Hill)
29. Wyard, S. J. and Cook, J. B. (1969). in *Solid State Biophysics*. (S. J. Wyard, editor). Chapter 3. (New York: McGraw-Hill)
30. Van de Vorst, A. (1970). *Acad. Roy. de Belgique, Classe des Sciences, Memoires*, **39**, Fascicule 2
31. Copeland, E. S. (1972). in *Biological Applications of Electron Spin Resonance*. (H. M. Swartz, J. R. Bolton and D. C. Borg, editors). (New York: John Wiley and Sons)
32. Muller, A. (1968). in *Energetics and Mechanism in Radiation Biology*. (G. O. Phillips, editor). (New York: Academic Press)
33. Gordy, W., Ard, W. B. and Schields, H. (1955). *Proc. Nat. Acad. Sci. U.S.*, **41**, 983
34. ten Bosch, J. J. (1967). *Int. J. Rad. Biol.*, **13**, 93
35. Milvy, P. and Farcasiu, M. (1970). *Rad. Res.*, **43**, 320
36. Copeland, E. S., Sanner, T. and Pihl, A. (1970). *Int. J. Rad. Biol.*, **18**, 85
37. Pullman, I., Blumenthal, N. C. and Handler-Bernich, E. (1971). *Rad. Res.*, **45**, 476
38. Kenny, P. and Commoner, B. (1969). *Nature (London)*, **223**, 1229
39. Dodd, N. J. F. and Ebert, M. (1970). *Int. J. Rad. Biol.*, **18**, 451
40. Bennett, J. E., Gibson, J. F. and Ingram, D. J. E. (1957). *Proc. Roy. Soc. (London)*, **A240**, 67
41. Kotani, M. (1968). *Advan. Quantum Chem.*, **4**, 227
42. Blumberg, W. E. and Peisach, J. (1971). *Advan. Chem. Ser.*, **100**, 271
43. Peisach, J., Blumberg, W. E., Ogawa, S., Rachmilewitz, E. A. and Oltzik, R. (1971). *J. Biol. Chem.*, **246**, 3342
44. Blumberg, W. E. and Peisach, J. (1971). in *Structure and Function of Macromolecules and Membranes*. (B. Chance, C.-P. Lee and T. Yonetani, editors) (New York: Academic Press)
45. Peisach, J. and Blumberg, W. E. (1969). *Electron Spin Resonance of Metal Complexes*, 71. (New York: Plenum Press)
46. Yonetani, T. and Leigh, Jr., J. S. (1971). *J. Biol. Chem.*, **246**, 4174 and references cited therein
47. Scholes, C. P. (1970). *J. Chem. Phys.*, **52**, 4890
48. Banerjee, R., Alpert, Y., Leterrier, F. and Williams, R. J. P. (1969). *Biochemistry*, **8**, 2862
49. Peisach, J., Blumberg, W. E., Wittenberg, B. A., Wittenberg, J. B. and Kampa, L. (1969). *Proc. Nat. Acad. Sci. U.S.*, **63**, 934
50. Henry, V. and Banerjee, R. (1970). *J. Molec. Biol.*, **50**, 99
51. Isomoto, A., Watari, H. and Kotani, M. (1970). *J. Phys. Soc. Japan*, **29**, 1571
52. Buchanan, B. B. (1966). *Structure and Bonding*, **1**, 109
53. Malkin, B. and Rabinowitz, J. C. (1967). *Ann. Rev. Biochem.*, **36**, 113
54. Kimura, T. (1968). *Structure and Bonding*, **5**, 1
55. Hall, D. O. and Evans, M. C. W. (1969). *Nature (London)*, **223**, 1342
56. Tsibris, J. C. M. and Woody, R. W. (1970). *Coord. Chem. Rev.*, **5**, 417
57. Beinert, H. (1972). in *Biological Applications of Electron Spin Resonance*. (H. M. Swartz, J. R. Bolton and D. C. Borg, editors) (New York: John Wiley and Sons)
58. Tsibris, J. C. M., Tsai, R. L., Gunsalus, I. C., Orme-Johnson, W. H., Hansen, R. E. and Beinert, H. (1968). *Proc. Nat. Acad. Sci. U.S.*, **59**, 959

59. Der Vartanian, D. V., Orme-Johnson, W. H., Hansen, R. E., Beinart, H., Tsai, R. L., Tsibris, J. C. M., Bartholomaus, R. C. and Gunsalus, I. C. (1967). *Biochem. Biophys. Res. Commun.*, **26**, 569

60. Tsibris, J. C. M., Tsai, R. L., Gunsalus, I. C., Orme-Johnson, W. H., Hansen, R. E. and Beinert, H. Unpublished results cited in reference 56

61. Orme-Johnson, W. H., Hansen, R. E., Beinert, H., Tsibris, J. C. M., Bartholomaus, R. C. and Gunsalus, I. C. (1968). *Proc. Nat. Acad. Sci. U.S.*, **60**, 368

62. Gibson, J. P., Hall, D. O., Thornley, J. H. M. and Whatley, F. R. (1966). *Proc. Nat. Acad. Sci. U.S.*, **56**, 987

63. Johnson, C. E., Cammack, R., Rav, K. K. and Hall, D. O. (1971). *Biochem. Biophys. Res. Commun.*, **43**, 564

64. Johnson, C. E., Bray, R. C., Cammack, R. C. and Hall, D. O. (1969). *Proc. Nat. Acad. Sci. U.S.*, **63**, 1234

65. Fritz, J., Anderson, R., Fee, J., Palmer, G., Sands, R. H., Orme-Johnson, W. H., Beinert, H., Tsibris, J. C. M. and Gunsalus, I. C. (1971). *Biochem. Biophys. Acta*, **253**, 110

66. Vänngård, T. (1972). in *Biological Applications of Electron Spin Resonance*. (H. M. Swartz, J. R. Bolton and D. C. Borg, editors) (New York: John Wiley and Sons)

67. Malkin, R. and Malmström, B. G. (1970). *Advan. Enzymol.*, 177

68. Gould, D. C. and Ehrenberg, A. (1968). in *Physiology and Biochemistry of Haemocyanins*. (F. Ghiretti, editor). 95. (New York: Academic Press)

69. Malmström, B. G. and Rydén, L. (1968). in *Biological Oxidations*. (T. P. Singer, editor). 415. (New York: Interscience)

70. Malmström, B. G., Reinhammar, B. and Vänngård, T. (1968). *Biochim. Biophys. Acta*, **156**, 67

71. Fee, J., Malkin, R., Malmström, B. and Vänngård, T. (1969). *J. Biol. Chem.*, **244**, 4200

72. Peisach, J., Levine, W. G. and Blumberg, W. E. (1967). *J. Biol. Chem.*, **242**, 2847

73. Malmström, B. G., Agrò, A. F. and Antonini, E. (1969). *European J. Biochem.*, **9**, 383

74. Bränden, R., Malmström, B. G. and Vänngård, T. (1971). *European J. Biochem.*, **18**, 238

75. Lazdunski, C., Chappelet, D., Petitclerc, C., Leterrier, F., Douzou, P. and Lazdunski, M. (1970). *European J. Biochem.*, **17**, 239

76. Nagel, R. L., Bemski, G. and Pincus, P. (1970). *Arch. Biochem. Biophys.*, **137**, 428

77. Joyce, B. K. and Cohn, M. (1969). *J. Biol. Chem.*, **244**, 811

78. Hill, H. A. O., Pratt, J. M. and Williams, R. J. P. (1965). *Proc. Roy. Soc. (London)*, **A288**, 352

79. Hamilton, J. A., Blakley, R. L., Looney, F. D. and Winfield, M. E. (1969). *Biochim. Biophys. Acta*, **177**, 374

80. Yonetani, T. and Asakura, T. (1969). *J. Biol. Chem.*, **244**, 4580 and references cited therein

81. Cohn, M. and Reuben, J. (1971). *Accounts Chem. Res.*, **4**, 214

82. Reed, G. H. and Ray, W. J., Jr. (1971). *Biochemistry*, **10**, 3190

83. Bray, R. C., Knowles, P. F., Meriwether, L. S. (1967). in *Magnetic Resonance in Biological Systems*. (A. Ehrenberg, B. G. Malmström and T. Vänngård, editors), 249. (Oxford: Pergamon Press)

84. Bray, R. C. and Knowles, P. F. (1968). *Proc. Roy. Soc. (London)*, **A302**, 351

85. Bray, R. C. and Meriwether, L. S. (1966). *Nature (London)*, **212**, 467

86. Knowles, P. F., Gibson, J. F., Pick, F. M. and Bray, R. C. (1969). *Biochem. J.*, **111**, 53

87. Bray, R. C., Pick, F. M. and Samuel, D. (1970). *European J. Biochem.*, **15**, 352

88. Commoner, B., Heise, J. J. and Townsend, J. (1956). *Proc. Nat. Acad. Sci. U.S.*, **42**, 710

89. Weaver, E. C. (1968). *Ann. Rev. Plant Phys.*, **19**, 283

90. Kohl, D. H. (1972). in *Biological Applications of Electron Spin Resonance*. (H. M. Swartz, J. R. Bolton, and D. C. Borg, editors). (New York: John Wiley and Sons)

91. Weaver, E. C. (1970). *Nature (London)*, **226**, 183

92. Bolton, J. R., Clayton, R. K. and Reed, D. W. (1969). *Photochem. Photobiol.*, **9**, 209

93. Loach, P. A. and Walsh, K. (1969). *Biochemistry*, **8**, 1908

94. McElroy, J. D., Feher, G. and Mauzerall, D. C. (1969). *Biochim. Biophys. Acta*, **172**, 180

95. Beinert, H., Kok, B. and Hoch, G. (1962). *Biochem. Biophys. Res. Commun.*, **7**, 209

96. Borg, D. C., Fajer, J., Felton, R. H. and Dolphin, D. (1970). *Proc. Nat. Acad. Sci. U.S.* **67**, 813

97. Norris, J. R., Uphaus, R. A., Crespi, H. L. and Katz, J. J. (1971). *Proc. Nat. Acad. Sci. U.S.*, **68**, 625

98. Kohl, D. H. and Wood, P. M. (1969). *Plant Physiol.*, **44**, 1439
99. Kohl, D. H., Wright, J. R. and Weissman, M. (1969). *Biochim. Biophys. Acta*, **180**, 536
100. Feher, G. (1971). *Photochem. Photobiol.*, **14**, 373
101. Malkin, R. and Bearden, A. J. (1971). *Proc. Nat. Acad. Sci. U.S.*, **68**, 16
102. Weaver, E. C. (1968). *Photochem. Photobiol.*, **7**, 93
103. Weaver, E. C. and Weaver, H. E. (1969). *Science*, **165**, 906
104. Tu, Shu-I. and Wang, Jui-H. (1969). *Biochem. Biophys. Res. Commun.*, **36**, 79
105. Stone, T. J., Buckman, T., Nordio, R. L. and McConnell, H. M. (1965). *Proc. Nat. Acad. Sci. U.S.*, **54**, 1010
106. Hamilton, C. L. and McConnell, H. M. (1968). in *Structural Chemistry and Molecular Biology*. (A. Rich and N. Davidson, editors). 115. (New York: W. H. Freeman and Co.)
107. Ohnishi, S. (1968). *Seibutsu Butsuri*, **8**, 118
108. Griffith, O. H. and Waggoner, A. S. (1969). *Accounts Chem. Res.*, **2**, 17
109. McConnell, H. M. and McFarland, B. G. (1970). *Quart. Rev. Biophys.*, **3**, 91
110. Jost, P. and Griffith, O. H. (1971). *Methods in Pharmacology*, **2**
111. Smith, I. C. P. (1972). in *Biological Applications of Electron Spin Resonance*. (H. M. Swartz, J. R. Bolton and D. Borg, editors. (New York: John Wiley and Sons)
112. Weiner, H. (1969). *Biochemistry*, **8**, 526
113. Hoffman, B. M., Schofield, P. and Rich, A. (1969). *Proc. Nat. Acad. Sci. U.S.*, **62**, 1195
114. Raison, J. K., Lyons, J. M., Mehlhorn, R. J. and Keith, A. D. (1971). *J. Biol. Chem.*, **246** 4036
115. Jost, P., Libertini, L. J., Herbert, V. C. and Griffith, O. H. (1971). *J. Molec. Biol.*, **59**, 77
116. Calvin, M., Wang, H. H., Entine, G., Gill, D., Ferruti, P., Harpold, M. A. and Klein, M. P. (1969). *Proc. Nat. Acad. Sci. U.S.*, **63**, 1
117. Butler, K. W., Dugas, H., Smith, I. C. P. and Schreider, H. (1970). *Biochem. Biophys. Res. Commun.*, **40**, 770
118. Kornberg, R. D. and McConnell, H. M. (1971). *Biochemistry*, **10**, 1111
119. Tonomura, Y., Watanabe, S. and Morales, M. (1969). *Biochemistry*, **8**, 2171
120. Seidel, J. C., Chopek, M. and Gergely, J. (1970). *Biochemistry*, **9**, 3265
121. McConnell, H. M., Deal, W. and Ogata, R. T. (1969). *Biochemistry*, **9**, 2580
122. Ohnishi, S., Maeda, T., Ito, T., Hwang, K. and Tyumu, I. (1968). *Biochemistry*, **7**, 2662
123. Ho, C., Baldassare, J. J. and Charache, S. (1970). *Proc. Nat. Acad. Sci. U.S.*, **66**, 722
124. Asakura, T and Drott, H. R. (1971). *Biochem. Biophys. Res. Commun.*, **44**, 1199
125. Hsia, J. C. and Piette, L. H. (1969). *Arch. Biochem. Biophys.*, **129**, 296
126. Jost, P. C., Waggoner, A. S. and Griffith, O. H. (1971). in *Structure and Function of Biological Membranes*. (L. I. Rothfield, editor). (New York: Academic Press)
127. Leute, R. K., Ullman, E. F., Goldstein, A. and Herzenberg, L. A. (1971). *Clinical Chem.*, **17**, 639
128. Keith, A. D., Waggoner, A. S. and Griffith, O. H. (1968). *Proc. Nat. Acad. Sci. U.S.*, **61**, 819
129. Hemmerich, P. (1968). *Proc. Roy. Soc. (London)*, **A302**, 335
130. Borg, D. C. (1972). in *Biological Applications of Electron Spin Resonance*. (H. M. Swartz, J. R. Bolton and D. C. Borg, editors). Chapter 7. (New York: John Wiley and Sons)
131. Beinert, H. (1972). in *Biological Applications of Electron Spin Resonance*. (H. M. Swartz, J. R. Bolton and D. C. Borg, editors). Chapter 8. (New York: John Wiley and Sons)
132. Ehrenberg, A. and Hemmerich, P. (1968). in *Biological Oxidations*. (T. Singer, editor), 238. (New York: Wiley-Interscience)
133. Ehrenberg, A. (1970). *Vit. and Horm.*, **28**, 489
134. Walker, W. H., Ehrenberg, A. and Lhoste, J-M. (1970). *Biochim. Biophys. Acta*, **215**, 166
135. Lhoste, J-M. (1968). *C. R. Acad. Sci. Paris*, **266**, 1059; (1968). *Studia Biophysica, Berlin*, **12**, 135
136. Wang, J. H. (1970). *Accounts Chem. Res.*, **3**, 90
137. Mukherjie, D. C., Cho, D. H. and Tollin, G. (1969). *Photochem. Photobiol.*, **9**, 273 and references cited therein
138. Wang, J. H. (1969). *Proc. Nat. Acad. Sci. U.S.*, **62**, 653
139. Tu, S.-I. and Wang, J. H. (1970). *Biochemistry*, **9**, 4505